Pesticide residues in food – 2004

FAO PLANT PRODUCTION AND PROTECTION PAPER

178

Report of the Joint Meeting of the FAO Panel of Experts on Pesticide Residues in Food and the Environment and the WHO Core Assessment Group on Pesticide Residues
Rome, Italy, 20 – 29 September 2004

WORLD HEALTH ORGANIZATION
FOOD AND AGRICULTURE ORGANIZATION OF THE UNITED NATIONS
Rome, 2004

The designations employed and the presentation of material in this information product do not imply the expression of any opinion whatsoever on the part of the Food and Agriculture Organization of the United Nations concerning the legal or development status of any country, territory, city or area or of its authorities, or concerning the delimitation of its frontiers or boundaries.

ISBN 92-5-105242-5

All rights reserved. Reproduction and dissemination of material in this information product for educational or other non-commercial purposes are authorized without any prior written permission from the copyright holders provided the source is fully acknowledged. Reproduction of material in this information product for resale or other commercial purposes is prohibited without written permission of the copyright holders. Applications for such permission should be addressed to the Chief, Publishing Management Service, Information Division, FAO, Viale delle Terme di Caracalla, 00100 Rome, Italy or by e-mail to copyright@fao.org

© FAO 2004

iii

CONTENTS

List of participants v

Abbreviations x

Use of JMPR reports and evaluations by registration authorities xi

Report of the 2004 JMPR FAO/WHO Meeting of Experts 1

1. Introduction 1

2. General considerations 3
    2.1   Guidance on the establishment of acute reference doses 3
    2.2   Definition of 'overall NOAEL' 9
    2.3   Interim acute reference dose 9
    2.4   Progress report on the JMPR work-sharing pilot project on trifloxystrobin ... 10
    2.5   Comparison of JMPR recommendations and interim MRL recommendations
          from the CCPR pilot project 13
    2.6   Estimation of maximum residue levels of pesticides in or on spices on the basis
          of monitoring results 19
    2.7   Revisited: MRLs for fat-soluble pesticides in milk and milk products 24
    2.8   Revisited: Dietary burden of animals for estimation of MRLs for animal
          commodities 26
    2.9   Statistical methods for estimating MRLs 29
    2.10 Application of the recommendations of the OECD project on minimum data
          requirements to the work of the JMPR. 30
    2.11 Alignment within one year of toxicological and residue evaluations for new and
          periodically reviewed compounds 33

3. Dietary risk assessments for pesticide residues in foods 35

4. Evaluation of data for establishing values for acceptable daily intakes and acute reference
    doses for humans, maximum residue limits and supervised trial median residue levels 39
    4.1   Bentazone (T, D) 39
    4.2   Captan (T, D) 40
    4.3   Carbofuran / Carbosulfan (R, D) 42
    4.4   Chlorpyrifos (R, D) 44
    4.5   Dimethipin (T, D) 48
    4.6   Dithiocarbamates (R) 49
    4.7   Ethoprophos (R, D) 50
    4.8   Fenitrothion (R, D) 62
    4.9   Fenpropimorph (T, D) 70
    4.10 Fenpyroximate (T, D) 72
    4.11 Fludioxonil (T, R, D)* 74

Contents

4.12	Folpet (T, D)	96
4.13	Glyphosate (T, D)**	98
4.14	Malathion (R, D)	103
4.15	Metalaxyl M (R)	106
4.16	Methamidophos (R)	115
4.17	Methomyl (R)	115
4.18	Oxydemeton methyl (R, D)	117
4.19	Paraquat (R, D)	124
4.20	Phorate (T, D)**	150
4.21	Pirimicarb (T, D)**	154
4.22	Pirimiphos methyl (R, D)	161
4.23	Prochloraz (R, D)	163
4.24	Propiconazole (T, D)**	180
4.25	Propineb (T, R, D)	185
4.26	Pyraclostrobin (R, D)	194
4.27	Spices (R, D)	212
4.28	Spinosad (R)	226
4.29	Triadimenol and triadimefon (T, D)**	231
4.30	Trifloxystrobin (T, R, D)*	241

5. Recommendations 271

6. Future work 273

Annexes 275
    Annex 1 Acceptable daily intakes, short-term dietary intakes, acute reference doses, recommended maximum residue limits and supervised trials median residue values recorded by the 2004 Meeting 275
    Annex 2 Index of reports and evaluations of pesticides by the JMPR 293
    Annex 3 International estimated daily intakes of pesticide residues 302
    Annex 4 International estimated short-term dietary intakes of pesticide residues 331
    Annex 5 Dietary risk assessment for spices 361
    Annex 6 Reports and other documents resulting from previous Joint Meetings of the FAO Panel of Experts on Pesticide Residues in Food and the Environment and the WHO Expert Group on Pesticide Residues 365
    Annex 7 Corrections to the report of the 2003 Meeting 373

LIST OF PARTICIPANTS

2004 Joint FAO/WHO Meeting on Pesticide Residues
Rome, 20–29 September 2004

FAO Members

Dr Ursula Banasiak, Federal Institute for Risk Assessment, Thielallee 88-92, D-14195 Berlin, Germany
    Tel.: (49 30) 8412 3337; Fax: (49 30) 8412 3260; E-mail: u.banasiak@bfr.bund.de

Dr Eloisa Dutra Caldas, University of Brasilia, College of Health Sciences, Pharmaceutical Sciences Department, Campus Universitàrio Darci Ribeiro, 70919-970 Brasília/DF, Brazil
    Tel.: (55 61) 307 3671; Fax: (55 61) 273 0105; E-mail: eloisa@unb.br

Dr Stephen Funk, Health Effects Division (7509C), United States Environmental Protection Agency, 1200 Pennsylvania Ave NW, 7509C, Washington DC 20460, USA (*FAO Chairman*)
    Tel.: (1 703) 305 5430; Fax: (1 703) 305 0871; E-mail: funk.steve@epa.gov

Mr Denis J. Hamilton, Principal Scientific Officer, Biosecurity, Department of Primary Industries and Fisheries, PO Box 46, Brisbane, QLD 4001, Australia (*FAO Rapporteur*)
    Tel.: (61 7) 3239 3409; Fax: (61 7) 3211 3293; E-mail: denis.hamilton@dpi.qld.gov.au

Mr David Lunn, Programme Manager (Residues–Plants), Dairy and Plant Products Group, New Zealand Food Safety Authority, PO Box 2835, Wellington, New Zealand
    Tel.: (644) 463 2654; Fax: (644) 463 2675; E-mail: dave.lunn@nzfsa.govt.nz

Dr Dugald MacLachlan, Australian Quarantine and Inspection Service, Australian Department of Agriculture, Fisheries and Forestry, Edmond Barton Building, Kingston, ACT 2601, Australia
    Tel.: (61 2) 6272 3183; Fax: (61 2) 6271 6522; E-mail: dugald.maclachlan@aqis.gov.au

Dr Bernadette C. Ossendorp, Centre for Substances and Integrated Risk Assessment, National Institute of Public Health and the Environment (RIVM), Antonie van Leeuwenhoeklaan 9, PO Box 1, 3720 BA Bilthoven, Netherlands
    Tel.: (31 30) 274 3970; Fax: (31 30) 274 4475; E-mail: bernadette.ossendorp@rivm.nl

Dr Yukiko Yamada, Research Planning and Coordination Division, National Food Research Institute, 2-1-12 Kannondai, Tsukuba 305-8642, Japan
    Tel.: (81 3) 3502 2319; Fax: (81 3) 3597 0389; E-mail: yukiko.yamada@affrc.go.jp

WHO Members

Professor Alan R. Boobis, Experimental Medicine & Toxicology, Division of Medicine, Faculty of Medicine, Imperial College London, Hammersmith Campus, Ducane Road, London W12 0NN, England (*WHO Chairman*)
    Tel.: (44 20) 8383 3221; Fax: (44 20) 8383 2066; E-mail: a.boobis@imperial.ac.uk

List of participants

Dr Les Davies, Science Strategy and Policy, Office of Chemical Safety, Australian Government Department of Health and Ageing, PO Box 100, Woden, ACT 2606, Australia
Tel.: (61 2) 6270 4378; Fax: (61 2) 6270 4353; E-mail: les.davies@health.gov.au

Dr Vicki L. Dellarco, United States Environmental Protection Agency, Office of Pesticide Programs (7509C), Health Effects Division, 401 M Street SW, Washington DC 20460, USA (*WHO Rapporteur*)
Tel.: (1 703) 305 1803; Fax: (1 703) 305 5147; E-mail: dellarco.vicki@epa.gov

Dr Helen Hakansson, Institute of Environmental Medicine, Karolinska Institutet, Unit of Environmental Health Risk Assessment, Box 210, Nobels väg 13, S-171 77 Stockholm, Sweden
Tel.: (46 8) 524 87527; Fax: (46 8) 34 38 49; E-mail: helen.hakansson@imm.ki.se

Dr Angelo Moretto, Dipartimento Medicina Ambientale e Sanità Pubblica, Università di Padova, via Giustiniani 2, 35128 Padova, Italy
Tel.: (39 049) 821 1377 / 2541; Fax: (39 049) 821 2550 / 2542; E-mail: angelo.moretto@unipd.it

Dr Roland Solecki, Pesticides and Biocides Division, Federal Institute for Risk Assessment, Thielallee 88-92, D-14195 Berlin, Germany
Tel.: (49 30) 8412 3232; Fax: (49 30) 8412 3260; E-mail: r.solecki@bfr.bund.de

Dr Maria Tasheva, Laboratory of Toxicology, National Center of Hygiene, Medical Ecology and Nutrition, 15 Dim. Nestorov Str., 1431 Sofia, Bulgaria
Tel.: (3592) 954 11 97; Fax: (3592) 954 11 97; E-mail: mtasheva@aster.net

Secretariat

Dr Arpàd Ambrus, Central Service for Plant Protection and and Soil Conservation, Plant and Soil Protection Directorate, Budaörsi ut 141-145, H-1118 Budapest, Hungary (*FAO Consultant*)
Tel.: (36 1) 309 1003; Fax: (36 1) 246 2955; E-mail: ambrus.arpad@ontsz.hu

Dr Andrew Bartholomaeus, Therapeutic Goods Administration, Commonwealth Department of Health and Ageing, MDP 122, PO Box 100, Woden, ACT 2602, Australia (*WHO Temporary Adviser*)
Tel.: (61 2) 62 32 8345; Fax: (61 2) 62 32 8355; E-mail: andrew.bartholomaeus@health.gov.au

Dr Lourdes Costarrica, Food and Nutrition Division, Food and Agriculture Organization of the United Nations (FAO), Viale delle Terme di Caracalla, 00100 Rome, Italy (*FAO Staff Member*)
Tel.: (39 06) 570 56060; Fax: (39 06) 570 54593; E-mail: lourdes.costarrica@fao.org

Mr Bernard Declercq, 13 impasse du court Riage, 91360 Epinay sur Orge, France (*FAO Consultant*)
Tel.: (33 1) 64488369; Fax: (33 1) 64488369; E-mail: bernard-declercq@wanadoo.fr

Dr Ghazi Dannan, Office of Pesticide Programs (7509C), United States Environmental Protection Agency, 1200 Pennsylvania Avenue NW, Washington DC 20460, USA (*WHO Temporary Adviser*)
Tel.: (1 703) 308 9549; Fax: (1 703) 305 5529; E-mail: dannan.ghazi@epa.gov

Dr Ian C. Dewhurst, Pesticides Safety Directorate, Mallard House, King's Pool, 3 Peasholme Green, York YO1 7PX, England (*WHO Temporary Adviser*)
Tel.: (44 1904) 455 890; Fax: (44 1904) 455 711; E-mail: ian.dewhurst@psd.defra.gsi.gov.uk

List of participants

Dr Salwa Dogheim, Central Laboratory of Residue Analysis of Pesticides and Heavy Metals in Food, Agriculture Research Center, Ministry of Agriculture, 7 Mahalla Street, Heliopolis, Cairo, Egypt (*FAO Consultant*)
Tel.: (2012) 215 5201; Fax: (2012) 418 2814; E-mail: s.dogheim@link.net

Professor P.K. Gupta, Toxicology Consulting Services Inc., C-44 Rajinder nagar, Bareilly 243 122 (UP), India (*WHO Temporary Adviser*)
Tel.: (91 581) 2300 628; mobile (91 581) 310 4922; E-mail: drpkg_brly@sancharnet.in

Dr Yibing He, Pesticide Residue Division, Institute for the Control of Agrochemicals, Ministry of Agriculture, Building 22, Maizidian Street, Cheoyang District, Beijing 100026, China (*FAO Consultant*)
Tel.: (86 10) 659 36997; Fax: (86 10) 641 94017; E-mail: heyibing@agri.gov.cn

Dr H. Jeuring, Chairman, Codex Committee on Pesticide Residues, Senior Public Health Officer, Food and Consumer Product Safety Authority, PO Box 19506, 2500 CM The Hague, Netherlands (*WHO Temporary Adviser*)
Tel.: (31 70) 448 4848; Fax: (31 70) 448 4747; E-mail: hans.jeuring@vwa.nl

Mr Antony F. Machin, Boundary Corner, 2 Ullathorne Road, London SW16 1SN, England (*FAO Editor*)
Tel. & Fax: (44 208) 769 0435; E-mail: afmachin@clara.net

Dr Timothy C. Marrs, Food Standards Agency, Room 504C, Aviation House, 125 Kingsway, London WC2B 6NH, England (*WHO Temporary Adviser*)
Tel.: (44 207) 276 8507; Fax: (44 207) 276 8513; E-mail: tim.marrs@foodstandards.gsi.gov.uk

Dr Jeronimas Maskeliunas, Joint FAO/WHO Food Standards Programme, Food and Nutrition Division, Food and Agriculture Organization of the United Nations (FAO), Viale delle Terme di Caracalla, 00100 Rome, Italy (*FAO Staff Member*)
Tel.: (39 06) 570 53967; Fax: (39 06) 570 54593; E-mail: jeronimas.maskeliunas@fao.org

Dr Heidi Mattock, 9 rue Jules Verne, 67400 Illkirch-Graffenstaden, France (*WHO Editor*)
Tel.: (33 3) 88 66 07 34; E-mail: heidimattock@yahoo.com

Dr Douglas B. McGregor, 38 Shore Road, Aberdour, KY3 0TU, Scotland (*WHO Temporary Adviser*)
Tel.: (44 1383) 860901; Fax: 44 1583 860901; E-mail: mcgregortec@btinternet.com

Dr Rudolf Pfeil, Pesticides and Biocides Division, Federal Institute for Risk Assessment, Thielallee 88-92, D-14195 Berlin, Germany (*WHO Temporary Adviser*)
Tel.: (49 30) 8412 3828; Fax: (49 30) 8412 3260; E-mail: r.pfeil@bfr.bund.de

Mr Tsuyoshi Sakamoto, Agricultural Chemicals Inspection Station, 2-772 Suzuki-cho, Kodaira, Tokyo 187-0011, Japan (*FAO Consultant*)
Tel.: (81 42) 383 2151 (switchboard); Fax: (81 42) 385 3361; E-mail: t-sakamoto@acis.go.jp

Dr Atsuya Takagi, Division of Toxicology, Biological Safety Research Centre, National Institute of Health Sciences, 1-18-1 Kamiyoga, Setagaya-ku, Tokyo 158-8501, Japan (*WHO Temporary Adviser*)
Tel.: (81 3) 3700 1141 (ext 406); Fax: (81 3) 3700 9647; E-mail: takagi@nihs.go.jp

List of participants

Dr Amelia Tejada, Pesticide Management Group, Plant Protection Service, Plant Production and Protection Division, Food and Agriculture Organization of the United Nations (FAO), Viale delle Terme di Caracalla, 00100 Rome, Italy (*FAO Joint Secretary*)
Tel.: (39 06) 570 54010; Fax: (39 06) 570 56347; E-mail: amelia.tejada@fao.org

Dr Angelika Tritscher, International Programme on Chemical Safety, World Health Organization, 1211 Geneva 27, Switzerland (*WHO Joint Secretary*)
Tel: (41 22) 791 3569; Fax: (41 22) 791 4848; E-mail: tritschera@who.int

Dr Gero Vaagt, Pesticide Management Group, Food and Agriculture Organization of the United Nations (FAO), Viale delle Terme di Caracalla, 00100 Rome, Italy (*FAO Staff Member*)
Tel.: (39 06) 570 55757; Fax: (39 06) 570 56347; E-mail: gero.vaagt@fao.org

Dr Gerrit Wolterink, Centre for Substances & Risk Assessment, National Institute for Public Health and the Environment (RIVM), Antonie van Leeuwenhoeklaan 9, PO Box 1, 3720 BA Bilthoven, Netherlands (*WHO Temporary Adviser*)
Tel.: (31 30) 274 4531; Fax: (31 30) 274 4401; E-mail: gerrit.wolterink@rivm.nl

Dr Jürg Zarn, Swiss Federal Office of Public Health, Food Toxicology Section, Stauffacherstrasse 101, CH-8004 Zürich, Switzerland (*WHO Temporary Adviser*)
Tel.: (41 43) 322 21 93; Fax: (41 43) 322 21 99; E-mail: juerg.zarn@bag.admin.ch

ABBREVIATIONS

(Well-known abbreviations in general use and SI units are not included.)

*	at or about the limit of quantification
ADI	acceptable daily intake
ai	active ingredient
AMPA	ammonethylphosphonic acid
ARfD	acute reference dose
bw	body weight
CCN	Codex classification number (for compounds or commodities)
CCPR	Codex Committee on Pesticide Residues
CSAF	chemical-specific adjustment factor
CXL	Codex level
DT_{50}	time to 50% decomposition
DT_{90}	time to 90% decomposition
F_1	first filial generation
F_2	second filial generation
FAO	Food and Agricultural Organization of the United Nations
FOB	functional observational battery
GAP	good agricultural practice
GEMS/Food	Global Environment Monitoring System–Food Contamination Monitoring and Assessment Programme
GLP	good laboratory practice
HPLC	high-performance liquid chromatography
HR	highest level of residue in the edible portion of a commodity found in trials to estimate a maximum residue limit in the commodity
HR-P	highest residue in a processed commodity calculated by multiplying the HR of the raw commodity by the corresponding processing factor
IEDI	international estimated daily intake
IESTI	international estimated short-term dietary intake
ILO	International Labour Organisation
IUPAC	International Union of Pure and Applied Chemists
JECFA	Joint Expert Committee on Food Additives
ISO	International Standards Organization
JMPR	Joint Meeting on Pesticide Residues
LC_{50}	median lethal concentration
LD_{50}	median lethal dose
LOAEL	lowest-observed-adverse-effect level
LOQ	limit of quantification
MRL	maximum residue limit
NOAEC	no-observed-adverse-effect concentration
NOAEL	no-observed-adverse-effect level
OECD	Organisation for Economic Co-operation and Development
PHI	post-harvest interval
Po	the recommendation accommodates post-harvest treatment of the food commodity
PoP	the recommendation accommodates post-harvest treatment of the primary food commodity

Abbreviations

P_{ow}	octanol–water partition coefficient
ppm	parts per million
RAC	raw agricultural commodity
STMR	supervised trials median residue
STMR-P	supervised trials median residue in a processed commodity calculated by multiplying the STMR of the raw commodity by the corresponding processing factor
THPI	1,2,3,6-tetrahydrophthalimide
TMDI	theoretical maximum daily intake
TRR	total radiolabelled residue
UNEP	United Nations Environment Programme
US	United States of America
W	The previous recommendation is withdrawn.
WHO	World Health Organization

Use of JMPR reports and evaluations by registration authorities

Most of the summaries and evaluations contained in this report are based on unpublished proprietary data submitted for use by JMPR in making its assessments. A registration authority should not grant a registration on the basis of an evaluation unless it has first received authorization for such use from the owner of the data submitted for the JMPR review or has received the data on which the summaries are based, either from the owner of the data or from a second party that has obtained permission from the owner of the data for this purpose.

PESTICIDE RESIDUES IN FOOD
REPORT OF THE 2004 JOINT FAO/WHO MEETING OF EXPERTS

1. INTRODUCTION

A Joint Meeting of the FAO Panel of Experts on Pesticide Residues in Food and the Environment and the WHO Core Assessment Group (JMPR) was held at FAO Headquarters, Rome (Italy), from 20 to 29 September 2004. The Panel Members of FAO and WHO met in preparatory sessions on 15–19 September.

The Meeting was opened by Dr Mahmoud Solh, Director, Plant Production and Protection Division, Department of Agriculture, FAO. On behalf of FAO and WHO, Dr Solh welcomed and thanked the participants for providing their expertise and for the significant time and effort put into this important activity. He noted that on the Meeting agenda there were a number of important issues for consideration, that would result in recommendations to the Codex Committee on Pesticide Residues (CCPR), as well as to Member States. He emphasized the importance of the JMPR and the commitment of FAO/WHO to its continuous support to this very important statutory body which gives scientific advice to FAO, WHO, their member countries and Codex.

Dr Solh noted that the CCPR had once again referred to the JMPR in order to speed up the establishment of maximum residue limits (MRLs) and to refine dietary intake as a result of a critical review by the JMPR, evaluation of the Codex process and the work of the OECD Working Group on Pesticides. He highlighted some of the important activities: the pilot project on work-sharing on the basis of national and regional evaluations; refinements of dietary risk assessment to approximate a more realistic situation at national and international levels; implementation of the recommendations on the York and zoning meetings to conserve resources; establishment of templates for estimating long- and short-term dietary intake; and development of a probabilistic approach to intake assessment.

The Meeting was held in pursuance of recommendations made by previous Meetings and accepted by the governing bodies of FAO and WHO that studies should be undertaken jointly by experts to evaluate possible hazards to humans arising from the occurrence of residues of pesticides in foods. The reports of previous Joint Meetings (see Annex 6) contain information on acceptable daily intakes (ADIs), acute reference doses (ARfDs), MRLs and the general principles that have been used for evaluating pesticides. The supporting documents (residue and toxicological evaluations) contain detailed monographs on these pesticides and include evaluations of analytical methods.

During the Meeting, the FAO Panel of Experts was responsible for reviewing residue and analytical aspects of the pesticides under consideration, including data on their metabolism, fate in the environment and use patterns, and for estimating the maximum levels of residues that might occur as a result of use of the pesticides according to good agricultural practice (GAP). Maximum residue levels and supervised trials median residue (STMR) values were estimated for commodities of animal origin. The WHO Core Assessment Group was responsible for reviewing toxicological and related data and for estimating ADIs and, if necessary, ARfDs, when possible.

The Meeting evaluated 31 pesticides, including two new compounds and 12 compounds that were re-evaluated within the periodic review programme of the CCPR, for toxicity or residues, or both.

The Meeting allocated ADIs and ARfDs, estimated MRLs and recommended them for use by the CCPR, and estimated STMR and highest residue levels as a basis for estimating dietary intake.

The Meeting devoted particular attention to estimating the dietary intakes (both short-term and long-term) of the pesticides reviewed in relation to their ADIs or ARfDs. In particular, for compounds undergoing a complete evaluation or re-evaluation, it distinguished between those for which the estimated intake is below the ADI and those for which the intake might exceed the ADI. Footnotes are used to indicate those pesticides for which the available information indicates that the ADI might be exceeded, and to denote specific commodities in which the available information indicates that the ARfD of the pesticide might be exceeded. A proposal to make this distinction and its rationale are described in detail in the reports of the 1997 JMPR (Annex 6, reference 80, section 2.3) and the 1999 JMPR (Annex 6, reference 86, section 2.2). Additional considerations are described in the report of the 2000 JMPR (Annex 6, reference 89, sections 2.1–2.3).

2. GENERAL CONSIDERATIONS

2.1 Guidance on the establishment of acute reference doses

This item summarizes a document drafted by a Working Group of the JMPR WHO Core Assessment Group[1], and the reader is referred to that document for a more detailed consideration of the issues outlined below. The document provides guidance on the issues to be considered when determining whether it is necessary to establish an acute reference dose (ARfD) on the basis of the hazard profile of a chemical as well on particular end-points that may be particularly relevant to acute effects. Note that it is intended that the detailed document be updated on a regular basis as further experience in establishing ARfDs is gained.

Introduction

In 1998, the JMPR WHO Core Assessment Group published brief guidance on procedures for setting ArfDs[1]. At its meeting in 2001, the JMPR established an international Working Group to compile a table of all available ARfDs and to collate information from different national agencies about their approaches to setting ARfDs. On the basis of an analysis of this inventory and a comparison of the technical policy approaches to setting ARfDs in different countries, further guidance was drafted by the Working Group and published by the 2002 JMPR[2]. At the request of the 2003 JMPR, the Working Group elaborated further guidance, including: (1) detailed advice on interpretation of certain toxicological end-points (haematotoxicity, immunotoxity, neurotoxicity, kidney toxicity, liver effects, endocrine effects and developmental toxicity) that might be particularly relevant to acute exposure to pesticides; and (2) recommendations about the design and conduct of a suitable single-dose study that might be useful in identifying a more appropriate no-observed-adverse-effect level (NOAEL) for an acute end-point. This guidance is summarized below.

In 2002, the JMPR recognized that databases on consumption are available for daily intakes but that this information generally cannot be further divided into individual meals. Thus, the original definition of the ARfD was reworded from 'over a short period of time, usually during one meal or one day' to 'the amount that can be ingested in a period of 24 h or less' (see definition below).

The necessity for setting an ARfD should be considered for other chemicals to which the population may be exposed, such as non-agricultural pesticides, veterinary compounds and contaminants. If it is considered that an ARfD is necessary for such a compound, the guidance provided in this and the more detailed document should be of value. It is hoped that a harmonized approach will be followed in establishing acute health values for these other types of compounds.

Analysis of the inventory of ARfDs set by different agencies (1995–2002)

An analysis of the ARfDs set for pesticides by several regulatory agencies between 1995 and 2002 was summarized in a working paper used by the Working Group in developing the guidance document published by the 2002 JMPR. The purpose of this inventory was to identify any obvious differences in the derivation of toxicology threshold values, to identify different approaches to the selection of safety or uncertainty factors, and to give more detailed guidance on harmonization of acute risk assessments. This analysis indicated significant variations in decisions about the need to set an ARfD for a particular chemical, in the selected NOAELs and in safety factors, and hence ARfDs, for the same pesticide among different agencies. Most agencies have based some ARfDs on studies in humans, particularly for inhibitors of

---

[1] Annex 6, reference 83
[2] Annex 6, reference 95

acetylcholinesterase activity[3]. The database also indicated that the practice employed by the Meeting, of using safety factors of less than 100 (or 10) for carbamate insecticides, had not been followed by any other agency in 1995–2002. It was apparent from the analysis of the database that the current data package of toxicological studies generated for the active ingredients of pesticides is poorly suited to the establishment of ARfDs.

Acute dietary risk assessment

At the international level, only limited data on acute consumption are available. Adults and children aged 1–6 years are the only population groups for which acute dietary intakes can be estimated. Therefore, during the Meeting, the models for calculating acute dietary intake of pesticide residues were based on these two populations. Since it is generally not possible to model separately the intake of women of childbearing age, the establishment of separate ARfDs is not particularly useful. Therefore, the ARfD used for modelling the dietary intake of the general population should protect this subgroup adequately.

Except for those chemicals that would be very unlikely to leave residues in foodstuffs, it was concluded that the trigger for performing an acute dietary intake assessment should be toxicity concern.

If it is considered unnecessary to establish an ARfD, it can be concluded that short-term intake is unlikely to present a public health concern. Therefore, it is not necessary to estimate the short-term intake of such substances.

General guidance on the derivation of ARfDs

Most of the scientific concepts applying to the establishment of acceptable dietary intakes (ADIs) apply equally to the establishment of ARfDs. This section is mainly based on the recommendations of the 2002 JMPR but expands on that guidance in a number of areas. The following key issues are highlighted:

Definition of the ARfD

The following definition of the ARfD was adopted by the 2002 JMPR: "The ARfD of a chemical is an estimate of the amount of a substance in food and/or drinking-water, normally expressed on a body-weight basis, that can be ingested in a period of 24 h or less, without appreciable health risk to the consumer, on the basis of all the known facts at the time of the evaluation".

General considerations in setting an ARfD

The establishment of an ARfD should be considered for all pesticides whose uses may lead to residues in food and drinking-water. The suggested numerical cut-off for setting ARfDs for pesticides was about 5 mg/kg bw; i.e. if calculations indicated that an ARfD would be greater than this value, then it would not be necessary on practical grounds to set an ARfD, as residue levels necessary to achieve this intake would be highly unlikely to occur in practice.

Biological and toxicological considerations

The appropriateness or otherwise of using doses and end-points from short- and long-term studies to establish ARfDs, needs to be carefully considered. The pertinent biology of the system affected should be considered in order to determine whether an acute exposure might compromise the ability of the organ to compensate and maintain homeostasis. Particular weight should be given to observations and investigations at the beginning of studies of repeated doses. Isolated findings showing no specificity or clear pattern are not necessarily indications of toxicity. In the absence of information to the contrary, all toxic effects seen in repeat-dose studies should be evaluated for their relevance in establishing an ARfD. In studies on compounds showing acute toxicity after repeated doses, the adequacy of investigations (including the duration of the

---

[3] Until 2002, the United States Environmental Protection Agency had based only one value in their database on a study in humans, but there has been a policy change since that date.

follow-up of the animals to see if there are any delayed effects) must be shown to be adequate before it can be concluded that the compound does not have any acute toxic potential. The NOAEL from the most sensitive species should be used, unless there is evidence to demonstrate it is not appropriate for a human risk assessment.

Stepwise process in setting an ARfD

The first step is to consider whether, on the basis of the acute toxicity profile and potency of effects, an acute health value is really necessary. The following process for setting ARfDs is suggested:

1. Evaluate the total database on the substance and establish a toxicological profile for the active substance.

2. Consider the principles for not setting an ARfD:
   - No findings indicative of effects elicited by an acute exposure are seen at doses up to about 500 mg/kg bw per day; and/or
   - No substance-related mortalities are observed at doses of up to 1000 mg/kg bw in single-dose studies after oral administration.
   - If mortality is the only trigger, the cause of death should be confirmed as being relevant to human exposures.
   - If the above criteria do not exclude the setting of an ARfD, then further consideration should be given to setting a value, using the most appropriate end-point.

3. Select appropriate end-points for setting an ARfD
   - Select the toxicological end-points that are most relevant for a single (day) exposure.
   - Select the most relevant or adequate study in which these end-points have been adequately determined.
   - Identify the NOAELs for these end-points.
   - Select the most relevant end-point providing the lowest NOAEL.
   - Use an end-point from a repeat-dose study of toxicity if the critical effect of the compound has not been adequately evaluated in a single-dose study. This is likely to be a more conservative approach and should be stated. This does not mean that a safety factor other than the default value should be applied. A refinement of such a NOAEL (e.g. in a special single-dose study) may be necessary, if the acute intake estimation exceeds such a conservatively-established ARfD.
   - If after consideration of all the end-points, an ARfD is not set, justify and explain the reasons.

4. Select appropriate safety factors for setting an ARfD.

   Derive the ARfD using an appropriate safety factor(s).

Safety factors

The process of deriving ARfDs involves the determination of the most appropriate NOAEL and a safety factor (also called an uncertainty or assessment factor). These factors are used to extrapolate from data in animals to the average human and to allow for interindividual variation within the human population.

In determining the appropriate safety factor, a stepwise approach is proposed:

- Determine whether the database is adequate to support the derivation of a chemical-specific adjustment factor (CSAF)[4].
- If a specific factor cannot be derived, consider if there is any information to indicate reduced or increased uncertainty. If not, the 100-fold (or 10-fold) default should be used.

Whenever a safety factor other than a default is used, a clear explanation of the derivation of the factor must be provided.

Different ARfDs for different population subgroups

It is preferable to set a single ARfD to cover the whole population. It is important to ensure that any ARfDs established are adequate to protect the embryo and fetus from possible effects in utero. While an ARfD based on developmental effects (on the embryo or fetus) would necessarily apply to women of childbearing age, it is recognized that such an ARfD may be conservative and not relevant to other population subgroups. This is the case for children aged 1–6 years, for whom specific data on acute consumption are available (and thus can be modelled separately with respect to acute dietary intake of residues); the use of such a conservative ARfD for children who generally have a higher intake of food commodities per unit body weight compared with adults could lead to an unreasonably conservative acute dietary risk assessment. Thus, in those situations in which a developmental end-point drives a very conservative ARfD, it may be necessary to set a second value based on another non-developmental end-point.

Alternatively, if the prenatal developmental end-point is the only acute effect of concern, there may be no need to perform an acute dietary intake risk assessment for children. The single ARfD based on a developmental end-point (in the embryo or fetus) would only be used to model the acute dietary intake of the adult population, which includes women of childbearing age.

Use of human data

Human data from accidental or deliberate poisonings, biomarker monitoring studies, epidemiology studies, volunteer studies, and clinical trials on the same or structurally similar compounds can provide useful data to help establish ARfDs. The use of data from human volunteers in chemical risk assessment is a controversial issue, with a range of views held by different countries and individuals. However, it is recognized that the use of such data can reduce the level of uncertainty inherent in extrapolating from animal models. There needs to be adequate consideration of both scientific and ethical issues. The JMPR has considered human data at many of its meetings. The 2002 JMPR reaffirmed the principle that end-points from studies in human volunteers could be used to set intake standards if they had been conducted in accordance with relevant ethical guidelines. The study should be assessed for the quality and integrity of the data and the adequacy of the documentation of methods (including statistics and control values) and results. A poorly designed or conducted study in humans should not be used for risk assessment. Additionally, because the designs of studies in humans have some limitations in comparison with those in experimental animals, their use should always be considered in the context of the overall toxicological database.

ARfDs based on studies in humans should provide a sufficient margin of safety for toxicological end-points that cannot readily be addressed by such studies (e.g. developmental toxicity).

When an initial estimate of the ARfD is less than the established ADI

If an ARfD derived using the principles outlined in this guidance document has a numerical value that is lower than the existing ADI, then the ADI should be reconsidered.

---

[4] Draft guidance document for the use of chemical-specific adjustment factors (CSAFs) for interspecies differences and human variability in dose/concentration–response assessment. WHO/UNEP/ILO International Programme on Chemical Safety, July 2001 (WHO/PCS01.4) (http://www.who.int/ipcs/methods/harmonization/areas/uncertainty/en/print.html, accessed 15 October 2004)

Specific guidance on the derivation of ARfDs

Particular toxicology end-points that are relevant to the establishment of ARfDs are considered below. Note that this summary is not intended to comprehensively cover all potentially relevant end-points, but focuses on the interpretation of those that have proved to be problematic in reaching a decision as to whether an effect is relevant to an acute exposure.

Haematotoxicity

The induction of methaemoglobinaemia is considered to be a critical effect in acute responses to chemical exposure. For acute exposure to methaemoglobin-inducing xenobiotics, a level of 4% methaemoglobin (or higher) above the background level in dogs, or a statistically significant increase relative to the background level in rodents is considered to be relevant in setting an ARfD. Haemolytic anaemias induced by mechanical damage, immune mediated anaemia, oxidative injury to erythrocytes and non-oxidative damage are considered to be less relevant for the derivation of ARfDs since the severity of such effects appears generally to depend on prolonged exposure. If changes in haematological parameters are observed early in a repeat-dose study and do not appear to progress during the course of the study, then such effects can be considered to relate to acute exposure to the substance. In assessing whether effects observed in repeat-dose studies should be used for setting an ARfD, the mechanism of action must be evaluated. If known, this could provide arguments for selecting, or not selecting, the end-point for setting an ARfD.

Immunotoxicity

Data on immunotoxicity derived from short-term studies are not likely to be appropriate for setting a reference dose for limits on acute exposure. It is unlikely that an acute exposure will produce persistent effects on immune function because the cells of the immune system are constantly replaced and because of the inherent redundancy in the system (e.g. alternative mechanisms exisit to resist infection).

Neurotoxicity

The nervous system has a limited capacity for repair and regeneration. Therefore, any neurotoxicity seen in repeat-dose studies could be the result of a single exposure and may not be reparable, i.e. any evidence of neurotoxicity should be considered relevant to an ARfD assessment, unless it can be demonstrated that the effects are produced only after repeated exposures. In addition to long-term or irreversible effects associated with acute exposure, attention should be paid to transient effects, as these could be considered to be adverse under some circumstances.

Delayed neurotoxicity after single chemical exposures can occur, and thus any acute exposure study should have an adequate period of investigation.

In functional observation batteries (FOBs), a large amount of data is produced; interpretation of such studies should include a consideration not only of the statistical significance of results, but alos of the nature, severity, persistence, dose–response relationship and pattern of the effects. Isolated findings showing no specificity or clear pattern do not necessarily indicate neurotoxicity.

The most common neurotoxic end-point used to date in the derivation of ARfDs for pesticides is inhibition of acetylcholinesterase activity. The Meeting has previously defined criteria for the assessment of cholinesterase inhibition[5]; these apply equally to the setting of ADIs and ARfDs. For inhibition of acetylcholinesterase activity, a specific cut-off (20%) is used routinely to differentiate between adverse and non-adverse effects.

Effects on the kidney and liver

If effects on these organs cannot be discounted as being either adaptive or the result of prolonged exposure, an ARfD can be derived on the basis of these effects. Such an ARfD is likely to be conservative

---

[5] Annex 6, reference 83

and it may be possible to subsequently refine it using the results of an appropriately designed single-dose study. When interpreting data on liver and kidney toxicity in repeat-dose studies, one has to consider two important aspects; firstly, the type of effect observed and secondly, any information on correlations between exposure duration and effect.

For liver toxicity, it is considered that findings of increased serum cholesterol, cirrhosis, induced activity of metabolizing enzymes, regenerative hyperplasia, hepatocyte hypertrophy, fibrosis, or sclerosis in repeat-dose studies are, in isolation, either adaptive or the result of prolonged exposure and are therefore are not applicable for deriving an ARfD.

For kidney toxicity, it is considered that the following findings of toxicity targeted to the kidney in repeat-dose studies are, in isolation, the result of prolonged exposure and are not applicable for deriving an ARfD: increased organ weight; regenerative hyperplasia; altered serum calcium and phosphate concentrations.

All other findings of liver and kidney toxicity should be considered to be potentially relevant to the derivation of an ARfD.

Endocrine effects

In general, effects on the endocrine system, other than those affecting female reproduction, are considered to be unlikely to arise as a consequence of acute exposure. However, because the process of evaluating endocrine toxicity and mechanisms is an evolving one, the guidance on endocrine disruptor chemicals is considered to be interim.

Developmental effects

Any treatment-related adverse effect on fetuses or offspring that has resulted from exposure during any phase of development should be considered to be potentially appropriate to use in acute dietary risk assessment, despite the fact that the period of treatment typically consists of repeated dosing. ARfDs based on reductions in fetal body-weight gain may be conservative and should be evaluated in the context of all pertinent data, including other developmental effects. Consideration should be given to the degree of maternal toxicity when considering whether fetal effects may be occurring as a direct effect of the chemical; the presence of severe maternal toxicity means that a direct effect is less likely.

Single-dose study protocol

ARfDs established using existing databases of repeat-dose studies may be overly conservative and it may be possible to refine the ARfD by conducting an appropriately-designed single-dose study. The protocol for a proposed single-dose study of toxicity is outlined in the detailed document. Unlike a classical median lethal dose (LD_{50}) study, this protocol is not intended to investigate mortality and significant morbidity. On the contary, it is aimed at investigating a range of more subtle end-points which may arise after a single exposure, or during 1 day of dietary exposure to a test substance. In particular, it is tailored to see whether toxic effects observed in the standard package of repeat-dose studies may occur after single doses, and if so, at what level of exposure.

It is important to note that a specific study designed to enable an accurate ARfD to be set should only be undertaken after the toxicology profile of an active substance has been reasonably well documented and understood, and it is apparent that the available database is not adequate to allow the establishment of a reasonable ARfD. Furthermore, it is not intended that this test guideline should become a routine data requirement. It can, however, be utilized for those chemicals with limited databases that do not allow a clear determination of an appropriate reference dose.

The draft test protocol for the single oral dose study of toxicity will be submitted to the Organization for Economic Co-operation and Development (OECD) Working Group of National Co-ordinators to the Test Guidelines Programme. If it is considered appropriate to incorporate into the OECD test guidelines, it will be useful in directing the conduct of studies directly relevant to the establishment of ARfDs for those chemicals which, because of their use pattern, have the potential to lead to residues in food and drinking-water.

Future considerations

It is intended that the detailed document drafted by the Working Group of the JMPR WHO Core Assessment Group be updated on a regular basis as further experience in establishing ARfDs is gained. For example, further more detailed guidance on specific end-points (e.g. emesis and diarrhoea in dogs) will be added as these issues are researched and assessed.

The introduction of a specific test protocol for a single-dose study of oral toxicity will aid the process of establishing ARfDs. Once introduced, the utility of such a test guideline will be evaluated by the Meeting. The detailed guidance document will be available from the JMPR Secretariat.

2.2 Definition of 'overall NOAEL'

During the toxicological evaluation of a compound, the Meeting often has available more than one study in which the same end-points have been addressed. In such situations, the dose spacing may be different, resulting in different NOAELs and lowest-observed-adverse-effect levels (LOAELs). The Meeting agreed that in such circumstances it might be appropriate to consider the studies together. When they are comparable, including consideration of study design, end-points addressed, and strain of animal, the 'overall NOAEL' should be the highest value identified in the available studies that provides a reasonable margin (≥ 2) over the lowest LOAEL, provided that due consideration is given to the shape of the dose–response curve.

2.3 Interim acute reference dose

The WHO Core Assessment Group is occasionally asked by the FAO Panel of Experts to establish an ARfD for compounds that were not scheduled for toxicological evaluation. In such cases, the ARfD is based on data available from previous evaluations and is used to perform an acute dietary risk assessment. The Meeting decided to call these values 'interim ARfDs' (in order to distinguish them from the ARfDs established for compounds that were scheduled for toxicological evaluation) and to publish the derivation of the interim ARfDs under general considerations. These interim ARfDs can be used in dietary risk assessments until they are replaced by a full evaluation, if this is considered necessary.

Setting an ARfD for propineb

At the present Meeting, the FAO Panel of Experts asked the WHO Core Assessment Group if an ARfD for propineb could be established on the basis of the data available to the 1993 JMPR. Propineb is metabolized to propylenethiourea, which forms part of the residue, and then to carbon disulfide. Propineb is of low acute toxicity (oral LD_{50}s in rats and mice are about 5 g/kg bw). The main effects of propineb are on the thyroid and on the nervous system.

The effects of propineb on the thyroid are mediated by propylenethiourea, which is similar to ethylenethiourea; these compounds block the accumulation of iodine by the thyroid gland. The 1993 JMPR took the view that those thyroid changes that were thought to be reversible, e.g. those observed in a study in rat could be considered adaptive. Furthermore, this effect is not appropriate for establishing an ARfD because of the buffering of thyroid hormone levels by homeostatic mechanisms. The existing ADI established by the JMPR (0–0.007 mg/kg bw per day) is based on the NOAEL (10 ppm, equal to 0.74 mg/kg bw per day) for decreased thyroxin and changes in thyroid weights in a short-term study of thyroid function in rats, with a 100-fold safety factor.

Propylenethiourea: The limited database on propylenethiourea was reviewed by the Meeting in 1993 when a temporary ADI of 0–0.0002 mg/kg bw was established. At this Meeting in 1993, no information was available on the metabolism of propylenethiourea. The temporary ADI was based on the LOAEL for liver tumours in a 2-year study in mice, with the application of a 1000-fold safety factor because of the inadequate database. The lowest dose in that study was considered to be a marginal effect level.

Carbon disulfide: Exposure to carbon disulfide is associated with cardiovascular disease, and carbon disulfide is also neurotoxic, producing demyelination, cerebellar atrophy and peripheral nerve abnormalities. Some or all of these effects are secondary to angiopathy. The effects on long axons might be caused by protein cross-linking. There is evidence from the clinical toxicology literature that carbon disulfide produces effects in humans after single, high-level exposures

Critical NOAELs for propineb: The neurotoxicity observed after repeated doses of propineb is considered to be potentially relevant to the establishment of an ARfD. The lowest NOAEL for neurotoxicity was identified in the multigeneration study in rats; the NOAEL in this study was 60 ppm, equivalent to 3 mg/kg bw per day on the basis of hind limb paralysis, and the LOAEL was 200 ppm, equivalent to 10 mg/kg bw per day. In the study of developmental toxicity in rats, the NOAEL for maternal toxicity was 10 mg/kg bw per day. Mild central neurotoxic effects were seen at 30 mg/kg bw per day in the dams. At the highest dose (100 mg/kg bw day), more serious central neurotoxicity was observed in the dams. Because of the shorter duration of dosing, the study of developmental toxicity in rats was considered to be more appropriate for the establishment of an ARfD than the multigeneration study in rats. Therefore the Meeting established an ARfD of 0.1 mg/kg bw, based on the NOAEL of 10 mg/kg bw per day from the study of developmental toxicity in rats, with a safety factor of 100. This should be considered to be an interim ARfD, until a full evaluation of the toxicological database has been conducted. Furthermore, this ARfD is probably conservative and could be refined using the results of an appropriate single-dose study.

2.4 Progress report on the JMPR work-sharing pilot project on trifloxystrobin

A FAO/WHO/OECD pilot project on work-sharing was conducted to test whether national and regional evaluations of pesticide residues and toxicology could be used as a basis for JMPR evaluations. The 2003 CCPR selected trifloxystrobin as the first compound for the project because it had been evaluated in Australia, Canada and the USA and by the European Commission and was scheduled for evaluation by the JMPR in 2004. Unfortunately, a Japanese evaluation was not available for the pilot project. Original data were also provided by the manufacturer.

The objective of work-sharing is to facilitate and expedite reviews by using national and regional evaluations, while maintaining independence and incorporating global perspectives. The pilot project was intended to promote, facilitate and formalize this practice, although it was not expected to save cost or time. Those would be the main goals in the future, with greater efficiency and more transparent evaluations.

Differences in the global applicability of evaluations of toxicological and residue data were noted: Studies on residues are often specific to particular registered uses, which differ from one country to another, while toxicological studies are generally globally relevant.

The data on the toxicology and residues of trifloxystrobin were evaluated by the 2004 JMPR, and experiences in work-sharing were assessed.

Experience in work-sharing on toxicological evaluations

The Australian monograph served as the 'master' evaluation and thus as the main basis for the JMPR monograph, because a preliminary review indicated that it was closest to the JMPR format. The sections on toxicokinetics, human data and metabolites were derived from the European Commission monograph, which was prepared by the United Kingdom.

In the first step, it was found that the manufacturer had submitted the same key toxicological studies on trifloxystrobin to all four agencies and to JMPR. In general, the national and regional reports described the methods and results of the studies in a reasonable level of detail.

The formats of the monographs and the amount of detail given in the study reviews differed substantially. The differences in the formats of the national evaluations made comparisons difficult, obviating importation of tables into the JMPR monograph. Tables in which the toxicological data from individual studies were summarized and the statistical analyses that supported the conclusions varied significantly in lay-out and detail, creating a great deal of work for the WHO temporary advisor.

In the next step, the study descriptions and selection of end-points in the Australian monograph were compared with those in the Canadian, US and European Commission monographs. This comparison revealed both similarities and differences in interpretations. When differences were found, JMPR experts made an independent evaluation of the original data and presented it to the JMPR. In only a few instances was it necessary to review the original data.

The process facilitated preparation of a transparent JMPR monograph, as the WHO evaluator could judge the results independently, usually without having to refer to the original studies. The evaluator could thus concentrate on areas of disagreement, helping to focus the JMPR deliberations. The evaluator also benefitted from others' insights into the significance of the effects seen in a study.

As trifloxystrobin represents relatively little hazard and has minimal toxicity, the national evaluations tended to differ on interpretation of the adversity of minor effects, such as changes in body weight and increases in liver weight.

The evaluations agreed reasonably well with regard to the selection and evaluation of key studies. Three studies with similar NOAELs were selected as pivotal for setting three ADI values, ranging from 0.038 to 0.1 mg/kg bw. Derivation of an ARfD was considered unnecessary in all the evaluations except for that of the US Environmental Protection Agency, which has a common practice of setting a separate ARfD for women of child-bearing age.

Experience in work-sharing on residue evaluations

The generic studies for residue reviews are: pesticide identity, physical and chemical properties, metabolism, environmental fate in soil and water–sediment systems, analytical methods, freezer storage stability, and fate of residues in processing and in feeding to farm animals. National and regional summaries and assessments of data on these subjects were taken into account. Supervised trials, which constitute the major part of a residue evaluation, were not included in this pilot project.

No one national or regional monograph could be used as the 'master' because of differences in the studies on trifloxystrobin. Furthermore, the studies in the dossier provided to JMPR were not all the same as those assessed at national and regional levels. The set of studies provided to the four national and regional authorities and to JMPR was identical only for metabolism in farm animals.

In the first step, a JMPR evaluation and an appraisal were prepared on the basis of the original studies provided by the company. In the second step, the evaluations of the studies in the national and regional reviews were compared with each other and with the JMPR evaluation. A 129-page report on the comparisons was prepared, which will be available on the FAO website.

The assessments of trifloxystrobin submitted by Australia and the European Commission were prepared in monograph format. In this format, excerpts from the applicant's dossier can be used (e.g. tables, metabolism schemes, some text), although the monograph does not include the applicant's proposals or recommendations. Independent, complete, final evaluations of a compound are made without considering registration purposes.

General considerations

The assessments of trifloxystrobin submitted by Canada and the USA were presented in review format. In this approach, the scientific results are summarized and then the applicant's dossier is subjected to a critical review which includes discussions of the applicant's proposals and recommendations with regard to registration of specific plant protection products.

The trifloxystrobin example showed some similarities and some differences in the procedures and approaches used by national and regional agencies for evaluating residues. These resulted in some divergence on conclusions, such as those for residue definitions and processing factors. As JMPR considers the worldwide use of pesticides when recommending MRLs for food commodities in international trade, its approach is not necessarily the same as those of national and regional organizations, which operate within registration systems.

Conclusions and recommendations

Issues relevant for both toxicological and residue evaluations

- The availability of several national and regional evaluations was useful for both the WHO and the FAO evaluators, despite the problems encountered. FAO, WHO and OECD should thus consider means to facilitate the provision of national or regional evaluations to JMPR evaluators.

- Consideration of multiple national and regional evaluations should aid progress towards international harmonization of dossiers and evaluations.

- The evaluation process, including standardization of formats and guidelines, should be harmonized further internationally. Good progress has been attained in the toxicological evaluations, while more work is necessary to improve work-sharing for residue evaluations.

- A further JMPR pilot work-sharing project should have more flexible procedures, which should be reviewed when the formats and evaluation procedures have been harmonized in guidance documents.

Issues more closely relevant to toxicological evaluations

- The national and regional evaluations differed considerably in their level of detail in describing toxicological studies, and more guidance and harmonization in this area could improve work-sharing.

- More guidance is needed on the interpretation of specific toxicological data, such as on changes in body and organ weights, and on interpretation of effects on the liver as adaptive or adverse.

- The key advantage of considering multiple national and regional evaluations in the JMPR process is the rapid identification of agreements and disagreements in data interpretation, thus focusing resources on areas of disagreement.

Issues more closely relevant to residue evaluations

- The project requires further work, incorporating changes based on the experience with trifloxystrobin, before work-sharing, with its anticipated benefits, can be implemented routinely. The experience with trifloxystrobin indicates that JMPR cannot comprehensively accept national or regional conclusions or recommendations for the residue topics included in the pilot project.

- Further progress in international work-sharing could be made by separating the summaries of key data from the conclusions in the submitted studies in the national or regional evaluations for risk assessment and management. This could facilitate mutual exchange and acceptance of study summaries.

- Work-sharing should focus on mutual use of summaries of data validated at national, regional and international level, covering all aspects of the residue evaluation, including data from supervised trials. This would allow exchange of a valid database, thus saving time and potentially reducing the workload.

- Specific assessment results, such as the definition of residue, could be used on a case-by-case basis.

General considerations

- The current workload of the JMPR precludes additional work by FAO Panel Members on this project, as it would be at the expense of normal residue evaluation commitments, which are regarded by Panel Members as the priority.

2.5 Comparison of JMPR recommendations and interim MRL recommendations from the CCPR pilot project

The 2004 CCPR agreed on the main steps of the procedure for establishing interim maximum residue limits (MRLs)[6]. These included a request to JMPR to compare its recommendations with the suggested interim MRLs for trifloxystrobin and fludioxonil and to comment on discrepancies, for the purposes of the pilot project; this would not be considered part of any normal procedure that might eventually evolve. The US Delegation to the CCPR, as Chair of the Pilot Project Working Group, provided FAO with detailed information on the chemicals nominated for interim MRLs. These included summaries of the proposed interim MRLs for fludioxonil and trifloxystrobin, as well as information on toxicology, residue chemistry and dietary risk assessment.

The table shows the proposed interim MRL and the corresponding JMPR recommended MRLs.

Commodity, CCN	Interim MRL recommendation (mg/kg)	JMPR MRL recommendation (mg/kg)	Comment on difference
Trifloxystrobin			
Interim definition: Plant and animal, trifloxystrobin + CGA321113 or (E,E)-methoxyimino{2-[1-(3-trifluoromethylphenyl)-ethylideneaminooxymethyl]phenyl}acetic acid			
JMPR definition: Plant, trifloxystrobin; animal, trifloxystrobin + CGA321113 or (E,E)-methoxyimino{2-[1-(3-trifluoromethyl-phenyl)ethylideneaminooxymethyl]phenyl}acetic acid			
Barley, GC640	0.3	0.5	High value from European Commission, 0.19 mg/kg. European Commission established 0.3 mg/kg (Tables B.7.49 and B.7.50)
			High value from JMPR, 0.40 mg/kg, based on trial in Germany, 1999
Grapes, FB269	3	3	
Grapes, dried, DF269	5	5	
Pome fruit, FP9	1	0.7	High value from European Commission, 0.44 mg/kg
Edible offal (mammalian), MO105	0.05	Kidney, MO98, 0.04*	Interim, wide scope
		Liver, MO99, 0.05	JMPR, narrow scope
Eggs, PE112	0.04*	0.04*	
Meat (mammalian), MM95	0.04*	0.05 (fat)	Interpretation of feeding study: JMPR determined a maximum residue in fat of 0.038 mg/kg; 0.04 mg/kg is the LOQ (0.02 trifloxystrobin + 0.02 metabolite). Trifloxystrobin present in fat (0.05 mg/kg) at a feeding level of 20 ppm is approximately twice the US and JMPR calculated dietary intake of cattle. The 0.04 value is based on one feed item, barley, as compared with the much greater intake of the total treated commodities considered by the US and JMPR.
Milks, ML106	0.02*	0.02*	
Poultry, edible offal of	0.04*	0.04*	

---

[6] ALINORM 04/27/24, paragraphs 220–234

Commodity, CCN	Interim MRL recommendation (mg/kg)	JMPR MRL recommendation (mg/kg)	Comment on difference
Poultry meat	0.04*	0.04* (fat)	Interpretation of feeding study

Fludioxonil

Interim definition: Plant, fludioxonil. Animal, fludioxonil + metabolites determined as 2,2-difluorobenzo[1,1]dioxole-4-carboxylic acid, calculated as fludioxonil

JMPR definition: Same

Commodity, CCN	Interim MRL recommendation (mg/kg)	JMPR MRL recommendation (mg/kg)	Comment on difference
Herbs (fresh), HH726	10	Basil, HH722, 10; Chives, HH727, 10	JMPR restricted MRLs to specific herbs and did not extend them to all herbs.
Herbs (dry), HH726	65	Basil, dry, DH722, 50; Chives, dry, HH727, 50	JMPR considered two trials and a drying factor of 8, yielding 15 and 24 mg/kg. The interim approach considered one trial each on dried basil and dried chives, with residues of 23 and 31 mg/kg.
Blackberry, FB264	5	5	
Blueberry, FB20	2	2	
Broccoli, VB400	2	0.7	Same data set. US interim value based on US brassica head and stem subgroup 5A (with higher residues for cabbage).
Cabbage, head, VB41	2	2	
Carrot, VR577	1	0.7	Same data set. Interim MRL based on US tolerance of 0.75 mg/kg rounded up under JMPR rules; highest residue value, 0.46 mg/kg. JMPR reported highest residue of 0.42 mg./kg from same data set.
Wheat, GC643	0.02*	Cereal grains GC80, 0.05*	Interim based on seed treatments in the USA, with an LOQ of 0.02 mg/kg. JMPR based on 71 trials in Europe and the USA, with LOQs ranging from 0.01 to 0.05 mg/kg.
Rye, GC650	0.02*		
Spelt, GC4673	0.02*		
Triticale, GC653	0.02*		
Barley, GC640	0.02*		
Oats, GC647	0.02*		
Maize, GC645	0.02*		
Popcorn, GC656	0.02*		
Sorghum, GC651	0.02*		
Cotton-seed, SO691	0.05*	0.05*	
Currants, FB21	2		Interim based on translation of blueberry field trials (< 0.05–1.4 mg/kg) to the US bushberry subgroup 13B. JMPR did not make this translation.
Grapes, FB269	2	2	
Longan, FI342	1	None	Based on lychee
Lychee, FI343	1	None	Same data set (US). JMPR considered the three trials to be in excess of GAP.
Mustard greens, VL485	20	10	Same data set. Interim maximum residue value, 7.7 mg/kg; JMPR maximum residue, 7.1 mg/kg
Onion, VA385	0.2	0.5	Interim based on US data only, with maximum value of 0.11 mg/kg (0.06 mg/kg average for high field trial). Interim included European data, with maximum value of 0.34 mg/kg
Pistachio, TN675	0.1	0.2	Same data set (US). Although highest residue was 0.08 mg/kg, JMPR estimated 0.2 mg/kg on basis of small size of set (n = 3).

Commodity, CCN	Interim MRL recommendation (mg/kg)	JMPR MRL recommendation (mg/kg)	Comment on difference
Potato, VR589	0.02*	0.02	
Pulasan, FI357	1	None	Based on lychee
Rambutan, FI358	1	None	Based on lychee
Rape-seed, SO495	0.01*	0.02*	Interim based on translation of data on treatment of other seed (wheat), with an LOQ of 0.01 mg/kg. JMPR based on 15 trials in Europe, with an LOQ of 0.02 mg/kg.
Raspberry, FB272	5	5	
Soya, SO4723	0.01*	None	JMPR received no data. Interim based on seed treatment use and translation of data for wheat, lettuce, pea, cucumber and radish (all below LOQ)
Spanish lime, FI366	1	None	Based on lychee
Stone fruit, FS12	5	5	
Strawberry, FB275	2	3	Interim based on US data only, with maximum value of 1.3 mg/kg. JMPR included European data, with maximum of 1.9 mg/kg.
Sunflower seed, SO702	0.01*	None	JMPR received no data. Interim based on seed treatment use and translation of data for wheat, lettuce, pea, cucumber and radish (all < LOQ)
Sweet corn (corn-on-the-cob), VO447	0.02*	0.01*	Interim based on LOQ for cereal grain group. Codex does not consider sweet corn in the cereal grain group and evaluated data separately.
Watercress, VL473	10	10	
Meat (from mammals other than marine), MM95	0.01*	0.01*	
Edible offal (mammalian), MO105	0.05*	0.05	
Milks, ML106	0.01*	0.01	
Poultry meat, PM110	0.01*	0.01*	
Poultry, edible offal of, PO111	0.05*	0.05*	
Eggs, PE112	0.05*	0.05*	

LOQ, limit of quantification; GAP, good agricultural practice

Trifloxystrobin

The interim MRL recommendations are based on information and data compiled by the USA from both the European Union and the USA. The maximum residue levels recommended by the JMPR are based on an independent review of studies supplied by the manufacturer and include field trial data from more diverse sources: Australia, Brazil, Canada, Colombia, Costa Rica, Ecuador, Europe (Belgium, Denmark, France, Germany, Italy, The Netherlands, Spain, Switzerland, United Kingdom), Guatemala, Honduras, Mexico, South Africa, and the USA.

The interim MRLs are based on the residue definition in the USA, which is parent plus an acid metabolite. This differs fundamentally from the JMPR definition for enforcement for plant commodities, which is trifloxystrobin only. The residue definitions for dietary intake and for animal commodities are identical.

The manufacturer proposed, via the Pilot Project Working Group on the Interim MRL, only three commodities for this test exercise: grapes, barley and pome fruit, while the JMPR evaluation included an extensive array of plant commodities. Thus, only a limited comparison is possible.

The recommended interim MRLs and JMPR MRLs are similar for some commodities. The exceptions are mammalian meat (interim, 0.04 mg/kg; JMPR, 0.05 mg/kg), pome fruit (interim, 1 mg/kg; JMPR, 0.7 mg/kg) and barley (interim, 0.3 mg/kg; JMPR, 0.5 mg/kg). The discrepancy in the values for meat may be due to the fact that only barley was considered in the animal diet, as this was the only feed commodity suggested by the manufacturer for an interim MRL. The value of 0.04 mg/kg represents the limit of detection of the analytical method. JMPR estimated the intake of cattle at 6.3 ppm, and the USA estimated it at 12 mg/kg. The USA maintains a tolerance of 0.5 mg/kg on the basis of a long list of trifloxystrobin-treated commodities. The value was reduced to 0.04 mg/kg because barley was the only feed item being considered for an interim MRL. The reviews of both the USA and the JMPR considered the maximum residue level in beef fat to be 0.05 mg/kg, derived from a 20 ppm dietary burden of trifloxystrobin. Thus, the difference can be attributed to the limited interim MRL list; the values would both be 0.05 mg/kg with the longer list of animal feed commodities. The value of 0.05 mg/kg is appropriate on the basis of the animal feed items for which JMPR is proposing MRLs.

The interim MRL value of 1 mg/kg for pome fruit is based on the highest residue value of 0.44 mg/kg reported from Europe. The US tolerance is 0.5 mg/kg, based on a maximum residue value of 0.23 mg/kg. Under the JMPR system, the MRL would be either 0.7 mg/kg or 1 mg/kg. The Meeting selected 0.7 mg/kg in view of the large number of trials available (74) and the median of 0.11 mg/kg. These factors suggest that the MRL may be set near the maximum residue value, i.e. 0.7 mg/kg and not 1 mg/kg.

The interim MRL for barley is based on data from field trials considered by the European Commission in establishing a European Union MRL of 0.03 mg/kg. The JMPR considered additional data from field trials. A trial in Germany yielded a maximum residue value of 0.40 mg/kg. The 0.5 mg/kg value is appropriate when this additional information is considered.

The dietary risk assessments for the interim MRLs and for the JMPR MRLs indicated no concern. Using supervised trials median residue (STMR) values and an ADI of 0.04 mg/kg bw per day, the JMPR estimated the intake to be 1–2% of the ADI in regional diets. Using MRL values (for the limited list of commodities) and a reference dose of 0.038 mg/kg bw per day for long-term exposure (USA), the Interim Pilot Project estimated the intake in regional diets to be 0–6% of the ADI.

A calculation of dietary intake during short-term exposure similar to that of JMPR was performed within the Pilot Project, taking the proposed interim MRLs as residue levels and the ARfD determined in the USA (2.5 mg/kg bw per day). It should be noted that the end-point applies only to the subpopulation of females aged 13–49 years and is based on developmental toxicity. The calculation was performed for the general population, because the Codex system does not have data on the consumption of this subpopulation The ARfD was not exceeded for any commodity. The maximum exposure of the general population (used as a surrogate for women of child-bearing age) was 1% of the ARfD, from apples. The JMPR determined that an ARfD need not be established.

Fludioxonil

The interim MRLs recommended for fludioxonil are based solely on data from the USA, as the European Commission has not completed its evaluation of fludioxonil. The JMPR made an independent evaluation of studies provided by the manufacturer, which included field trials from Chile, Europe (Denmark, France, Greece, Hungary, Italy, Spain, Sweden, Switzerland), South Africa and the USA.

The residue definitions in the interim proposal and the JMPR recommendation are identical for plants and animals for both enforcement and dietary intake. The interim and the JMPR MRL recommendations differed for currants, lychee, longan, pulasan, rambutan, Spanish lime, strawberry, sweet corn, broccoli, carrot, mustard greens, bulb onion, basil (dry), chives (dry), pistachio and cereal grains.

The interim MRL for *currants* is based on a translation of data from US blueberry field trials, in which the residue values ranged from < 0.05 to 1.4 mg/kg. Blueberries are the representative commodity for the US bushberry subgroup 13B. The JMPR had no data from trials conducted under GAP and did not consider a translation of data on blueberries to currants.

The interim MRLs for *lychee, longan, pulasan, rambutan* and *Spanish lime* are based on three US field trials with lychee. The JMPR received the same data set but considered that the three trials were conducted substantially in excess of US GAP. US GAP specifies a maximum of four applications, each at 0.25 kg ai/ha, with a 0 day PHI; however, in the trials, five or seven applications were made at the maximum rate, with an approximate 7-day re-treatment interval. On the basis of the results of decline studies conducted with other fruit crops, JMPR concluded that the extra applications might make a significant (25%) contribution to the residue.

The interim MRL proposal for *strawberry* is based on US data only, with a maximum value of 1.3 mg/kg. The JMPR considered both European and US data; the maximum value in the former was 1.9 mg/kg.

The interim MRL proposal for *sweet corn* is based on US data for cereal grains, with an LOQ of 0.01 mg/kg. The Codex System does not place sweet corn in the cereal grains group, and the JMPR evaluated data for sweet corn and field corn, with an LOQ of 0.02 mg/kg.

The same US data were used to derive the interim and the JMPR MRL recommendations for *broccoli*. The interim MRL is based on a high residue value of 0.53 mg/kg, found on one of two samples in one field trial. The value is also based on a US tolerance for the brassica head and stem subgroup 5A, which allows higher residue values for cabbage. The JMPR MRL was based on a high residue value of 0.36 mg/kg, which is the average for the two samples in the same field trial ([0.53 mg/kg + 0.19 mg/kg]/2). Under JMPR procedures, 0.5 mg/kg is an appropriate MRL.

The same US data were used to derive the interim and the JMPR MRL recommendations for *carrot*. The interim MRL is based on the US tolerance of 0.75 mg/kg, rounded under JMPR rules to 1 mg/kg. Residue values ranged from 0.04 to 0.42 mg/kg in the nine trials. JMPR considered 0.7 mg/kg to be an appropriate MRL estimate.

The same US data were used to derive the interim and the JMPR MRL recommendations for *mustard greens*. The residue values ranged from 0.06 to 7.1 mg/kg in seven trials. JMPR considered 10 mg/kg to be an appropriate MRL estimate. The proposed interim MRL value of 20 mg/kg does not seem appropriate, given the number of trials and the range of residue values.

The same US data were used to derive the interim and the JMPR MRL recommendations for *dried chives* and *dried basil*. The interim MRL of 65 mg/kg for dried herbs is based on the results of one trial each with basil and chives, with residues of 23 and 31 mg/kg, respectively. The recommended JMPR maximum residue level of 50 mg/kg for dried basil is based on two trials and a drying factor of 8 (from one trial), yielding values of 15 and 24 mg/kg. For dried chives, the recommended JMPR maximum residue level is based on two trials and a drying factor of 8 (from one trial), yielding values of 14 and 31 mg/kg. The recommended JMPR maximum residue level of 50 mg/kg is appropriate for both dried basil and dried chives. The JMPR considered that the limited data available on basil (two trials, one including drying) and chives (two trials, one including drying) were inadequate to recommend an MRL for the general category of dried herbs.

The same US data were used to derive the interim and the JMPR MRL recommendations for *pistachio* nuts. Three trials were conducted in the USA, with the highest residue being 0.08 mg/kg. The interim MRL recommendation is based on the US tolerance of 0.1 mg/kg. The JMPR considered that the number of data points was minimal and therefore estimated an MRL of 0.2 mg/kg.

The interim MRL recommendation for soya (oil seed) of 0.01 mg/kg is based on a translation of data from treatment of seeds for wheat, lettuce, pea, cucumber and radish. JMPR did not receive any data on soya seed treatment and did not translate data on rape or cereal grain.

The interim MLR for rape-seed is based on a translation of data from various seed treatments, with an LOQ of 0.01 mg/kg. The JMPR value is based on the results of trials on the treatment of rape-seed, with an LOQ of 0.02 mg/kg.

The various interim MRLs for cereal grains are based on field trials of seed treatment for wheat, maize and other crops in the USA and are set at the LOQ of 0.02 mg/kg. The JMPR recommendation for the maximum residue level for *cereal grain* is based on 71 seed treatment trials in Europe and the USA, with LOQs ranging from 0.01 to 0.05 mg/kg. The highest value was selected by the JMPR.

The dietary risk assessments with the interim and the JMPR MRLs found no dietary concern. Using STMR values and an ADI of 0.4 mg/kg bw per day, the JMPR estimated the intake at 0–1% of the ADI for regional diets. Using MRL values and a reference dose of 0.03 mg/kg bw per day for long-term exposure (US), the Interim Pilot Project estimated the intake to be 0–13% of the ADI for regional diets. Neither calculation indicates concern associated with long-term dietary intake by the general populations in the five regions of the Global Environment Monitoring System–Food Contamination Monitoring and Assessment Programme (GEMS/Food).

A calculation of dietary intake for short-term exposure similar to that of JMPR was performed in the USA, using the proposed interim MRLs as residue values and the ARfD as determined in the USA (1.0 mg/kg bw per day). It should be noted that the end-point applies only to the subpopulation of females aged 13–49 years. The calculation was performed for the general population, because the Codex system does not have data on the consumption of this subpopulation. The maximum exposure of the general population (used as a surrogate for women of child-bearing age) was 10% of the ARfD, from stone fruits. The JMPR considered that an ARfD need not be established. Both the Pilot Group for the Interim MRL and the JMPR concluded that there was no concern associated with short-term dietary intake from the uses considered.

General comments

Several significant differences in the approaches used by the Pilot Group for the Interim MRL and the JMPR become apparent during consideration of the maximum residue levels, because the JMPR uses the average for replicate samples or from replicate analyses, whereas the interim proposals are based on maximum values. This difference usually has only a minimal effect. The JMPR has access to a larger database of field trials and can thus make recommendations on the basis of wider use. The larger database might give rise to different estimates, even when the residue values are below the LOQ. Consideration of full studies from the manufacturer allows inclusion of results that might be considered 'outliers' by the European Commission, thereby giving rise to significantly different estimates of maximum residue levels. As crop groupings and extrapolations differ among classification systems, an interim proposal for a maximum residue level in a commodity that is based on a national crop group might be significantly different from that based on a single commodity by the JMPR.

The JMPR agrees with the goals of the project, namely to accelerate the process for establishing MRLs within the Codex system and the introduction of standards for pesticides that are safer alternatives for use on food and feed. The Meeting expressed concern over the meaning of the word 'safer', which could mean either less toxic (hazard) or a lower residue level (exposure), and the extent of the 'safer' designation, which could be either for human health (dietary intake) or for environmental effects. The CCPR has included a number of safeguards in the process, the most important being analyses of long-term and short-term dietary intake based on JMPR methods and consideration by the Codex Alimentarius Commission.

The JMPR considers that extensive use of the interim MRL process might create a serious problem. As the interim MRLs are limited to a period of 4 years, pesticides nominated in this process must be scheduled for and reviewed by the JMPR within this period. If there are many nominations for interim MRLs for pesticides, the currently limited resources of the JMPR might mean that the evaluations of some of the pesticides might not be completed within the 4 years or that other priorities, such as periodic reviews and evaluations, might have to be severely curtailed.

2.6 Estimation of maximum residue levels of pesticides in or on spices on the basis of monitoring results

The setting of MRLs for spices on the basis of monitoring results was discussed by the CCPR on several occasions. The 2002 JMPR prepared guidelines for the submission of monitoring data for evaluation (xx). The CCPR at its Thirty-sixth Session proposed that commodity group A028 be subdivided into groups on the basis of the parts of plants from which they are obtained—seeds, fruits or berries, roots or rhizomes, bark, buds, arils and flower stigmas—and that MRLs for pesticides that had been evaluated within the Codex system should be set for these sub-groups rather than for each of the pesticide–spice combinations.[7]

The CCPR proposed that the MRL should cover at least 95% of the residue population at the 99% confidence level.[8]

The Meeting evaluated the data on residues provided but emphasized that estimating maximum, high and median residue values does not necessarily mean that the use of those compounds for use on spices is approved by the JMPR.

The Meeting considered the nature of monitoring results and identified the following principal differences from residue data derived from supervised field trials:

- The origin and treatment of the lots sampled are not known.
- The sampled commodity might be composed of the produce of several small fields.
- The residues in spice samples were determined by multi-residue procedures, and their contents were usually screened for organochlorine and organophosphorous compounds, which have relatively high LOQs.
- When residue values are reported as being below the LOQ, it is not necessarily true that the sampled commodity was not treated with or exposed to the pesticide.

Consequently, estimation of maximum residue levels for pesticides on the basis of monitoring results requires a different approach from that used for the evaluating the results of supervised field trials.

Basic principles for evaluating monitoring data

The Meeting assumed that the laboratories reported only valid results. Therefore, all residue data were taken into account as there was no scientific ground for excluding any value as an outlier due to the fact that no information was available on registered or approved uses or application conditions of the pesticides.

It is unlikely that all the sampled commodities were treated with the pesticides in the multi-residue screening procedure; therefore, the proportion of commodities treated with or exposed to a given pesticide was calculated from the ratio of samples containing detectable residues and the total number of samples analysed.

The distributions of residues were scattered or skewed upwards, and no distribution fitting appeared to be appropriate. Consequently, distribution-free statistics was used in estimating the maximum residue level, covering the 95th percentile of the population at the 95% confidence level. Thus, the estimated

---

[7] ALINORM 04/24, para 236
[8] CX/PR 04/13

maximum residue level encompasses at least 95% of the residues with 95% probability (in 95% of cases). To satisfy this requirement, a minimum of 58–59 samples is required.

STMR and the highest residue values can be calculated only from supervised trials. The corresponding values from the monitoring data are indicated as median and high residue values, and these can be used like the STMR and highest residue values for estimating short-term and long-term intake of residues.

In accordance with the recommendation of the CCPR, maximum, high and median residue values were estimated for pesticide residues in the spice subgroups shown in Table 1 if at least 58 samples belonging to one subgroup were analysed for the given pesticide. The minimum sample size of 58 provides 95% assurance of finding at least one residue value above the 95th percentile of the residue population in the sampled object. It is not, however, known how many of the measured values are above the 95th percentile and what percentile (95.1th, 99th or 99.9th) the highest residue represents. The 95th percentile of the sample does not necessarily represent the 95th percentile of the residue population in the sampled commodity, especially when the sample size is small.

The procedure used depended on the number of samples containing detectable residues.

- When no sample contained detectable residues, the highest reported LOQ value was used as the maximum residue level and the high residue value. The median residue value was calculated from the reported LOQ values.

- When a large number of samples (> 120) contained detectable residues, the sample size was sufficiently large to calculate the upper 95% one-tailed confidence limit of the population of residues, on the basis of binominal (distribution free) probability calculation.[9]

The confidence limits calculated for pirimicarb residues in anise seeds (number of positive samples, 129) are illustrated in Figure 1. The data set contained 129 residue values with maximum and 95th percentile values of 1.4 and 0.68 mg/kg, respectively. The upper confidence limit of the 95th percentile is between 0.93 and 1.2 mg/kg. This would require an MRL of 1.2 mg/kg, but it is rounded up to 2 mg/kg according the general practice of the JMPR.

When more detectable residue values are available (n = 343; max 10 values: 11, 9.4, 7, 6.9, 6.8, 6.8, 6.7, 6, 5.9, 4.4, mg/kg; 95th percentile of sample, 3.2 mg/kg; median, 0.5 mg/kg), the upper 95% confidence limit is lower than the maximum of residues observed, and the residue values above the upper confidence limit need not be considered for the maximum residue level. The situation is illustrated in Figure 2.

As the upper 95% confidence limit is between 5.9 and 6 mg/kg, an MRL of 7 mg/kg would be appropriate. The highest three residue values (7, 9.4 and 11 mg/kg) were above the upper 95% confidence limit and did not influence the estimated limit.

3. A few samples (≤ 120) contained detectable residues. In such cases, the upper 95% confidence limit cannot be calculated from the 95th percentile value of the residues. Sufficient allowance should be given when the maximum residue level is estimated to be above the highest residue value observed. The situation is illustrated in Figure 3 by the example of dimethoate residues in anise (n = 61; maximum, 3 mg/kg; 95th percentile for all samples, 0.9 mg/kg). For the given residue population, the Meeting estimated a maximum residue level of 5 mg/kg.

---

[9] Hamilton DJ, Ambrus Á, Dieterle RM, Felsot Á, Harris C, Petersen B, Racke K, Wong SS, Gonzalez R, Tanaka K. Pesticide residues in food—Acute dietary intake. Pest Manag Sci 2004;60:311–39.

General considerations

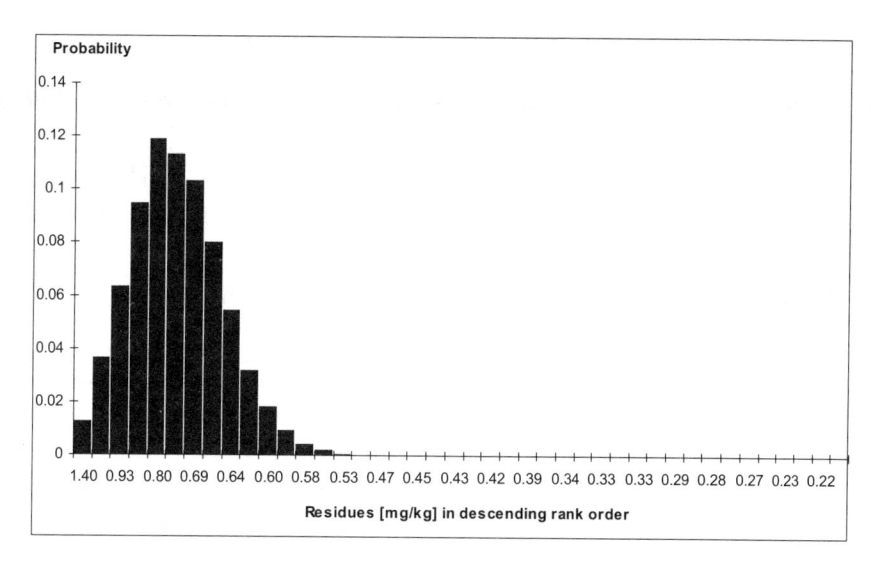

Figure 1. Upper 95% confidence limits for the 95th percentile of pirimicarb residues in anise seed (*n* = 129)

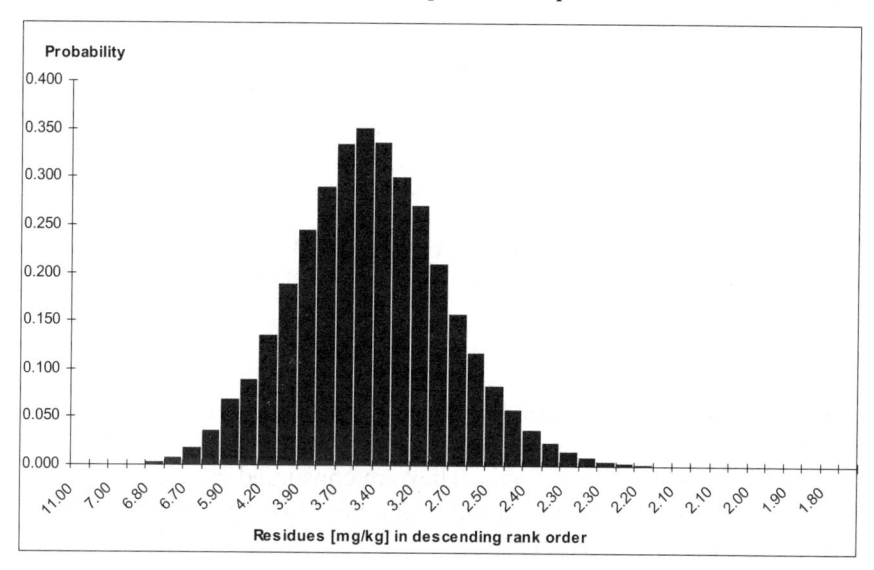

Figure 2. Upper 95% confidence limits for the 95th percentile of chlorpyrifos residues in cumin seed (*n* = 343)

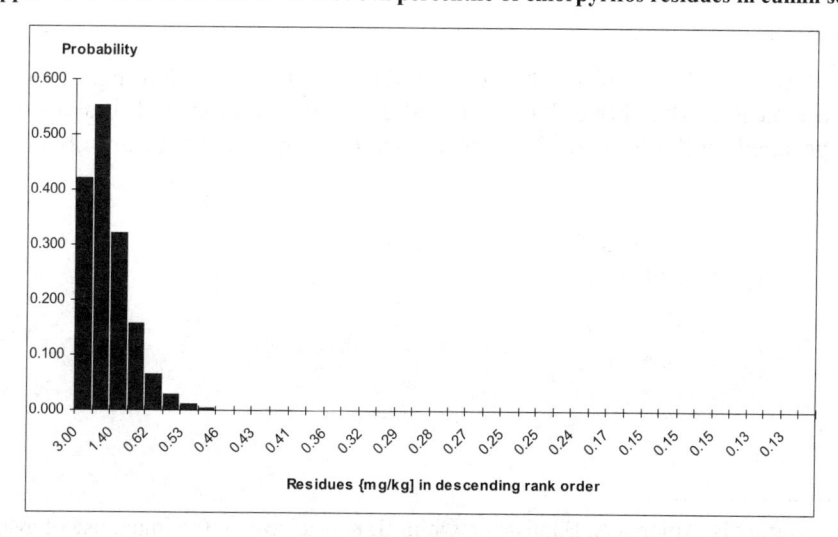

Figure 3. Distribution of estimated 95th percentile estimates of dimethoate residues in anise seed (*n* = 61)

Calculation of international estimated short-term dietary intake (IESTI) and international estimated daily intake (IEDI)

The revised subgroups of spices in group A28 do not all correspond to the consumption figures used for intake calculations. Therefore, the calculations were made with the combined amounts listed for spices in the GEMS/Food tables, in grams per day.

The calculations of short-term intake were based on the highest value of residues of the given pesticide measured in any spice sample.

The IEDI was calculated only from the residue levels detected in the pesticide–spice commodity combination that made the highest contribution to intake of any of the subgroups. A factor for the proportion of treated commodities was calculated from the ratio of the number of samples containing detectable residues and that containing undetectable residues. Adjustment of the median residue value by this factor was one of the recommended procedures for refining intake calculations.[10]

Recommendations

The Meeting recommended that the CCPR accept the principle of setting MRLs on the basis of monitoring results covering the 95th percentile of the residue population at the 95% confidence level. This decision would facilitate use of the statistical procedures for estimating maximum residue levels and acceptance of recommended limits. It should be noted, however, that when MRLs are set at the 95th percentile with 95% confidence, the residue levels might exceed the MRLs in 5% of cases.

Monitoring results should not be used for estimating maximum residue levels that reflect post-harvest use, which results in much higher residue values than foliar application or exposure to spray drift.

Definition of spices

Spices (Group 028) are dried aromatic seeds, buds, roots, bark, pods, berries or other fruits from a variety of plants, rhizomes and flowers, or parts thereof, which are used in relatively small quantities for seasoning, flavouring or imparting aroma to foods. They are consumed in dried form after being added to or sprinkled on foods. The portion of the commodity to which the MRL applies (and which is analysed) is only the dried commodity as it moves in international trade.

Spices of interest and proposed subgroups

The Committee agreed that spices should be collected into smaller groups to facilitate the setting of group MRLs. On the basis of the above definition and after consultation with the spice trade industry, the spices of interest, grouped on the basis of the parts of plants from which they are obtained, are as shown in the table below.

---

[10] Hamilton D, Murray B, Ambrus Á, Baptista G, Ohlin B, Kovacicova J. Optimum use of available residue data in the estimation of dietary intake of pesticides. Pure Appl Chem 1997;69:1373–410.

General considerations

Group 028—Spices (as modified by the CCPR at its Thirty-sixth Session, 2004; CX PR 04/13)

CCN[a]	Common name	Botanical name	Industry associations consulted
Seeds	Ajowan	*Carum ajowan*	ESA
	Bishop's weed	*Trachspermum ammi*	ISO, India
HS 771	Anise	*Pimpinella anisum*	ESA, ASTA (fruit in ISO, India)
	Black caraway	*Nigella satia*	ASTA
HS 624	Celery seed	*Apium gravveolens*	ASTA, ESA, (fruit in India, ISO)
HS 779	Coriander	*Coriandrum sativum*	ASTA, India (fruit in ESA)
HS 780	Cumin	*Cuminum cyminum*	ESA, ASTA (fruit in ISO, India)
HS 730	Dill	*Anethum graveolens*	ESA, ASTA, India (fruit in ISO)
HS 731	Fennel	*Foeniculum vulgare*	ESA, ASTA (fruit in India)
HS 782	Fenugreek	*Trigonella foenum graecum*	ESA, ASTA, India, ISO
HS 789	Nutmeg	*Myristica fragrans*	ESA, ASTA, India (kernel in ISO)
HS 4783	Poppy seed	*Papaver somniferum*	India
HS 4785	Sesame seed	*Sesamum indicum* or *orientale*	India
	Mustard seed		Proposed addition
Fruits and berries			
HS 4769	Allspice	*Pimenta dioica*	ESA, ASTA, ISO, India
	Anise pepper; Japan pepper	*Zanthoxylum piperitum* DC	ESA
	Candlenut	*Aleyrites moluccana*	Indonesia
HS 774	Caraway	*Carum carvi*	ESA, India, ISO (seed in ASTA)
HS 775	Cardamom	*Elettaria cardamomum*	ESA, ASTA, India, ISO
HS 786	Juniper	*Juniperus communis*	ASTA, ESA, India, ISO
HS 735	Lovage	*Levisticum officinale* Koch	ISO, India (root in ESA)
HS 790	Pepper (black, white, green)	*Piper nigrum*	ESA, ASTA, India, ISO
HS 791	Pepper, long	*Piper longum*	India, ISO
HS 792	Pimento	*Pimenta officinalis*	ESA, ASTA
	Pink pepper	*Schinus terebinthifolius*, *S. molle*	ASTA, ESA
	Sichuan pepper	*Zanthoxylum bungei* Planch	ESA
	Star anise	*Illicium verum*	ASTA, ESA, ISO, India
HS 795	Vanilla bean	*Vanilla planifolia*; or *tahitensis*	ESA, ASTA, ISO, India
Barks			
HS 4775	Cassia bark	*Cinnamomum aromaticum*	ASTA, ESA, India, ISO
HS 777	Cinnamon	*Cinnamomum zeylanicum*, *C. verum*	ESA, ASTA, India
	Tejpat	*Cinnamomum tamala*	India
Roots and rhizomes			
	Asafoetida	*Ferula assafoetida*	ESA, India, ISO
HS 781	Elecampane	*Inula helenium*	ESA
HS 784	Ginger	*Zingiber officinale*	ESA, ASTA, India, ISO
	Greater galangal; Siamese ginger	*Apinia galanga* Willd	ESA, India, ISO

CCN[a]	Common name	Botanical name	Industry associations consulted
	Lesser galangal	*Apinia officinarum* Hance	ESA, ISO
	Galangal	*Kaempferia galanga*	ESA, ISO
HS 772	Sweet flag	*Acorus calamus*	India, ISO
HS 794	Turmeric	*Curcuma longa*	ESA, ASTA, India, ISO
Buds			
HS 7743	Capers	*Capparis spinosa*	ESA, India, ISO
HS 778	Cloves	*Syyzgiunm aromaticum*; *Caryophyllus aromaticus*	ESA; ASTA, India, ISO
Flower stigmas			
	Saffron	*Crocus sativus*	ESA, India, ISO, ASTA, Spain, Malta
Arils			
HS 788	Mace	*Myristica fragrans.*	ISO, India, ASTA, ESA

CCN, Codex classification number; ASTA, American Spice Trade Association; ESA, European Spice Association; India, Indian Spices Board; ISO, International Standards Organization (specifically, ISO 676 Standard for Spices and Condiments)

[a] Those with a CCN are currently in the list of spices; others are additions to the list.

2.7 Revisited: MRLs for fat-soluble pesticides in milk and milk products

The 2003 JMPR reiterated its explanation of the rationale behind the expression of MRLs for fat-soluble pesticides in milk and milk products.[11] Nevertheless, the 2004 CCPR indicated that the procedure is still complex and considered that it was bound to introduce errors. For instance, the JMPR assumes that adding the suffix 'F' to an MRL for milk clearly implies that analysis for compliance testing must be conducted on the milk fat. The 2004 CCPR noted that the suffix 'F' means only that the specific rule for application of the MRL to the fat in milk products applies. It stated that the MRL applies to the milk as such and might therefore be determined for whole milk.

Currently, the JMPR follows the Codex convention of expressing the MRL for fat-soluble compounds in milk on the basis of the calculated whole product, assuming that all milks contain 4% fat.[12] The residue concentration is calculated for the whole product from the concentration measured in fat, and the MRL would be 1/25th of the residue concentration estimated for the milk fat. Many pesticides are, however, have intermediate solubility in fat; even if they are regarded as fat-soluble, they can be distributed equally between the fat and non-fat portions of milk. For example, if the ratio of residue concentrations between the fat and aqueous phases is 15:1 in milk with 4% fat, the ratio of residue mass distribution is about 2:3, meaning that most of the residue remains in the aqueous phase.

The Meeting decided that, for fat-soluble pesticides, two maximum residue levels will be estimated, if the data permit: one for whole milk and one for milk fat. For enforcement purposes, a comparison can be made either of the residue in milk fat with the MRL for milk (fat) or of the residue in whole milk with the MRL for milk. When needed, maximum residue levels for milk products can then be calculated from the two values, by taking into account the fat content of the milk product and the contribution from the non-fat fraction.

---

[11] Annex 6, reference 98
[12] FAO/WHO 2000: Codex Alimentarius Volume 2B, Pesticide Residues in Food—Maximum residue limits. Joint FAO/WHO Food Standards Programme, p. 4. Rome, 2000.

General considerations

Analytical methods should be made available for both milk and milk fat (both with a practical LOQ). The fat should preferably be separated from the milk by physical means, not by chemical solvent extraction, because in solvent extraction residues are extracted from both the aqueous and the lipid phase. As in this way cream (containing 40–60% fat) and not 100% milk fat is obtained, the lipid content of the cream should also be reported. The Meeting requested the CCPR ad-hoc Working Group on Methods of Analysis to give further guidance on analytical methods for measuring residues of fat-soluble pesticide in milk. Estimation of maximum residue levels for spinosad in milk is given here as an example (see also this report).

The 2001 JMPR reported that residues were measured in 119 samples of milk and cream after direct treatment of dairy cows with spinosad, and that the mean quotient of the concentration in cream divided by the concentration in milk was 4.2. A plot of residue levels in whole milk versus residue levels in cream (Figure 4) showed that the residue level in milk was approximately 24% of that in cream (line of best fit through the origin). Data on spinosad residues in milk and cream from a study of feeding to dairy cows are summarised in Table 79 of the JMPR Residue Evaluations of 2001[13]. The mean quotient of the concentration in cream divided by the concentration in milk after feeding at levels of 1, 3 and 10 ppm was 4.0, in good agreement with the results of direct treatment.

The 2001 JMPR estimated a maximum residue level of spinosad in cattle milk of 1 mg/kg on the basis of the highest value in milk, 0.65 mg/kg, after direct treatment. The calculated concentration in cream would be $0.65 \times 4.2 = 2.7$ mg/kg. On the assumption that the cream in the feeding study was approximately 50% fat, the concentration in fat would be about 5 mg/kg.

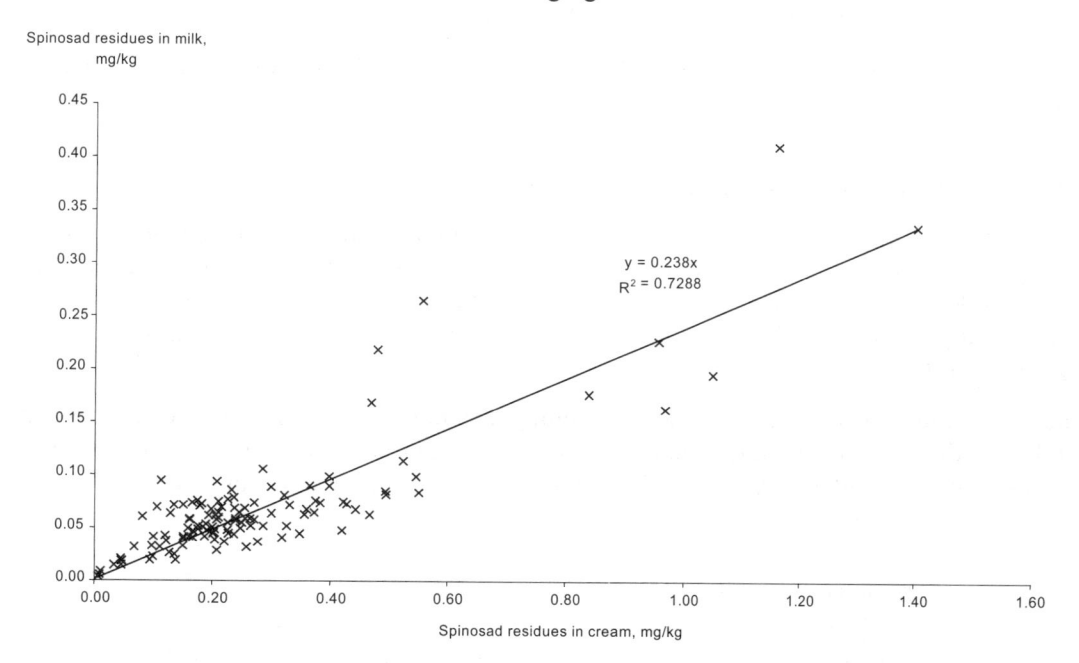

Figure 4. Residues of spinosad in milk and cream after direct treatment of dairy cows

The Meeting estimated a maximum residue level for spinosad residues in cattle milk fat of 5 mg/kg. The Meeting noted that the residue of this pesticide of intermediate fat solubility was estimated to be 8.4 times higher in milk fat than in milk, while the default factor would have been 25 when milk MRLs for fat-soluble pesticides are expressed as previously.

---

[13] Annex 6, reference 93

2.8 Revisited: Dietary burden of animals for estimation of MRLs for animal commodities

The 1997 JMPR developed guidance for estimating maximum residue and STMR levels for products of animal origin when residues are transferred from feed items. As a result of experience gained since that time, the Meeting agreed that animals could be exposed for extended periods to certain commodities such as fodder, grain and feeds treated post-harvest which contain residues at the highest level. The Meeting was informed that this situation pertains in Australia and probably in other regions of the world.

For example, on a farm on which 20 ha of an animal feed (forage, fodder or grain) were grown per year with a yield of 10 t/ha on a dry weight basis, enough would be produced to feed 333 head of cattle for 1 month. If the feed constituted less than 100% of the diet, more head of cattle could be fed for 1 month, or the duration of feeding might be longer.

In addition, in the experience of the Meeting, the residue levels of many pesticides on animal feed commodities decrease to only a limited extent during storage.

The Meeting agreed that the assumption of the 1986 JMPR, which was used in developing guidance in this area, that "it was unrealistic to assume that the theoretical maximum residue level would be achieved and maintained in the rations of food-producing animals receiving feeds produced on the farm" should no longer apply. Hence, the concept of the time required for residue levels to reach a plateau is also no longer required, thus simplifying estimation of dietary burden. It is recognized that it is unlikely that the individual ingredients of mixed feeds produced from commercially available ingredients would all contain residues at the theoretical maximum level; in these cases, the STMR should be applied to each of the components.

A revision of the relevant text in the *FAO Manual*, taking the above into account, will appear on the FAO website.

Residues arising from consumption of feed items

The 1986 JMPR explained its use of feeding studies with farm animals in estimating MRLs for foods of animal origin. It made the point that a sensible judgement must be made about the expected level of ingestion. The 2004 JMPR noted that it is realistic to assume that the theoretical maximum residue level would be achieved and maintained in the rations of food-producing animals receiving feeds grown on the farm for extended periods.

The estimation of residues that will arise in animal commodities is a two-pronged process, involving feeding studies with farm animals and calculations of dietary burden. These two sets of information are combined to estimate the residue levels that will be found in animal commodities in practice:

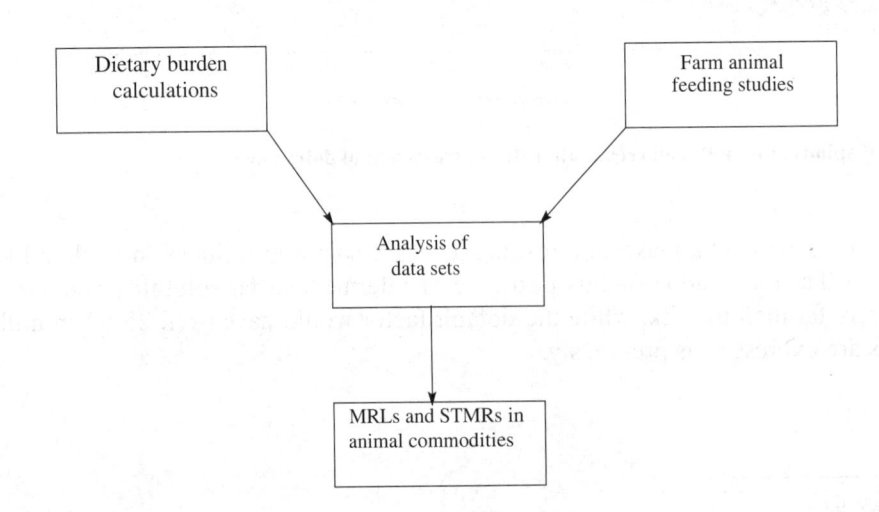

General considerations

The following decision matrix is recommended for use in estimating maximum residue levels and STMR values:

Maximum residue level	STMR
Choose:	Choose:
feed commodity, highest residue or STMR-P (for dietary burden calculation)	feed commodity STMR or STMR-P (for dietary burden calculation)
highest residue level[1] (from feeding study in farm animals)	mean residue level[1] (from feeding study in farm animals)

STMR-P, supervised trials median residue in a processed commodity calculated by multiplying the STMR of the raw commodity by the corresponding processing factor

[1] Residue levels in tissues and eggs of the relevant group of animals in the feeding study. For milk, choose the mean residue in milk from the relevant group of animals in all cases.

The residue contribution of a feed commodity for estimating maximum residue levels is calculated from the percentage of the total diet and the highest residue level in raw agricultural commodity feed items. For processed commodities that are likely to originate from a number of farms, e.g. apple pomace, the STMR-P of the processed commodity is chosen as the likely highest residue value that will occur in practice.

Maximum residue levels in animal commodities are derived from the highest residue values in feed commodities, and STMRs for animal commodities are derived from the STMRs for feed commodities. Separate tables are made for each MRL and STMR estimate, in which all feed items, their Codex commodity group and the residue levels found in crop residue trials are listed. The basis of the residue level is provided; i.e. the basis of the maximum residue level estimate is the highest level for raw agricultural commodities and the STMR-P for processed commodities. The percentage dry matter in the feeds is derived from Appendix IX of the *FAO Manual*, 'Maximum proportion of agricultural commodities in animal feed', except when the data stipulate 100% dry matter. The residue of each feed commodity on a dry weight basis is then calculated.

Starting with the feed item with the highest residue level, the percentage of each feed in the livestock diet is allocated from Appendix IX. Usually, only one feed commodity from each Codex commodity group is used; if more than one is used, it is only up to the full percentage feed allocation for that group. Feeds are allocated a percentage of the diet for each animal until no more than 100% of the diet is used.

The residue contribution (mg/kg) of each feed is then calculated from the dry weight of the residue and the corresponding percentage of the diet. All residue contributions for each animal are then added to determine the total dietary burden.

The procedure illustrated in the diagram and described above is demonstrated in the following worked example for spinosad.

Maximum dietary burden of farm animals estimated for spinosad

Commodity	Codex commodity group	Residue (mg/kg)	Basis of residue value	Dry matter (%)	Residue dry weight (mg/kg)	Feed allocation in total diet (%) Beef cattle	Feed allocation in total diet (%) Dairy cattle	Feed allocation in total diet (%) Poultry	Residue contribution of feeds (mg/kg) Beef cattle	Residue contribution of feeds (mg/kg) Dairy cattle	Residue contribution of feeds (mg/kg) Poultry
Apple pomace, wet	AB	0.064	STMR-P	40	0.16	10			0.016		
Citrus pulp	AB	0.12	STMR-P	91	0.13						
Maize forage	AF	3.1	Highest	100	3.1	40	50		1.24	1.55	
Maize fodder	AS	2.1	Highest	100	2.1						

Commodity	Codex commodity group	Residue (mg/kg)	Basis of residue value	Dry matter (%)	Residue dry weight (mg/kg)	Feed allocation in total diet (%)			Residue contribution of feeds (mg/kg)		
						Beef cattle	Dairy cattle	Poultry	Beef cattle	Dairy cattle	Poultry
Wheat straw and fodder, dry	AS	0.83	Highest	100	0.83						
Sorghum	GC	0.68	Highest	86	1.2	40	40	80	0.27	0.27	0.544
Almond hulls	AM	1.1	Highest	90	2.2	10	10		0.11	0.11	
Cotton-seed hulls		0.0020	STMR-P	90	0.0022						
Cotton-seed meal		0.0017	STMR-P	88	0.0019			20			0.0004
					Total	100	100	100			
					Maximum dietary burden				1.64	1.93	0.54

STMR dietary burden of farm animals estimated for spinosad

Commodity	Codex commodity group	Residue (mg/kg)	Basis of residue value	Dry matter (%)	Residue dry weight (mg/kg)	Feed allocation in total diet (%)			Residue contribution of feeds (mg/kg)		
						Beef cattle	Dairy cattle	Poultry	Beef cattle	Dairy cattle	Poultry
Apple pomace wet	AB	0.064	STMR-P	40	0.16	10			0.016		
Citrus pulp	AB	0.12	STMR-P	91	0.13						
Maize forage	AF	0.70	STMR	100	0.70	40	50		0.28	0.35	
Maize fodder	AS	0.46	STMR	100	0.46						
Wheat straw and fodder, dry	AS	0.215	STMR	100	0.22						
Sorghum	GC	0.165	STMR	86	0.19	40	40	80	0.08	0.08	0.15
Almond hulls	AM	0.56	STMR	90	0.62	10	10		0.062	0.062	
Cotton seed hulls	SO	0.0020	STMR-P	90	0.0022						
Cotton seed meal	SO	0.0017	STMR-P	88	0.0019			20			0.00039
					Total	100	100	100			
					STMR dietary burden				0.43	0.49	0.15

The poultry dietary burdens are not discussed further in this example.

Use of the results of farm animal feeding studies and dietary burdens to estimate maximum residue levels and STMR values for commodities of animal origin

The calculations of dietary burden are compared with the feeding levels in studies of farm animals to estimate maximum residue levels and STMR values on the basis of the following guidelines.

- When a feeding level in a feeding study matches the dietary burden, the residue levels reported in the study can be used directly as estimates of residue levels in tissues, milk and eggs resulting from the dietary burden.

- When a feeding levels in a feeding study differs from the dietary burden, the resulting residues in tissues, milk and eggs can be estimated by interpolation between the closest feeding levels.

General considerations

- When the dietary burden is below the lowest feeding level in the study, the resulting residues in tissues, milk and eggs can be estimated by applying the transfer factor (residue level in milk or tissue ÷ residue level in diet) at the lowest feeding level to the dietary burden.

- When the dietary burdens of beef and dairy cattle are different, the higher value should be used for calculating the residues in muscle, fat, liver and kidney.

- For estimating maximum and highest residue levels in meat, fat, liver, kidney and eggs, the highest residue level found in an animal in the relevant feeding group of the study should be used.

- For estimating STMR values in meat, fat, liver, kidney and eggs, the mean residue levels in animals in the relevant feeding group of the study are used.

- For estimating maximum residue levels and STMRs in milk, the mean residue levels in the animals in the relevant feeding group of the study are used.

- No more than about 30% above the highest feeding level can be extrapolated to a dietary burden.

The feeding levels in the farm animal feeding studies are entered into a table with the calculations of dietary burden and analysed on the basis of the above guidelines.

In the example of spinosad, the maximum dietary burdens of beef and dairy cattle are 1.6 and 1.9 mg/kg, respectively. Therefore, the levels of residues in tissues and milk are taken by interpolation from the 1 and 3 ppm feeding levels in the study.

The STMR dietary burdens (0.43 and 0.49 mg/kg) are below the lowest feeding level, 1 ppm. Therefore, the resulting residues in tissues and milk are calculated by applying the transfer factors at the lowest feeding level to the STMR dietary burdens.

The highest residue level in an individual tissue from an animal in the relevant feeding group is used with the highest dietary burden of residue to calculate the probable highest residue level in animal commodities. The mean tissue residue in the animals in the relevant feeding group is used with the STMR dietary burden to estimate the STMR values for animal commodities. For milk, the mean plateau residue level in milk from the relevant feeding group was used to estimate both the maximum residue level and the STMR.

The estimated dietary burden of spinosad was 1.6 mg/kg for beef cattle and 1.9 mg/kg for an actual feeding level of 3 ppm in the transfer study. The STMR value was 0.43 mg/kg for beef cattle and 0.49 mg/kg for dairy cattle for an actual feeding level of 1 ppm. The mean spinosad residue level in milk from dairy cattle was 0.13 mg/kg, while the value interpolated from the dietary burden and the feeding levels and the residue levels found in the transfer study was 0.087. The highest tissue residue levels in individual dairy cattle in the relevant feeding group were 1.7 mg/kg in fat (interpolated value, 1.1 mg/kg), 0.069 mg/kg in muscle (0.044 mg/kg), 0.44 mg/kg in liver (0.28 mg/kg) and 0.26 mg/kg in kidney (0.16 mg/kg). The mean residues in dairy cattle tissue (or milk) in the relevant feeding group were 0.044 mg/kg in milk (interpolated value, 0.022 mg/kg), 0.65 mg/kg in fat (0.32 mg/kg), 0.020 mg/kg in muscle (0.010 mg/kg), 0.13 mg/kg in liver (0.064 mg/kg) and 0.065 mg/kg in kidney (0.032 mg/kg).

As the STMR burden of dairy cattle exceeds that of beef cattle, it is used as the maximum residue level and the estimated STMR for fat, muscle, liver and kidney.

The highest residue levels expected in tissues and milk are: 1.1 mg/kg in fat, 0.16 mg/kg in kidney, 0.28 mg/kg in liver and 0.087 mg/kg in milk. The proposed STMR values are: 0.010 mg/kg in cattle meat, 0.032 mg/kg in cattle kidney, 0.064 mg/kg in cattle liver and 0.022 mg/kg in milk.

2.9 Statistical methods for estimating MRLs

The Meeting was informed of developments in the use of statistics to estimate the maximum residue level. A group of experts was formed in the USA, consisting of persons involved in MRL setting in the countries of the North American Free Trade Association and the European Food Safety Authority. They investigated the available procedures, primarily those used in the European Union, and decided to pursue a procedure in which a log normal distribution of data is assumed and the lower 95% upper confidence limit on the 95th percentile or the point estimate of the 99th percentile from normal Q–Q plots is used as an approximation of the maximum residue level. The plots give visual evidence of the degree of log normal behaviour of particular residue values. Some comparisons were made of the results obtained from application of this procedure and of the procedure used currently by the JMPR for data sets on fludioxonil. In most cases, there was good agreement. Small data sets often result in estimates of maximum residue levels that are considerably above the maximum. Statistically, this results from the uncertainty associated with few data points.

It was emphasized that any statistical system must be easy to use (spreadsheet) and be based on accepted statistical principles. Such systems might assist evaluators but could never replace professional judgement.

The method is still being developed. The Meeting expressed interest in receiving spreadsheets and documentation for evaluation, when available, and will await further developments.

2.10 Application of the recommendations of the OECD project on minimum data requirements to the work of the JMPR

The 2003 JMPR noted that some of the recommendations of the OECD York Workshop[14] and the FAO/OECD Zoning Steering Group were used routinely. The JMPR first considered the activities of the York Workshop at its 1999 Meeting, and subsequent Meetings have incorporated the Workshop recommendation into their work where practical. For example, as a result of the York Workshop, the JMPR has considered glasshouse trials conducted worldwide at the same GAP equivalent. Likewise, post-harvest trials conducted throughout the world at the same GAP are considered equivalent. The York Workshop also provided limited recommendations on crop translations, and the JMPR has been using that guidance routinely. It is noted, however, that the Zoning Steering Group addressed only foliar uses and that the York workshops considered primarily foliar and post-harvest uses. There is no guidance on soil treatment, seed treatment or use of herbicides. It was agreed that a pilot study would be conducted for the 2004 JMPR, in which the effect of full implementation of the recommendations would be considered for a new compound evaluation. Fludioxonil was selected.

The York Workshop and the follow-up Zoning Steering Group offered guidance that is potentially useful in two areas. First, they assigned the minimum number of trials required for evaluating a given pesticide–commodity combination for a maximum residue level, on the basis of three criteria: (1) the significance of the item in the diet of the general population; (2) the significance of the item in trade; and (3) the number of zones in which the commodity is grown. Unfortunately, the Workshop failed to define two of the three criteria adequately. It suggested that an item is 'significant' if its consumption represents $\geq 0.5\%$ of the total diet in any region. Importance in trade and the criteria for designating zones were not defined, and the subsequent Zoning Steering Group did not find a correlation between climatic growing regions and crop residue levels. The Zoning Steering Group suggested that a definition of zones based on climate, e.g. temperate–wet, temperate–dry, cold and tropical, was not relevant.

---

[14] *Minimum data requirements for establishing Maximum Residue Limits (MRLs) and import tolerances*, Doc. 2734/SANCO/99

General considerations

In view of the undefined criteria, the Meeting found it difficult to implement the scheme in determining the minimum number of trials required for a given pesticide–commodity combination. In the absence of guidance, the Meeting based the required number of trial on one or several zones where the likelihood of the crop being grown in substantial quantity in very different geographic and agricultural practice areas was considered. Likewise, the Meeting used its best collective opinion in deciding on the significance of the commodity in trade. The Meeting considered significance relative to international trade and not local or regional trade; thus, a commodity might be of substantial economic importance to a nation or region but not significant on the global scale.

The second area of assistance was in defining representative commodities for crop groups and listing some acceptable translations of data. For example, the York Workshop decided that field trial data for orange or grapefruit and mandarin or lemon could be used to represent all citrus fruit and that data for tomato and pepper could be extrapolated to okra and eggplant. This aspect was not, however, completed; for example, no recommendations were made for leafy vegetables. The Meeting noted that the Codex classification system for food and feed items is currently undergoing a limited revision, and this may be of assistance.

The Meeting compiled a list of commodities for which the following information was available: recommendations on maximum residue levels for fludioxonil, the number of trials available at GAP, the decision of the JMPR, the apparent requirements under the York Workshop criteria, and the result of application of the York Workshop criteria. The list is shown below.

Maximum residue level recommendations for fludioxonil

Commodity	No. of trials at GAP	JMPR MRL recommendation	OECD minimum database			Recommendation
			No. of zones	Significant in trade (T), diet (D)	No. of trials	
Pears	7	Yes	2–3	D	8	No
Stone fruit	11	Yes (Po)	Po	T, D	8	Yes
Grapes	18	Yes	2–3	T, D	12	Yes
Strawberry	25	Yes	2–3	D	8	Yes
Raspberry	4	Yes	2–3		4	Yes
Blackberry	0	Yes, translation of raspberry				Yes (translation)
Dewberry	0	Yes, translation of raspberry				Yes (translation)
Blueberry	8	Yes	2–3		4	Yes
Spring onions	3	Yes	2–3		4	No
Bulb onions	13	Yes	2–3	D	8	Yes
Broccoli	7	Yes	2–3	D	8	No
Cabbage	6	Yes	2–3	D	8	No
Cucumber	13	Yes	2–3	D	8	Yes
Summer squash	2	Yes (support from cucumber)	2–3		4	No
Tomato	16	Yes	2–3	T, D	12	Yes
Bell pepper	10	Yes	2–3	T, D	12	Yes
Eggplant	4	Yes	2–3		3	Yes
Sweet corn (on-the-cob)	8	Yes	2–3	D	8	Yes
Lettuce, head	17	Yes	2–3	T, D	12	Yes
Watercress	2	Yes (support from mustard greens)	2–3		3	No

Commodity	No. of trials at GAP	JMPR MRL recommendation	OECD minimum database			Recommendation
			Assumptions			
			No. of zones	Significant in trade (T), diet (D)	No. of trials	
Mustard greens	7	Yes (support form watercress)	2–3		3	Yes
Bean pod with seed	22	Yes	2–3	T, D	12	Yes
Pea with pod	0	Translation from bean with pod				Yes (translation)
Peas (succulent)	6	Yes	2–3	T, D	12	No
Pea and bean (dry)	5	Yes	2–3	D	8	No
Potato	17	Yes	2–3	T, D	12	Yes
Carrot	7	Yes	2–3	D	8	No
Asparagus	2	No	2–3		3	No
Cereal grains	71	Yes (wheat, rye, barley, maize, sorghum)	> 3	T, D	16	Yes
Pistachio nuts	3	Yes	1		3	Yes
Rape-seed	15	Yes	2–3	T	8	Yes
Cotton-seed	8	Yes	2–3	T, D	12	No
Chives	2	Yes	1		3	No
Basil	2	Yes	1		3	No
Maize forage	7	Yes	> 3	N/A	10	No
Fodder and straw of cereal grains	50	Yes	> 3	N/A	10	Yes

Po, the recommednation accommodates post-harvest treatment of the food commodity; N/A, not applicable

[1] Crops with no trials at GAP are not listed.

The Meeting noted that it had extended estimates of maximum residue levels to commodities for which use of the OECD guidance would not have so indicated (cotton-seed, basil, chives, peas, peas and beans (dry), watercress, summer squash, broccoli, cabbage, green onions, pears). These assignments were based, however, on application of criteria that are not defined or only partially described. For example, if one zone were assumed (rather than two or three), use of the OECD guidance would have led to maximum residue level recommendations for all commodities except the herbs, summer squash and watercress.

These evaluations of additional commodities for maximum residue levels are the result of independent scientific judgement by the JMPR. For example, the Meeting considered six or seven trials (in cabbage, pear, broccoli) conducted under GAP and properly documented to be acceptable.

Although the Meeting usually requires a minimum of three or four trials for a minor commodity, it considered two fully documented trials under GAP adequate for the herbs chives and basil. As these commodities are herbs, their consumption is extremely low and pesticide residues on them would not usually pose a hazard. Because the maximum residue level was based on only two trials, the Meeting estimated a considerably higher level than the highest residue level: The high residue level for chives was 3.9 mg/kg, and the Meeting estimated a maximum residue level of 10 mg/kg.

The Meeting considered five trials to be adequate for establishing maximum residue levels for dry beans and dry peas, as the residues were below the LOQ. The OECD recommendations suggest eight trials.

A similar situation existed for cotton-seed. The GAP is for seed treatment use, and all the residue levels were below the LOQ. Metabolism studies after seed treatment and other seed treatment trials predict residues below the LOQ. Thus, the Meeting considered five trials to be acceptable, although the York Workshop would have required eight.

The Meeting also noted that, while the York Workshop provided some information on the translation of residue values for one commodity to another, it did not provide for the mutual support of other commodities with similar use patterns. In the present example, the Meeting agreed to use data on cucumber to support the MRL for summer squash (zucchini) and to use data on mustard greens to support the value for watercress. This is common practice by many national governments and is based on similarities in crop morphology. The York Workshop did provide for the translation of data on raspberry to blackberry and dewberry and the translation of data on bean with pod to pea with pod.

The Meeting recalled the conclusion of the Zoning Steering Group, that trials on a given commodity conducted at the same GAP with similar residues on day 0 be considered equivalent regardless of geographic location. The Group suggested that application method, crop type and local agricultural practices are major contributors to differences in residue levels among trials conducted under the same GAP. Climate has only a minor direct effect. The JMPR suggested, therefore, that hypothetical zones (not geographical zones) be developed on the basis of crop type and variations in agricultural practice. For example, wheat is grown in a relatively uniform manner worldwide (one zone), while grapes are grown under a variety of conditions (of crop height, leaf numbers and plant density; multiple zones).

The Meeting concluded that some of the recommendations of the York Workshop and Zoning Steering Group used by the JMPR will continue to be considered as auxiliary advice but that substantial additional work is required to make the recommendations generally applicable as guidance. Areas in which additional effort is needed include: (1) defining significance in trade, perhaps with a table of commodities and their classifications; (2) determining the criteria for zones and the number of zones for each commodity; (3) extending the list of commodity translations for the purpose of recommending a maximum residue level for one commodity on the basis of available field trial data for another commodity; and (4) completing the list of representative crops for the purpose of recommending group maximum residue levels.

2.11 Alignment within one year of toxicological and residue evaluations for new and periodically reviewed compounds

When a new compound or one undergoing periodic review is evaluated, it is generally preferable to conduct the toxicological and residue reviews in the same year. Practical problems may, however, arise; e.g. when the residue definition is uncertain, the residue evaluation cannot proceed satisfactorily or efficiently. In such cases, it is preferable that the toxicological evaluation precede the residue evaluation.

The Meeting recommended that the toxicological and residue evaluations of new compounds or those undergoing periodic review be scheduled for the same year, when practical. When the residue definition is problematic (e.g. different residue definitions in different national and regional registration systems), the toxicological evaluation should be scheduled 1 year ahead of the residue evaluation.

3. DIETARY RISK ASSESSMENT FOR PESTICIDE RESIDUES IN FOOD

Assessment of risk from long-term dietary intake

Risks associated with long-term dietary intake were assessed for compounds for which MRLs were recommended and STMRs estimated at the present abd previous Meetings. Dietary intakes were calculated by multiplying the concentrations of residues (STMR or STMR-P values or recommended MRLs) by the average daily per capita consumption estimated for each commodity on the basis of the GEMS/Food diet[1,2]. International estimated daily intakes (IEDIs) were derived when STMR or STMR-P values were available, and theoretical maximum daily intakes (TMDIs) were calculated when only existing or recommended MRLs were available. Codex MRLs that have been recommended by the JMPR for withdrawal were not included in the estimates.

Long-term dietary intakes are expressed as a percentage of the ADI for a 60-kg person, with the exception of the intake calculated for the Far Eastern diet, in which a body weight of 55 kg is used. The estimates are summarized in Table 1. Percentages are rounded to one whole number up to 9 and to the nearest 10 above that. Percentages above 100 should not necessarily be interpreted as giving rise to a health concern because of some conservative assumptions used in the assessments. The detailed calculations of long-term dietary intake are given in Annex 3.

The Meeting drew attention to the use of residue levels in muscle tissue for estimating dietary intake of residues in meat of fat-soluble compounds. Previously, residue levels in trimmable fat, with adjustment by a default factor, were used to estimate dietary intake from meat.

Assessments of long-term risk for betazone, captan, carbofuran, fenpropimorph, fenpyroximate, folpet, dimepthipin, methamidophos and pirimiphos-methyl were considered by previous meetings. For these compounds, the recommendations made at this Meeting will not significantly affect the previous assessments.

Calculations of dietary intake can be further refined at the national level by taking into account more detailed information, as described in the guidelines for predicting intake of pesticide residues.[1]

Table 1. Summary of risk assessments of long-term dietary intake conducted by the 2001 JMPR

Codex code	Name	ADI (mg/ kg bw)	Intake range (% of maximum ADI)	Type of assessment
017	Chlorpyrifos	0–0.01	3–30	IEDI
149	Ethoprophos	0–0.0004	5–10	IEDI
037	Fenitrothion	0–0.05	110–330	IEDI
211	Fludioxonil	0–0.4	0–1	IEDI
158	Glyphosate	0–1	1	TMDI
049	Malathion	0–0.3	0	IEDI
212	Metalaxyl/ Metalaxyl M	0–0.08	2–10	TMDI
094	Methomyl	0–0.02	1–20	IEDI
166	Oxydemeton methyl	0–0.0003	3–30	IEDI
057	Paraquat	0–0.005	2–5	IEDI
112	Phorate	0–0.0007	40–200	TMDI
101	Pirimicarb	0–0.02	3–20	TMDI

---

[1] WHO (1997) *Guidelines for predicting dietary intake of pesticide residues.* 2nd revised edition, GEMS/Food Document WHO/FSF/FOS/97.7, Geneva

[2] WHO (1997) *Food consumption and exposure assessment of chemicals.* Report of a FAO/WHO Consultation. Geneva, Switzerland, 10–14 February 1997, Geneva

Compound		ADI	Intake range (% of	Type of assessment
Codex code	Name	(mg/kg)	maximum ADI	
142	Prochloraz	0–0.01	7–10	IEDI
160	Propiconazole	0–0.07	0–1	TMDI
105	Propineb [1]	0–0.007	4–30	IEDI
210	Pyraclostrobin	0–0.03	0–3	IEDI
203	Spinosad	0–0.02	9–30	IEDI
133	Triadimefon	0–0.03	1–6	TMDI
168	Triadimenol	0–0.03	1–20	TMDI
213	Trifloxystrobin	0–0.04	1–2	IEDI

Spices and dried chili peppers

Long-term dietary intake (IEDI) from the consumption of spices was calculated for 28 pesticides for which recommendations were made at the present Meeting on the basis of monitoring data (General item 2.6; Annex 5). For these compounds, the additional contributions to the IEDI from the consumption of spices were < 1% of the respective ADIs for all GEMS/Food regional diets.

The Meeting also calculated TMDIs from the consumption of dried chili peppers using maximum residue levels estimated at this Meeting for 46 pesticides, on the basis of MRLs for peppers (Annex 5). The TMDIs for carbaryl and dimethoate were > 100 % of the respective ADIs in at least one regional diet.

The TMDIs were > 5–100% of the respective ADIs in at least one regional diet for acephate, azinphos-methyl, carbendazim, chlorothalonil, chlorpyrifos, chlorpyrifos-methyl, cyhexatin, cyromazine, dicofol, ethephon, ethoprophos, fenarimol, methamidophos, methomyl, monocrotophos, oxamyl, phosphamidon, procymidone, profenofos, tebufenozide and vinclozolin. For these compounds, a long-term intake calculation based on all uses should be performed before the risk assessment can be finalized.

The TMDIs were ≤ 5% of the respective ADIs in all regional diets for abamectin, benalaxyl, cyfluthrin, cypermethrin, diazinon, dichlofluanid, dinocap, dithiocarbamates, fenpropathrin, fenvalerate, imidacloprid, metalaxyl, methoxyfenozide, permethrin, piperonyl butoxide, propamocarb, pyrethrins, quintozene, spinosad, tebuconazole, tolylfluanid, triadimefon and triadimenol. The Meeting agreed that the intake of these compounds through consumption of dried chili peppers would not significantly affect the risk assessment based on all other uses of the compounds.

Assessment of risk from short-term dietary exposure

Risks associated with short-term dietary intake were assessed for compounds for which STMR and highest residue values were estimated at the present and previous Meetings and for which acute reference doses (ARfDs) has been established, in commodities for which data on consumption were available. The procedures used for calculating the short-term intake are described in detail in sections 2.10 and 3 of the report of the 2003 JMPR.[3]

A risk assessment for short-term dietary intake was conducted for each commodity–compound combination by assessing the IESTI as a percentage of the ARfD. When the maximum residue level was estimated for a Codex commodity group (e.g. citrus fruits), intakes were calculated for individual commodities within the group.

On the basis of data received by the present or previous Meetings, the establishment of an ARfD for bentazone, metalaxyl M, fludioxinil, glyphosate, spinosad, trifloxystrobin was considered unnecessary. The intake of these compounds was therefore not estimated.

An ARfD might be necessary for pirimiphos-methyl but has not yet been established. Therefore, the short-term risk assessment could not be finalized. The Meeting recommended that this compound be evaluated for the establishment of ARfDs in the near future.

---

[3] Annex 6, reference 98

Dietary risk assessment

ARfDs were established for phorate, pirimicarb, propiconazole, triadimefon and triadimenol, but short-term intakes were not calculated as STMR and highest residue values were not available for these compounds.

On the basis of data received by the present Meeting, the establishment of ARfDs for captan and folpet was considered necessary only for women of child-bearing age. The corresponding IESTI could, however, be calculated only from the consumption data available for the general population.

The recommendation made for methamidophos at the present Meeting does not affect the previous assessment for this compound.

The short-term intakes, as percentages of the ArfDs, of the general population and of children are summarized in Table 2. The percentages are rounded to one whole number up to 9 and to the nearest 10 above that. Percentages above 100 should not necessarily be interpreted as giving rise to a health concern because of some conservative assumptions used in the assessments. The detailed calculations of short-term dietary intakes are given in Annex 4.

Table 2. Summary of risk assessments of short-term dietary intake conducted by the 2004 JMPR

Compound Codex code	Name	ARfD (mg/kg bw)	Commodity	Percentage of ARfD General population	Children ≤ 6 years
007	Captan	0.3[1]	Grape	150	–
			Other 19 commodities	0–100	–
096	Carbofuran	0.009	All 20 commodities	1–50	0–100
017	Chlorpyrifos	0.1	All 7 commodities	0–10	0–40
151	Dimethipin	0.2	All 12 commodities	0	0–1
149	Ethoprophos	0.05	All 11 commodities	0–1	0–3
037	Fenitrothion	0.04	Maize	80	160
			Wheat bran, unprocessed	90	150
			Wholemeal bread	50	120
			Other 22 commodities	0–100	0–80
188	Fenpropimorph	0.2	All 15 commodities	0–7	0–10
193	Fenpyroximate	0.01	Apple	50	130
			Grape	120	310
			Other 7 commodities	0–10	0–30
041	Folpet	0.2[1]	Lettuce, head	190	–
			Other 9 commodities	1–100	–
049	Malathion	2	6 commodities	0–4	0–10
094	Methomyl	0.02	Pepper	20	20
166	Oxydemeton methyl	0.002	Apple	50	130
			Cabbage, head	50	120
			Grape	80	220
			Orange	30	120
			Other 16 commodities	0–30	0–90
057	Paraquat	0.006	All 44 commodities	0–20	0–50
142	Prochloraz	0.1	Mushroom	130	150
			Other 46 commodities	0–30	0–70
105	Propineb	0.1[2]	Pepper	110	120
			Other 12 commodities	0–30	0–90
210	Pyraclostrobin	0.05	All 37 commodities	0–30	0–90

[1] Applies only to women of child-bearing age
[2] Interim ARfD

Spices and dried chili peppers

Short-term dietary risk assessments from the consumption of spices were performed for 12 pesticides for which recommendations were made at the present Meeting on the basis of monitoring data (item 2.6, Annex 5), when ARfD values were available. The IESTI ranged from 0 to 30 % of the ARfD for both the general population and children for azinphos-methyl, chlorpyrifos, diazinon, dimethoate, disulfoton, endosulfan, fenitrothion, malathion, methamidophos, parathion-methyl and phosalone. The intake of mevinphos represented 170% of the ARfD for the general population and 160% for children.

Short-term dietary risk assessments from the consumption of dried chilli peppers were performed for 14 pesticides for which ARfD values were established, using the maximum residue level estimated by the present Meeting. The intakes ranged from 0 to 100% of the respective ARfDs for the general population and children for acephate, carbaryl, chlorpyrifos, diazinon, ethephon, ethoprophos, imidacloprid, methamidophos, methoxyfenozide, pyrethrins and tolyfluanid. The intakes were 120% of the ARfD for both populations for dimethoate. They were 260% of the ARfD for the general population and 270% for children for oxamyl.

4. EVALUATION OF DATA FOR ESTABLISHING VALUES FOR ACCEPTABLE DAILY INTAKES AND ACUTE REFERENCE DOSES FOR HUMANS, MAXIMUM RESIDUE LIMITS AND SUPERVISED TRIAL MEDIAN RESIDUE LEVELS

4.1 BENTAZONE (172)

TOXICOLOGY

Evaluation for an acute reference dose

Bentazone is a herbicide that was first evaluated by the JMPR in 1991, when an ADI of 0–0.1 mg/kg bw was allocated on the basis of the NOAEL of 9 mg/kg bw per day in a long-term study in rats and a safety factor of 100. Further data were made available to the 1998 JMPR, including observations in humans and a 90-day study in rats fed with 6-hydroxybentazone, a metabolite of bentazone. Data from studies of genotoxicity with 6-hydroxybentazone were also supplied. It was concluded that 6-hydroxybentazone was less toxic than bentazone, and the ADI of 0–0.1 mg/kg bw was maintained. Data were not evaluated to establish an ARfD. The present Meeting re-evaluated some of the previously evaluated data in order to establish an ARfD, and two cases of acute human poisoning were described.

The oral median lethal dose (LD_{50}) for bentazone was 1200–2500 mg/kg bw in rats. In 13-week studies in mice, rats and dogs, interference with blood clotting was a consistently observed effect. There was prolongation of the prothrombin and partial thromboplastin times in mice and rats, and prolongation of the prothrombin and bleeding time in dogs. Additionally, extramedullary haematopoiesis, haemorrhage and haemosiderosis were found in mice at autopsy. Toxicological effects in rats were less dramatic, the NOAEL being identified on the basis of clinical chemistry changes. In dogs, clinical effects, such as hyperactivity, ataxia, prostration and tremor, were seen. At the highest dose in dogs, at histopathological examination of tissues post mortem there was congestion and necrosis in the liver, together with fatty change and, in the spleen, evidence of extramedullary haematopoiesis. Fatty change in the myocardium and cloudy swelling of the renal tubular cells were also observed. The NOAEL for the study in mice was 400 ppm (equal to 90 mg/kg bw per day) on the basis of prolonged prothrombin and partial thromboplastin times at higher dietary concentrations. The NOAEL for the study in rats was 400 ppm (equal to 25.3 mg/kg bw per day) on the basis of clinical chemistry changes observed at the next highest dietary concentration. In the study in dogs, the NOAEL was 300 ppm (equal to 12.0 mg/kg bw per day) on the basis of clinical effects observed at higher dietary concentrations. Three deaths were observed at the highest dose in weeks 11 and 12 of the study. In a 1-year study in dogs, clinical signs (emaciation, dehydration, hyperaemia, alopecia and diarrhoea, which was occasionally bloody) were seen at the highest dietary concentration. The NOAEL for the study was 400 ppm (equal to 13.1 mg/kg bw per day) on the basis of clinical signs, weight loss and anaemia at the highest dietary concentration. It was not considered appropriate to set an ARfD on the clinical signs, reduced body weight or haematological changes occurring in dogs, since significant clinical effects were not seen early in these two studies.

Three studies of developmental toxicity in rats treated by gavage (two studies) or by dietary administration (one study) were evaluated by the Meeting. In the earlier study of rats treated by gavage, neither maternal nor fetal toxicity was seen at any dose; the NOAEL for both maternal and fetal toxicity was thus 200 mg/kg bw per day, the highest dose tested. In the later study in rats treated by gavage, in which higher doses were administered, the NOAEL was 100 mg/kg bw per day on the basis of maternal toxicity (decreased food consumption) and fetal toxicity (postimplantation loss, reduced fetal weight and incompletely ossified fetal skeletons). In the study of developmental toxicity in rats given diets containing bentazone, the NOAEL for maternal toxicity was 4000 ppm (equal to 324 mg/kg per day) on the basis of decreased weight gain and food consumption at 8000 ppm. The NOAEL for fetal toxicity was also 4000 ppm (equal to 324 mg/kg bw per day) on the basis of decreased fetal weights and fetal liver petechiae at 8000 ppm. Bentazone was not teratogenic in any of the studies of developmental toxicity in rats. The Meeting assessed two studies of developmental toxicity in rabbits. In one study, the NOAEL for maternal and fetal toxicity was 150 mg/kg bw per day, the highest dose tested; neither maternal nor fetal toxicity was observed at any dose. In the second study, which was conducted using higher doses than the earlier study, the NOAEL for maternal toxicity was 150 mg/kg bw per day on the basis of reduction in maternal food consumption at 375 mg/kg bw per day. Postimplantation loss was increased at 375 mg/kg bw per day and there was total implantation loss in one dam. Bentazone was not teratogenic in either study of developmental toxicity in rabbits.

Two case reports of fatal self-poisoning in humans were characterized by vomiting, diarrhea, drowsiness and death from cardiac arrest.

Toxicological evaluation

The Meeting concluded that the establishment of an ARfD was unnecessary. An addendum to the toxicological monograph was prepared

Estimate of acute reference dose

Unnecessary

Studies that would provide information useful for continued evaluation of the compound

Further observations in humans

DIETARY RISK ASSESSMENT

Short-term intake

The 2004 JMPR decided that it was not necessary to establish an ARfD for bentazone. The Meeting therefore concluded that the short-term dietary intake of bentazone residues is unlikely to present a public health concern.

4.2 CAPTAN (007)

TOXICOLOGY

Evaluation for an acute reference dose

Captan is a fungicide used for the control of fungal diseases in crops. The JMPR evaluated captan in 1963, 1965, 1969, 1973, 1978, 1982, 1984, 1990 and 1995. Toxicological monographs were prepared in 1963, 1965 and 1969 and addenda to the monographs were prepared in 1973, 1977, 1978, 1982, 1984, 1990 and 1995. In 1984, an ADI of 0–0.1 mg/kg bw was established on the basis of a NOAEL of 12.5 mg/kg bw per day in studies of reproductive toxicity in rats and monkeys. The present Meeting considered the requirement for an ARfD, based on data from the previous evaluations for JMPR and from new studies.

In mice treated orally with captan, the captan molecule is largely degraded to 1,2,3,6-tetrahydrophthalimide (THPI) and thiophosgene (via thiocarbonyl chloride) in the stomach before reaching the duodenum. No captan was detected in blood or urine. Studies of metabolism in vitro with human blood revealed that captan is rapidly degraded to THPI, with a calculated half-life of 1–4 s. Thiophosgene is detoxified by reaction with, e.g. cysteine or glutathione and is ultimately rapidly excreted.

The acute oral toxicity of captan in rats is low (LD_{50}, > 5000 mg/kg bw). Mice fed diets containing captan at a concentration of 3000 ppm, equal to 440 mg/kg bw per day, for 28 days showed an initial reduction in food consumption of about 37%. Food consumption gradually recovered over the first week of treatment, although it remained lower that that of controls throughout the 4-week treatment period. After 1 day, no treatment-related macroscopic and microscopic changes were observed in the duodenum or any other tissue examined. From day 3 onwards, the duodenum showed crypt cell hyperplasia, shortening of villi and a general disorganization of the villus enterocytes. From day 7 onward, immature cells were seen at the villus tips.

In a 28-day range-finding study in which dogs were given captan at doses of 30 to 1000 mg/kg bw per day, dose-related emesis, reduced body-weight gain and food consumption were observed in all treatment groups. No other clinical signs were observed. Haematological parameters and histopathology of the duodenum were within normal limits.

In a study from the published literature, the teratogenic effects of a number of phthalimide derivatives, including captan, were tested in pregnant golden hamsters. The Meeting noted that this study had major limitations (e.g. small number of animals per dose, limited reporting of the data) and is therefore of limited value. It does, however, suggest that developmental effects may occur after a single exposure to captan, albeit at maternally toxic doses.

In a study of developmental toxicity in rats treated by gavage, captan was not teratogenic. The NOAEL for maternal toxicity was 18 mg/kg bw per day on the basis of a reduction in body weight and food consumption. The NOAEL for offspring toxicity was 90 mg/kg bw per day on the basis of the reduction in fetal body weight and an increased incidence of skeletal variations.

In a study in rabbits treated by gavage, the NOAEL for maternal toxicity was 10 mg/kg bw per day on the basis of a markedly reduced body-weight gain and reduced food consumption at 30 mg/kg bw per day. The NOAEL for embryo- and fetotoxicity was 10 mg/kg bw per day on the basis of increases in skeletal variations at 30 and 100 mg/kg bw per day. At 100 mg/kg bw per day, increased incidences of early and late intra-uterine deaths were observed, as were increased incidences of several malformations. The NOAEL for these effects was 30 mg/kg bw per day. Multiple malformations observed in two fetuses in the group receiving the intermediate dose were considered to be incidental. In another study in rabbits treated by gavage, the NOAEL for maternal toxicity was 10 mg/kg bw per day on the basis of reduced body-weight gain and food consumption at 40 mg/kg bw per day. On the basis of the increase in postimplantation losses and the increase in incidence of minor skeletal variations at 160 mg/kg bw per day, the NOAEL for embryo- and fetotoxicity was 40 mg/kg bw per day. In a third study in rabbits treated by gavage, the NOAEL for maternal toxicity was 12 mg/kg bw per day on the basis of reductions in body-weight gain during the initial phase of treatment. The NOAEL for embryo- and fetotoxicity was 25 mg/kg bw per day on the basis of a reduction in fetal body weight at 60 mg/kg bw per day. The Meeting considered that maternal toxicity and the associated increases in skeletal variations and fetal body-weight reductions observed were likely to be caused by high local concentrations of captan produced by administration by gavage and are not considered to be relevant to dietary exposure.

While few data on humans are available, captan is known to have caused allergic dermatitis and eye irritation in humans. After ingesting 7.5 g of Captan 50 WP, which is a suspension of captan mixed with water (ratio, 50%), a 17-year-old woman (body weight not reported) experienced some clinical signs, which started 3 h after ingestion and recovered within 72 h. Assuming a body weight of 50–60 kg, this intake equates to a dose of 62.5–75 mg/kg bw.

Toxicological evaluation

Other than developmental effects, captan produced no toxicological effects that might be considered to be a consequence of acute exposure. The Meeting concluded that it was not necessary to establish an ARfD for the general population, including children aged 1–6 years for whom separate data on dietary intake are available. The Meeting concluded that it might be necessary to establish an ARfD to protect the embryo or fetus from possible effects in utero. Such an ARfD would apply to women of childbearing age.

The maternal toxicity and associated increases in skeletal variations and fetal body-weight reductions observed in studies of developmental toxicity in rabbits are likely to be caused by high local concentrations of captan and are not considered to be relevant to dietary exposure. However, the observed intra-uterine deaths and fetal malformations could not, with confidence, be attributed to maternal toxicity.

The Meeting concluded that the database was insufficient (in particular, with regard to the absence of studies on the developmental effects of THPI) to establish the mode of action by which the increased incidences of intra-uterine deaths and of fetuses with malformations, observed at 100 mg/kg bw per day (NOAEL, 30 mg/kg bw per day) in rabbits, were induced. As a consequence, their relevance for deriving an ARfD could not be dismissed. Therefore the Meeting established an ARfD of 0.3 mg/kg bw, based on a NOAEL of 30 mg/kg bw per day for increased incidences of intra-uterine deaths and malformations at 100 mg/kg bw per day in the study in rabbits and a safety factor of 100. The use of a safety factor of 100 was considered to be conservative; although the mode of action by which the developmental effects were induced is uncertain, they are possibly secondary to maternal toxicity. The ARfD also covers the effects observed in the case report in humans. The Meeting noted that it might be possible to refine the ARfD using the results of an appropriately designed study.

An addendum to the toxicological monograph was prepared.

Estimate of acute reference dose

0.3 mg/kg bw for women of childbearing age

Unnecessary for the general population

DIETARY RISK ASSESSMENT

Short-term intake

The Meeting established an ARfD (0.3 mg/kg bw) for captan for women of childbearing age and decided that an ARfD was unnecessary for the general population including children aged 1–6 years. Women of childbearing age are also part of the general population.

In the absence of relevant studies on the developmental effects of THPI, the Meeting was unable to determine whether or not THPI should be excluded from the residue for dietary risk assessment. The Meeting was not able to finalize the risk assessment before an evaluation of the residue definition for risk assessment and associated residue values for dietary intake estimation had been completed.

4.3   CARBOFURAN (096)/ CARBOSULFAN (145)

RESIDUE AND ANALYTICAL ASPECTS

Carbofuran, resulting from use of carbosulfan, was reviewed for residue levels within the periodic review programme in 1997. When the compound was re-evaluated by the JMPR in 2002 and 2003, short-term risks were assessed for commodities for which recommendations had been made at those Meetings: rice, sweet corn, maize and potato. The CCPR at its Thirty-fifth Session, taking into account concerns expressed by the Delegation of Australia and the Observer from the European Commission, requested GEMS/Food to perform a full short-term intake assessment of carbofuran, to include all the commodities not evaluated previously for which recommendations existed. The assessment was presented to the CCPR at its Thirty-sixth Session (CX/PR 03/4). Except for the consumption of oranges (sweet and sour) by children, none of the IESTI value exceeded the ARfD of 0.009 mg/kg bw. The assessment was conducted with the highest residue (HR) level in the edible portion of 0.5 mg/kg, recommended by the 1997 JMPR for oranges, sweet, sour, from a residue data set in whole orange derived from 53 supervised trials conducted with carbosulfan according to GAP. A maximum residue level of 0.5 mg/kg and a STMR of 0.1 mg/kg were also recommended.

At its Thirty-sixth Session, the Committee noted (ALINORM 04/27/24) that the European Commission had established an ARfD 10 times lower than that established by JMPR. The Committee decided to return to Step 6 the draft MRLs for cantaloupe; cucumber; mandarin; oranges, sweet, sour; squash, summer; and sweet corn (corn-on-the-cob) to address concern about short-term intake.

For the purposes of dietary intake, the residue definition for carbofuran arising from use of carbosulfan and carbofuran is carbofuran + free and conjugated 3-OH carbofuran, expressed as carbofuran. The analytical methods include an acid hydrolysis step to release the conjugate.

At the present Meeting, data on residues in orange pulp in supervised trials conducted with carbosulfan and submitted to the 1997 JMPR were evaluated.

Results of supervised trials on crops

Six trials in Spain conducted according to GAP with carbosulfan in orange, evaluated by the 1997 JMPR, showed residue levels of carbamates (carbosulfan, carbofuran or 3-OH carbofuran) of < 0.05 mg/kg at 0 days (four trials), 45 days (six trials) and 105 days (four trials). The post-harvest interval (PHI) is 110 days. The residue levels in peel ranged from 0.21 to 0.84 mg/kg 45 days after the last application (two trials) and from < 0.05 to 0.25 mg/kg within 110 days' PHI (four trials).

The Meeting agreed that it is unlikely that residues of carbamates arising from the use of carbosulfan will be present in orange pulp at levels higher than the LOQ (0.05 mg/kg). The Meeting estimated an STMR and a highest residue level of 0.05 mg/kg for carbofuran in orange, sweet, sour.

This estimate is supported by a study on metabolism evaluated by the 1997 JMPR, in which the pulp of oranges treated with ^{14}C-carbosulfan contained no more than 0.3% of the total radioactive residues 30 days after treatment.

DIETARY RISK ASSESSMENT

Long-term intake

The Meeting agreed that the STMR in the edible portion of orange estimated by the present Meeting would result in a lower IEDI than that calculated by the 2002 JMPR. The Meeting confirmed the previous conclusion that the long-term intake of carbofuran residues arising from use of carbosulfan and/or carbofuran on the commodities considered by the JMPR is unlikely to present a public heath concern.

Short-term intake

The IESTI of carbofuran, was calculated for all the commodities not evaluated previously, including citrus. The IESTI represented 1–50% of the ARfD for the general population and 1–100% of that for children. The Meeting concluded that the short-term intake of carbofuran residues in the commodities evaluated at this Meeting is unlikely to present a public health concern.

4.4 CHLORPYRIFOS (017)

RESIDUE AND ANALYTICAL ASPECTS

At its Twenty-fifth Session in 1993 (ALINORM 93/24A para. 251), the CCPR identified chlorpyrifos as a candidate for periodic review. At its Twenty-ninth Session in 1997, it scheduled periodic reviews for toxicology in 1999 and for residue chemistry in 2000. The 1999 toxicology review confirmed the ADI of 0.01 mg/kg bw and also established an ARfD of 0.1 mg/kg bw. In the 2000 residue chemistry review, information was supplied on the identity and physical properties of the active ingredient and technical material, metabolism in plants and animals, environmental fate, storage stability, animal feeding studies, field trials, GAP (national labels) and fate of residues in processing.

Chlorpyrifos was scheduled for re-evaluation in 2004 for consideration of maximum residue levels in cotton, potato, rice and soya bean. No MRLs were recommended by the 2000 JMPR on these commodities because of lack of relevant GAP labels or insufficient residue data. Relevant GAP labels and additional residue data to support proposed Codex MRLs in these commodities were submitted to the Meeting for evaluation. The CCPR at its Thirty-sixth Session in 2004 agreed that JMPR would review data from India to support establishment of a maximum residue level on tea. Information was submitted by the Government of India for this purpose.

Methods of analysis

Methods for enforcement, data collection and monitoring of chlorpyrifos in different matrices were submitted and evaluated by JMPR in 2000. Additional methods were submitted to the present Meeting for the analysis of cotton-seed, potato, rice and soya bean. Various extraction and clean-up methods are followed by gas chromatography with flame photometric detection. The LOQ for chlorpyrifos is 0.01 mg/kg in all matrices.

The method of analysis for chlorpyrifos in tea was submitted by the Government of India. Residues in the cleaned extract were quantified by gas chromatography with flame-photometric or electron-capture detection. The LOQ is 0.02 mg/kg tea.

Results of supervised trials on crops

Cotton-seed

Supervised field trials on cotton were conducted at three sites in Australia in 2000 to provide additional residue data for consideration of MRLs for this crop. Bridging studies were conducted with a water-dispersible granule containing 750 g ai/kg and with an emulsifiable concentrate containing 300 g ai/l. Three foliar broadcast applications of each formulation were made at a rate of 1.5 kg ai/ha per application at a 5–8-day interval. A twofold rate of 3.0 kg ai/ha per application was also included. The applications were made at growth stages from late flowering to 15% of the bolls opened, with the last application 17 or 28 days before harvest. Although two sites in Australia have an early PHI (17 days), the residue levels in cotton-seed in these trials were included for MRL consideration because they were within the range of those in all trials. The trials conformed with Australian GAP (0.15–1.5 kg ai/ha in three applications and a 28-day PHI with either formulation). The highest residue level in replicate plots was chosen for estimating the STMR as a worst-case scenario.

Two supervised field trials on cotton were conducted at two sites in Brazil in 1992. Data on residues from the trials were submitted to JMPR in 2000 and accepted as meeting GAP. Briefly, the trials were conducted with an emulsifiable concentrate containing 480 g ai/l at a rate of either 0.96 or 1.92 kg ai/ha (twofold rate) per application. Foliar broadcast applications were made twice in one trial and three times in the other. In 2000, three additional supervised field trials on cotton were conducted at three sites in Brazil, with three foliar broadcast applications of the same emulsifiable concentrate at a rate of 0.96 or 1.92 kg ai/ha per application. The applications were made at growth stages from plant emergence to 30% of the bolls opened. Two replicate plots were used per treatment in each field site. Samples were hand-harvested from control and treated plots at 0, 7 (or 5), 14 (or 15), 21 (or 22) and 28 days after the last application and ginned to generate the cotton-seed (undelinted) samples. The trials conformed with Brazilian GAP (0.14–0.96 kg ai/ha in one to three applications and a 21-day PHI).

Supervised field trials on cotton were conducted in the USA in 1973, 1974 and 1986. Data on residue levels from these trials were submitted to JMPR 2000, and three of the trials were accepted as meeting GAP. Briefly, the trials were conducted with an emulsifiable concentrate containing 480 g ai/l applied as foliar broadcast applications. At a site in Mississippi in 1973, nine applications were made at a rate of either 1.12 or 2.24 kg ai/ha per application. Samples were hand-harvested 0, 3, 7 and 14 days after the last application and ginned to generate undelinted cotton-seed. At another site in Mississippi in 1974, two applications were made at a rate of 0.28 kg ai/ha per application, followed by 12 applications each at 1.12 kg ai/ha. Undelinted cotton-seed samples were generated similarly 15 days after the last application. At a site in California in 1986, five applications were made at 1.12 kg ai/ha, and undelinted cotton-seed samples were generated 14 days after the last application.

On the basis of the trials that conformed to GAP, the chlorpyrifos residue levels, in ranked order (median underlined), was: 0.01 (two), 0.02, 0.03 (two), 0.05, <u>0.07</u> (four), 0.11, 0.12, 0.16 and 0.18 mg/kg. The Meeting estimated a maximum residue level of 0.3 mg/kg for cotton-seed, an STMR of 0.07 mg/kg and a highest residue level of 0.18 mg/kg from the supervised trial in the USA in 1986.

Potato

The 2000 JMPR considered data on residues from supervised trials on potatoes, but did not establish MRLs because there were insufficient trials at GAP to estimate the STMR or maximum residue level. Additional data from supervised trials after at-plant application (soil in-furrow) on potatoes were generated in Brazil and submitted to the present Meeting. The proposed MRL from the data on residues after at-plant application would also cover residue levels after foliar application. The data previously summarized for at-plant application were provided again to allow evaluation of residue levels in all supervised trials conducted in Brazil. Four supervised field trials on potatoes were conducted in 1993–94 in Brazil. The data from these trials were submitted to the JMPR in 2000 and accepted as meeting GAP. The trials were conducted with either a granular formulation containing 100 g ai/kg or an emulsifiable concentrate containing 450 g ai/l. A single application was made to soil in-furrow at planting, at a rate of 1.5, 3.0 or 6.0 kg ai/ha for the granular formulation and 2.9 or 5.9 kg ai/ha for the emulsifiable concentrate.

In 1999–2000, seven supervised field trials on potatoes were conducted in Brazil to provide additional data on residue levels for MRLs. Four trials were conducted with the granular formulation at a rate of 3 or 6 kg ai/ha, and three trials were conducted with the emulsifiable concentrate at a rate of 2.7 or 5.4 kg ai/ha. Three replicate plots were maintained for each treatment. A single application of each formulation was made to soil in-furrow at planting. Three samples of potatoes were collected manually at normal harvest from the control and treated plots 100–124 days after application. The highest residue level in replicate plots was chosen for estimating the STMR as a worst-case scenario.

On the basis of at-plant treatment in trials conforming to GAP, the chlorpyrifos residue levels, in ranked order, were: 0.02, 0.03, 0.10, 0.13, 0.29, <u>0.51</u>, 0.57, 0.58, 0.65, 0.69 and 0.87 mg/kg. The Meeting estimated a maximum residue level of 2 mg/kg for potato, an STMR of 0.51 mg/kg and a highest residue level of 0.87 mg/kg.

Rice

The 2000 JMPR considered data on residues from supervised trials on rice, but did not establish MRLs because no trials were conducted at the relevant GAP. Relevant GAP in Colombia, the Philippines and Viet Nam was made available to support the results of the supervised trials submitted to the JMPR in 2000. Some additional data from supervised trials on rice, generated in India and Thailand since 2000, were submitted to the present Meeting. Supervised trials were thus conducted in Colombia, India, the Philippines, Thailand and Viet Nam.

In Colombia, two supervised field trials on rice were conducted in 1998 at two sites. Data on residue levels from these two trials were submitted to the JMPR in 2000. Briefly, the trials were conducted with an emulsifiable concentrate containing 480 g ai/l, applied three times to upland rice. The first application was made at a rate of 0.96 kg ai/ha after germination, followed by a second application at 0.72 kg ai/ha when the plants were at tillering crop growth stage; the final application was made at a rate of 0.34 kg ai/ha 20–21 days before harvest, when 100% of pinnacles were present. Three replicate plots were maintained for each trial. Rice samples from the control and treated plots were hand-harvested 20–21 days after the last application. The supervised trials conformed to Colombian GAP (0.34–0.96 kg ai/ha in a maximum of three applications and 20-day PHI).

In the Philippines, two supervised field trials on rice were conducted in 1998 at two sites. An emulsifiable concentrate containing 300 g ai/l was applied three times at a rate of 0.3 kg ai/ha per application. Data on residue levels in these two trials were submitted to JMPR 2000. The first two applications were made 25 and 40 days after transplantation, and the last application was made 25 days before harvest. Rice grain and straw samples from the control and treated plots were harvested 25 days after the last application. The supervised trials conformed to GAP in the Philippines, which is 0.30 kg ai/ha in a maximum of three applications and 25 days' PHI.

In Viet Nam, two supervised field trials on rice were conducted in 1998 at two sites. Data on residue levels in these two trials were submitted to the JMPR in 2000. Briefly, the trials were conducted with an emulsifiable concentrate containing 300 g ai/l, applied at a rate of 0.42 kg ai/ha. Rice grain and straw samples from the control and treated plots were harvested 10 days after the last application. The supervised trials conformed to GAP in Viet Nam, which is 0.3–0.42 kg ai/ha in one application and 28 days' PHI.

In Thailand, three supervised field trials on rice were conducted in 2002. An emulsifiable concentrate of chlorpyrifos containing 400 g ai/l was applied three times as a foliar application at a rate of 400 g ai/ha per application. The first application was made at mid-booting, and the final one at seed formation. Samples of rice grain and straw from the control and treated plots were hand-harvested 7, 14 and 21 days after the last application. The GAP in Thailand is 0.40 kg ai/ha, with an unspecified number of applications and 7–14 days' PHI.

In India, three supervised field trials on rice were conducted at three sites in 2002. An emulsifiable concentrate of chlorpyrifos containing 200 g ai/l was applied as a foliar application at a single rate of 375 g ai/ha. Samples of grain and straw from the control and treated plots were taken 14 (or 15), 21 and 30 days after application. As the labels were not available in English, the Meeting did not evaluate the data from India.

On the basis of trials on rice conforming to GAP, the chlorpyrifos residue levels, in ranked order, were: 0.02 (two), 0.08 (two), <u>0.09</u>, <u>0.15</u>, 0.16, 0.19 and 0.28 mg/kg. The Meeting estimated a maximum residue level of 0.5 mg/kg for rice, an STMR of 0.12 mg/kg and a highest residue level of 0.28 mg/kg.

Soya bean

The 2000 JMPR considered data from supervised trials on soya beans, but did not establish MRLs because the data from accepted GAP trials were insufficient for estimating the STMR or maximum residue level. Additional data from supervised trials conducted in 1994–96 on soya beans in Brazil were submitted to the Meeting. The data from the five trials conducted in the USA according to GAP were provided again to the Meeting.

In Brazil, two field trials were conducted in 1992–93 with an emulsifiable concentrate containing 480 g ai/l, applied three times at either 0.48 or 0.96 kg ai/ha (twofold label rate) per application. Soya beans from the control and treated plots were hand-harvested 20–21 days after the last application. In 1992–93, an additional supervised field trial was conducted with the emulsifiable concentrate applied three times at 0.48 or 0.96 kg ai/ha per application. Soya beans from the control and treated plots were hand-harvested 21 days after application. In 1995, one supervised field trial was conducted in Brazil with the emulsifiable concentrate applied at either 0.34 or 0.77 kg ai/ha. Soya beans from the control and treated plots were harvested 21 days after application. Brazilian GAP is 0.12–0.48 kg ai/ha with one to two applications and 21 days' PHI. The supervised trials represent the worst-case scenario. The residue levels were below the LOQ (< 0.01 mg/kg) in all trials conducted at either the maximum or twice the maximum label rate or in single or triple applications.

In the USA, supervised field trials on soya beans were conducted in 1975–76. Data on residue levels from these trials were submitted to the JMPR in 2000, and five trials were accepted as meeting GAP. Briefly, the trials were conducted with an emulsifiable concentrate containing 480 g ai/l applied once as a directed broadcast spray at crop emergence, followed by three to four foliar broadcast applications during the growing season. The application rates were 0.56–2.2 kg ai/ha at emergence and 0.56–1.1 kg ai/ha at each foliar application. Soya beans were collected from the control and treated plots at normal harvest, 28–31 days after the last application. For replicate plots, the highest residue level was chosen for consideration of the MRL, as a worst-case scenario.

On the basis of trials conforming to GAP, the chlorpyrifos residue levels, in ranked order, were: ≤ 0.01 (six), 0.01 (two) and 0.05 mg/kg. The Meeting estimated a maximum residue level of 0.1 mg/kg for soya bean, an STMR of 0.01 mg/kg and a highest residue level of 0.05 mg/kg.

Tea

Six supervised field trials were conducted in 1995, 1996, 1998 and 1999 at various sites in India. A chlorpyrifos emulsifiable concentrate containing 200 g/l was applied once at a rate of 0.20 kg ai/ha (0.05 kg ai/hl, 400 l/ha water), which complied with GAP for chlorpyrifos on tea as submitted by the Government.

On the basis of trials conforming to GAP in India, the chlorpyrifos residue levels in tea, in ranked order, were: 0.03, 0.15, 0.19, 0.57, 0.77 and 1.13 mg/kg. The Meeting estimated a maximum residue level of 2.0 mg/kg for tea, an STMR of 0.34 mg/kg and a highest residue level of 1.13 mg/kg.

Fate of residues during processing

Studies on processing of cotton-seed, rice and soya beans were submitted but not evaluated by the JMPR in 2000 because no MRLs were established for the raw agricultural commodities of these crops. The processing studies were resubmitted to the present Meeting for evaluation of residue levels in processed products of these raw agricultural commodities. The processing factors and estimated STMR-Ps for cotton-seed, rice and soya bean are summarized below:

Processed commodity	Processing factor	STMR (mg/kg) (RAC)	STMR-P (mg/kg)
Cotton hulls	0.7	0.07	0.05
Cotton-seed meal	0.1	0.07	< 0.01
Cotton-seed oil, crude	1.4	0.07	0.10
Cotton-seed oil, refined	0.2	0.07	0.01
Rice hulls	2.44	0.12	0.29
Rice bran	1.80	0.12	0.22
Rice husked	0.13	0.12	0.016

Processed commodity	Processing factor	STMR (mg/kg) (RAC)	STMR-P (mg/kg)
Polished rice	0.07	0.12	0.008
Soya bean meal	< 0.2	0.01	< 0.002
Soya bean crude oil	0.4	0.01	0.004
Soya bean refined oil	0.4	0.01	0.004
Soya bean refined	0.5	0.01	0.005

STMR-P, STMR of raw agricultural commodity × processing factor of processed product

Residues in animal commodities

The 2000 JMPR estimated the dietary burden of chlorpyrifos in farm animals and poultry in cases in which calculations from the MRLs yielded maximum theoretical dietary intakes, and calculations from STMR values for feed allowed estimation of STMR values for animal commodities. The present Meeting concluded that the contribution of residues to feed, calculated for the uses considered this year, would not increase the dietary burden assessed by the 2000 JMPR. The Meeting maintained the recommendations of the 2000 JMPR.

DIETARY RISK ASSESSMENT

Long-term intake

IEDIs for chlorpyrifos were calculated for the five GEMS/Food regional diets from the STMRs and STMR-Ps estimated by this Meeting, in addition to those for 61 commodities from the JMPR 2000 evaluation. The IEDIs were 3–30% of the maximum ADI (0–0.01 mg/kg bw), as shown in Annex 3. The Meeting concluded that the intake of residues of chlorpyrifos resulting from uses that have been considered by the JMPR is unlikely to present a public health concern.

Short-term intake

The IESTI for chlorpyrifos was calculated for the commodities for which MRLs, STMR values and highest residue values were estimated and for which data on consumption (large portion and unit weight) were available. The results are shown in Annex 4.

The ARfD for chlorpyrifos is 0.1 mg/kg bw. The short-term intakes were calculated for commodities for which highest residues or HR-Ps were estimated by the present Meeting. The calculated short-term intakes were < 100% of the ARfDs for children (0–40%) and for the general population (0–10%). The Meeting concluded that the intake of residues of chlorpyrifos resulting from uses that have been considered by the JMPR is unlikely to present a public health concern for consumers.

4.5 DIMETHIPIN (151)

TOXICOLOGY

Evaluation for an acute reference dose

The 1999 JMPR established an ARfD for dimethipin of 0.02 mg/kg bw on the basis of a NOAEL of 20 mg/kg bw per day and a LOAEL of 40 mg/kg bw per day for skeletal malformations (increased incidence of fetuses and of litters containing fetuses with scoliosis and 27 presacral vertebrae) in a study of developmental toxicity in rabbits, and using a safety factor of 1000 in consideration of the nature of the effects caused. The 2002 JMPR concluded that the 1000-fold safety factor might be excessive and that the ARfD of dimethipin should be reconsidered on the basis of appropiate data.

The present Meeting reconsidered the ARfD for dimethipin. The study of developmental toxicity in rabbits was re-evaluated in the light of a larger set of historical control data for Dutch belted rabbits, provided by the sponsor.

In a study of developmental toxicity in rabbits, does treated with dimethipin at a dose of 40 mg/kg bw per day showed body-weight loss on days 6–12 of gestation and decreased body-weight gain on days 6–28 of gestation. Fetal and litter incidences of scoliosis were 0% and 0% in the controls and 4.0% and 23.1% at 40 mg/kg bw per day, respectively. The observed incidence of scoliosis at 40 mg/kg bw per day was at the upper bound of that for historical controls (i.e. fetal and litter incidences of 4.1 and 20%, respectively). The NOAEL for both maternal and developmental toxicity was 20 mg/kg bw per day.

Toxicological evaluation

The Meeting established an ARfD of 0.2 mg/kg bw on the basis of the NOAEL of 20 mg/kg bw per day in the study of developmental toxicity in rabbits and a safety factor of 100. The Meeting considered that a safety factor of 100 was adequate, since the observed developmental toxicity was at the upper range of the historical control incidence and was possibly secondary to maternal toxicity.

An addendum to the toxicological monograph was prepared.

Estimate of acute reference dose

0.2 mg/kg bw

DIETARY RISK ASSESSMENT

Short-term intake

The Meeting established an ARfD (0.2 mg/kg bw) for dimethipin. The 2001 JMPR had calculated the IESTI for dimethipin for 12 food commodities (and their processed fractions) for which MRLs were estimated and for which consumption data were available, using the previous ARfD of 0.02 mg/kg bw. The IESTI represented 0% of the ARfD for the general population and 0–1% of the ARfD for children. The Meeting concluded that the short-term intake of residues of dimethipin from uses that have been considered by the JMPR is unlikely to present a public health concern.

4.6 DITHIOCARBAMATES (105)

RESIDUE AND ANALYTICAL ASPECTS

Propineb was evaluated by the present Meeting within the CCPR periodic review programme, and the recommendations for MRLs, STMRs and highest residue levels are discussed in the appraisal of that compound. Recommended MRLs for dithiocarbamates arising from use of propineb are consolidated here.

The 1996 JMPR recommended MRLs for dithiocarbamates in almond hulls, almond, pecan, pome fruits and stone fruits which were based on residue data for ziram. The 1996 JMPR also recommended that estimates of maximum residue levels for dithiocarbamates, which relied primarily on data for ziram, should be temporary until the relevant data on environmental fate had been evaluated.

50 Dithiocarbamates

In view of the decision of the 2003 JMPR that data on environmental fate need be reviewed only when they directly affect estimation of maximum residue levels, the Meeting decided to withdraw its requirement for information on the environmental fate of ziram in soil and in water–sediment systems.

4.7 ETHOPROPHOS (149)

RESIDUE AND ANALYTICAL ASPECTS

Ethoprophos, a nematicide and soil-insecticide, was evaluated for residues in 1984 and 1987. The toxicology of ethoprophos was reviewed within the periodic review programme by the 1999 JMPR. Ethoprophos was listed as a priority by the the CCPR at its Thirtieth Session (Alinorm 99/24 App VII) for for periodic review of residues by the 2001 JMPR. The manufacturer requested postponement of the residue evaluation.

The Meeting received information on identity; metabolism and environmental fate; analysis of residues; use pattern; residues resulting from supervised trials on strawberry, banana, cucumber, melon, pepper, tomato, potato, sweet potato and sugar-cane; fate of residues during storage and in processing; residues in food in commerce or at consumption; and national maximum residue limits.

Metabolism

Animals

The Meeting received information on the fate of [1-ethyl-^{14}C]ethoprophos rats, lactating goats and laying hens dosed orally.

Studies on metabolism in laboratory animal (rats) were evaluated by the WHO Expert Group of the 1999 JMPR, which concluded that ^{14}C-ethoprophos is rapidly and virtually completely absorbed, metabolized and excreted after oral administration to rats. The main route of excretion was urine (51–56%), but significant proportions were excreted in expired air (about 15%) and faeces (10–14%). Little radiolabel was found in tissues at 168 h, representing less than 2.5% of the dose, and the highest concentrations were found in excretory organs (liver, kidneys and lungs). There was no evidence that bioaccumulation would occur after repeated doses. Ethoprophos was metabolized by dealkylation of one or both *S*-propyl groups, followed by conjugation.

Lactating goats given feed containing ^{14}C-ethoprophos at a concentration of 32 ppm excreted 78% of the administered radiolabel in urine (including cage rinse), 3.6% in faeces (including the gastrointestinal tract and contents) and 1.7% in milk; 3.9% of the administered dose was found in tissues. During the 7-day dosing period, 2% of the applied radiolabel was found in expired air. The highest concentration of radioactive residues was found in liver (8.8 mg/kg), while kidney contained 0.93 mg/kg, milk 0.49 mg/kg, muscle 0.095 mg/kg and fat 0.051 mg/kg. The total recovery of the administered dose was 88%.

The majority of the radiolabel in liver and kidney remained in the post-extraction solids, and enzyme and acid digests of these solids co-chromatographed with amino acid standards. Radiolabelled amino acids can be formed by hydrolysis of ethoprophos to ethanol and subsequently to acetaldehyde, acetate, acetyl coenzyme A and amino acids (tricarboxylic acid cycle). Thin-layer chromatography of the polar liver extract showed three radioactive spots, representing 1.1%, 1.4% and 0.45 % of the total radioactive residues (TRR). The first spot co-chromatographed with *O*-ethyl-*S*-propyl phosphorothioate and ethyl phosphate, while the other two spots did not co-chromatograph with any of the reference markers used. The parent compound was not found. Radioactivity in the kidney extract was not characterized.

Most of the radioactivity in muscle was released from the post-extracted solids by acid or base treatment, while that in fat was distributed approximately equally between the extracted and unextracted fractions. The radiolabel in the post-extracted solids could be released by enzyme digestion. No further characterization of muscle or fat fractions was attempted owing to the low levels of radioactivity.

The residue levels in milk reached a plateau on the first day of treatment, with an average level over days 0–7 of 0.49 mg/kg (maximum, 0.68 mg/kg). The radioactivity in the chloroform extract of milk (55% TRR) co-chromatographed with standards of the fatty acids palmitic acid, oleic acid and stearic acid, which were poorly resolved. Radiolabelled fatty acids can be formed by hydrolysis of ethoprophos to ethanol. No parent compound was found.

When laying hens were given feed containing ^{14}C-ethoprophos at a concentration of 2.1 ppm for 7 days, 48% of the total administered radioactivity was recovered in excreta (including the gastrointestinal tract and contents), 1.0% in egg whites, 9.3% in egg yolks, 3.6% in expired volatiles and 3.2% in tissues and blood. The total recovery of the administered radioactivity was 64%. The highest concentration of radioactive residues was found in liver, at 1.2 mg/kg, followed by kidney at 0.42 mg/kg; 0.069 mg/kg radioactive residue was found in fat and 0.010 mg/kg in muscle. A maximum residue level of 0.64 mg/kg was found in egg yolk and 0.029 mg/kg in egg white.

As in goats, most of the radioactivity in liver and kidney remained in the post-extracted solids. Enzyme and acid digests of these solids co-chromatographed with amino acid standards. Thin-layer chromatography of the polar extract of liver contained three radioactive zones, representing 1.9%, 0.95% and 2.0% TRR. The first zone contained *O*-ethyl-*S*-propyl phosphorothioate or ethyl phosphate, the second zone did not co-chromatograph with any of the reference markers used, and the third zone co-chromatographed with *O*-ethyl-*O*-methyl-*S*-propyl phosphorothioate or *O*-ethyl-*S*-methyl-*S*-propyl phosphorodithioate. No parent compound was found.

Most of the radioactivity in muscle was released from the post-extracted solids by acid or base treatment, while that in fat was present mainly in the organic extract. The radiolabel could be released from the post-extracted solids by enzyme digestion. No further characterization of muscle or fat fractions was attempted owing to the low levels of radioactivity.

The radioactive residue level reached a plateau in egg whites on the third day of treatment, but no plateau was reached in egg yolks during the 7-day treatment. The average concentrations found were 0.021 mg/kg in egg whites (average over days 3–7; maximum, 0.029 mg/kg) and 0.30 mg/kg in egg yolks (average over days 0–7; maximum, 0.64 mg/kg). In egg yolks, 84% was extractable in hexane and 11% in chloroform. The hexane fraction of egg yolks co-chromatographed with the fatty acids palmitic, myristic, oleic and stearic acid, which were poorly resolved. No parent compound was found.

The metabolism of ethoprophos in laboratory animals was similar to that in farm animals.

Plants

The Meeting received information on the fate of ethoprophos labelled with ^{14}C in the ethyl or the propyl group after soil treatment before planting of pulses or oil seeds (French beans), cereals (maize), root and tuber vegetables (potatoes) and leafy crops (cabbage).

In a greenhouse, *French bean* bedding plants (variety Contender) were planted in clay pots filled with steam-sterilized soil treated with [α-^{14}C-ethyl]- or [α-^{14}C-propyl]ethoprophos. The compound was applied as a granule formulation at 14.3 mg ai/kg soil. The plants were grown for 63 days and were sampled at weekly intervals from day 7 onwards. The residue levels in soil extracts decreased with time, while the total residues in the bean plants increased with time, from 2.2% of the total applied radioactivity to 13% with the ethyl label and from 0.58% to 8.3% with the propyl label between days 7 and 63. Mainly extractable residues were found early in the study, while unextracted residues predominated (> 57%) from day 21 onwards. In mthanol:water extracts of the bean plants, the main compounds were *O*-ethyl-*S*-propyl phosphorothioate and ethyl phosphate. In dichloromethane extracts, the main compounds were the parent (maximum, 13%) and ethyl propyl sulfide (maximum, 9.2%). The amount of parent compound decreased with time after application and contributed < 10% from day 28 onwards. Minor amounts of propyl disulfide, ethyl propyl sulfoxide (plus methyl propyl sulfoxide) and ethyl propyl sulfone (plus methyl propyl sulfone) were present at some sampling times.

In a greenhouse, *maize seeds* were planted in clay pots filled with steam-sterilized soil treated with [α-^{14}C-ethyl]- or [β-^{14}C-propyl]ethoprophos. Ethoprophos was applied as a granule formulation at 14.3 mg ai/kg soil. Maize plants were grown for 100 days and were sampled at 10-day intervals from day 18 onwards. The residue levels in soil extracts were constant, while those in maize plants increased from 0.96% of the applied radiolabel to 74% for the ethyl label and from 0.26% to 34% with the propyl label between days 18 and 100. Most of the extractable residues in the maize plants were found early in the study, while unextracted residues predominated (> 67%) from day 38 onwards. In methanol:water extracts of the maize plants, the main compounds were *O*-ethyl-*S*-propyl phosphorothioate and ethyl phosphate. In dichloromethane extracts, the main compounds were the parent (maximum, 40% TRR) and ethyl propyl sulfide (maximum, 7.6% TRR). The amount of parent compound decreased over time and contributed < 10% from day 48 onwards. Small amounts of propyl disulfide, ethyl propyl sulfoxide (plus methyl propyl sulfoxide) and ethyl propyl sulfone (plus methyl propyl sulfone) were present at some sampling times. The ethyl label was found mainly on ethyl phosphate.

In a second study on *maize*, silt loam was treated with [1-ethyl-^{14}C]ethoprophos (emulsifiable concentrate formulation) at a rate of 13 kg ai/ha in plastic-lined wooden boxes placed in the field. The actual concentration in the soil was 10 mg ai/kg. The application mixture was incorporated to a depth of 10 cm. Sweet maize seeds (variety Early extra sweet) were planted 3 days after soil treatment and were sampled at the green forage stage (soil, whole plant), at maturity (shanks, husks, silks, grain, empty cobs) and at the fodder stage (soil, senescent stalks without cobs). The TRR was 2.2 mg/kg in maize forage, 0.27 mg/kg in maize cobs, 0.25 mg/kg in grain, 0.79 mg/kg in husks and 1.4 mg/kg in fodder. Most of the TRR in these matrices was solvent-extractable. Acid or base hydrolysis released a further 6–14% TRR from forage, grain, cobs and fodder; however, 13% TRR in forage and 40% TRR in grain, cobs and fodder remained unextracted. Ethyl phosphate was the main metabolite detected in green forage, grain and fodder (10%, 35% and 8.9%, respectively). Parent ethoprophos and its metabolite *O*-ethyl-*S*-propyl phosphorothioate were also present in small amounts in forage and fodder. The extracts of forage and fodder further tentatively contained < 1% each of *O*-ethyl-*O*-methyl-*S*-propyl phosphorothioate and *O*-ethyl-*S*-methyl-*S*-propyl phosphorodithioate.

Silt loam was treated with [1-ethyl-^{14}C]ethoprophos (emulsifiable concentrate formulation) at a rate of 13 kg ai/ha in plastic-lined wooden boxes placed in the field. The actual concentration in the soil was 15 mg ai/kg. The mixture was incorporated to a depth of 10 cm. *Potatoes* (variety Kenebeck) were planted 3 days after soil treatment, and soil and plants were sampled at the 'new potato' stage and at maturity. The TRR was 0.24–0.54 mg/kg in tubers and 1.1–3.8 mg/kg in vines. Most of the TRR was extracted with aqueous methanol. Acid or base hydrolysis solubilized a further 17% of the radioactivity in the vines, while 31% TRR in vines and 23% TRR in tubers remained unextracted. In both vines and the tubers, the main metabolite was ethylphosphate (12% and 38% TRR, respectively). Parent ethoprophos, *O*-ethyl-*S*-propyl phosphorothioate and *O*-ethyl-*O*-methyl-*S*-propyl phosphorothioate (the latter tentatively) were present in small amounts in the vines but were not detected in tubers.

To determine the nature of the unextracted residues in potatoes, sandy loam was treated with [1-ethyl-^{14}C]ethoprophos (emulsifiable concentrate formulation) at a dose rate of 13 kg ai/ha in plastic-lined wooden boxes placed in the field. The mixture was incorporated to a depth of 10 cm; the actual concentration in the soil was 5.9 mg ai/kg. Potatoes (minituber variety Kennebec) were planted 3 days after soil treatment and were harvested 118 days (new potato tubers) or 167 days after treatment (mature potatoes). The TRR and extractability were comparable with those in the first study. A sequential extraction scheme showed that 41% TRR in new potato tubers consisted of solvent-extractable residues, 11% TRR was present in starch, 8.5% TRR in protein, 4.4% TRR in pectin, 3.7% TRR in lignin, 8.2% in hemicellulose and 8.8% TRR in cellulose. The unextracted radioactive residue associated with starch was shown to be ^{14}C-glucose.

Ethoprophos

Silt loam was treated with [1-ethyl-^{14}C]ethoprophos (emulsifiable concentrate) at a rate of 11 kg ai/ha in plastic-lined wooden boxes placed in the field. The actual concentration in the soil was 7.6 mg ai/kg. The mixture was incorporated to a depth of 7.6 cm. *Cabbage* bedding plants (variety Stonehead) were planted 2 days after soil treatment, and soil and plants were sampled at the leafy stage and at maturity. The TRR was 16 mg/kg in leafy cabbage and 3.1 mg/kg in head cabbage. Most of the TRR was extractable, and ethylphosphate was the main metabolite found in both leafy and head cabbage extracts (21% and 24%, respectively). Ethoprophos and *O*-ethyl-*S*-propyl phosphorothioate were present at 0.3–4% in both types of cabbage, and *O*-ethyl-*O*-methyl-*S*-propyl phosphorothioate and *O*-ethyl-*S*-methyl-*S*-propyl phosphorodithioate were tentatively identified at 0.4–1.7%. A supplementary characterization study showed that most of the unextractable radioactive residues in cabbage were incorporated into plant structural components, mainly in lignin (38%).

The metabolism of ethoprophos in plants appears to be qualitatively similar to that in animals; however, the toxicologically significant metabolites *O*-ethyl-*O*-methyl-*S*-propyl phosphorothioate and *O*-ethyl-*S*-methyl-*S*-propyl phosphorodithioate were tentatively identified in hen liver, maize green forage and fodder, potato vines and cabbage heads, but not in rats or goats.

Environmental fate

Soil

The Meeting received information on aerobic degradation in soil and studies on rotational crops (confined and field).

The route and rate of degradation of [1-^{14}C-propyl]ethoprophos was investigated in three studies in different soils under aerobic conditions in the dark at 10 °C and 20–25 °C. On the basis of an application rate of 10.5 kg ai/ha in the field, the test substance was applied at a nominal concentration of 10–14 mg ai/kg dry weight of soil. The main degradation product in soil under aerobic conditions was ^{14}CO$_2$, which accounted for 54–60% of the applied radioactivity after 90 days at 22–25 °C and 43–50% after 110 days at 10 °C. Most of the radioactivity in the extracts was associated with unchanged ethoprophos, representing 90–94% on day 0 and 7.2–9.4% on day 90 at 22 °C. One major metabolite was identified as *O*-ethyl-*S*-propyl phosphorothioate (maximum, 3.6–7.9% of the applied radioactivity); two minor metabolites were *O*-ethyl-*O*-methyl-*S*-propyl phosphorothioate (maximum, 0.7%) and *O*-ethyl-*S*-methyl-*S*-propyl phosphorodithioate (maximum, 0.3%). The half-life of ethoprophos at ambient temperature was 10–25 days, while that at 10 °C was two to three times longer.

In a study of a confined rotational crop, a sandy loam soil was sprayed with [1-ethyl-^{14}C]ethoprophos as an emulsifiable concentrate at a rate equivalent to 13.4 kg ai/ha and thus incorporated into the top 10 cm of soil. The soil was placed in boxes inside a screened enclosure, which was heated and covered with plastic during the winter months. The soil was left fallow for 30–365 days after treatment. *Wheat* (variety Anza), *spinach* (variety Polka) and *radish* (variety Cherry Belle) were each planted 30, 120 and 365 days after treatment, and immature and mature crops were harvested at each planting interval. Soil samples were collected at application, at each planting and at each harvest. The TRR in rotational crops was generally much lower after a plant-back interval of 365 days than that after a plant-back interval of 30 days; e.g. mature wheat straw contained a radioactive residue level of 47 mg/kg after a plant-back interval of 30 days and 0.65 mg/kg after a plant-back interval of 365 days. Crops harvested when immature showed similar extractability, while the total extractability from mature wheat was generally lower than that from mature spinach or radish. Some of the remaining solids could be hydrolysed by acid or alkaline treatment; however, 1.9–42% TRR remained unextractable, with the highest portion in wheat chaff 365 days after treatment.

Parent ethoprophos was present in extracts of immature and early maturing crops (radish) at both the 120-day and the 30-day plant-back interval. The parent compound was not found in mature wheat or spinach at the 120-day plant-back interval or in any crop at the 365-day plant-back interval. The main component in each crop matrix was ethyl phosphate, but *O*-ethyl-*S*-propyl phosphorothioate was also found. Many unidentified compounds were found, some at levels > 10% TRR or 0.05 mg/kg. After hydrolysis of immature spinach extracts from the 120-day plant-back interval, two of the unknown compounds (12% and 8.5% TRR) were found to be conjugates of ethyl phosphate. The levels of the remaining unknown compounds were not sufficient for structural identification. The main component in acid and base hydrolysates of the crops was the parent compound (0.13 mg/kg in mature wheat straw at the 120 day plant-back interval). Most of the remaining radiocarbon was associated with *O*-ethyl-*O*-methyl-*S*-propyl phosphorothioate (0.02 mg/kg). Unextractable residues were characterized in mature wheat straw. General incorporation into cellular components was 40% TRR in extractable residues, 7.7% TRR in starch, 1.5% TRR in protein, 1.9% TRR in pectin, 11% TRR in lignin, 14% TRR in hemi-cellulose, 10% TRR in cellulose and 22% TRR in insoluble residue; the overall recovery was 105%.

In a field study of rotational crops, unlabelled ethoprophos was applied once as an emulsifiable concentrate to sandy loam before planting at an actual rate of 13.5 kg ai/ha. The rotational crops were root vegetables (radish roots), leafy crops (radish leaves, red leaf lettuce, collards), cereals (forage, grain and straw from winter wheat, spring wheat and sorghum) and pulses or oil seeds (forage, grain and straw from cow peas, wando peas, green peas and soya beans and mustard forage). The crops were planted 1, 4, 8 and 12 months after application at two sites. Sample extracts were analysed for ethoprophos and *O*-ethyl-*S*-propyl phosphorothioate by gas chromatography with flame photometry detection. The residue levels were below the LOQ of 0.01 mg/kg in all treated samples from both test sites, except in radish root and radish leaves. The highest level of parent compound found in radish root was in samples taken at the plant-back interval of 31 days with harvest 32 days after planting, at 0.023 mg/kg; in the same samples, the highest level of *O*-ethyl-*S*-propyl phosphorothioate was 0.039 mg/kg. The presence of ethoprophos and *O*-ethyl-*S*-propyl phosphorothioate in radish root and tops was confirmed by gas chromatography–mass spectrometry but at levels at least an order of magnitude lower than those measured by gas chromatography with flame photometry detection.

Methods of analysis

The Meeting received information on enforcement and monitoring methods for the determination of ethoprophos in foodstuffs of plant and animal origin and on the analytical methods used in studies of rotational crops, supervised trials and studies of storage stability, processing and monitoring for determination of ethoprophos and the metabolite *O*-ethyl-*S*-propyl phosphorothioate in foodstuffs of plant origin.

Five enforcement methods were submitted. Ethoprophos can be determined by the Dutch multi-residue method MRM-1 (validated for non-fatty matrices, quantification by gas chromatography with nitrogen–phosphorus or mass spectrometry detection; LOQ, 0.01–0.05 mg/kg) and with the German multi-residue methods DFG-S8 (validated for fruits and vegetables, quantification by gas chromatography with electron capture or alkali flame ionization detection; LOQ, 0.02 mg/kg) and DFG-S19 (validated for foodstuffs of plant and animal origin, quantification with gas chromatography with flame photometry, mass spectrometry or PFP detection, depending on the module used; LOQ, 0.01 mg/kg). Ethoprophos could not be determined by the multi-residue protocols of the US Food and Drug Administration. Method AR 271-01 was proposed as an enforcement method for determination of ethoprophos in milk, egg, meat, fat and liver and is considered valid in the range 0.01–0.1 mg/kg (quantification by gas chromatography with flame photometric detection).

All the methods used in the various studies were based on extraction with hexane, methanol, acetone, ethyl acetate, acetonitrile or petroleum ether:acetone, followed by a clean-up and determination by gas chromatography with MC, flame photometry, nitrogen–phosphorus, flame photometric, mass spectrometry, electron capture or tandem mass spectrometry detection. The LOQs ranged from 0.005 mg/kg to 0.05 mg/kg, 0.01 mg/kg being the most common.

Ethoprophos

Stability of residues in stored analytical samples

The Meeting received data on the stability of residues in crops with a high water content (pineapple, broccoli, cabbage, potato, sweet potato, tomato), in dry crops with starch and protein (maize), in dry crops with fat or oil, starch and protein (peanut), in special cases (sugar-cane, tobacco (green, cured)), in processed commodities (pineapple juice, peanut crude oil, peanut refined oil, maize crude oil, maize refined oil, maize starch, refined cane sugar) and in feed remains (pineapple bran, pineapple feed pulp, peanut hulls, peanut meal, peanut vine, dry peanut hay, maize meal, maize forage, maize fodder, maize grain dust, sugar-cane molasses) stored frozen. Crops with a high water and a high acid content (citrus fruits) were not investigated.

The freezer storage stability of ethoprophos depends on the matrix. Parent ethoprophos was found to be stable at –20 °C for a maximum of 9 months in broccoli and pineapple fruit, but for at least 9–12 months in other crops with a high water content (cabbage, sweet potato, potato, peanut vine, maize forage). In another study, the parent compound was stable for at least 19 months in tomato and potato. It was stable for at least 12 months in dry crops with starch and protein (maize grain) and for a maximum of 12 months in tobacco and peanut nutmeat. Ethoprophos was not stable in peanut hay. Ethoprophos and *O*-ethyl-*S*-propyl phosphorothioate were stable at –20 °C for at least 15 months in sugar-cane and its processed commodities. No general conclusions can be drawn for processed commodities and remains.

The results showed that, in general, *O*-ethyl-*S*-propyl phosphorothioate is not stable at –20 °C for < 1 month, although longer storage times are possible for some crops.

Definition of the residue

Ethoprophos is metabolized rapidly in rats and livestock and was not found in edible tissues. In metabolism studies with labelled compounds, most of the radioactivity was found to be incorporated into natural components, such as fatty acids and amino acids. Low levels of *O*-ethyl-*S*-propyl phosphorothioate or ethyl phosphate were identified in goat and hen liver, and *O*-ethyl-*O*-methyl-*S*-propyl phosphorothioate and *O*-ethyl-*S*-methyl-*S*-propyl phosphorodithioate were tentatively identified in hen liver at low levels. The main route of metabolism in livestock is hydrolysis of the P–S bond, yielding *O*-ethyl-*S*-propyl phosphorothioate and propyl sulfide, with hydrolysis of *O*-ethyl-*S*-propyl phosphorothioate to ethyl phosphate; the ethyl moiety can be split off and become incorporated into natural components like amino acids and fatty acids.

Although ethoprophos is not found in edible tissues, the Meeting agreed that, in the absence of a better indicator, the parent should be considered the compound of interest in animal commodities, both for enforcement and for dietary risk assessment. The log octanol–water partition coefficient (P_{ow}) of 2.99 indicates that the residue is not fat-soluble.

The main route of metabolism is similar in plant and animals, although other routes differ. Propylsulfide in plants can react with a parent molecule to yield ethylpropyl sulfide and propyl disulfide, while propylsulfide is methylated in rats.

In edible plant parts (mature maize grain, potato and cabbage), the major residue is ethyl phosphate, which is considered not to be toxicologically relevant and is thus not included in the residue definition for dietary risk assessment. Furthermore, ethyl phosphate is formed by several other organophosphate pesticides (e.g. parathion) and can therefore not be used for enforcement purposes. Ethylpropyl sulfide, which was found in amounts similar to the parent compound in French beans, was not found in rats; however, it behaves similarly to methylpropyl sulfide, which was detected in rats. It is not expected that this metabolite will be toxicologically significant.

Possible candidates for the residue definition are the parent, *O*-ethyl-*S*-propyl phosphorothioate, *O*-ethyl-*O*-methyl-*S*-propyl phosphorothioate and *O*-ethyl-*S*-methyl-*S*-propyl phosphorodithioate.

Ethoprophos

As reported by the 1999 JMPR, *O*-ethyl-*S*-propyl phosphorothioate, *O*-ethyl-*O*-methyl-*S*-propyl phosphorothioate and *O*-ethyl-*S*-methyl-*S*-propyl phosphorodithioate were tested for toxicity and for their ability to inhibit cholinesterase activity in female rats given single oral doses. The last two metabolites had approximately the same cholinergic toxicity as the parent compound, while the first was less toxic than the parent. As *O*-ethyl-*S*-propyl phosphorothioate is less toxic than the parent compound in rats, is rapidly converted to ethyl phosphate and is not found in mature maize grain, potato tubers or mature cabbage heads, the Meeting decided not to include this metabolite in either residue definition. The two remaining metabolites were not detected in mature maize grain or potato tubers but were tentatively identified in mature cabbage heads. These metabolites were also tentatively identified in animal feedstuffs (maize forage and fodder), although they were not identified in rats. It is possible that the molecules are artefacts formed during extraction with methanol. In view of the low levels found in the metabolism studies, the Meeting decided not to include *O*-ethyl-*O*-methyl-*S*-propyl phosphorothioate or *O*-ethyl-*S*-methyl-*S*-propyl phosphorodithioate in either residue definition.

Definition of the residue for compliance with MRLs and for estimating dietary intake: ethoprophos, for both plant and animal commodities.

Results of supervised residue trials on crops

Supervised trials were available for stawberry, banana, cucumber, melon, pepper, tomato, potato and sugar-cane, but none were provided for the remaining commodities that currently have a Codex level (CXL). Therefore, the Meeting decided to withdraw the current recommendations for beetroot, cabbage head, gherkin, grape, lettuce head, maize, maize fodder, onion bulb, peanut, peanut fodder, pea (pods and succulent or immature seeds), pineapple, pineapple fodder, pineapple forage, soya bean and soya bean fodder.

Berries and other small fruit

Strawberry

Ethoprophos is registered in Austria and Spain for use on strawberry with granule and emulsifiable concentrate formulations at the pre-planting or planting stages. Four trials on strawberry were conducted in Italy in 1996–98 at two sites. Application was by drip irrigation with emulsifiable concentrates throughout the growing season but before the fruits had formed. Although drip irrigation is usually the critical GAP and no residues were detected in the trials, the application rate stated on the available labels was 6 kg ai/ha, while that used in the trials was only 1.8–3.5 kg ai/ha: None of the trials was conducted according to GAP.

The Meeting agreed that the available data were insufficient to estimate a maximum residue level for ethoprophos in strawberry.

Assorted tropical and sub-tropical fruits minus inedible peel

Banana

Trials on bananas were reported to the Meeting from Brazil (GAP: 3.0 g ai/tree, two applications, 3-day PHI), Costa Rica (GAP for Central America: 2.9–3.0 g ai/tree, one application, 30-day PHI), Côte d'Ivoire (GAP: 4.0–8.0 g ai/tree, two to three applications, PHI not specified) and the Phillipines (GAP: 4.0–5.0 g ai/tree, two applications, PHI not specified).

In one trial in the Côte d'Ivoire, the residue levels in banana were below the LOQ (< 0.02 mg/kg).

None of the 20 Costa Rican trials was conducted according to GAP in Central America, as 15 involved overdosing, 14 involved more than one treatment or residues were measured at a PHI of < 30 days. As residue levels above the LOQ were not measured in any of the trials, the Meeting decided that the six trials with a PHI of ≤ 30 days could be considered for estimating the MRL. The residue levels in banana were < 0.02 mg/kg in all six trials.

The two trials in Brazil did not comply to GAP (overdosing, with only one application), and no residues were found. The five trials in the Philippines were also not conducted according to GAP (underdosing). In two of the trials, residue levels of 0.0065 mg/kg and 0.013 mg/kg were found in pulp.

The Meeting decided to combine the results of the trial in the Côte d'Ivoire and of the six trials in Costa Rica to estimate the maximum residue level for banana. The levels in all seven trials were < 0.02 mg/kg. The Meeting agreed to withdraw the previously recommended maximum residue level for banana of 0.02* mg/kg and to replaced it by a recommendation of 0.02 mg/kg. The Meeting estimated an STMR and a highest residue level for ethoprophos in banana of 0.02 mg/kg.

Fruiting vegetables, cucurbits

Cucumber

Indoor trials on cucumber in which soil received overall treatment before planting or at transplanting with a granule formulation were reported from Canada and The Netherlands. No GAP was reported for either country. In the five Canadian trials, conducted according to US GAP (12–15 kg ai/ha), the residue levels were < 0.01 mg/kg. In four of the six Dutch trials conducted according to Italian, Portuguese or Spanish GAP (3–10 kg ai/ha, 30–60-day PHI), the residue levels in cucumber were < 0.01 mg/kg.

Seventeen outdoor trials on cucumber, with overall soil treatment with a granule formulation before planting or at transplanting, were reported from the USA. In the five conducted according to US GAP, the residue levels were < 0.005 and < 0.02 (four) mg/kg.

Nine indoor trials on cucumber in which soil received spray treatment with emulsifiable concentrate formulations pre- and post-planting were reported from southern Europe (France, Greece, Italy, Portugal and Spain). The trials were evaluated against Spanish GAP (6.0 kg ai/ha, one application, 60-day PHI). All the trials involved overdosing. In one trial in which ethoprophos was applied after planting, an actual residue level of 0.0090 mg/kg was found at 21 days PHI; however, all trials at the correct PHI showed residue levels of < 0.005 mg/kg.

Ten outdoor trials on cucumber in which soil received spray treatment with emulsifiable concentrate formulations before or at planting were reported from the USA. All the trials were conducted according to US GAP, but the results of one trial was excluded from evaluation as the samples were purportedly mislabelled. The residue levels were < 0.01 (nine) mg/kg.

Eight indoor trials on cucumber in which soil was treated with emulsifiable concentrate formulations by drip irrigation after planting or transplanting were reported from southern Europe (France, Italy and Spain). The trials complied with Spanish GAP (0.6 kg ai/ha, 1–10 applications, maximum total of 6 kg ai/ha, 60-day PHI), except that the latest PHI for which residue levels were reported was 14–15 days. On these days, the residue levels were < 0.005 (six), < 0.01 and 0.012 mg/kg.

The Meeting concluded that, irrespective of the method of application and the site (indoors or outdoors), the residue levels would not be expected to exceed the enforcement LOQ of 0.01 mg/kg. The Meeting agreed to withdraw the previously recommended maximum residue level for cucumber of 0.02* mg/kg and to replace it by a recommendation of 0.01 mg/kg. The Meeting estimated an STMR and a highest residue level for ethoprophos in cucumber of 0.01 mg/kg.

Melon

Nine outdoor trials on melon involving overall soil treatment with granule formulations before, at and after planting were reported from southern Europe (France, Italy and Spain). The trials were compared with Portuguese GAP (8 kg ai/ha, 56-day PHI). The residue levels were < 0.005 (seven), 0.0055 and 0.010 mg/kg. The levels in melon pulp (edible portion) were ≤ 0.005 (eight) and 0.012 mg/kg.

Eight outdoor trials on melon involving post-transplanting drip irrigation with emulsifiable concentrate formulations were reported from southern Europe (France, Italy and Spain); however, there is no GAP for drip irrigation on melon in southern Europe.

On the basis of the trials of overall soil treatment, the Meeting agreed to withdraw the previously recommended maximum residue level for melon, except watermelon, of 0.02* mg/kg and to replace it by a recommendation of 0.02 mg/kg. The Meeting estimated an STMR of 0.005 mg/kg and a highest residue level of 0.012 mg/kg for ethoprophos in the edible portion of melon.

Fruiting vegetables other than cucurbits

Pepper

Eleven indoor trials on sweet pepper involving overall soil treatment with granule formulations before or at planting were reported from southern Europe (France, Greece, Italy and Spain). The trials were evaluated against Spanish GAP (6.0–8.0 kg ai/ha, one application, 60-day PHI). The residue levels were: < 0.005 (nine), 0.007 and 0.027 mg/kg.

A further 12 trials from southern Europe on green and sweet pepper involved application of ethoprophos as an emulsifiable concentrate formulation by post-planting drip irrigation. Ten could be evaluated against Italian GAP (1.7–3.5 kg ai/ha, three to four applications, 30-day PHI). The residue levels were: < 0.005 (four), < 0.01 (two), 0.006, 0.0068, 0.007 and 0.044 mg/kg. Two trials on green pepper yielded higher residue levels, but the latest sampling was at a PHI of 14–15 days.

The Meeting decided to combine the results of all the trials, yielding residue levels, in ranked order, of: ≤ 0.005 (13), < 0.01 (two), 0.006, 0.0068, 0.007 (two), 0.027 and 0.044 mg/kg.

The Meeting agreed to withdraw the previously recommended maximum residue level for pepper of 0.02* mg/kg and to replace it by a recommendation of 0.05 mg/kg for sweet pepper. The Meeting estimated an STMR of 0.005 and a highest residue level of 0.044 mg/kg for ethoprophos on sweet peppers.

Tomato

Six trials on tomato fruit involving overall soil treatment with granule formulations before or after planting were reported from Brazil (two, no GAP), The Netherlands (indoors) and the USA (three, no GAP). The dose used in the Dutch trial was twice that of Spanish GAP, but no residue level above the LOQ was found (< 0.01 mg/kg).

In 20 trials in southern Europe on tomato fruit, ethoprophos was applied as an emulsifiable concentrate formulation by post-planting drip irrigation or band spraying. The 13 trials conducted according to Italian GAP (1.7–3.5 kg ai/ha, three to four applications, total maximum of 8.6 kg ai/ha, 30-day PHI) or Spanish GAP (0.8–2.0 kg ai/ha, several applications, total maximum of 6 kg ai/ha, 60-day PHI) yielded residue levels of < 0.005 (four) and < 0.01 (nine) mg/kg.

On the basis of the trials conducted in southern Europe, the Meeting estimated a maximum residue level of 0.01* mg/kg, an STMR of 0.005 mg/kg and a highest residue level of 0.01 mg/kg for ethoprophos on tomato.

Root and tuber vegetables

Potato

The results of 62 trials were available in which ethoprophos was applied to potatoes after overall soil treatment or band application with granule formulations before or at planting. Ware potatoes are normally harvested within 90–120 days after application at or a few days before planting. Early maturing potatoes can be harvested before 90 days, while late maturing ones (such as Russet Burbank or Maris Piper varieties) are usually harvested after 120 days. The PHI therefore depends on the crop variety. On most labels, no PHI is indicated, as treatment is made before or at planting, and the potatoes are harvested when they are ready. In

Ethoprophos

trials in which the time of maturity of the potataoes was not indicated, the residue level measured at the shortest PHI was used for evaluation.

Three Dutch trials were evaluated against Dutch GAP (overall application, pre-planting: 4–10 kg ai/ha; band application at planting, 2.5 kg ai/ha), all yielding < 0.02 mg/kg. In 19 German trials evaluated against Dutch GAP, the residue levels were: < 0.01 (10), 0.0076, 0.012 (two), 0.014, 0.016, 0.017 (two), 0.02 and 0.03 mg/kg.

Three of four trials in the United Kingdom in 1995 suffered from abnormal weather conditions, resulting in retarded growth of the tubers. As residues were found in control samples (up to 0.096 mg/kg in one trial), these trials were excluded from evaluation. The remaining 14 trials in the United Kingdom could not be evaluated against that country's GAP (overall application, pre-planting: 6.6–11 kg ai/ha; band application pre-planting, 4.0–6.0 kg ai/ha; 56-day PHI) because the PHI was longer. Eleven of the trials could be evaluated against Dutch GAP, yielding residue levels of: < 0.005 (seven), 0.005 and < 0.01 (three) mg/kg.

Five French trials could be compared to French GAP (overall application, pre-planting: 6–10 kg ai/ha), yielding residue levels of: < 0.005 (two), < 0.01 (two) and 0.011 mg/kg.

One Spanish trial was evaluated against Spanish GAP (overall application, pre-planting: 6–8 kg ai/ha), yielding a residue level of < 0.01 mg/kg.

Two Greek trials could not be compared with Greek GAP (overall application, pre- or at planting: 8–10 kg ai/ha; 60-day PHI) because of the specified PHI. When they were evaluated against Spanish GAP, the residue levels were < 0.02 mg/kg in both.

Fourteen trials in the USA with a granule formulation were compared with US GAP (pre-planting until prior to crop emergence: overall application, 4.5–13 kg ai/ha; band application, 10 kg ai/treated ha = 3.4 kg ai/ha). In the 12 that complied with GAP, the residue levels were: < 0.01 and < 0.02 (11) mg/kg. In three trials in the USA with an emulsifiable concentrate formulation, which complied with US GAP, the residue levels were < 0.01 (two) and < 0.02 mg/kg.

The Meeting decided to combine the residue levels from all the studies, yielding, in ranked order: < 0.005 (nine), 0.005, ≤ 0.01 (19), < 0.02 (17), 0.0076, 0.011, 0.012 (two), 0.014, 0.016, 0.017 (two), 0.02 and 0.03 mg/kg. The Meeting estimated a maximum residue level of 0.05 mg/kg, an STMR of 0.01 mg/kg and a highest residue level of 0.03 mg/kg.

For assessing the risk to consumers of short-term intake, the possible residue level in single units is more important than the average residue level in a lot, which is the residue level in a representative composite sample. The concept of a variability factor was introduced to describe the relationship between the level in a high-residue unit and the typical or average level in the whole batch. The concept was refined to a more precise definition: the residue level in the 97.5th percentile unit divided by the mean residue level for the lot. There is a relation between the number of data from field trials, the proportion (percentile) of the population covered and the confidence level. The 2003 Meeting determined a method for calculating the variability factor on the basis of probabilities of random sampling from a population, making no assumptions as to the type of distribution.

In four of the trials on potato, residue levels were measured in individual units. Two were among the trials conducted in the United Kingdom in 1995 that were considered unreliable (see above). In a trial in France in 2001, 48 of 50 samples had undetectable residues, making the result unsuitable for calculation of a variability factor. In the fourth trial, conducted in the United Kingdom in 1999, 88 of 100 samples contained finite residue levels, so that a variability factor could be calculated. Applying the method referred to above to the 100 individual values available and using the 97.5th percentile in the calculation, the best estimate of the variability factor is 4.1 when the 12 data points below the LOQ are assumed to be at the LOQ, and 4.2 when those values are assumed to be 0. The 95% confidence limits on these estimates are 2.63 – > 5.6 and 2.75 – > 5.6, respectively. The Meeting decided to use the default variability factor of 3 in calculating the short-term intake of ethoprophos, as this value was within the confidence interval of the calculated factor, and the default factor was based on a much larger database.

Sweet potato

The Meeting received the results of four trials on sweet potato in the USA, three of which complied to US GAP (3.3–4.4 kg ai/ha). The residue levels were < 0.01 (two) and 0.014 mg/kg. Three trials is insufficient for recommending a maximum residue level, but the Meeting decided to extrapolate the data on potato to sweet potato, because GAP is similar for the two crops.

On the basis of the trials on potato, the Meeting estimated a maximum residue level of 0.05 mg/kg, an STMR of 0.01 mg/kg and a highest residue level of 0.03 mg/kg for sweet potato.

Grasses for sugar or syrup production

Sugar-cane

Fourteen trials were available in which ethoprophos in granule formulations was applied to sugar-cane in the open furrow at planting. Of these, nine trials from the USA complied to US GAP (band application: 2.2–4.6 kg ai/ha; 10–27 kg ai/treated ha). In all cases, the residue level was below the LOQ (< 0.02 mg/kg). Three trials in India were evaluated against Indonesian GAP (band application, pre-planting: 1.0–2.0 kg ai/ha), yielding residue levels of < 0.01 mg/kg. The two trials in Brazil with application after planting had no matching GAP.

The Meeting decided to combine the results of the trials in India and the USA, yielding residue levels, in ranked order, of: < 0.01 (three) and < 0.02 (nine) mg/kg. The Meeting estimated a maximum residue level of 0.02 mg/kg, an STMR of 0.02 mg/kg and a highest residue levelof 0.02 mg/kg for ethoprophos on sugar-cane.

Miscellaneous fodder and forage crops (group 052)

Sugar-cane leaves

In the three trials on sugar-cane in India, the residue levels in leaves were < 0.01 mg/kg. In three of the trials in the USA, the residue levels in leaves were < 0.02 mg/kg.

The Meeting estimated an STMR and a highest residue level of 0.02 mg/kg for ethoprophos on sugar-cane forage.

Fate of residues during processing

The Meeting received information on the fate of residues during commercial storage of bananas. After successive storage at 10 °C and 20 °C for fruit ripening, the residue level remained within 80–130% of the original level.

The Meeting received information on the fate of incurred residues of ethoprophos during the processing of potatoes and sugar-cane.

In the first study on potato, the raw agricultural commodity and the processed fractions (washed tubers, wash water, peeled tuber, wet peel, dry peel, flakes, chips) did not contain residues (< 0.01 mg/kg ethoprophos and < 0.01 mg/kg *O*-ethyl-*S*-propyl phosphorothioate). In the second study, no residues were found (< 0.005 mg/kg ethoprophos) in the raw agricultural commodity or in processed fractions (peeled and baked potato). Nevertheless, residues were found in potato peel, at 0.022 and 0.062 mg/kg (average, 0.042 mg/kg) in variety Maris Peer and 0.009 and 0.011 mg/kg (average, 0.010 mg/kg) in variety Desiree. As

Ethoprophos

the raw agricultural commodity did not contain residues, no processing factors for potato could be calculated; however, it can be concluded tentatively that the residue concentrates in peel and not in potato pulp.

After treatment with ethoprophos at planting, a 2-t batch of sugar-cane was processed into bagasse, mixed juice, clarified juice, mud, syrup, raw sugar and molasses according to commercial practices in a pilot plant. No quantifiable residues (< 0.02 mg/kg) were found in the raw agricultural commodity or its processed fractions. In a second study, in which ethoprophos was applied at planting, no residues (< 0.01 mg/kg) were detected in sugar-cane stalks or juice. Even when an exaggerated dose rate was used, in a third study, no residues (< 0.01 mg/kg) were found in stalks, bagasse, mixed juice, clarified juice, clarifier mud, syrup, molasses or sugar. Therefore, no processing factors for sugar-cane could be calculated.

The Meeting also received information on the distribution of residues in peel and pulp fractions of banana and melon. The results of two trials on banana in which residues were found at harvest indicate that ethoprophos tends to concentrate in the pulp fraction of banana. The results of six trials on melon in which residues were found at harvest indicate that ethoprophos is present in both peel and pulp fractions. Generally, the peel fractions contained slightly higher residue levels.

Residues in animal commodities

Dietary burden of farm animals

The Meeting estimated the dietary burden of ethoprophos residues in farm animals from the diets listed in Appendix IX of the *FAO Manual*. Only one feed commodity from each Codex Commodity Group was used, in this case potato culls (group VR). Calculation from highest residue values provides the concentrations in feed suitable for estimating MRLs for animal commodities, while calculation from the STMR values for feed is suitable for estimating STMR values for animal commodities. In the case of processed commodities, the STMR-P value is used for both intake calculations.

On the basis of a highest residue value of 0.03 mg/kg and 20% dry matter in potato culls, the maximum contribution of residue to the dietary burden would be 0.11 mg/kg for beef cattle given feed containing 75% potato culls and 0.06 mg/kg for dairy cattle given feed containing 40% potato culls.

On the basis of an STMR of 0.01 mg/kg and 20% dry matter, the mean dietary burden of beef cattle given feed containing 75% potato culls would be 0.038 mg/kg, and that of dairy cattle given feed containing 40% culls would be 0.02 mg/kg.

Maximum residue levels

The results of the metabolism study in lactating goats given feed containing 32 ppm ethoprophos indicate that no residues are to be expected in mammalian commodities at a maximum dietary burden of 0.11 mg/kg. As laying hens are not exposed to ethoprophos, no maximum residue levels for poultry commodities are required.

The Meeting estimated a maximum residue level of 0.01* mg/kg in mammalian meat, offal and milks, and STMR and highest residue values of 0.

DIETARY RISK ASSESSMENT

Long-term intake

The IEDIs of ethoprophos, on the basis of the STMRs estimated for 11 commodities, for the five GEMS/Food regional diets represented 5–10% of the maximum ADI (0–0.0004 mg/kg bw), see Annex 3. The Meeting concluded that the long-term intake of residues of ethoprophos resulting from uses that have been considered by JMPR is unlikely to present a public health concern.

Short-term intake

The IESTIs for ethoprophos were calculated for 11 food commodities for which maximum residue levels had been estimated and for which consumption data were available. The results are shown in Annex 4.

The IESTI represented 0–1% of the ARfD (0.05 mg/kg bw) for the general population and 0–3% of the ARfD for children. The Meeting concluded that the short-term intake of residues of ethoprophos resulting from uses that have been considered by the JMPR is unlikely to present a public health concern.

4.8 FENITROTHION (037)

RESIDUE AND ANALYTICAL ASPECTS

Fenitrothion was evaluated for residues within the periodic review programme of the CCPR by the 2003 JMPR, which recommended an MRL of 10 mg/kg (accommodating post-harvest treatment, Po) for cereals. The Meeting indicated that additional information on metabolism in cereals (including rice) after pre-harvest treatment, a validated analytical method for the determination of fenitrothion in animal commodities, freezer storage stability of residues in animal commodities, farm animal transfer studies and a processing study on rice were desirable. At the present Meeting, an undertaking was given to submit additional data to support uses on apple, pear and soya bean (dried seed), as these are important traded commodities.

The CCPR at its Thirty-sixth Session agreed to retain the existing MRLs for meat (0.05* mg/kg, fat), milks (0.002* mg/kg), unprocessed rice bran (20 mg/kg), polished rice (1 mg/kg), processed wheat bran (2 mg/kg), unprocessed wheat bran (20 mg/kg), wheat flour (2 mg/kg) and wheat wholemeal (5 mg/kg).

The data submitted to address the outstanding points are summarized below.

Metabolism

Animals

No additional data were submitted.

Plants

The 2003 JMPR evaluated the fate of fenitrothion after spray application to grapes and tomato (crop category fruits) and the fate of fenitrothion in stored rice. The Meeting received two additional studies on the fate of fenitrothion after pre-harvest spray application to rice under simulated paddy growing conditions.

In the first study, rice plants (variety Hatsushimo) were transplanted to a paddy field which had been prepared in a vinyl tent. The rice plants were sprayed once at 0.375 kg ai/ha with a ^{32}P-fenitrothion emulsifiable concentrate formulation. At various intervals after application, rice plants were sampled at random by cutting down at the base and separating the leaf sheath and leaf blade. At normal harvest, mature grains were harvested and separated into bran and polished rice.

In mature rice grain, the main metabolite was phosphoric acid, while phosphorothionic acid, dimethylphosphorothioic acid and parent were also found. The residue levels were 7–33 times higher in rice bran than in polished rice. Fenitrooxon, desmethylfenitrothion, desmethylfenitrooxon, dimethylphosphoric acid, monomethylphosphorothioic acid and monomethylphosphoric acid were not detected. In rice grains from plants (varieties Kin-nampu, Ginga, Aichi-asahi) treated with unlabelled fenitrothion, the parent compound could not be detected in rice bran or in polished rice (< 0.1 mg/kg), but 3-methyl-4-nitrophenol was found in rice bran of the varieties Kin-nampu and Aichi-asahi.

In a second study, rice (variety Nihonbare) was grown in a glasshouse under conditions simulating growth in a rice paddy. Rice seedlings were planted in pots filled with soil and flooded with 3–5 cm water throughout the study. The rice was treated with four spray applications of an emulsifiable concentrate formulation containing [phenyl-^{14}C]fenitrothion at a nominal dose rate of 0.75 kg ai/ha per application 81 (2 months after planting), 28, 21 and 14 days before mature harvest. Twelve days before harvest, water was withheld from the plants. Rice plants, soil and root samples were collected, and the plants were separated into unhulled whole grain and straw (leaf and stem). The unhulled whole grain was further processed to brown rice and chaff (hulls). The brown rice was further processed to give 90% (w/w) polished rice and 11% (w/w) bran.

A high proportion of the radioactivity was extractable (65–89% TRR). Fenitrothion was detectable in all rice fractions, at 0.003–0.3 mg/kg. The major metabolite in unhulled whole grain, brown rice, polished rice, chaff, bran and straw was 3-methyl-4-nitrophenol, at 0.09–3.9 mg/kg fenitrothion equivalents, and this compound was present in a mixture of free and conjugated forms. The conjugates were hydrolysed to the free form by enzymatic (β-glucosidase) and acid hydrolyses. Fenitrothion was present at levels of 0.30 mg/kg in unhulled whole grain rice, 0.027 mg/kg in brown rice and 0.003 mg/kg in polished rice.

The percentage transference would be equivalent to 8.9% for brown rice and 1.0% for polished rice. Fenitrooxon was present at levels of 0.14 mg/kg in unhulled whole grain rice and 0.009 mg/kg in brown rice; it was not detected in polished rice. Fenitrooxon was also found in bran at 0.042 mg/kg, chaff at 0.84 mg/kg and straw at 0.27 mg/kg. Most of the radioactivity in the post-extracted solids was released by acid or base hydrolysis. 3-Methyl-4-nitrophenol was detected in all the hydrolysates except that of polished rice. Minor amounts (< 2.5% TRR) of aminofenitrothion, fenitrooxon, fenitrothion S-isomer and fenitrothion were also released on hydrolysis. In polished rice, only fenitrothion S-isomer was found.

Environmental fate

No additional data were submitted.

Methods of analysis

The 2003 JMPR evaluated enforcement and monitoring methods for plant commodities. The Meeting received two additional analytical methods intended for enforcement and monitoring for animal commodities (RRC 78-32 and RRC 78-32A).

Method RRC 78-32 is used for the determination of fenitrothion (parent), fenitrooxon, aminofenitrothion and 3-methyl-4-nitrophenol in milk, cream and cattle tissues. Method RRC 78-32A is a modification for eggs and poultry tissues. In this appraisal, the focus is on parent compound. Cattle milk, cream and tissue, and eggs and poultry tissues were extracted with acetone or acetonitrile:methanol:water and cleaned, and fenitrothion was determined by gas chromatography–flame photometry detection. The reported LOQ was 0.01 mg/kg for cattle milk and cream and 0.05 mg/kg for eggs and cattle and poultry tissues. The method is considered to be outdated because of the use of glass gas chromatography columns.

The Meeting received information on analytical methods for the determination of parent in foodstuffs of plant, animal or environmental origin as used in various additionally submitted study reports (residue trials, storage stability of analytical samples, processing studies, feeding studies). Fenitrothion was determined by gas chromatography with flame photometry, nitrogen–phosphorus or FT detection. The reported LOQs ranged from 0.001 mg/kg to 0.1 mg/kg.

Stability of residues in stored analytical samples

The 2003 JMPR evaluated the stability in storage of residues in dry crops with starch and protein (grain of wheat, barley, rice) and in rice straw. The Meeting received additional data on the stability of residues in crops with a high water content (apple, green soya bean, green broad bean) and dry crops with starch and protein (dry seeds of soya bean, kidney bean, pea).

The Meeting concluded that fenitrothion residues are stable at –20 °C for at least 192 days in apples, for at least 155 days in legumes, for at least 149 days in cereals and for at least 98 days in pulses. In several trials, however, samples were stored at –15 °C, –10 °C, +5 °C or +10 °C. Information on the stability of the samples under these conditions was not available.

Definition of the residue

The 2003 JMPR proposed fenitrothion as the residue definition for compliance with MRLs and for dietary intake, for both plant and animal commodities. The residue is not fat-soluble.

The supported uses of fenitrothion at that time were pre-harvest application on cereals and post-harvest application on stored cereal grains. The 2003 Meeting concluded that the available studies were adequate only for post-harvest uses on stored cereal grains; to support pre-harvest uses on cereals, relevant metabolism studies were required.

The manufacturer now wishes to support use on pome fruit and soya bean. The 2003 JMPR evaluated studies of metabolism in grape and tomato after pre-harvest use. The main metabolites were 3-methyl-4-nitrophenol and its conjugates. Two studies on metabolism in rice after pre-harvest use of fenitrothion were made available to the present Meeting.

The studies confirm that the main metabolite is 3-methyl-4-nitrophenol (free and conjugated). The parent compound was found in the raw agricultural commodity and in all processed fractions (unhulled whole rice grain, brown rice, polished rice, bran, hulls and rice straw). In addition, fenitrooxon was found in unhulled whole rice grain (0.14 mg/kg), brown rice (0.009 mg/kg), bran (0.042 mg/kg), hulls (0.84 mg/kg) and straw (0.27 mg/kg). Fenitrooxon should be considered for inclusion in the residue definition for dietary intake, as the 2000 JMPR concluded that it is the most important metabolite with respect to toxicity. As it is found in small amounts in brown rice and not at all in polished rice (the edible portions for humans), however, the Meeting decided not to include fenitrooxon in the residue definition for dietary intake.

Fenitrothion S-isomer and aminofenitrothion, metabolites that were found but not discussed previously, were present in such small amounts that the Meeting concluded that their inclusion in the residue definition was unnecessary.

Therefore, the Meeting concluded that the residue definition as proposed by the 2003 JMPR do not require alteration.

Definition of the residue for compliance with MRLs and for estimating dietary intake: fenitrothion, for both plant and animal commodities.

Results of supervised trials on crops

The Meeting received information on supervised trials on apple, pear, green broad bean, green soya bean, dry soya bean, dry beans and dry peas.

In 2003, the Meeting summarized the results of supervised trials of pre-harvest treatment on cereal grains (rice, wheat, barley, triticale) in Japan and Australia. In some trials, pre-harvest treatment was combined with seed treatment before planting. The trials were not evaluated at that time, because information on metabolism after pre-harvest treatment was lacking.

Pome fruit

Nineteen trials on apple and two trials on pear were conducted in Canada in 1972–73. As there is no GAP in Canada or in the USA, the trials could not be evaluated.

In six of 24 trials on apple conducted in Japan in 1989–95, the analytical samples excluded hulls, pedicels, stylar scars and cores. As the Codex commodity for which the maximum residue level is estimated is the whole fruit, the trials could not be evaluated. In the remaining 18 trials, only pedicels, stylar scars and cores were removed. The Meeting decided that this deviation from the Codex commodity definition was acceptable, and it evaluated the trials. GAP in Japan is three applications of 0.025–0.050 kg ai/ha, with a PHI of 30 days. In two trials, residues were found in control samples, and those trials could not be evaluated.

The residue levels in trials meeting GAP were, in ranked order: < 0.01, 0.01 (three), 0.02 (two), 0.04, 0.08, 0.10 (two), 0.11, 0.12 and 0.41 mg/kg. The Meeting decided to withdraw the currently recommended maximum residue level for apple and replace it by 0.5 mg/kg, with an STMR of 0.04 mg/kg and a highest residue value of 0.41 mg/kg.

There were only two trials on pear, and neither complied with GAP. Furthermore, Japanese GAP allows six applications on pear and only three on apple. The Meeting decided to maintain withdrawal of the recommendation for pear.

Legume vegetables

Three trials on green broad beans (seeds only) and seven trials on green soya beans (seeds or beans with pods) were conducted in 1971–95 in Japan. There is no Japanese GAP for broad beans. Japanese GAP for soya beans (dry and green) is foliar spray treatment at 0.025–0.050 kg ai/ha, four applications, 21-day PHI. Two trials in which the portion analysed was green soya bean in the pod and which were conducted according to GAP yielded residue levels of 0.12 mg/kg and 0.18 mg/kg. Two trials is insufficient to estimate a MRL for soya bean, immature seeds.

Pulses

Nineteen trials on dry harvested soya beans, four trials on beans and two indoor trials on peas were carried out in 1971–90 in Japan. Japanese GAP is described above, except that when fenitrothion is applied by foliar spray from an unmanned helicopter, the dose is 0.50 kg ai/ha. In all but four trials, the PHI exceeded that specified for GAP. In the four trials, the residue levels were: 0.004 (two) and < 0.01 (two) mg/kg. The Meeting decided that four trials is insufficient to estimate a MRL for soya bean, dry.

Cereal grains

The 2003 Meeting received information from supervised trials on pre-harvest treatment of cereal grains (rice, wheat, barley, triticale) in Australia and Japan. Because of lack of data on metabolism in cereal grains after pre-harvest treatment, the trials could not be evaluated at that time. As the present Meeting received the requested data, the trials can now be evaluated.

Sixteen trials on rice were conducted in Japan in 1993–96 (see 2003 JMPR). In the trials conducted in 1993–95, seeds were soaked in a fenitrothion solution for 24 h before planting and then given four applications of fenitrothion from a knapsack sprayer during the growing season. In the trials performed in 1996, fenitrothion was applied four times from an unmanned helicopter, without prior soaking of the seeds. Fenitrothion is registered in Japan for pre-harvest use on rice at a maximum of four foliar spray applications of 0.375–0.90 kg ai/ha and a PHI of 21 days. The residue levels in the trials complying with Japanese GAP were: < 0.01 (four), 0.01 (two), 0.02, 0.03, 0.04, 0.05, 0.06, 0.07, 0.08, 0.09, 0.10 and 0.12 mg/kg.

Four trials on *winter wheat* were conducted in Australia in 2001 and in Japan in 1993. In these trials, fenitrothion was applied three times from a knapsack sprayer or a boom sprayer. Fenitrothion is registered in Australia for pre-harvest use on cereals with a maximum of three applications of 0.27–0.55 kg ai/ha (interval ≥ 14 days) and a PHI of 14 days; in Japan, it is registered for pre-harvest use on wheat with a single application of 0.45–0.60 kg ai/ha and a PHI of 7 days. The Australian trials were at Australian GAP and yielded residue levels of 0.10 and 0.21 mg/kg. The trials in Japan did not comply completely with Japanese GAP (three applications instead of one), but the results were evaluated. The residue levels were < 0.01 and 0.30 mg/kg.

Three trials on *winter barley* were conducted in Australia in 2001 and in Japan in 1993. Fenitrothion was applied three times from a knapsack sprayer or a hand-held boom sprayer. In Australia, fenitrothion is registered for pre-harvest use on cereals with a maximum of three applications of 0.27–0.55 kg ai/ha (interval ≥ 14 days) and a PHI of 14 days. In Japan, fenitrothion is not registered for use on barley. The Australian trial was at Australian GAP and yielded a residue level of < 0.06 mg/kg.

One trial on *winter triticale* was conducted in Australia in 2001, in which fenitrothion was applied three times from a hand-held boom sprayer. The trial was at the Australian GAP for cereal and yielded a residue levek of 0.08 mg/kg.

The Meeting decided to combine the results of all the trials on cereals residue, to yield residue levels, in ranked order, of: < 0.01 (five), < 0.06, 0.01 (two), 0.02, <u>0.03</u>, <u>0.04</u>, 0.05, 0.06, 0.07, 0.08 (two), 0.09, 0.10 (two), 0.12, 0.21 and 0.30 mg/kg. As the residue levels resulting from pre-harvest treatment were lower than those after post-harvest treatment, the Meeting decided to maintain the current recommendations for cereal grain of 10 mg/kg (Po), an STMR of 5 mg/kg and a highest residue of 7.6 mg/kg.

Straw, fodder and forage of cereal grains and grasses

Two trials on *rice* were conducted in Japan in 1993 according to Japanese GAP. The residue levels in straw were 0.31 and 0.44 mg/kg.

Two trials on *winter wheat* were conducted in Australia in 2001 according to Australian GAP. The residue levels in straw were 1.2 and 4.1 mg/kg.

One trial on *winter barley* was conducted in Australia in 2001 according to Australian GAP. The residue level in straw was 0.41 mg/kg.

One trial on *winter triticale* was conducted in Australia in 2001 according to Australian GAP. The residue level in straw was 2.0 mg/kg.

The Meeting decided that there were insufficient data to estimate a maximum residue level in cereal straw.

Fate of residues during processing

The 2003 JMPR evaluated the fate of fenitrothion during simulated processing, in stored rice during polishing and cooking and in stored wheat during milling and baking. The Meeting received three additional studies on the fate of fenitrothion after post-harvest treatment of rice during polishing and cooking and in barley during malting.

Rice stored for up to 12 months after post-harvest treatment with fenitrothion at 2–15 g ai/t was separated into polished rice and bran. The polished rice was washed and cooked for 10–15 min at normal to 2.5 atm pressure. Parent compound was analysed in all processed products. The calculated processing factors were 0.11–0.15 for polished rice (mean, 0.14), 6.6–7.2 for bran (mean, 6.9), 0.041–0.049 for washed polished rice (mean, 0.046) and 0.0060–0.033 for cooked washed polished rice (mean, 0.020).

Paddy rice stored for up to 6 months after post-harvest treatment with fenitrothion at 15 g ai/t was processed into husked rice, polished rice and bran. Husked and polished rice were cooked for 25 min. Parent compound was analysed in all processed products. The calculated processing factors were 0.18 for husked rice, 0.08 for polished rice, 0.11 for cooked husked rice and 0.04 for cooked polished rice.

Paddy rice was processed into husked rice, polished rice, hulls and bran. Parent compound was analysed in all processed products. The calculated processing factors were 0.031–0.64 for husked rice (mean, 0.17), < 0.002–0.087 for polished rice (mean, 0.018), 0.12–10 for hulls (mean, 4.3) and 0.018–2.0 for bran (mean, 0.61).

Fenitrothion

Barley stored for up to 6 months after post-harvest treatment with 15 g ai/t fenitrothion was processed into malt and was analysed for parent. The calculated processing factors were 0.16-0.24 for malt (mean 0.20).

The table below summarizes the processing factors for wheat, barley and rice commodities. The information on wheat was discussed by the 2003 JMPR. Three studies were available on the processing of rice, yielding different results. As details of the growing conditions, treatments, storage and processing were absent or partial, the Meeting could not judge which study was the most representative of household processing. Therefore, maximum processing factors were used. Processing factors used for the calculation of maximum residue levels and STMR-Ps are underlined.

Commodity	Processing factor (range)	No. of trials	Processing factor (mean)	STMR-P (mg/kg)	HR-P (mg/kg)
Wheat bran	3.9–4.0	2	3.95	19.75	30.02
Wheat flour	0.21–0.26	2	0.235	1.175	1.786
White bread	0.089–0.11	2	0.10	0.5[a]	0.76
Wholemeal bread	0.33–0.43	2	0.38	1.9	2.888
Barley malt	0.16–0.24	2	0.20	1	1.52
Husked rice	0.031–0.64	22	0.17	3.2	4.864
Polished rice	< 0.002–0.15	26	0.039	0.75	1.14
Rice hulls	0.12–10	21	4.3	50	76
Rice bran	0.018–7.2	23	1.7	36	54.72
Cooked husked rice	0.11	1		0.55	0.836
Cooked polished rice	0.04	1		0.2	0.304
Washed polished rice	0.041–0.049	4	0.046	0.23	0.3496
Cooked, washed, polished rice	0.0060–0.033	13	0.020	0.1	0.152

[a] The Meeting noted that the 2003 Report incorrectly reported an STMR-P of 0.05 mg/kg in white bread.

Using the highest residue level for cereal grains (7.6 mg/kg) and the processing factors indicated above, the Meeting estimated maximum residue levels of 30 mg/kg in wheat bran and 60 mg/kg in rice bran. The Meeting decided to withdraw the current recommendations for 1 mg/kg in polished rice, 2 mg/kg in wheat flour, 1 mg/kg in white bread and 3 mg/kg in wholemeal bread (accommodating post-harvest treatment, PoP), because the MRL would be lower than that of the raw agricultural commodity.

Using the STMR for cereal grains (5 mg/kg), the Meeting estimated STMR-Ps for wheat bran, wheat flour, white bread, wholemeal bread, barley malt, husked rice, polished rice, rice hulls, rice bran, cooked husked rice, cooked polished rice, washed polished rice and cooked washed polished rice, as shown in the table above.

Furthermore, using the highest residue level for cereal grains (7.6 mg/kg), the Meeting estimated HR-Ps for wheat bran, wheat flour, white bread, wholemeal bread, barley malt, husked rice, polished rice, rice hulls, rice bran, cooked husked rice, cooked polished rice, washed polished rice and cooked washed polished rice, as shown in the table above.

The Meeting decided to use the STMR-P and HR-P values for cooked husked rice and cooked polished rice in calculating the dietary intake of fenitrothion from rice.

Fenitrothion

Residues in animal commodities

Dietary burden of farm animals

The Meeting estimated the dietary burden of fenitrothion residues in farm animals from the diets listed in Appendix IX of the *FAO Manual*. Only one feed commodity from each Codex Commodity Group is used; therefore, the calculation includes wheat grain, but no other cereals, and rice bran, but not rice hulls. Calculation from the highest residue values provides the concentrations in feed suitable for estimating MRLs for animal commodities, while calculation from the STMR values for feed is suitable for estimating STMR values for animal commodities. In the case of processed commodities, the STMR-P value is used for both intake calculations.

Estimated maximum dietary burden of farm animals

Commodity	Codex code	Residue (mg/kg)	Basis	Dry matter (%)	Residue, dry wt (mg/kg)	Dietary content (%) Beef cattle	Dairy cattle	Poultry	Residue contribution of feeds (mg/kg) Beef cattle	Dairy cattle	Poultry
Rice bran	CM	36	STMR-P	90	40						
Rice hulls	CM	50	STMR-P	90	55.6						
Wheat milled by-products[a]	CF	19.75	STMR-P	88	22.44	40	50	50	8.98	11.22	11.22
Wheat grain	GC	7.6	Highest residue	89	8.54						
Total						40	50	50	8.98	11.22	11.22

[a] Use of wheat bran

Estimated mean dietary burden of farm animals

Commodity	Codex code	Residue (mg/kg)	Basis	Dry matter (%)	Residue, dry wt (mg/kg)	Dietary content (%) Beef cattle	Dairy cattle	Poultry	Residue contribution of feeds (mg/kg) Beef cattle	Dairy cattle	Poultry
Rice bran	CM	36	STMR-P	90	40						
Rice hulls	CM	50	STMR-P	90	55.6						
Wheat milled by-products[a]	CF	19.75	STMR-P	88	22.44	40	50	50	8.98	11.22	11.22
Wheat grain	GC	5	STMR	89	5.62						
Total						40	50	50	8.98	11.22	11.22

[a] Use of wheat bran

Residues in grazing animals and feeding studies

The 2003 JMPR evaluated the fate of fenitrothion residues in cattle grazing on fenitrothion-treated grass and in cattle fed fenitrothion-treated corn. The Meeting received an additional study on cattle grazing on pasture treated with fenitrothion, a feeding study in lactating dairy cattle and a feeding study in laying hens.

Fenitrothion

Fenitrothion as technical material was applied by air to a 120-ha paddock by the ultra-low volume technique at the rate of 0.508 kg ai/ha. There were 28 *cattle* on the pastures when the application was made, and 38 were brought to the treated pastures immediately after spraying. Four cattle were maintained as controls, with no exposure to fenitrothion. The cattle were allowed to graze on the sprayed pasture for varying periods and were slaughtered within 1 day after removal from the paddock, 2, 4, 7, 14 or 21 days after treatment. Samples of subcutaneous and perirenal fat, meat (neck muscle) and liver were taken for analysis. Some cattle were removed from the treated pasture and were grazed on clean pasture for 2, 4 or 7 days before slaughter to determine the effect on the residue levels of this procedure. Soil and pasture were sampled on days –3, 0, 1, 2, 4, 7, 10, 14 and 21 and analysed for fenitrothion residues. The worst-case estimate of total exposure (pasture + soil + deposition) of the cattle on the day of spraying was 800–900 mg of fenitrothion per animal. As the total intake of dry matter from the pasture was assumed to be 10 kg/day, the exposure corresponds to a fenitrothion content of approximately 90 ppm.

Data on residues found in the pasture indicate a half-life for fenitrothion of 1–2 days, which declined with time; the decay during the first period appeared to be faster than that during later periods on a per hour basis. About 50% of the fenitrothion was gone within 1 day, 75% within 3 days and 90% within 6 days. Data on residues in the soil indicated a half-life for fenitrothion of 2–3 days. The residue levels in all animal commodities were below the LOQ (< 0.2 mg/kg for fat, < 0.02 mg/kg for muscle and liver).

Groups of three *lactating cows* received feed containing fenitrothion at a concentration of 0, 10, 30 or 100 ppm for 28 days. All animals were milked twice daily, and composite milk samples from the morning and afternoon milkings on days –1, 0, 3, 7, 14, 21 and 28 were analysed for residues. Cream was analysed separately; the cream content of the milk was 5–8% and the butterfat content was 3–5%. The levels of residues of fenitrothion in milk were below the LOQ of 0.01 mg/kg for all groups. Residues were measured in cream at some times (scattered among groups), but the levels were never higher than 0.01 mg/kg. Samples of liver, kidney, muscle (cardiac, hind-quarter and front-quarter) and fat (omental and perirenal) were taken. Residue levels above the LOQ of 0.05 mg/kg were not found in any sample at any feed level.

Layer and broiler *hens* received feed containing fenitrothion at 0, 10, 30 or 100 ppm for 28 days. Egg samples were collected twice a week from each group and frozen until ready for analysis. Eggs from birds in the same dose group were combined. Half of the hens at each dietary concentration were killed on day 14, and the remaining hens were killed on day 28 of the study. Samples of red muscle, white muscle, liver and fat were taken at both times. The residue levels in all tissue samples taken on days 14 and 28 were below the limit of determination of 0.05 mg/kg. No residues of fenitrothion were found in eggs taken over the 28-day period.

Maximum residue levels

The calculated dietary burden of dairy cattle is 11.22 mg/kg, and that of beef cattle is 8.98 mg/kg. In the feeding study with cattle described above, no residues were found at levels above the LOQ (0.05 mg/kg) in muscle, fat, liver or kidney at dietary concentrations of 10, 30 and 100 mg/kg. Therefore, no residues at levels above the LOQ are to be expected at the calculated dietary burden. The levels of residues of fenitrothion in milk were below the LOQ of 0.01 mg/kg.

The calculated dietary burden of poultry is 11.22 mg/kg. In the feeding study in poultry, no residues were detected in muscle, liver, fat or eggs (< 0.05 mg/kg) at dietary concentrations of 10, 30 and 100 mg/kg.

The Meeting recommended maximum residue levels of 0.05* mg/kg in meat (from mammals other than marine mammals), in edible offal (mammalian), in poultry meat and in eggs. The Meeting also recommended a maximum residue level of 0.01 mg/kg in milks. The STMR and the highest residue values for muscle, fat, liver, kidney, poultry meat and fat are estimated to be 0 mg/kg.

Fenitrothion

DIETARY RISK ASSESSMENT

Long-term intake

The IEDIs of fenitrothion, on the basis of the STMRs estimated for 12 commodities, for the five GEMS/Food regional diets represented 110–330% of the maximum ADI (0–0.005 mg/kg bw), see Annex 3. The information provided to the JMPR precludes an estimate that the dietary intake would be below the ADI.

The Meeting noted that the calculations of long-term intake were conservative, as they did not take into account the reduction in residue levels obtained by processing cereal grains, except for processing of wheat, barley and rice. The Meeting extrapolated processing data on wheat to rye. Information on processing of barley (uses besides beer), maize, millet and sorghum would be particularly useful for refining the intake calculations.

Short-term intake

The IESTIs for fenitrothion were calculated for 25 food commodities for which maximum residue levels had been estimated and for which consumption data were available. The results are shown in Annex 4.

The IESTI represented 0–100% of the ARfD (0.04 mg/kg bw) for the general population and 0–160% of the ARfD for children. The values 120%, 150% and 160% represent the estimated short-term intake of wholemeal bread, wheat bran (unprocessed) and maize (fresh, flour), respectively, by children. The Meeting concluded that the short-term intake of residues of fenitrothion from uses other than on these three commodities that have been considered by the JMPR is unlikely to present a public health concern.

The Meeting noted that the intake calculations were conservative, as they did not take into account the reduction in residue levels obtained by processing cereal grains, except for processing of wheat, barley and rice. Information on processing of maize would be particularly useful for refining the intake calculations. Nevertheless, the exceedences found for children in the consumption of wheat bran and wholemeal bread cannot be further refined.

The Meeting noted that the ADI and the ARfD for fenitrothion were established by the 2000 JMPR. At that time, the concepts of overall NOAEL (see item 2.3) and of compound-specific adjustment factors (see item 2.1) had not been fully developed. In addition, the process of establishing ARfDs was at an early stage of development. In view of these considerations, the Meeting concluded that a review of the toxicological database of fenitrothion, taking into account the new concepts, could lead to a refinement of the ADI and the ARfD.

4.9 FENPROPIMORPH (188)

TOXICOLOGY

Evaluation for an acute reference dose

Fenpropimorph is a morpholine fungicide with systemic activity, interfering with sterol biosynthesis. It was first evaluated by the 1994 JMPR, which established an ADI of 0–0.003 mg/kg bw on the basis of a NOAEL of 10 ppm, equal to 0.3 mg/kg bw per day, in a 2-year study of toxicity and carcinogenicity in rats. At the 2001 JMPR, an ARfD of 1 mg/kg bw was established on the basis of a NOAEL of 100 mg/kg bw per day in an acute neurotoxicity study in rats. In 2002, the Government of Germany asked the Meeting to reconsider the ARfD established for fenpropimorph by the 2001 JMPR, because this Government considered the NOAEL of 15 mg/kg bw per day for teratogenicity in the rabbit to be a more appropriate basis for the ARfD. The present review was undertaken to determine the appropriate end-point and NOAEL for

Fenpropimorph

establishing an ARfD and to evaluate a new screening study of pre- and postnatal developmental toxicity in rats, which was submitted to the present Meeting.

Fenpropimorph is of low acute toxicity; in rats, the oral LD_{50} was 1500–3500 mg/kg bw, the dermal LD_{50} was 4300 mg/kg bw, and the inhalation LC_{50} was 2.9 mg/l of air.

In a study of acute neurotoxicity in rats, the NOAEL was 100 mg/kg bw per day on the basis of clinical and behavioural signs observed at doses of 500 and 1500 mg/kg bw per day.

In a screening study of pre- and postnatal developmental toxicity in rats, the NOAEL for maternal toxicity was < 5 mg/kg bw per day on the basis of decreased food consumption during pregnancy and lactation and reduced body-weight gain during pregnancy, at all doses. The NOAEL for developmental toxicity was < 5 mg/kg bw per day on the basis of reduced body weight or body-weight gain in pups at all doses during lactation.

In a study of prenatal developmental toxicity in rats, an increased incidence of cleft palate (14 fetuses from seven litters) was observed at the highest dose of 160 mg/kg bw per day. At this dose, severe maternal toxicity, including mortality, was found. The NOAEL for developmental toxicity was 40 mg/kg bw per day, while the NOAEL for maternal toxicity was 10 mg/kg bw per day.

In a study of prenatal developmental toxicity in Himalayan rabbits, severe maternal toxicity, including mortality, was found at 60 mg/kg bw per day, the highest dose tested. The number of early resorptions and dead fetuses was increased such that only one fetus survived. This fetus had several abnormalities, including syndactyly of the forelimbs, an anomalous position of the hindlimbs and micromelia, fusion of individual sternebrae and reduced weight and length. At 36 mg/kg bw per day, the clinical signs in dams (diarrhoea, salivation) were less severe and occurred at a lower incidence than at the highest dose. Two animals aborted and three were killed *in extremis*. The number of postimplantation losses was slightly increased at this dose. Six fetuses from two litters had pseudoankylosis, a skeletal variation. The NOAEL was 12 mg/kg bw per day for both maternal toxicity and developmental toxicity.

In a study of prenatal developmental toxicity in Russian rabbits, maternal toxicity (swelling of the anus, reduction of food consumption and of body weight or body-weight gain, weight loss) was observed only at the highest dose tested, 30 mg/kg bw per day. There were no effects on pre- or postimplantation loss, number of live or dead fetuses per litter and sex ratio. Mean gravid uterus weight and weights of male fetuses were significantly reduced at the highest dose. There was an increase in the total number of malformations (21 fetuses from four litters) and in findings described as 'anomalies' (36 fetuses from 13 litters). The malformations occurred mainly in the litters of three dams that showed marked signs of toxicity during treatment. Twenty fetuses from three litters had shortened fore- and hindlimbs, and four fetuses from two litters had a cleft palate. Furthermore, position anomalies were observed in the forelimbs in twenty-five fetuses from seven litters and in the hindlimbs in eight fetuses from three litters. The NOAEL was 15 mg/kg bw per day for both maternal toxicity and developmental toxicity.

Toxicological evaluation

The Meeting established an ARfD of 0.2 mg/kg bw on the basis of an overall NOAEL of 15 mg/kg bw per day for embryo- and fetotoxicity and teratogenicity in two studies of prenatal developmental toxicity in rabbits and using a safety factor of 100. The LOAEL of 5 mg/kg bw per day for decreased body-weight gain in pups in the screening study of pre- and postnatal developmental toxicity in rats is related to repeated pre- and postnatal exposure and is therefore not considered to be an appropriate basis for establishing an ARfD.

Fenpropimorph

Levels relevant to risk assessment

Species	Study	Effect	NOAEL	LOAEL
Rat	Acute neurotoxicity	Neurotoxicity	100 mg/kg bw per day	500 mg/kg bw per day
	Screening for pre- and postnatal developmental toxicity	Maternal toxicity	< 5 mg/kg bw per day	5 mg/kg bw per day
		Developmental toxicity	< 5 mg/kg bw per day[a]	5 mg/kg bw per day[a]
	Prenatal developmental toxicity	Maternal toxicity	10 mg/kg bw per day	40 mg/kg bw per day
		Developmental toxicity	40 mg/kg bw per day	160 mg/kg bw per day
Rabbit	Prenatal developmental toxicity	Maternal toxicity	12 mg/kg bw per day	36 mg/kg bw per day
		Developmental toxicity	12 mg/kg bw per day	36 mg/kg bw per day
	Prenatal developmental toxicity	Maternal toxicity	15 mg/kg bw per day	30 mg/kg bw per day
		Developmental toxicity	15 mg/kg bw per day	30 mg/kg bw per day

[a] Effect considered not relevant for a single exposure

Estimate of acute reference dose

0.2 mg/kg bw

DIETARY RISK ASSESSMENT

Short-term intake

The Meeting estimated an ARfD of 0.2 mg/kg bw for fenpropimorph. The 2001 JMPR had calculated the IESTI for fenpropimorph for 16 food commodities (and their processed fractions) for which MRLs were estimated and for which consumption data were available using the previous ARfD of 1 mg/kg bw.

The IESTI represented 0–7% of the ARfD for the general population and 0–10% of the ARfD for children.

The Meeting concluded that the short-term intake of residues of fenpropimorph from uses that have been considered by the JMPR is unlikely to present a public health concern.

4.10 FENPYROXIMATE (193)

TOXICOLOGY

Evaluation for an acute reference dose

Fenpyroximate is a phenoxypyrazole acaricide. Fenpyroximate was evaluated by the 1995 JMPR, when an ADI of 0–0.01 mg/kg bw was established on the basis of a NOAEL of 1 mg/kg bw per day in a 104-week study in rats and a safety factor of 100. The critical effect in that study was a reduction in body-weight gain. Fenpyroximate was re-evaluated by the present Meeting in order to determine an ARfD.

The acute oral LD_{50} of fenpyroximate was 245 and 480 mg/kg bw in male and female rats, respectively. The toxic effects of fenpyroximate include diarrhoea, failure to gain weight and haematological and clinical chemistry changes. In short-term (range-finding) dietary studies in mice, the effects of fenpyroximate were mainly limited to decreases in food consumption and reduced body-weight gain. The

Fenpyroximate

NOAELs for the two studies were 20 ppm (equal to 2.58 and 3.07 mg/kg bw per day in males and females, respectively) and 80 ppm (equal to 10.8 mg/kg bw per day in males and 11.7 mg/kg bw per day in females). In a 13-week dietary study in rats, effects were seen on body-weight gain, food consumption and haematological and clinical chemistry parameters. The NOAEL for the study was 20 ppm, equal to 1.30 mg/kg bw per day in males and 1.65 mg/kg bw per day in females. In a 13-week study in dogs given capsules containing fenpyroximate, weight loss or decreased weight gain, decreased food consumption and diarrhoea were seen. No NOAEL was identified in this study and the LOAEL was 2 mg/kg bw per day, on the basis of diarrhoea occurring at all doses. This was considered to be a minimal effect level and occurred from the beginning of the study. In a dietary study of reproductive toxicity in rats, parental toxicity comprised reduced body-weight gain and food consumption in both sexes and increased testicular and epididymal weights in males. Offspring toxicity consisted of reduced body-weight gain. A reduction in conception rate and fertility index was observed in one generation only. The NOAELs for parental, offspring and reproductive toxicity were all 30 ppm (equal to 1.99 mg/kg bw per day in parental males, 2.44 mg/kg bw per day in parental females, 2.33 mg/kg bw per day in F_1 males and 2.82 mg/kg bw per day in F_1 females). In a study of developmental toxicity in rats treated by gavage, effects were seen on maternal body weight, while increases in thoracic ribs were found in fetuses. The NOAEL for both effects was 5 mg/kg bw per day. In preliminary and substantive studies of the developmental toxicity of fenpyroximate in rabbits treated by gavage, reduced maternal body weight and food consumption was seen at the higher doses. These effects were not considered relevant for establishing an ARfD. The NOAEL for maternal toxicity was 1 mg/kg bw day. Significant elevations in the frequency of unilateral and bilateral slightly folded retinas were observed in the fetuses at 5 mg/kg bw per day. No other evidence of fetotoxicity was observed and the biological significance of this finding is unclear.

Toxicological evaluation

The Meeting established an ARfD of 0.01 mg/kg bw on the basis of the a minimal LOAEL of 2 mg/kg bw per day for the induction of diarrhoea at the beginning of a 13-week study of toxicity in dogs. It was unclear whether the diarrhoea was the result of a direct irritant or pharmacological effect of fenpyroximate. A safety factor of 200 was used since no NOAEL was identified. This ARfD is probably conservative and could be refined using the results of an appropriately designed study.

An addendum to the toxicological monograph was prepared

Estimate of acute reference dose

0.01 mg/kg bw

Studies that would provide information useful for continued evaluation of the compound

Appropriately designed single-dose study

DIETARY RISK ASSESSMENT

Short-term intake

The Meeting established an ARfD of 0.01 mg/kg bw for fenpyroximate. The 1999 JMPR had calculated the IESTI for fenpyroximate for 10 food commodities (and their processed fractions) for which MRLs were estimated and for which consumption data were available, but was not able to finalize the risk assessment because an ARfD was not available.

The IESTI represented 0–120% of the ARfD for the general population and 0–310% of the ARfD for children. The value 120% represents the estimated short-term intake of grapes by the general population. The values 130% and 310% represent the estimated short-term intakes of apples and grapes by children.

4.11 FLUDIOXONIL (211)

TOXICOLOGY

Fludioxonil is the ISO approved name for a new phenylpyrrole fungicide, 4-(2,2-difluoro-1,3-benzodioxol-4-yl)pyrrole-3-carbonitrile (IUPAC), that interferes with glucose transport across fungal membranes. Fludioxonil has not been evaluated previously by the JMPR.

After oral administration of radiolabelled fludioxonil, the radiolabel is rapidly and extensively (approximately 80% of the administered dose) absorbed, widely distributed, extensively metabolized and rapidly excreted, primarily in the faeces (approximately 80%) via the bile (approximately 70%), with a small amount being excreted in the urine (approximately 20%). The maximum blood concentration is reached within 1 h after administration. Elimination is biphasic, with half-lives of between 2 and 5 h for the first phase and between 30 and 60 h for the second phase. Fludioxonil is rapidly cleared from the blood and tissues, and there is consequently negligible potential for accumulation. The metabolism of fludioxonil proceeds primarily through oxidation of the pyrrole ring, leading to one major (57% of the administered dose) and one minor (4% of the administered dose) oxo-pyrrole metabolite. Hydroxylation of the phenyl ring yields the corresponding phenol metabolite, which represents 2% of the administered dose. These phase I metabolites are subsequently excreted as glucuronyl and sulfate conjugates and, together with unabsorbed and unchanged fludioxonil excreted in faeces, account for approximately 75% of the administered dose. The dimerization of the hydroxy pyrole metabolite produces a metabolite of an intense blue colour.

The dermal absorption of fludioxonil, excluding material bound to the skin, is low in rats in vivo ($< 5\%$) and in human skin in vitro ($< 0.5\%$). In a study of dermal penetration in rats in vitro, values for dermal absorption at low levels of application were comparable to those obtained in a study performed in vivo ($< 2\%$), but at higher levels significantly overestimated absorption in vivo (38%).

Fludioxonil has low acute toxicity in rats when administered by oral, dermal or inhalation routes, producing no deaths at 5000 and 2000 mg/kg bw and 2.6 mg/l of air, respectively, the highest doses tested. There were also no deaths in mice given fludioxonil at 5000 mg/kg bw by gavage. Fludioxonil is a slight eye irritant in rabbits, but is neither a skin irritant in rabbits nor a skin sensitizer in guinea-pigs (Magnusson & Kligman maximization assay).

In studies with repeated doses in mice and rats, the liver (necrosis, centrilobular hypertrophy, increased serum cholesterol and 5′nucleotidase), the kidneys (nephropathy, inflammation, cysts) and haematopoietic system (mild anaemia) were the principal targets. Such effects often set the LOAELs for these studies, together with reduced body-weight gains. In mice, these effects were observed after 90 days of treatment at 450 mg/kg bw per day and at 590 mg/kg bw per day in one 18-month study, but not at 360 mg/kg bw per day in another such study. In rats, effects were seen at doses of ≥ 400 mg/kg bw per day in short-term studies and at 110 mg/kg bw per day in a 2-year study; lower body-weight gains were also observed at these doses. Liver toxicity was generally manifested by increased concentrations of serum cholesterol and bilirubin and centrilobular hypertrophy and/or necrosis. Anaemia in mice (at > 590 mg/kg bw per day for 18 months) and rats (at 1300 mg/kg bw per day for 3 months) was seen at doses greater than the LOAEL. In dogs, anaemia was observed at the LOAEL (at 290 mg/kg bw per day for 3 months, but only after 4 weeks of treatment; and at a dose of 300 mg/kg bw per day for 12 months). No haematological effects were observed in shorter studies in mice (at ≤ 1050 mg/kg bw per day for 90 days) or rats (at ≤ 2500 mg/kg bw per day for 20 days and at ≤ 1000 mg/kg bw per day for 28 days)

Blue discolouration of the urine, perineal fur, kidneys and gastrointestinal tract were common observations in all species. These effects were secondary to the formation of the blue metabolite in quantities that were sufficient, at high doses, to stain the various tissues. The effect is not toxicologically significant and was disregarded in identifying NOAELs from studies in which it was observed.

Fludioxonil gave negative results in assays for reverse mutation in *S. typhimurium* and *E. coli*, gene mutation in Chinese hamster V79 cells, unscheduled DNA synthesis in rat hepatocytes, micronucleus formation in bone marrow of rats and mice in vivo and chromosome aberration in Chinese hamsters in vivo. Fludioxonil was clastogenic in Chinese hamster ovary cells (CCL61) in vitro at non-cytotoxic concentrations. There was no evidence of heritable genetic damage in an assay for dominant lethal mutations in mice.

The Meeting concluded that fludioxonil is unlikely to be genotoxic in vivo.

The carcinogenicity potential of fludioxonil was examined in a study in rats and in two studies in mice. While the incidence of lymphomas was slightly increased in females in one study in mice receiving diets containing fludioxonil at a concentration of 3000 ppm (equal to 360 mg/kg bw per day), no increase was observed in a concurrent life-time study in mice given diets containing fludioxonil at dietary concentrations of up to 7000 ppm (1000 mg/kg bw per day). Lymphoma is a common finding in ageing female CD-1 mice, and the historical incidence at the laboratory conducting these studies was 13–32%, which encompasses the incidence noted in females at 3000 ppm (30%). Given the high background rate of this finding and the lack of any increase in the other study by the same authors using higher doses, the Meeting concluded that the apparent increase in lymphomas observed in one study in mice was incidental. There was no evidence of carcinogenic potential with fludioxonil in the study in rats. The overall NOAELs in the long-term studies were 112 and 37 mg/kg bw per day in mice and rats respectively.

On the basis of the above consideration and on the lack of genotoxic potential in vivo, the Meeting concluded that fludioxonil is unlikely to pose a carcinogenic risk to humans.

In a two-generation study of reproductive toxicity in rats, at a dose of 210 mg/kg bw per day, adult males had reduced body-weight gains and food consumption and pups had lower body-weight gains. The NOAEL for parental and pup toxicity was 21 mg/kg bw per day. The NOAEL for effects on reproductive performance was 21 mg/kg bw per day on the basis of reduced pup weights. As no effects on body-weight gain, or any other parameter, were seen in adult rats at 37 mg/kg bw per day in a 2-year study, the NOAEL for parental animals in the study of reproductive toxicity can also be considered to be > 37 mg/kg bw per day. The NOAEL of 21 mg/kg bw per day for pup toxicity was based on effects observed at 210 mg/kg bw per day. These effects were relatively mild and the dose–response relationship appears to be shallow (12% decrease in body-weight gain over the dose range of 190 mg/kg bw per day, or a 1%, or less, decrease in body-weight gain per dose increment of 16 mg/kg bw per day). Assuming a linear dose–response relationship between 21 and 210 mg/kg bw per day, then at the proposed overall NOAEL for rats of 37 mg/kg bw per day, a decrease in body-weight gain of ≤ 1% would be predicted in pups; this would not be interpreted as being an adverse effect. Consequently, the use of an overall NOAEL of 37 mg/kg bw per day is also appropriate for pup toxicity. In a study of developmental toxicity in rabbits and another in rats, fludioxonil was neither teratogenic nor fetotoxic and fetal weights were unaffected at doses of up to 1000 and 300 mg/kg bw per day, respectively. Maternal toxicity in these studies was limited to reduced body-weight gain at 1000 and 300 mg/kg bw per day in rats and rabbits, respectively.

The Meeting concluded that the existing database on fludioxonil was adequate to characterize the potential hazards to fetuses, infants and children.

Studies of acute oral toxicity and genotoxicity with a range of plant metabolites of fludioxonil showed that these metabolites are of low acute oral toxicity and are not genotoxic. The NOAEL in a 90-day study in rats given diets containing a photolytic/hydrolytic degradation product of fludioxonil found in soil and water was 800 ppm (equal to 58 mg/kg bw per day), on the basis of increased relative kidney weight and tubular casts at ≥ 2500 ppm (in males) and minimal to slight atrophy of the olfactory epithelium at ≥ 2500 ppm.

Fludioxonil

Toxicological evaluation

The Meeting established an ADI of 0–0.4 mg/kg bw based on a NOAEL of 37 mg/kg bw per day in a 2-year dietary study in rats and a 100-fold safety factor.

Although effects on the kidneys occurred after relatively short periods of exposure, the Meeting concluded that such effects were unlikely to result from a single exposure. Consequently, the Meeting concluded that an ARfD for fludioxonil was unnecessary.

A toxicological monograph was prepared.

Levels relevant to risk assessment

Species	Study	Effect	NOAEL	LOAEL
Mouse	18 month study of toxicity and carcinogenicity[a]	Toxicity	1000 ppm, equal to 112 mg/kg bw per day	3000 ppm, equal to 360 mg/kg bw per day
		Carcinogenicity	3000 ppm equal to 360 mg/kg bw per day[b]	—
Rat	2-year study of toxicity and carcinogenicity[a]	Toxicity	1000 ppm, equal to 37 mg/kg bw per day	3000 ppm, equal to 110 mg/kg bw per day
		Carcinogenicity	3000 ppm, equal to 110 mg/kg bw per day[b]	—
	Two-generation study of reproductive toxicity[a]	Parental toxicity	300 ppm, equal to 21 mg/kg bw per day[d]	3000 ppm, equal to 210 mg/kg bw per day
		Embryo- and fetotoxicity	300 ppm, equal to 21 mg/kg bw per day[d]	3000 ppm, equal to 210 mg/kg bw per day
	Developmental toxicity[c]	Maternal toxicity	100 mg/kg bw per day	1000 mg/kg bw per day
		Embryo- and fetotoxicity	1000 mg/kg bw per day[b]	—
Rabbit	Developmental toxicity[c]	Maternal toxicity	100 mg/kg bw per day	300 mg/kg bw per day
		Embryo- and fetotoxicity	300 mg/kg bw per day[b]	—
Dog	12-month study of toxicity[a,e]	Toxicity	1000 ppm, equal to 33 mg/kg bw per day	8000 ppm, equal to 300 mg/kg bw per day

[a] Diet
[b] Highest dose tested
[c] Gavage
[d] The NOAEL of 21 mg/kg bw per day in this study was adjusted upwards to 37 mg/kg bw per day on the basis of an absence of effects in 90-day and 2-year studies in rats at 64 and 37 mg/kg bw per day, respectively. Additionally, interpolation of the reduced weight gain in pups in the study of reproductive toxicity indicated a probably non-adverse, reduction in weight gain of 1% or less at the higher NOAEL of 37 mg/kg bw per day.
[e] The LOAEL for this study was 300 mg/kg bw per day. Owing to the wide dose spacing used for this study, the NOAEL is conservative; hence the slightly higher NOAEL obtained in the 2-year study in rats was selected for establishment of the ADI.

Estimate of acceptable daily intake for humans

    0–0.4 mg/kg bw

Estimate of acute reference dose

    Unnecessary

Studies that would provide information useful for continued evaluation of the compound

    Further observations in humans

Fludioxonil

Critical end-points forsetting guidance values for exposure to fludioxonil

Absorption, distribution, excretion and metabolism in animals	
Rate and extent of oral absorption	Rapid, approximately 80%
Dermal absorption	Poor, < 10% in the rat in vivo; 1% or less in human skin in vitro
Distribution	Extensive
Rate and extent of excretion	Largely complete within 24 h; approximately 10% in urine and 80% in the faeces; 70% of the administered dose was excreted in the bile
Potential for accumulation	Low, no evidence of accumulation
Metabolism in mammals	Extensively metabolized, involving primarily oxidation of the pyrrole ring leading to a major (57% of the administered dose) and a minor (4% of the administered dose) oxo-pyrrole metabolite, followed by glucuronyl- and sulfate conjugation
Toxicologically significant compounds (animals, plants and the environment)	Parent compound and metabolites
Acute toxicity	
Rat, LD_{50}, oral	> 5000 mg/kg bw (no deaths)
Rat, LD_{50}, dermal	> 2000 mg/kg bw (no deaths)
Rat, LC_{50}, inhalation	> 2.6 mg/l (no deaths)
Rabbit, dermal irritation	Not irritating
Rabbit, eye irritation	Slight irritant
Skin sensitization	Not sensitizing (Magnusson & Kligman test)
Short-term studies of toxicity	
Target/critical effect	Damage to liver (rats and dogs) and kidney (mice and rats)
Lowest relevant oral NOAEL	1000 ppm, equal to 33 mg/kg bw per day (12-month study in dogs)
Lowest relevant dermal NOAEL	200 mg/kg bw per day (rats)
Lowest relevant inhalation NOAEC	—
	Unlikely to pose a genotoxic risk in vivo
Genotoxicity	
Long-term studies of toxicity and carcinogenicity	
Target/critical effect	Reduced body-weight gains and liver necrosis in rats, liver and kidney damage in mice
Lowest relevant NOAEL	1000 ppm, equal to 37 mg/kg bw per day, in rats
Carcinogenicity	Not carcinogenic in rats or mice; unlikely to pose a carcinogenic risk to humans
Reproductive toxicity	
Reproductive target/critical effect	Reduced pup weight gains in rats at parentally toxic doses
Lowest relevant reproductive NOAEL	300 ppm, equal to 21 mg/kg bw per day
Developmental target/critical effect	None
Lowest relevant developmental NOAEL	300 mg/kg bw per day (the highest dose tested in rabbits)
Neurotoxicity/delayed neurotoxicity	No evidence of neurotoxicity in any study conducted
Other toxicological studies	Studies on plant metabolites and a photolytic/hydrolytic degradation product of fludioxonil indicated that these were of no greater toxicity than the parent compound
Medical data	Medical monitoring since 1992 of employees engaged in the manufacture of fludioxonil, or its formulation, into products has not revealed any adverse health effects

Summary	**Value**	**Study**	**Safety factor**
ADI	0–0.4 mg/kg bw	2-year study in rats (liver effects and reduced body-weight gains)	100
ARfD	Unnecessary	—	—

Fludioxonil

RESIDUE AND ANALYTICAL ASPECTS

Fludioxonil, or 4-(2,2-difluorobenzo[1,3]dioxol-4-yl)-1*H*-pyrrole-3-carbonitrile, is a fungicide that belongs to the chemical class phenylpyrroles. It functions by blocking the protein kinase which catalyses the phosphorylation of a regulatory enzyme of glycerol synthesis. It is specific for a limited number of fungi. It was evaluated for the first time by the 2004 Joint Meeting.

Metabolism

Animals

The metabolism of ^{14}C-pyrrole-labelled fludioxonil was studied in goats and laying hens. Two goats were given radiolabelled fludioxonil orally at a level equivalent to 100 ppm in the feed for 4 consecutive days. The levels of radioactive residue, calculated as fludioxonil, were: 0.07 mg/kg in tenderloin muscle, 0.19 mg/kg in fat, 5.8 mg/kg in liver, 2.9 mg/kg in kidney and 2.2 mg/kg in milk on day 4. Organic solvents released 35% of the TRR in liver, 76% in muscle, 50% in kidney, 35% in liver, 87% in fat and 90% in milk. Protease treatment of the solid residues from solvent extraction of liver, kidney and muscle released 75–91% of the remaining activity. Less than half of this released activity was characterized as proteins by derivatization with 2,4-dinitrofluorobenzene.

The main component identified in muscle was fludioxonil, representing 24% and 43% of the TRR in the two goats. Likewise, fludioxonil was the main component of the residue in omental fat, representing 83% TRR. The main identified metabolite in muscle was the sulfate conjugate of the 2-hydroxy or 5-hydroxy derivative of fludioxonil (22% or 2% TRR). Minor metabolites identified in muscle (< 10% TRR) included the 2-*O*-glucuronide derivative of fludioxonil and the 5-*O*-glucuronide derivative of fludioxonil. (The position numbers refer to the pyrrole ring.) About 50% of the residue in muscle and 83% of the residue in fat were identified.

Multiple components were found in kidney and liver. The following were identified in kidney: 2-*O*-glucuronide derivative of fludioxonil (23% TRR); 7′-*O*-glucuronide derivative (8% TRR); 5-*O*-glucuronide derivative (15% TRR); fludioxonil (2% TRR); and 2- or 5-*O*-sulfate ester (0.7% TRR), for a total identification of 48%. In liver, only fludioxonil was identified (14% TRR). Two labile compounds (24% TRR) were also encountered. No compounds without the pyrrole–phenyl linkage were identified.

On the basis of the identified and characterized residues, the Meeting concluded that the metabolism of fludioxonil via the oral route in goats involves oxidation of the pyrrole ring at the 2 and 5 positions, followed by rapid conversion to sulfate and glucuronide conjugates. A minor route involves oxidation of the benzodioxol ring at the 7′ position and conversion to the glucuronide conjugate. Evidence was also found for substantial incorporation into natural products, including proteins, in kidney and liver.

Five laying hens were given gelatin capsules containing [^{14}C-pyrrole]fludioxonil for 8 consecutive days at a rate equivalent to about 89 ppm in the feed. The vast majority of the radiolabelled residue was eliminated in the excreta (88–102% of the total administered dose). The levels of radioactive residues, calculated as fludioxonil, in the tissues and eggs were as follows: liver, 8.9 mg/kg; muscle, 0.12 mg/kg; skin with fat, 0.25 mg/kg; peritoneal fat, 0.17 mg/kg; egg yolk, 1.8 mg/kg (day 7); egg white, 0.054 mg/kg (day 7).

A series of organic solvent extractions released 61% TRR in liver, 33% in kidney, 62% in muscle, 42% in skin with fat, 74% in egg white and 83% in egg yolk. The solids remaining after solvent extraction of liver (33% TRR), kidney (54%) and muscle (34%) were solubilized with protease and characterized by treatment with 2,4-dinitrofluorobenzene. Protease solubilized 54% of the unextracted activity in liver, 63% of that in kidney and 67% of that in muscle. About 25% of the released radioactivity (< 10% TRR) was derivatized by 2,4-dinitrofluorobenzene at pH 2, indicating the terminal amino group of amino acids.

Alkaline hydrolysis (15% KOH, 95 °C) released all the remaining radioactivity from the solvent-extracted liver (33% TRR), but it could be characterized only as acidic, polar compounds.

About 69% of the TRR in eggs, 24% in liver, 14% in kidney, 44% in muscle and 29% in skin with fat were identified. The main metabolites identified in eggs were the sulfate conjugate of the 1-hydroxy derivative of fludioxonil (40% TRR), the succinamic acid derivative (10% TRR) and the sulfate conjugate of the 2-hydroxy or 5-hydroxy derivative (13% TRR). Fludioxonil was a minor component (2.1% TRR) in eggs. The succinamic acid derivative was the only significant metabolite identified in liver, at about 6% TRR. The metabolites identified in kidney were the glucuronide conjugate of the 2-hydroxy or 5-hydroxy derivative (4.7% TRR), fludioxonil (2.6% TRR) and the 7′-hydroxy derivative (2.8% TRR). The main components identified in breast muscle were fludioxonil (29% TRR) and the sulfate conjugate of the 1-hydroxy derivative. A similar situation existed for skin with attached fat, which contained fludioxonil (9.8%) and the sulfate conjugate of the 1-hydroxy derivative (14%).

On the basis of the characterizations and identifications made in the study of metabolism in hens, the Meeting concluded that metabolism in poultry involves oxidation at the C-2, C-5 and N-1 positions in the pyrrole ring and at the C-7′ of the benzodioxol ring. This is followed by the formation of sulfate or glucuronide conjugates. The C-2 hydroxypyrrole further oxidizes to the 2,5-dioxo-2,5-dihydro pyrrole and succinamic acid derivatives. The last two compounds are unique to poultry. The remaining metabolites found in the hen and all the metabolites in ruminants were also found in rats. The studies of metabolism in rats were reviewed by the WHO Expert Group of the 2004 JMPR.

Plants

The metabolism of radiolabelled fludioxonil resulting from its foliar application has been studied in grape, tomato, peach, green onion and head lettuce. Grape vines were sprayed three times at 3-week intervals with [pyrrole-4-C^{14}] fludioxonil at a rate of 500 g ai/ha per application. Samples of grapes and leaves were taken at intervals, immediately after the first application, up to grape maturity 35 days after the final application. Grapes at maturity contained 2.5–2.8 mg/kg of radiolabelled residue, calculated as fludioxonil. About 57% of the TRR was a surface residue, released by a methanol–water rinse; another 32% of the TRR was released by solvent extraction. The leaves at maturity contained 5.2 mg/kg of radiolabelled residue, 52% as a surface residue and 44% solvent extracted.

The residues in grapes and leaves were extensively identified. In grapes at maturity, seven compounds were identified, but only fludioxonil at 70% TRR exceeded 2% TRR. The metabolites included the succinamic acid derivative (< 1% TRR), the 3-hydroxy succinamic acid derivative (< 1% TRR), the glucose conjugate of 2-hydroxyacetamide benzodioxol (< 1% TRR), 2-hydroxyacetamide benodioxol (< 1% TRR), the 2-hydroxy-5-oxo derivative (2% TRR), the 2,5-dioxo derivative (< 1% TRR) and the 1-hydroxy-2,5-dioxo derivative (< 1% TRR). Similar metabolites were identified in leaves, fludioxonil representing 69% of the TRR; no other metabolite exceeded 6% TRR.

The metabolism of [pyrrole-4-^{14}C]fludioxonil was studied in greenhouse tomato plants that were sprayed three times at 2-week intervals with a wettable powder formulation at a single application rate of 750 g ai/ha. Forty days after the last application, leaves and tomatoes were sampled. The leaves contained a fludioxonil-equivalent radiolabelled residue level of 7.0 mg/kg, and the tomatoes contained 0.28 mg/kg. Of the residue on tomatoes, 41% was on the surface. Rinsing and solvent extraction released 95% of the residues in tomatoes and 95% of those in leaves. About 73% of the tomato residue and 69% of the leaf residue was fludioxonil. Five metabolites, representing 3.6% of the TRR in tomato, were identified. These were the same metabolites as in grapes, except that the 2,5-dioxopyrrole derivative was not found and the benzodioxole-4-carboxylic acid derivative was present but at below the LOQ (< 0.001 mg/kg).

The metabolism of [phenyl-U-^{14}C]fludioxonil was studied in peaches. Three foliar applications at 30-day intervals were made at 130 and 1300 g ai/ha, starting at petal fall. Mature fruit was collected 28 days after the second treatment. In a second trial, two applications, 950 and 2860 g ai/ha, were made at a 35-day interval, starting at petal fall. Samples of immature and mature fruits were taken 30 and 114 days after the second application. The radiolabelled residue level, calculated as fludioxonil, was 0.083 mg/kg and

0.98 mg/kg in the first trial after application at 130 and 1300 g ai/ha, respectively. The residue level on mature peaches in the second trial was 0.26 mg/kg (114-day PHI).

Extraction with acetonitrile:water:acetic acid (80:20:1, v/v) released ≥ 88% TRR in all cases. Analyses were conducted on extracts from 28-day peaches treated at 130 and 1300 g ai/ha in the first trial and on 114-day peaches from the second trial. Fludioxonil was the main component in all cases, ranging from 22% TRR to 62% TRR. Eight metabolites were identified in the 114-day PHI peaches, of which four are also grape or tomato metabolites (succinamic acid derivative, 3% TRR; 2-hydroxy-5-oxo derivative, 1.4% TRR; 2-hydroxy-5-oxo derivative, 1% TRR; benzodioxole-4-carboxylic, 1% TRR). The other metabolites included oxidized fludioxonil glucose conjugates at 7% TRR and an oxirane-2-carboxylic acid derivative at 3% TRR. About 54% of the TRR in peach was identified.

The metabolism of [phenyl-U-^{14}C]fludioxonil on green onions was studied after radiolabelled fludioxonil was applied twice at a 14-day interval at a rate of 560 or 680 g ai/ha and at 2800 or 3380 g ai/ha. Samples were taken at maturity (14-day PHI) and at other intervals. TRR as fludioxonil represented 1.6 mg/kg on the onions given the 560 or 680 kg ai/ha treatment and 10 mg/kg on those given the 2800 or 3380 g ai/ha treatment. After the 2800 or 3380 g ai/ha treatment, 51% of the TRR was souble in organic solvents and 21% in water.

The metabolitic profiles were qualitatively similar at the two treatment levels and at the various sampling intervals. In the onions treated at 2800 or 3380 g ai/ha at mature harvest (14-day PHI), fludioxonil comprised 49% of the TRR. Six metabolites were identified, but none represented > 2% TRR. These were the same metabolites identified in the studies of grape, tomato and peach metabolism.

The metabolism of [pyrrole-4-^{14}C]fludioxonil on head lettuce was studied after three foliar treatments at 10-day intervals at 200 g ai/ha. A second experiment was conducted at 600 g ai/ha per application. With a 6-day PHI, the TRR calculated as fludioxonil was 1.3 mg/kg after treatment at 200 g ai/ha and 5.8 mg/kg at 600 g ai/ha treated. Almost 100% of the radioactivity was extracted with methanol:water. Fludioxonil was the main component (68% TRR after the 200 g ai/ha treatment, 80% after the 600 g ai/ha treatment).

Six metabolites, four of which corresponded to metabolites in the studies of tomato, grape and peach, were identified. No metabolite exceeded 3% TRR. Metabolites unique to head lettuce were lactic acid conjugates of fludioxonil (1–2% TRR).

Several studies were also conducted on metabolism after seed treatment. Seed potatoes were treated with [pyrrole-4-^{14}C]fludioxonil at a rate of 2.5 g ai/100 kg seed. The pieces were planted, and mature potatoes were harvested after 95 days. The tuber contained 0.006 mg/kg radiolabelled residue, calculated as fludioxonil. Fludioxonil represented 21% of the TRR.

Rice seeds were soaked in a [pyrrole-4-^{14}C]fludioxonil solution equivalent to 6.5 kg ai/100 kg seed. Rice plants were grown in a glasshouse and harvested at maturity, 152 days after treatment. Stalks, hulls and seeds contained ≤ 0.002 mg/kg radiolabelled residue as fludioxonil equivalents.

The metabolism of [pyrrole-4-^{14}C]fludioxonil was studied in field-grown spring wheat plants treated at 7.4 g ai/100 kg seed. Plants were harvested 48 days (ear emergence), 83 days (milky stage) and 106 days (maturity) after treatment. At 48 days, stalks contained 0.005 mg/kg of radioactive residue (calculated as fludioxonil). At 83 days, stalks contained 0.004 mg/kg and ears contained 0.002 mg/kg. At maturity, stalks contained 0.015 mg/kg, husks contained 0.005 mg/kg, and grain contained 0.003 mg/kg.

Cotton-seed was treated at a rate of 2.5 or 5.0 g ai/100 kg seeds with [pyrrole-4-^{14}C]fludioxonil and then planted in sandy loam soil in pots. Plants were sampled at maturity, 186 days after treatment. Cotton-seed treated at 5.0 g ai/100 kg seed contained 0.003 mg/kg TRR, and those treated at 2.5 g ai/100 kg contained 0.012 mg/kg. Only 20–30% of the radioactivity could be extracted.

Soya bean seeds were treated with [pyrrole-4-^{14}C]fludioxonil at a rate of 5.0 g ai/100 kg seed and grown to mature plants in a greenhouse. The plants were sampled at intervals of 28 days after planting (sixth node stage), 38 days (mid- to full bloom stage) and 133 days (maturity). Soya bean forage (sixth node) contained 0.096 mg/kg TRR, calculated as fludioxonil. Soya bean hay (mid-flowering) contained 0.041 mg/kg. At maturity, stalks contained 0.005 mg/kg, dry beans contained 0.015 mg/kg, and dry hulls contained 0.012 mg/kg. The main tentatively identified component in forage and hay was 6-hydroxy-2*H*-chromeno[3,4-*c*]pyrrol-4-one, representing 2% TRR.

The metabolism of fludioxonil in and on plants after foliar and seed treatment is adequately understood. Generally, the residue concentrations resulting from seed treatment were too low to permit extraction and identification. The numerous studies of foliar application indicate a similar metabolic pathway, showing fludioxonil as the main component of the residue.

The pathway is characterized by the generation of a large number of metabolites and proceeds mainly through oxidation. Each metabolite represents < 10% TRR. With the exception of oxidation at the 7′-C of the benzoldioxol ring, the oxidations and conjugations occur at the C-2, C-5 and N-1 positions of the pyrrole ring. Ultimately, cleavage of the pyrrole ring, probably via the formation of succinamic acid derivatives, results in formation of 2,2-difluorobenzo[1,3]dioxole metabolites. In studies with pyrrole- or phenyl-labelled ^{14}C-fludioxonil, no metabolites were found, indicating cleavage of the bond between the phenyl and pyrrole ring.

No information was provided on the degradation of fludioxonil when applied post-harvest. Nevertheless, the use of both short and long PHIs in the trial on metabolism in peaches after foliar application provides some indication of the fate of fludioxonil when applied to fruit post-harvest. The study of metabolism in peach shows that the main constituent in the residue is fludioxonil.

Soil

The degradation of fludioxonil on soil exposed to light is rapid, with a half-life of < 1 day for the component of fludioxonil on the surface. On the basis of isolated and identified degradates in studies of radiolabelled compound, it would appear that fludioxonil degrades to 4-(2,2-difluorobenzo[1,3]dioxol-4-yl)-2,5-dioxo-2,5-dihyrdo-1*H*-pyrrole-3-carbonitrile or the 2,5-dioxo derivative of fludioxonil. This metabolite undergoes epoxidation at the C-3 to C-4 position and pyrrole ring opening to give 3-carbamoyl-2-cyano-3-(2,2-difluorobenzo[1,3]dioxol-4-yl)oxirane-2-carboxylic acid. The latter degrades to 2,2-difluoro-benzo[1,3]-dioxole-4-carboxylic acid, a compound found in the studies of rotational crops (see below).

The breakdown of fludioxonil in soil under aerobic conditions with no exposure to light is slow. Mineralization to carbon dioxide is the main route of breakdown (4–45% of applied radioactivity). Some unextractable residues (8–27%) also form. The half-life in sandy loam soil is approximately 250 days.

Four studies of confined rotational crops were conducted with [pyrrole-4-^{14}C]fludioxonil. In the first study, soil was sprayed with [pyrrole-4-^{14}C]fludioxonil at a rate of 750 g ai/ha, and lettuce, winter wheat, sugar beets and maize were planted after intervals of 90, 140, 320 and 345 days, respectively. Lettuce at maturity (152 days post-treatment) contained 0.006 mg/kg radiolabelled residue, winter wheat stems and grain contained 0.008 and 0.002 mg/kg 429 days post-treatment, sugar beet roots and tops contained 0.001 and < 0.001 mg/kg, respectively, and maize stalks and grain contained 0.005 and < 0.001 mg/kg, respectively, at maturity (519 days after treatment). The concentrations of residue were too low to pursue isolation and identification.

In a follow-up study, spring wheat, mustard and turnips were planted 33 days after application of [pyrrole-4-^{14}C]fludioxonil to bare ground at a rate of 120 g ai/ha. At maturity, the residue levels were < 0.01 mg/kg (TRR) in the turnips and mustard greens and 0.006 mg/kg in wheat grain. In 25% mature wheat forage (109 days post-treatment) and in wheat straw (175 days post-treatment), however, the residue levels were 0.058 and 0.12 mg/kg, respectively. The following components were identified in immature wheat forage: fludioxonil (2.4% TRR, 0.001 mg/kg), 6-hydroxy-2*H*-chromeno[3,4-*c*]pyrrol-4-one (11% TRR, 0.006 mg/kg, tentative identification), 4-hydroxy-2,5-dione derivative (4.2% TRR, 0.002 mg/kg), 2,5-dioxo derivative (< 0.001 mg/kg, tentative identification), 2,2-difluoro-benzo[1,3]dioxole-4-carboxylic acid (2.3%

TRR, 0.001 mg/kg, tentative identification) and 2-(2,2-difluorobenzo[1,3]dioxol-4-yl)-2-hydroxyacetamide (< 0.001 mg/kg, tentative identification). Fludioxonil (< 0.001 mg/kg) and similar metabolites at similar concentrations were detected in wheat straw. This work was confirmed by another experiment conducted at 60 g ai/ha.

In a final trial, [phenyl-U-^{14}C]fludioxonil was sprayed onto bare ground at a rate of 1120 g ai/ha. Rotational crops of spring wheat, mustard and radishes were planted 30, 90 and 210 days after treatment and grown to normal maturity. Radish tubers contained 0.14, 0.019 and 0.019 mg/kg of radiolabelled residue at plant-back intervals of 30, 90 and 210 days, about 50% of which could be extracted with organic solvents and water. Mustard greens contained 0.033, 0.044 and 0.050 mg/kg at 30, 90 and 120 days after treatment. Mature wheat straw contained 0.36, 0.14 and 0.11 mg/kg radiolabelled residues, and grain contained 0.058, 0.021 and 0.019 at 30, 90 and 120 days plant-back, of which about 40% from straw and 20% from grain was extractable.

The main metabolite identified in the various commodities was 2,2-difluorobenzo[1,3]dioxole-4-carboxylic acid, at levels ranging from 4.4% TRR in mature wheat straw (30-day plant-back) to 38% TRR (radish tuber, 90-day plant-back). Fludioxonil generally represented < 4% TRR (≤ 0.001 mg/kg) in all matrices except mature radish tuber (30-day plant-back), in which it represented 12% TRR or 0.016 mg/kg.

Field rotational crop studies were conducted in which fludioxonil was applied four times to bare soil at 280 g ai/ha per application, followed at plant-back intervals of 30, 90, 150 and 210 days by sowing of wheat, turnips and leaf lettuce. The mature crops contained no detectable residues of fludioxonil at any plant-back interval, with a LOQ of 0.01 mg/kg.

The nature and extent of the residue in rotational crops after use of fludioxonil on the primary crop is adequately delineated. Similar patterns were observed with pyrrole- and phenyl-labelled ^{14}C-fludioxonil, although somewhat greater concentrations of residue were encountered with the phenyl label. In these trials, fludioxonil was not taken up into rotational crops at plant-back intervals as short as 30 days. The metabolism of fludioxonil in the crops was apparently the same as that seen in target crop studies, but this conclusion is speculative as little or no residue was generally found. Primarily on the basis of the confined study with [phenyl-U-^{14}C]fludioxonil, the metabolism and degradation of this compound is characterized by oxidation and cleavage of the pyrrole ring. No metabolites of cleavage of the bond between the phenyl and the pyrrole ring were observed. The proposed metabolic and degradation pathway is that suggested for foliar application of fludioxonil.

The Meeting concluded that the presence of fludioxonil residues in succeeding (rotational) crops from foliar applications is unlikely.

Methods of analysis

The Committee concluded that adequate analytical methods exist for both monitoring and enforcing MRLs and for gathering data in supervised field trials and processing studies. Methods REM-133/AG631A and AG-597 are suitable for the determination of fludioxonil in samples of plant origin. The methods are fully validated for a range of crops and crop types. In addition, fludioxonil residues can be determined in samples of plant origin by European multi-residue method DG S17.

Method REM-133 involves high-performance liquid chromatography (HPLC) with ultraviolet detection (268 nm). Only fludioxonil is determined. Samples are extracted and then placed on a phenyl solid-phase extraction cartridge and eluted with the appropriate solvent. The samples are analysed by HPLC with column switching (C-18 and phenyl). The validated LOQ is 0.01–0.04 mg/kg. In some European field trials, method REM 133 was modified by the use of only one HPLC column (amino) with a fluorescence detector (excitation, 265 nm; emission, 312 nm). The method was radiovalidated. In this method, 89% of the total radioactivity was solubilized, and 66% of the fludioxonil determined in the metabolism study was identified.

Fludioxonil

Method AG-597 is another HPLC method with ultraviolet detection (268 nm). Only fludioxonil is determined. Samples are extracted and then cleaned-up by silica solid-phase extraction. Analysis is usually conducted on an amino or a C18 column. The method was validated with a wide array of commodities, with limits of determination of 0.01–0.02 mg/kg, except for sorghum grain, for which the limit was 0.05 mg/kg. The method was validated by the US Environmental Protection Agency. Liquid chromatography with mass spectrometry can be used for confirmation, with quantification on ion 247.

A European multi-residue method based on DFG S19 was developed for an array of plant commodities. Extracts are separated by gel permeation chromatography and analysed by capillary gas chromatography with a mass selective detector, monitoring ions 248, 154 and 127. The method was validated for fludioxonil only at 0.02 mg/kg for tomato, orange, wheat and rape and at 0.01 mg/kg for grape wine.

The Meeting concluded that an adequate method exists for the determination of fludioxonil and certain metabolites in livestock commodities (meat, milk, poultry, eggs). In the HPLC method, fludioxonil and metabolites are converted to 2,2-difluoro-1,3-benzodioxole-4-carboxylic acid. The resulting residue is quantified by external calibration against standards of this conversion product, with HPLC and a ultraviolet detector (230 nm). Column switching is used, and alternate columns are specified as a confirmatory procedure. The method was validated at 0.01 mg/kg for muscle and milk and at 0.05 mg/kg for eggs, fat, liver and kidney.

Stability of residues in stored analytical samples

The Meeting concluded that fludioxonil is stable in an array of stored frozen commodities. No degradation of fludioxonil was observed in any frozen commodity throughout the duration of the studies. Fludioxonil is stable for at least 24 months in frozen samples of the following commodities: cereal grains, cereal straw, apple, tomato, grape, pea, rape-seed, maize grain, maize meal, sorghum hay, potato tuber and potato flake. Fludioxonil is stable for at least 12 months in frozen broccoli, cabbage and carrots and for 9 months in frozen chives. Fludioxonil is also stable for at least 3 months in frozen peach, plum, cherry and blueberry.

The Meeting also concluded that fludioxonil and metabolites, determined as 2,2-difluoro-1,3-benzodioxole-4-carboxylic acid, are stable for at least 12 months in frozen muscle and for at least 18 months in frozen liver, milk and eggs.

Definition of the residue

The results of the studies of metabolism after both seed treatment and foliar treatment show that the main identified component of the radiolabelled residue is fludioxonil. The identified metabolites generally represent < 10% of the TRR. The toxicological evaluation did not reveal any metabolites of special concern relative to the parent. The Meeting concluded that the residue definition for plant commodities for compliance with MRLs and for estimation of dietary intake is fludioxonil.

In the analytical methods for plant commodities, HPLC with ultraviolet detection or gas chromatography with mass spectrometry detection, only fludioxonil is determined.

The results of the studies of metabolism in goats and hens were similar. In goats, the main identified metabolite in meat, fat and liver was fludioxonil, representing 33%, 83% and 14% TRR, respectively. The main metabolite in milk and kidney was the pyrrole carbonitrile-*O*-glucuronide, representing 65% and 31% TRR, respectively, and the parent was absent. In hens, fludioxonil was present in muscle (7.9–28.% TRR) and skin plus attached fat (9.8%). It accounted for 1.2% of the TRR in liver, 2.6% in kidney and 2.2% in egg yolk (equivalent to 2.1% egg TRR). The main identified component of the radioactive residue in eggs and fat was the sulfate conjugate of 4-(2,2-difluorobenzo[1,3]dioxol-4-yl)-1-hydroxy-1*H*-pyrrole-3-carbonitrile. The benzene–pyrrole linkage was intact in all the identified metabolites. The toxicological evaluation did not reveal any metabolites of particular concern relative to the parent.

The P_{ow} for fludioxonil is 4.1, suggesting that fludioxonil is fat-soluble. In goats, the radioactive residue represented 0.07 mg/kg TRR in muscle and 0.26 mg/kg in fat. The main component in muscle and fat was fludioxonil (24–43% TRR in muscle and 83% in fat). The Meeting concluded that the fludioxonil residue is fat-soluble, but it also noted the lack of information on milk fat from both the metabolism and the feeding study.

In the validated analytical method for fludioxonil, fludioxonil and pyrrole-derivative metabolites are converted to 2,2-difluorobenzo[1,3]dioxole-4-carboxylic acid.

The Meeting concluded that the residue definition of the residue for livestock commodities (for compliance with MRLs and for estimation of dietary intake) is the sum of fludioxonil and its benzopyrrole metabolites, determined as 2,2-difluoro-benzo[1,3]dioxole-4-carboxylic acid and expressed as fludioxonil.

Results of supervised trials on crops

Supervised trials were conducted with foliar treatment, seed treatment and post-harvest treatment of a variety of crops worldwide.

Citrus fruit

Citrus (orange, lemon, grapefruit) was treated by post-harvest dip (120 g ai/hl) or spray (1000 g ai/250 000 kg fruit) in 28 trials conducted in the USA. GAP specifies a maximum of two treatments, one on entering storage and a second on exit of storage for market distribution, at a single application rate of 500 g ai/250 000 kg fruit (2 mg/kg; 0.85 kg ai/hl for droplet-type applications with a low-volume concentrate, 0.24 kg ai/hl for high-volume jet-type sprays) and 0.06 kg ai/hl for 30-s dip treatments. All trials were conducted at twice GAP in a single-application dip or high-volume spray, and nine of the trials included a second application at twice GAP with a re-treatment interval of 0 days. In the absence of data on residue level decline during storage of citrus, the Meeting considered application at twice GAP an approximation of the practical situation of two treatments at GAP with a variable interval between applications.

The residue levels on orange (six trials; one treatment at twice GAP), in ranked order, were: 0.48, 0.90, 1, 1.4, 2.2 and 2.8 mg/kg. The levels on lemon (seven trials; one treatment at twice GAP) were: 0.46, 0.54, 1., 1.1 (two), 2.9 and 3.2 mg/kg, and those on grapefruit (six trials; one treatment at twice GAP) were: 0.51, 0.94, 0.95, 1.4, 3.8 and 5.2 mg/kg. The nine trials consisting of two sequential applications, each at twice the GAP application rate, were considered exaggerations and were not used; the residue levels ranged from 0.52 mg/kg to 6.0 mg/kg.

The Meeting decided to combine the data; the residue levels on citrus (19 trials; one treatment at twice GAP single rate), in ranked order, were: 0.46, 0.48, 0.51, 0.54, 0.90, 0.94, 0.95, 1 (two), 1.1 (two), 1.4 (two), 2.2, 2.8, 2.9, 3.2, 3.8 and 5.2 mg/kg. Data on residues in pulp were available from only one trial on oranges, in which flesh and peel contained approximately equal concentrations of fludioxonil. The Meeting estimated a maximum residue level for whole citrus of 7 mg/kg and an STMR of 1.1 mg/kg.

Pome fruit

Apples were treated by post-harvest dip or spray in the USA with a 50% wettable powder formulation. GAP specifies a maximum of two treatments, one on entering storage and a second on exit from storage for market distribution, at a single application rate of 500 g ai/250 000 kg fruit (2 mg/kg; 0.85 kg ai/hl for droplet-type applications with a low-volume concentrate, 0.24 kg ai/hl for high-volume jet-type sprays) and 0.06 kg ai/hl for dip treatments of approximately 30 s. Seven trials were conducted at approximately the GAP rate (single application), and two trials were conducted at the GAP rate with two sequential applications: dip at 0.06 kg ai/hl, followed by packing-line spray at 2.5 mg/kg (125% GAP). As GAP specifies two treatments, the Meeting regarded the two trials conducted with two applications as an approximation of GAP. The residue levels were 2.0 and 2.2 mg/kg. The Meeting considered two trials inadequate for estimating a maximum residue level.

Pears were treated by post-harvest dip or spray treatment in the USA with a 50% wettable powder formulation. GAP is identical to that for apples. Twelve trials were conducted, but only two were conducted with two applications: 0.048 kg ai/hl drench (80% dip GAP), followed by a packing-line spray at 0.2–0.6 kg ai/hl or 2.2–6.6 mg/kg fruit (110–300% GAP). As GAP specifies two treatments, the Meeting regarded the two trials conducted with two applications as an approximation of GAP. The residue levels (with an exaggerated rate for the second application) were 1.6 and 2.8 mg/kg. The Meeting considered two trials insufficient for estimating a maximum residue level.

The Meeting considered combining the post-harvest trials on pear and apple (same GAP) for mutual support, but considered four trials insufficient for these commodities.

Pears received foliar treatment with a 25% water-dispersible granule formulation in seven trials conducted at GAP (three in Italy, three in Spain and one in France). The GAPs are as follows: Italy, 0.02 kg ai/hl, 0.25 kg ai/ha, three applications, 14-day PHI; Spain, 0.025 kg ai/ha, 0.25 kg ai/ha, three applications, 7-day PHI. No GAP was available for France, and the GAP of Spain was applied to all trials (7-day PHI). The residue levels, in ranked order, were: 0.14, 0.15, 0.18, <u>0.21</u>, 0.28, 0.32 and 0.36 mg/kg. The Meeting estimated a maximum residue level of 0.7 mg/kg and an STMR of 0.21 mg/kg.

Stone fruit

In seven post-harvest treatment trials (spray or dip), *peaches* were treated at the GAP of 0.06 kg ai/hl with a 50% wettable powder formulation. The residue levels on peaches after treatment (no storage interval), in ranked order, were: 0.37, 0.42, 1.6, <u>2.2</u>, 2.8, 3.4 and 3.6 mg/kg

Trials of foliar application of fludioxonil (62.5% water-dispersible granules, 25% fludioxonil) were conducted in France, Italy and Spain. The relevant GAPs are: France, 0.015 kg ai/hl, 14-day PHI; Italy, 0.015 kg ai/hl, 0.25 kg ai/ha, two applications, 14-day PHI. No GAP was available for Spain, and the GAP of Italy was applied. The residue levels in 11 trials at GAP, in ranked order, were: 0.02, 0.04 (two), 0.08 (two), 0.11, 0.23 (two), 0.29 and 0.33 mg/kg. The data set on post-harvest treatment contained the highest residue values and was used to estimate the maximum residue level and the STMR.

Post-harvest treatment of *plums* was investigated in two trials in the USA. GAP is spray application at 0.06 kg ai/hl of a 50% wettable powder formulation. The results were 0.10 and 0.92 mg/kg. As the results for post-harvest treatment of plums were not statistically significantly different from those for peaches with the same GAP, the populations can be combined for mutual support.

Trials of foliar application of fludioxonil (22.5% water-dispersible granules, 25% fludioxonil) to plums were conducted in France, Italy, Germany and Switzerland. The relevant GAPs are: France, 0.012 kg ai/hl, 0.12 kg ai/ha, three applications, 14-day PHI; Italy, 0.025 kg ai/hl, 0.25 kg ai/ha, two applications, 14-day PHI; Switzerland, 0.3 kg ai/ha, two applications, PHI not specified. GAP in Germany was not available, and the GAP of Italy was applied. In 12 trials at GAP, the residue levels, in ranked order, were: < 0.02, 0.03, 0.04, 0.05, 0.06 (two), 0.065, 0.07, 0.09, 0.10, 0.11 and 0.17 mg/kg. The data set on post-harvest treatment contained the highest residue values and was used to estimate the maximum residue level and the STMR.

Post-harvest treatment of *cherries* was investigated in two trials in the USA. GAP is spray application at 0.06 kg ai/hl of a 50% wettable powder formulation. The reside levels were 0.19 and 0.68 mg/kg. As the results for post-harvest treatment of cherries were not statistically significantly different from those for peaches, with the same GAP, the populations can be combined for mutual support.

A 25% water-dispersible granule formulation of fludioxonil was applied as a foliar spray to cherries in Europe. In six trials, the residue levels ranged from 0.16 to 0.43 mg/kg after a treatment rate of 0.019 kg ai/hl and a PHI of 7 days. No GAP was provided for any country in Europe.

The results for post-harvest treatment (GAP, dip or spray at 0.06 kg ai/hl) of peaches, plums and cherries were combined. The residue levels in the 11 trials, in ranked order, were: 0.10, 0.19, 0.37, 0.42, <u>0.68</u>,

<u>0.92</u>, 1.6, 2.2, 2.8, 3.4 and 3.6 mg/kg. The Meeting estimated a maximum residue level of 5 mg/kg and an STMR of 0.80 mg/kg for stone fruit.

Berries and other small fruit

Grape

Trials on foliar treatment of grape vines were available from Chile, France, Germany, Greece, Italy, South Africa, Spain and Switzerland. The relevant GAPs (25% water-dispersible granules) are: Chile, 0.25 kg ai/ha, two applications, 7-day PHI; France, 0.3 kg ai/ha, two applications, 60-day PHI; Germany, 0.015 kg ai/hl, 0.24 kg ai/ha, two applications, 35-day PHI; Italy, 0.02 kg ai/hl, 0.2 kg ai/ha, two applications, 21-day PHI; Spain, 0.25 kg ai/ha, two applications, 21-day PHI; Switzerland, 0.3 kg ai/ha, one application, early season. The residue values at the GAP of Chile, in ranked order, were: 0.18, 0.24 and 0.28 (two) mg/kg. The trials in France (northern), Germany and Switzerland were evaluated against the GAP of Germany, resulting in six trials in Germany (0.17, 0.20, 0.21, 0.24, 0.28, 0.31 mg/kg) and five trials in Switzerland (0.90, 0.99, 1.4, 1.6 (two) mg/kg) at GAP and combined: 0.17, 0.20, 0.21, 0.24, 0.28, 0.31, 0.90, 0.99, 1.4 and 1.6 (two) mg/kg. The GAP of Spain was used to evaluate the trials in Greece, Italy and Spain. The residue levels in two trials in Spain and one in Italy at this GAP were 0.22, 0.41 and 0.43 mg/kg. The Meeting combined the 18 values for Chile, Germany and Switzerland, Spain and Italy (same population) and found a ranked order of: 0.17, 0.18, 0.20, 0.21, 0.22, 0.24 (two), <u>0.28</u> (three), 0.31, 0.41, 0.43, 0.90, 0.99, 1.4 and 1.6 (two) mg/kg. The Meeting estimated a maximum residue level of 2 mg/kg and an STMR of 0.28 mg/kg.

Strawberry

Foliar applications of a 50% wettable powder formulation were made to strawberries in the USA, and of a 25% water-dispersible granule formulation in Europe (glasshouse and outdoor). The relevant GAPs are: France, 0.25 kg ai/ha, one application, 3-day PHI; Germany, 0.125 kg ai/hl, 0.25 kg ai/ha, three applications, 7-day PHI; Italy, 0.02 kg ai/hl, 0.2 kg ai/ha, three applications, 7-day PHI; Spain, 0.25 kg ai/ha, three applications, 7-day PHI; Switzerland, 0.025 kg ai/hl, 0.3 kg ai/ha, two applications, 14-day PHI; USA, 0.25 kg ai/ha, four applications, 0-day PHI. The values from the eight trials in the USA in ranked order were: 0.22, 0.43, 0.54, 0.62, 1.0, 1.2 (two) and 1.3 mg/kg. At GAP of Spain and Germany (0.25 kg ai/ha, three applications, 7-day PHI), the values from outdoor trials in Germany were 0.04 and 0.05 (two) mg/kg; those in Switzerland were 0.13 (two) mg/kg; those in France were 0.09, 0.25, 0.61 and 0.77 mg/kg; that in Italy was 0.14 mg/kg; those in Spain were 0.64 and 0.83 mg/kg; and that in the United Kingdom was 0.11 mg/kg. These 13 values may be combined: 0.04, 0.05 (two), 0.09, 0.11, 0.13 (two), 0.14, 0.25, 0.61, 0.64, 0.77 and 0.83 mg/kg. When the European and US populations were combined, the residue levels, in ranked order, were: 0.04, 0.05 (two), 0.09, 0.11, 0.13 (two), 0.14, 0.22, 0.25, 0.43, 0.54, 0.61, 0.62, 0.64, 0.77, 0.83, 1.0, 1.2 (two) and 1.3 mg/kg.

Indoor trials were also conducted in France, Italy, Spain and Switzerland. The ranked order of residue values evaluated against the GAP of Italy was: 0.11, 0.21, 0.27 and 1.9 mg/kg.

When the results of the indoor and outdoor trials were combined, the residue levels in the 25 trials, in ranked order, were: 0.04, 0.05 (two), 0.09, 0.11 (two), 0.13 (two), 0.14, 0.21, 0.22, 0.25, <u>0.27</u>, 0.43, 0.54, 0.61, 0.62, 0.64, 0.77, 0.83, 1.0, 1.2 (two), 1.3 and 1.9 mg/kg. The Meeting estimated a maximum residue level of 3 mg/kg and an STMR of 0.27 mg/kg.

Raspberry

Foliar applications of a 25% water-dispersible granule formulation of fludioxonil were made to raspberries in Germany and the USA. The relevant GAPs are: Switzerland, 0.025 kg ai/hl, 0.32 kg ai/ha, two applications, 14-day PHI; and USA, 0.25 kg ai/ha, four applications, 0-day PHI. The residue levels, in ranked order, were: 0.19, 0.24 (two) and 0.30 mg/kg in Germany and 0.96, 1.0 (three) and 3.6 mg/kg in the USA.

Fludioxonil

The Meeting estimated a maximum residue level of 5 mg/kg and an STMR of 1.0 mg/kg for raspberries and extrapolated the values to blackberry and dewberry on the basis of the trials in the USA, which had the highest values.

Blueberry

Foliar applications of a 25% water-dispersible granule formulation of fludioxonil were made to blueberries in Germany and the USA. The relevant GAP is: USA, 0.25 kg ai/ha, four applications, 0-day PHI. No GAP was available for Germany or other European countries. The residue levels in ranked order at GAP in the USA were: < 0.05, 0.14, 0.26, <u>0.52</u>, <u>0.68</u>, 0.84, 0.90 and 1.4 mg/kg. The Meeting estimated a maximum residue level of 2 mg/kg and an STMR of 0.60 mg/kg.

Black and red currant

Foliar application of a 25% water-dispersible granule formulation of fludioxonil was made to black currants in four trials and to red currants in one trial in Germany. As no GAP is available for Germany or other European countries, the Meeting could not estimate an STMR or maximum residue level.

Assorted tropical and subtropical fruits

Lychee

Fludioxonil (25% water-dispersible granules) was applied as a foliar spray to lychee in the USA, where GAP is: 0.25 kg ai/ha, four applications, 0-day PHI. The residue levels in ranked order were: 0.81, 0.92 and 1.4 mg/kg. The Meeting noted that five or seven applications were made at about 7-day intervals and that the extra one or three applications would have been made ≤ 21 days before harvest. On the basis of studies of decline in other fruit crops, they might have made a significant contribution (about 25%) to the final residue level. Therefore, the Meeting did not estimate a maximum residue level or an STMR.

Kiwi

Kiwi fruit in the USA were treated post-harvest at 0.06 kg ai/hl with a wettable powder formulation. GAP specifies application of a 50% wettable powder formulation as a dip at 0.06 kg ai/hl for 30 s or as a low-volume application with a control droplet-type application at 0.24 kg ai/hl or 2.5 mg/kg fruit. Trials were conducted, with two methods (dip, spray) at two locations and a single method (dip) at a third. The ranked order of residue levels in the five trials was: 1.6, 5.2, <u>7.2</u>, 8.6 and 9.0 mg/kg. The Meeting estimated mg/kg a maximum residue level of 15 mg/kg and an STMR of 7.2 mg/kg for kiwi fruit.

Pomegranate

Pomegranate in the USA were treated post-harvest at 0.06 kg ai/hl with a wettable powder formulation. The residue levels were 0.65 and 0.95 mg/kg; however, there is no GAP, and the Meeting could not estimate a maximum residue level or an STMR.

Bulb vegetables

Green (spring) onions

Fludioxonil was applied as a foliar spray of a wettable powder formulation to green onions in the USA. The relevant GAP is 0.25 kg ai/ha, four applications, 7-day PHI. The residue levels in ranked order were 0.14, <u>0.59</u> and 3.0 mg/kg. The Meeting estimated a maximum residue level of 5 mg/kg and an STMR of 0.59 mg/kg.

Bulb onion

Fludioxonil (wettable powder formulation) was applied as a foliar spray to onions in France, Italy, Germany and Switzerland and in the USA. The relevant GAPs are: Austria, 0.25 kg ai/ha, three applications, 7-day PHI; Switzerland, 0.25 kg ai/ha, two applications, unspecified PHI; and USA, 0.25 kg ai/ha, four applications, 7-day PHI. GAP in Switzerland (assumed 0-day PHI) was applied to the other European

Fludioxonil

countries in the absence of a GAP for southern Europe. The residue levels in trials on bulb onions (fresh) at the Swiss GAP were: France, < 0.02, 0.05 and 0.06 mg/kg; and Italy, < 0.02, 0.04, 0.07 and 0.34 mg/kg. The levels in three trials on bulb onions (dry) in the USA at US GAP were: < 0.02 (three), 0.04 (two) and 0.06 mg/kg. The Meeting combined the data sets for Europe and the USA and found a ranked order of residue levels of: < 0.02 (five), 0.04 (three), 0.05, 0.06 (two), 0.07 and 0.34 mg/kg. The Meeting estimated a maximum residue level of 0.5 mg/kg and an STMR of 0.04 mg/kg.

Brassica vegetables

Broccoli

Fludioxonil (water-dispersible granule formulation) was applied as a foliar spray to broccoli in Canada and the USA. The relevant GAP is: 0.25 kg ai/ha, four applications, 7-day PHI. The residue levels in seven trials at US GAP, in ranked order, were: 0.07, 0.10, 0.18, 0.23, 0.26, 0.34 and 0.36 mg/kg. The Meeting estimated a maximum residue level of 0.7 mg/kg and an STMR of 0.23 mg/kg.

Cabbage

Fludioxonil (water-dispersible granule formulation) was applied as a foliar spray to cabbage in the USA. The relevant GAP is: 0.25 kg ai/ha, four applications, 7-day PHI. The residue levels in ranked order on cabbage with wrapper leaves in six trials at GAP were: 0.17, 0.17, 0.21, 0.27, 0.5 and 1.2 mg/kg. The Meeting estimated a maximum residue level of 2 mg/kg and an STMR of 0.24 mg/kg.

Fruiting vegetables

Cucumber

A 25% water-dispersible granule formulation of fludioxonil was applied as a foliar spray (glasshouse and field) to cucumbers in Greece, Spain and Switzerland. The relevant GAPs are: Italy, 0.02 kg ai/hl, 0.20 kg ai/ha, three applications, 7-day PHI; Spain, 0.025 kg ai/hl, three applications, 7-day PHI; Switzerland, 0.025 kg ai/hl, 3-day PHI. GAP for Greece was not available, and that of of Italy and Spain was used. The results from the 10 glasshouse trials (seven in Spain, one in Greece, two in Switzerland) in ranked order were: < 0.02, 0.02 (two), 0.06 (two), 0.07, 0.08 (two), 0.11 and 0.14 mg/kg. The results from the field trials (one in Greece, two in Spain) were: < 0.02, 0.02 and 0.03 mg/kg. The populations are not statistically significantly different, and the combined results are: < 0.02 (two), 0.02 (three), 0.03, 0.06 (two), 0.07, 0.08 (two), 0.11 and 0.14 mg/kg. The Meeting estimated a maximum residue level of 0.3 mg/kg and an STMR of 0.06 mg/kg.

Summer squash (zucchini)

Two indoor trials were conducted on zucchini in Italy. The relevant GAP is: 25% water-dispersible granule, 0.02 kg ai/hl, 0.20 kg ai/ha, three applications, 7-day PHI. The residue levels were 0.05 and 0.06 mg/kg. The Meeting agreed to use the results for cucumber as support for summer squash. The residue levels in ranked order were: < 0.02 (two), 0.02 (three), 0.03, 0.05, 0.06 (three), 0.07, 0.08 (two), 0.11 and 0.14 mg/kg. The Meeting estimated a maximum residue level of 0.3 mg/kg and an STMR of 0.06 mg/kg.

Cantaloupe

A 50% wettable powder formulation was applied to cantaloupe vines (three times 0.28 kg ai/ha, 0.84 kg ai/ha total, 14-day PHI) in the USA by drip irrigation. GAP specifies drip irrigation application of a 50% wettable powder formulation at a rate of 0.28 kg ai/ha. The total seasonal application is limited to 0.84 kg ai/ha, and the PHI is 14 days. The residue levels in ranked order were: < 0.02 (two) and 0.02 (two) mg/kg. The Meeting estimated a maximum residue level of 0.03 mg/kg and an STMR of 0.02 mg/kg for whole melon. No information was available on the residue in pulp.

Tomato

Fludioxonil (25% water-dispersible granules) was applied as a foliar spray in glasshouses (11 trials) and in the field (two trials) in Greece, Spain, Switzerland and the United Kingdom. The relevant GAPs are:

Italy, 0.02 kg ai/hl, 0.2 kg ai/ha, three applications, 7-day PHI; Spain, 0.025 kg ai/hl, three applications, 7-day PHI; Switzerland, 0.25 kg ai/ha, two applications, 3-day PHI. GAPs were not available for Greece or the United Kingdom. As three applications were used in all trials, the GAPs of Italy and Spain were used (7-day PHI). The residue levels in the 14 glasshouse trials at this GAP in ranked order were: 0.05, 0.08, 0.09 (two), 0.10 (two), 0.13 (two), 0.14 (two), 0.16, 0.21, 0.28 and 0.32 mg/kg. The levels in the outside trials in Switzerland were: 0.04 and 0.07 mg/kg. The two groups were statistically the same population, and the combined levels from the 16 trials were: 0.04, 0.05, 0.07, 0.08, 0.09 (two), 0.10 (two), 0.13 (two), 0.14 (two), 0.16, 0.21, 0.28 and 0.32 mg/kg. The Meeting estimated a maximum residue level of 0.5 mg/kg and an STMR of 0.12 mg/kg.

Bell pepper

Fludioxonil (25% water-dispersible granules) was applied as a foliar spray to bell (sweet) peppers in eight glasshouse trials and two field trials in Spain and Switzerland. The relevant GAPs are: Austria, 0.025 kg ai/hl, 0.25 kg ai/ha, three applications, 7-day PHI; Italy, 0.02 kg ai/hl, 0.2 kg ai/ha, three applications, 7-day PHI; Spain, 0.25 kg ai/ha, three applications, 7-day PHI. The GAP of Austria and Italy was used. The ranked order of residue levels in the eight glasshouse trials (six in Spain, two in Switzerland) at GAP was: 0.08, 0.10, 0.14, 0.22, 0.29, 0.46, 0.56 and 0.60 mg/kg. The ranked order in field trials at GAP (one in Italy, one in Spain) was: 0.06 and 0.13 mg/kg. As the two groups are from the same population, they were combined to give levels of: 0.06, 0.08, 0.10, 0.13, 0.14, 0.22, 0.29, 0.46, 0.56 and 0.60 mg/kg. The Meeting estimated a maximum residue level of 1 mg/kg and an STMR of 0.18 mg/kg.

Eggplant

Fludioxonil formulated as 25% water-dispersible granule was applied to eggplant as a foliar spray three times in glasshouse trials in Italy and Spain. The relevant GAPs are: Italy, 0.02 kg ai/hl, 0.2 kg ai/ha, three applications, 7-day PHI; Spain, 0.025 kg ai/hl, three applications, 7-day PHI. The results at GAP in ranked order were: 0.03, 0.06, 0.06 and 0.08 mg/kg. The Meeting estimated a maximum residue level of 0.3 mg/kg and an STMR of 0.06 mg/kg.

Sweet corn (corn-on-the-cob)

Fludioxonil (flowable concentrate) was applied to sweet corn seed in the USA before planting. The relevant GAP is 5 g ai/1000 kg seed. The residue levels were < 0.01 in three trials at three to five times GAP. The Meeting recognized the similarity of sweet corn and maize (see below) and decided to translate the field trial data for seed treatment of maize (same GAP as sweet corn) to sweet corn seed treatment. The residue levels in the eight trials were all < 0.01 mg/kg. The Meeting estimated a maximum residue level of 0.01 (*) mg/kg and an STMR 0.01 mg/kg.

Leafy vegetables

Lettuce, head

Fludioxonil (25% water-dispersible granules) was applied as a foliar spray to lettuce in 11 glasshouse and 17 field trials in France, Germany, Italy, Spain and Switzerland. The relevant GAPs are: France, 0.15 kg ai/ha, four applications, 14-day PHI; Italy, 0.018 kg ai/hl, 0.18 kg ai/ha, three applications, 14-day PHI; Spain, 0.15 kg ai/ha, three applications, 14-day PHI; Switzerland, 0.12 kg ai/ha, two applications, early season. No GAP was available for Germany. The GAP of Italy was used to evaluate the trials. The ranked order of residue levels in the glasshouse trials was: 0.72, 0.98, 1.1, 2.4, 2.5, 2.7 (two), 3.4 (two), 4.7 and 6.0 mg/kg. The ranked order of residue levels in the field trials was: < 0.02 (six), 0.02 (two), 0.04 (three), 0.07, 0.11, 0.17, 0.29, 1.2 and 1.2 mg/kg. The two sets are not from the same population. In the basis of the indoor trials, the Meeting estimated a maximum residue level of 10 mg/kg and an STMR of 2.7 mg/kg.

Watercress

Watercress was treated with fludioxonil (25% water-dispersible granules) as a foliar spray in the USA. The relevant GAP is: 0.25 kg ai/ha, four applications, 0-day PHI. In the two trials at GAP, the residue levels were 4.2 and 4.5 mg/kg. The OECD York Workshop recommended a minimum of three trials for commodities that are not significant in trade or in the diet. (See mustard greens.)

Mustard greens

Supervised trials were conducted on mustard greens in the USA. The relevant GAP is: 0.24 kg ai/ha, water-dispersible granules, four applications, 7-day PHI. The ranked order of residue levels in the seven trials at GAP was: 0.06, 0.49, 0.54, 0.76, 1.2, 6.6 and 7.1 mg/kg. The Meeting decided to combine the results of the trials on watercress and mustard greens (same GAP) for mutual support. The ranked order of levels in the nine trials was 0.06, 0.49, 0.54, 0.76, 1.2, 4.2. 4.5, 6.6 and 7.1 mg/kg. The Meeting estimated a maximum residue level of 10 mg/kg and an STMR of 1.2 mg/kg for both watercress and mustard greens.

Legume vegetables and pulses

Bean pod with seed (common bean, French bean, edible podded bean)

Fludioxonil (water-dispersible granules) was applied as a foliar spray to beans in pod in 22 field and glasshouse trials in France, Spain and Switzerland. The relevant GAPs are: France, 0.083 kg ai/hl, 0.25 kg ai/ha, number of applications not specified, 14-day PHI; Spain, 0.025 kg ai/hl, three applications, 14-day PHI. No GAP was available for Switzerland, and the GAP of France was applied. The ranked order of residue levels from the 15 field trials was: < 0.02, 0.02 (two), 0.03 (five), 0.04 (two), 0.06 (three), 0.09 and 0.13 mg/kg. The ranked order in the seven glasshouse trials was: 0.03, 0.04 (two), 0.06, 0.09, 0.17 and 0.20 mg/kg. The two groups are not from different populations and were therefore combined to give residue levels of: < 0.02, 0.02 (two), 0.03 (six), 0.04 (four), 0.06 (four), 0.09 (two), 0.13, 0.17 and 0.20 mg/kg. The Meeting estimated a maximum residue level of 0.3 mg/kg and an STMR of 0.04 mg/kg for beans (pods and/or immature seeds). This maximum residue level and STMR are extended to peas with pod.

Trials were also conducted on the seed treatment of broad bean and French bean seeds (flowable concentrate) at 5 g ai/100 kg seed in Denmark and Germany. The residue levels were < 0.02 mg/kg on bean seed in all six trials, but no GAP was available.

Peas (succulent)

Fludioxonil (25% water-dispersible granules) was applied as a foliar spray to pea vines in France and Switzerland. The relevant GAP is that of France for legume vegetables: 0.083 kg ai/hl, 0.25 kg ai/ha, number of applications not specified, 14-day PHI. No GAP was available for Switzerland, and the GAP of France was applied. The ranked order of residue levels in the trials at GAP was: ≤ 0.02 (10) and 0.02 mg/kg.

Trials were also conducted of seed treatment of peas with a flowable concentrate or water-dispersible granule formulation in France and the United Kingdom. The residue levels were < 0.02 mg/kg in the six trials conducted at the GAP of the United Kingdom (5% water-dispersible granules, 10 g ai/100 kg seed).

The Meeting estimated a maximum residue level of 0.03 mg/kg and an STMR of 0.02 mg/kg for peas, shelled (succulent seeds) on the basis of the trials with foliar application. Nevertheless, the maximum residue level and STMR also accommodate seed treatment use and are extended to succulent beans without pod.

Pulses (dry bean and dry pea)

A water-dispersible granule formulation of fludioxonil was applied as a foliar spray to pea and bean (kidney) vines in France. The relevant GAP is: Austria and Spain, water-dispersible granules, 0.25 kg ai/ha, two applications, 14-day PHI. No GAP was available for France, and the GAP of Spain was applied. The ranked order of residue levels in dry pea and bean was: < 0.02 (two), 0.04 (two) and 0.05 mg/kg.

Supervised trials on the seed treatment of pea seed in France were also considered. GAP in the United Kingdom is application of a 5% water-dispersible granule formulation of fludioxonil (w/w) at a rate of 10 g ai/100 kg pea seed. The residue levels in the seven trials at this GAP were < 0.02 mg/kg in dry seed at harvest.

The Meeting estimated a maximum residue level of 0.07 mg/kg and an STMR of 0.02 mg/kg for dry peas and for dry beans after foliar application of fludioxonil. The Meeting noted that this also accommodates use of fludioxonil for seed treatment.

Root and tuber vegetables

Potato

Fludioxonil (flowable concentrate, dustable powder) was applied to potato pieces as seed treatment in six trials in Australia, three in South Africa and 13 in the USA. The available GAPs are: Australia, flowable concentrate, 10%, 2.5 g ai/100 kg seed; USA, flowable concentrate, 2.5 g ai/100 kg seed. The ranked order of residue levels on mature potatoes in trials at GAP in Australia and the USA was: < 0.01 (16) and 0.01 (17) mg/kg. The Meeting estimated a maximum residue level of 0.02 mg/kg and an STMR of 0.01 mg/kg.

Yam

A 50% wettable powder formulation of fludioxonil was applied as post-harvest treatment to yams at 0.06 kg ai/hl in the USA. GAP specifies application of a 50% wettable powder formulation as a single dip application at a rate of 0.06 kg ai/hl for about 30 s. Two trials were conducted at GAP, and in each trial both whole tubers and tuber pieces (cut yams) were tested. The ranked order of residue levels was: 4.6 and 5.0 mg/kg. The Meeting regarded two independent trials as insufficient for estimating a maximum residue level or an STMR.

Carrot

Nine trials were conducted in the USA in which carrot plots were given four foliar applications of fludioxonil at 0.24 kg ai/ha. The relevant GAP is: water-dispersible granules, 0.25 kg ai/ha, four applications, 7-day PHI. The residue levels in seven trials at GAP were: 0.04, 0.16, 0.18, <u>0.20</u>, 0.20, 0.25 and 0.42 mg/kg. The Meeting estimated a maximum residue level of 0.7 mg/kg and an STMR of 0.20 mg/kg.

Asparagus

In two trials in Germany, asparagus plants were treated with a water-dispersible granule formulation after harvest. This gives a PHI of about 240 days. No GAP was available for Germany. GAP for Austria specifies use of a 25% water-dispersible granule formulation three times with a 14–21-day interval at 0.042 kg ai/hl or 0.25 kg ai/ha per application. No PHI is specified, but treatments are to be made at transplantation from the glasshouse to the field. The residue levels in the two German trials were < 0.02 mg/kg. The Meeting considered two trials insufficient for estimating a maximum residue level or an STMR.

Cereal grains

Fludioxonil formulations were applied to wheat in France, Germany and Switzerland as seed treatment. The relevant GAPs are: Austria, Belgium, United Kingdom, flowable concentrate formulation, 5 g ai/100 kg seed, one application. In the 48 trials conducted at or above GAP, the residue levels in ranked order were: < 0.02 (36) and < 0.04 (12) mg/kg.

One trial was reported from Denmark in which rye seed was treated. The relevant GAP is: Austria, flowable concentrate, 5 g ai/100 kg seed. No GAP is available for Denmark. The residue level was < 0.02 mg/kg.

Fludioxonil was applied as seed treatment to barley in 30 trials in France, Germany and Switzerland. The relevant GAPs are: Austria, Belgium, United Kingdom, flowable concentrate formulation, 5 g ai/100 kg seed. No GAP was available for France, Germany or Switzerland. The residue levels in six trials conducted at or above GAP were < 0.02 mg/kg.

Fludioxonil was applied as a seed treatment to maize (field corn) in 27 trials in France, Germany, Greece, Hungary, Italy and Spain, three trials in South Africa and five trials in the USA. The relevant GAP is: USA, flowable concentrate formulation, 5 g ai/100 kg seed. No GAPs were available for Europe or South Africa, and the GAP of the USA was applied. The ranked order of residue values in trials conducted at GAP and at three and five times the GAP rate were < 0.01 (five) and < 0.02 (seven) mg/kg.

Fludioxonil was applied as seed treatment to sorghum in four trials in the USA. The relevant GAP is flowable concentrate formulation, 5 g ai/100 kg seed. The residue level at three to five times the GAP rate was < 0.05 mg/kg.

All residue levels resulting from seed treatment of the five different cereal grains were below the LOQ. The combined results from all the trials, in ranked order, were: < 0.01 (five), ≤ 0.02 (50), < 0.04 (12) and < 0.05 (four) mg/kg. The Meeting estimated a maximum residue level of 0.05 (*) mg/kg and an STMR of 0.02 mg/kg for cereal grains.

Pistachio nut

Fludioxonil was applied as a water-dispersible granule formulation to pistachio trees in the USA. The relevant GAP is: 0.25 kg ai/ha, four applications, 7-day PHI. The residue levels, in ranked order, were: 0.04, 0.05 and 0.08 mg/kg. The Meeting estimated a maximum residue level of 0.2 mg/kg and an STMR 0.05 mg/kg.

Rape-seed

Fludioxonil was applied to rape as seed treatment in trials in France, Germany, Sweden and the United Kingdom. GAP in Germany is treatment of seed with a flowable concentrate formulation at 12 g fludioxonil per 100 kg seed. The residue levels in the 15 trials at this GAP were < 0.02 mg/kg. The Meeting estimated a maximum residue level of 0.02 (*) mg/kg and an STMR of 0.02 mg/kg.

Cotton-seed

Fludioxonil was applied as seed treatment (flowable concentrate and emulsion formulations) to cotton in Greece and the USA. The relevant GAP is: USA, flowable concentrate, 5 g ai/100 kg seed. No GAP was available for Greece (or any other European country), and the GAP of the USA was applied. The ranked order of residue levels in the trials was: < 0.02 (two) and ≤ 0.05 (six) mg/kg. The Meeting estimated a maximum residue level of 0.05 (*) mg/kg and an STMR of 0.05 mg/kg for cotton-seed.

Herbs

Fludioxonil was applied as a foliar spray (water-dispersible granules) to chives and basil in the USA. The relevant GAP is 0.25 kg ai/ha, four applications, 7-day PHI. The residue levels were: 1.8 and 3.9 mg/kg on chives and 1.9 and 3.0 mg/kg on basil. The Meeting estimated a maximum residue level of 10 mg/kg and an STMR of 2.4 mg/kg for fresh basil and a maximum residue level of 10 mg/kg and an STMR of 2.8 mg/kg for fresh chives.

For each herb, one trial included drying. The drying factor for chives is 8 (31/3.9), and that for basil is also 8 (23/3). Application of this factor to the data from field trials with chives yielded a revised ranked order of residue levels: 14 and 31 mg/kg. Therefore, the Meeting estimated a maximum residue level of 50 mg/kg and an STMR of 22 mg/kg for dried chives. Application of the drying factor for basil yield a revised ranked order of 15 and 24 mg/kg. Therefore, the Meeting estimated a maximum residue level of 50 mg/kg and an STMR of 20 mg/kg for dried basil.

Animal feedstuffs

Straw, fodder and forage of cereal grains and grasses

Trials of residue levels in forage, fodder and straw after application of fludioxonil as seed treatment were conducted with wheat, rye, barley, maize, sweet corn and sorghum. Trials on wheat were conducted in

Fludioxonil

Europe according to the following relevant GAP: Austria, Belgium, United Kingdom, flowable concentrate formulation, 5 g ai/100 kg seed, one application. In the 45 trials conducted at or above GAP, the ranked order of residue levels in straw was: < 0.02 (eight), < 0.04 (14) and < 0.05 (23) mg/kg. The ranked order of residue levels in forage was: < 0.02 (seven) and < 0.04 (11) mg/kg.

A trial on rye was conducted in Denmark according to the GAP for Austria: flowable concentrate, 5 g ai/100 kg seed. The residue level was < 0.05 mg/kg in straw and < 0.05 mg/kg in forage.

Trials on barley were conducted in Europe according to the GAP for Austria, Belgium and the United Kingdom: flowable concentrate formulation, 5 g ai/100 kg seed. In five trials conducted at or above GAP, the ranked order of residue levels were: < 0.02 and < 0.05 (four) mg/kg in straw and < 0.05 (three) mg/kg in forage.

Trials on maize and sweet corn were conducted Europe (maize only) and in the USA. The relevant GAP was: USA, flowable concentrate formulation, 5 g ai/100 kg seed, as no GAP was available for any country in Europe. There were no detectable residues in fodder (< 0.01 (five) mg/kg) or forage (< 0.01 (seven) mg/kg). Using the default moisture content value of 40% for maize forage (*FAO Manual*, Appendix IX), the Meeting estimated a maximum residue level of 0.03 (*) mg/kg (0.01/0.40) and an STMR of 0 mg/kg (0.00/0.40) for maize forage (dry).

Field trials were conducted on sorghum in the USA, the relevant GAP being flowable concentrate formulation, 5 g ai/100 kg seed. The residue levels after exaggerated application rates were < 0.01 (four) mg/kg on fodder and < 0.01 (four) mg/kg on forage.

The combined values for fodder and straw in ranked order were: < 0.01 (nine), < 0.02 (nine), < 0.04 (four) and < 0.05 (28). As no data were available on the moisture content, the default value of 83% for maize fodder was used (*FAO Manual*, Appendix IX). The Meeting estimated an STMR of 0 mg/kg (0.00/0.83) and a maximum residue level of 0.06 (*) mg/kg (0.05/0.83) for fodder (dry) and straw of cereal grains.

Rape forage and straw

Fludioxonil was applied to rape as seed treatment in trials in France, Germany, Sweden and the United Kingdom. The GAP in Germany is treatment of seed with an flowable concentrate formulation at 12 g fludioxonil per 100 kg seed. The residue levels were < 0.05 mg/kg in forage (12) and straw (six).

Fate of residues during processing

A study of hydrolysis with [pyrrole-4-^{14}C]fludioxonil showed that fludioxonil is stable under the typical conditions of pasteurization, baking, brewing, boiling and sterilization.

The processing (transfer) factors through commercial-type processes for plums, strawberries, grapes, citrus and tomato are summarized in the table below. Factors could not be calculated for cereal grains, cottonseed or potatoes because there were no quantifiable residues in the raw agricultural commodities, even in trials with exaggerated treatment rates.

Processing factors and STMR-P values for various commodities

Raw agricultural commodity				Processed commodity			
Commodity	MRL (mg/kg)	STMR (mg/kg)	HR (mg/kg)	Commodity	Processing factor	STMR-P (mg/kg)	HR-P (mg/kg)
Plum[1]	5	0.80	3.6	Prunes (dried plums)	1.91[2]	0.96	4.3
				Juice	0.10	0.080	
				Preserves	0.50	0.40	

Raw agricultural commodity				Processed commodity			
Commodity	MRL (mg/kg)	STMR (mg/kg)	HR (mg/kg)	Commodity	Processing factor	STMR-P (mg/kg)	HR-P (mg/kg)
				Puree	0.80	0.64	
Strawberry	3	0.38	2.2	Juice	0.16	0.061	
				Preserves	0.62	0.24	
				Jam	0.34	0.13	
Grapes	2	0.28	1.6	Raisins (dried grapes)	1.1[3]	0.31	1.8
				Juice	0.92[4]	0.26	
				Wine (< 100 days)	0.30[5]	0.08	
				Wine (>100 days	0.036[6]	0.010	
Lemons				Juice	0.031		
				Oil	61		
				Pulp	2.1		
Tomato	0.5	0.12	0.32	Juice	0.22[7]	0.026	
				Paste	1.4[8]	0.17	
				Pomace (wet)	3.3[9]		

[1] Stone fruit, includes field trial data for cherries and peaches
[2] Four trials, range 1.8–2.7, mean 1.91, median 1.6
[3] 15 trials, range 0.58–1.7, mean 1.1, median 1.1
[4] 12 trials, range 0.58–1.0, mean 0.92, median 0.86
[5] 17 trials, range 0.012–0.86, mean 0.30, median 0.24
[6] 11 trials, range 0.0086–0.11, mean 0.036, median 0.029
[7] Four trials for pasteurized juice, range 0.20–0.24, average 0.22, median 0.22
[8] Four trials for pasteurized paste, range 1.1–1.6, average 1.4, median 1.35
[9] Two trials for wet pomace, 3.0 and 3.6

Residues in animal commodities

A feeding study was conducted in which three groups of three dairy cows received 0.55 ppm, 1.6 ppm or 5.5 ppm fludioxonil in the diet for 28–30 days. Residues of fludioxonil and metabolites, determined as CGA-192155 (2,2-difluorobenzo[1,1]dioxole-4-carboxylic acid), were quantifiable only at the highest feeding level (5.5 ppm). Residues were found in the milk of two of three cows, with maximum values of 0.019 mg/kg and 0.014 mg/kg on days 14 and 21, respectively. At the lowest feeding level, residues were detected in milk on days 3–21 at levels of 0.001–0.004 mg/kg, with maximum detection on day 3.

Only tissue samples from cows fed the 5.5 ppm diet were analysed. No residues of fludioxonil or metabolites were found. The LOQ was 0.01 mg/kg in muscle and 0.05 mg/kg in liver, kidney and fat (perirenal and omental).

The dietary intake of ruminants and poultry can be calculated from the recommended STMRs or HRs and consideration of possible animal feed items. The table below shows the bases for the dietary intake calculation.

# Fludioxonil

Commodity	Group	Maximum or highest residue level (mg/kg)	STMR or STMR-P (mg/kg)	Dry matter (%)	Dietary content (%) Beef cattle	Dietary content (%) Dairy cows	Dietary content (%) Poultry	Residue contribution (mg/kg) Beef cattle	Residue contribution (mg/kg) Dairy cows	Residue contribution (mg/kg) Poultry
Barley grain	GC	0.05		88	50	40	75			
Barley straw (dry)	AS	0.06			10	60				
Cotton-seed	SO	0.05		88	25	25				
Cotton-seed meal		0.05		89	15	15	20			
Maize grain	GC	0.05		88	80	40	80	0.02	0.03	0.05
					25	40				
Maize forage (dry)	AF	0.03			40	50				
Maize fodder (dry)	AS	0.06			25	15				
Oat grain	GC	0.05		89	50	40	80			
Oat straw (dry)	AS	0.06		90	10	10				
Potato waste	AB		0.01	15	75	40		0.05	0.03	
Rape meal		0.02		88	15	15	15			
Rape forage	AM	0.05		30	30	30				
						20				
Rye grain	GC	0.05			40	40	50			
Rye straw (dry)	AS	0.06		88	10	10				
Wheat grain	GC	0.05		89	50	40	80			
Wheat forage	AF	0.05		25						
Wheat straw (dry)	AS	0.06		88	10	10				
Pea seed	VD	0.07		90	20	20	20			0.02
Total					**100**	**100**	**100**	**0.07**	**0.06**	**0.07**

The calculated dietary intakes of beef cattle, dairy cows and poultry are 0.07, 0.06 and 0.07 mg/kg, respectively.

No quantifiable residue was found in the tissues of ruminants at levels 60 times (cows) and 80 times (beef cattle) the calculated dietary burden. Fludioxonil and metabolites were detected in liver and kidney at concentrations of 0.014–0.017 mg/kg and 0.022–0.025 mg/kg, respectively, at the 5.5 ppm feeding level. None was detected in fat or muscle. The Meeting concluded that the maximum residue level is the LOQ, 0.05 (*) mg/kg, for offal and 0.01 (*) mg/kg for muscle and that the STMR values for edible offal and muscle are both 0 mg/kg.

In milk, the highest residue level found was 0.019 mg/kg with the 5.5 ppm diet (60 times). The Meeting concluded that the maximum residue level is the LOQ, 0.01 (*) mg/kg, and that the STMR value for milk is 0 mg/kg.

No feeding study was available with poultry. The dietary intake calculation shows a possible burden of 0.07 ppm. The study of the nature of the residue in poultry was conducted at 89 ppm (1300 times) for 8 consecutive days. While short of the normal 30-day feeding study, the extreme exaggeration provides some idea of the likelihood of residues of fludioxonil and metabolites occurring in poultry commodities. The identified residue levels in eggs, liver, kidney, muscle and skin with fat were 0.26, 0.046, 0.020, 0.036 and 0.036 mg/kg, respectively. This strongly suggests that residues will not be quantifiable in these commodities at a feeding level of 0.07 mg/kg. Therefore, the Meeting estimated MRLs at the LOQ of 0.05 (*) mg/kg for eggs, 0.01 (*) mg/kg for poultry meat and 0.05 (*) mg/kg for poultry offal and STMRs of 0 mg/kg for eggs, poultry meat and poultry offal.

96   Fludioxonil

DIETARY RISK ASSESSMENT

Long-term intake

The IEDIs of fludioxonil based on the STMRs estimated for 45 commodities for the five GEMS/Food regional diets were 0–1% of the ADI (Annex 3). The Meeting concluded that the long-term dietary intake of residues of fludioxonil is unlikely to present a public health concern.

Short-term intake

The 2004 JMPR decided that an ARfD for fludioxinil is unnecessary. The Meeting therefore concluded that the short-term dietary intake of fludioxonil residues is unlikely to present a public health concern.

4.12   FOLPET (041)

TOXICOLOGY

Evaluation for an acute reference dose

Folpet is a fungicide used for the control of fungal diseases in crops. The Meeting prepared toxicological monographs on folpet in 1969 and 1995, and addenda to the monographs were prepared in 1973, 1984, 1986 and 1990. In 1995, an ADI of 0–0.1 mg/kg bw was established on the basis of a NOAEL of 10 mg/kg bw per day in a 2-year study of toxicity and carcinogenicity in rats, a 1-year study of toxicity in dogs, and studies of reproductive toxicity in rats and rabbits, and using a safety factor of 100. The present Meeting considered the requirement for an ARfD for folpet, based on data from the previous evaluations for JMPR and from new studies.

In rodents treated orally, folpet is rapidly degraded to phthalimide and thiophosgene (via thiocarbonyl chloride). Studies of metabolism in vitro with human blood revealed that folpet is rapidly degraded to phthalimide, with a calculated half-life of 4.9 s. Thiophosgene is rapidly detoxified by reaction with cysteine or glutathione, for example, and is ultimately rapidly excreted.

The acute oral toxicity of folpet in rats is low (LD$_{50}$, > 2000 mg/kg bw). In a study in pregnant hamsters, mortality occurred after a single dose of 400 mg/kg bw. In groups of pregnant New Zealand white rabbits treated for 3 days with folpet at a dose of 60 mg/kg bw per day, mortality was observed that may have been related to treatment.

Mice fed diets containing folpet at 5000 ppm, equal to 845 or 1060 mg/kg bw, for 24 h showed a reduction in food consumption of 10–20%. Immediately after the 24 h of treatment, minimal to moderate epithelial erosions and degeneration of the proximal duodenum were observed in some of these animals. Microscopy revealed moderate loss of villi and slight mucosal congestion in some animals. Microscopy was not performed at later time-points.

Studies of developmental toxicity with folpet have been carried out in hamsters, rats and rabbits.

In a study from the published literature, the teratogenic effects of a number of phthalimide derivatives, including folpet, were tested in pregnant golden hamsters. The Meeting noted that this study has major limitations (e.g. small number of animals per dose, limited reporting of the data) and is therefore of limited value. It does, however, suggest that developmental effects may occur after a single exposure to folpet, albeit at maternally toxic doses.

Folpet

Folpet has been tested in a number of studies of developmental toxicity in rats. In a study in Sprague-Dawley rats treated by gavage, the NOAEL for maternal toxicity was 60 mg/kg bw per day on the basis of reduced body-weight gain and food consumption and increased incidence of clinical signs at 360 mg/kg bw per day. The NOAEL for embryo- and fetotoxicity was 360 mg/kg bw per day, the highest dose tested. In another study in Sprague-Dawley rats treated by gavage, pregnant females received folpet at a dose of 0, 150, 550, or 2000 mg/kg bw on days 6–15 of gestation. On the basis of effects on body weight and food consumption, the NOAEL for maternal toxicity was 150 mg/kg bw per day. There was a slightly increased incidence of angulated ribs and the reduction in ossification of the interparietal bone at 150 mg/kg bw per day. The maternal toxicity and the associated fetal effects are likely to be caused by high local concentrations of folpet and are not considered to be relevant to dietary exposure. In a third study in Sprague-Dawley rats treated by gavage, the NOAEL for maternal toxicity was 100 mg/kg bw per day on the basis of a reduced body-weight gain in the group receiving the highest dose. As treatment had no effect on fetal growth and development, the NOAEL for developmental toxicity was 800 mg/kg bw per day, the highest dose tested. Therefore, in these three studies of developmental toxicity in rats, the overall NOAEL for maternal toxicity was 150 mg/kg bw per day on the basis of reduction of body-weight gain and food consumption. In two out of three studies, no fetal developmental anomalies were found at doses of up to 800 mg/kg bw per day. In one study, however, a possible slight increase in developmental anomalies was reported at 150 mg/kg bw per day.

Folpet has been tested in a number of studies of developmental toxicity in rabbits treated by gavage. In a study in which New Zealand white rabbits were given folpet at a dose of 0, 10, 20, or 60 mg/kg bw per day on days 6–28 of gestation, the NOAEL for maternal toxicity was 10 mg/kg bw per day on the basis of reduced body-weight gain and food consumption. The NOAEL for fetal toxicity was 10 mg/kg bw per day on the basis of reduced fetal body weights. The maternal toxicity and the associated reduction in fetal body weight are likely to be caused by high local concentrations of folpet and are not considered to be relevant to dietary exposure. At 60 mg/kg bw per day, there was a significant increase in the incidence of hydrocephaly in four fetuses out of three litters. In these same fetuses, skull, gastric, and pulmonary abnormalities were also observed. As the observation of hydrocephaly and cleft palate in one fetus at the intermediate dose was considered to be within the historical control range, the NOAEL for these effects was 20 mg/kg bw per day.

In a second study, HY/CR New Zealand white rabbits were given folpet at a dose of 0, 10, 40, or 160 mg/kg bw per day on days 7–19 of gestation. The NOAEL for maternal toxicity was 10 mg/kg bw per day on the basis of reductions in body-weight gain and in gravid uterine weight. The NOAEL for fetal toxicity was 10 mg/kg bw per day on the basis of an increased incidence of bilateral lumbar ribs and delayed skeletal maturation.

In a pulse-dose study, pregnant D1A Hra: (New Zealand white) rabbits were given folpet at a dose of 60 mg/kg bw per day by gavage on days 7–9, 10–12, 13–15, or 16–18 of gestation. There were occasional occurrences of abortion, but it was not clear whether these abortions were related to treatment with folpet. Maternal body weight and food consumption were significantly reduced in all treated animals. Two fetuses with hydrocephalus were observed, one in the group treated on days 10–12 of gestation and one in the group treated on days 16–18 of gestation. These incidences were considered to be within the historical control range. A significantly increased incidence (12.1%) of fetuses with an irregularly shaped fontanelle was observed in the group treated on days 13–15 of gestation; the incidence in controls was 4.5%. The significance of these effects was not clear.

Toxicological evaluation

Other than developmental effects, folpet produced no toxicological effects that might be considered to be a consequence of acute exposure. The Meeting concluded that it was not necessary to establish an ARfD for the general population, including children aged 1–6 years for whom separate data on dietary intake are available. The Meeting concluded that it might be necessary to establish an ARfD to protect the embryo or fetus from possible effects in utero. Such an ARfD would apply to women of childbearing age.

The maternal toxicity and the associated reductions in fetal body weight, delayed ossification and increased incidences in skeletal variations observed in studies of developmental toxicity in rabbits are likely to be caused by high local concentrations of folpet and are not considered to be relevant to dietary exposure. However, the increased incidence of hydrocephalus observed could not be attributed with confidence to maternal toxicity.

The Meeting concluded that the database was insufficient (in particular, with regard to the absence of studies on the developmental effects of phthalimide) to establish the mode of action by which the increased incidence of hydrocephalus, observed in rabbits at 60 mg/kg bw per day (NOAEL, 20 mg/kg bw per day) was induced, and as a consequence, their relevance for deriving an ARfD could not be dismissed. Therefore the Meeting established an ARfD of 0.2 mg/kg bw based on a NOAEL of 20 mg/kg bw per day for the increased incidence of hydrocephalus at 60 mg/kg bw per day in rabbits and a safety factor of 100. The use of a safety factor of 100 was considered to be conservative; although the mode of action by which the developmental effects are induced was uncertain, they are possibly secondary to maternal toxicity. The Meeting noted that it might be possible to refine this ARfD using the results of an appropriately designed study.

An addendum to the toxicological monograph was prepared.

Estimate of acute reference dose

0.2 mg/kg bw for women of childbearing age

Unnecessary for the general population

DIETARY RISK ASSESSMENT

Short-term intake

The Meeting set an ARfD of 0.2 mg/kg bw for folpet for women of childbearing age and decided that an ARfD was unnecessary for the general population, including children aged 1–6 years. Women of childbearing age are also part of the general population.

In the absence of relevant studies on the developmental effects of phthalimide (metabolite of folpet), the Meeting was unable to determine whether phthalimide should be excluded from the residue definition for dietary risk assessment. The Meeting was not able to finalize the risk assessment before an evaluation of the residue definition for risk assessment and associated residue values for dietary intake estimation had been completed.

4.13   GLYPHOSATE (158)

TOXICOLOGY

Glyphosate (*N*-(phosphonomethyl)glycine) is a non-selective systemic herbicide that was last evaluated by the JMPR in 1986, when an ADI of 0–0.3 mg/kg bw was established on the basis of a NOAEL of 31 mg/kg bw per day, the highest dose tested in a 26-month study of toxicity in rats. In 1997, JMPR evaluated aminomethylphosphonic acid (AMPA), the main metabolite of glyphosate, and concluded that AMPA was of no greater toxicological concern than its parent compound. A group ADI of 0–0.3 mg/kg bw was established for AMPA alone or in combination with glyphosate. Glyphosate was re-evaluated by the present Meeting within the periodic review programme of the CCPR. The Meeting reviewed new data on glyphosate that had not been reviewed previously and relevant data from the previous evaluations.

After oral administration to rats, [^{14}C]glyphosate was only partially absorbed (about 30–36%) from the gastrointestinal tract. Absorption was not significantly dose-dependent over the range of 10 to 1000 mg/kg bw. Peak plasma concentrations of radiolabel were observed at 0.5–1 h after dosing in rats and hens, respectively, and at 6–8 h after dosing in goats. The highest tissue concentrations were found in bone, with lower concentrations being found in bone marrow, kidney and liver. After oral administration, about 60–70% was eliminated in the faeces. Of the glyphosate that was absorbed, most was excreted in the urine and < 0.2% in expired air. After intravenous application, faecal excretion via bile was only about 2–8% of the administered dose. Whole body clearance (about 99% of an orally administered dose) occurred within approximately 168 h. The estimated half-life for whole-body elimination of the radiolabel was 2.1–7.5 h for the alpha phase and 69–337 h for the beta phase. Repeated dosing did not alter absorption, distribution and excretion. There was very little biotransformation of glyphosate; the only metabolite, AMPA, accounted for ≤ 0.7% of the administered dose in excreta; the rest was unchanged glyphosate.

Glyphosate has low acute oral toxicity in mice (LD$_{50}$, > 2000 mg/kg bw; no deaths at this dose) and rats (LD$_{50}$, > 5000,mg/kg bw), low acute dermal toxicity in rats (LD$_{50}$, > 2000 mg/kg bw) and rabbits (LD$_{50}$, > 5000 mg/kg bw), and low acute inhalation toxicity in rats (LC$_{50}$, > 4.43 mg/l of air). Clinical signs after acute oral exposure included reduced activity, ataxia and convulsions.

Glyphosate was not irritating to the skin, but produced moderate to severe eye irritation with irreversible corneal opacity in one study. Glyphosate salts were slightly irritating to the eye, with minimal to moderate conjunctival irritation and slight iritis that usually disappeared within 48 h after exposure. Glyphosate was not a skin sensitizer in guinea-pigs.

In short-term studies of toxicity in different species, the most important effects were clinical signs related to gastrointestinal irritation, salivary gland changes (hypertrophy and increase in basophilia of cytoplasm of acinar cells) and hepatotoxicity. In mice, reduced body-weight gain was seen at dietary concentrations of 25 000 ppm. Alterations of the salivary glands were present in mice in one of two short-term studies at dietary concentrations of ≥ 6250 ppm; the NOAEL for this finding was 3125 ppm (equal to 507 mg/kg bw per day). In rats, findings included soft faeces, diarrhoea, reduced body-weight gain, decreased food utilization and slightly increased plasma enzyme activities (alkaline phosphatase, alanine aminotransferase) at dietary concentrations of ≥ 20 000 ppm. Additionally, in two out of four 90-day studies in rats, increased incidences of alterations of the salivary glands were observed. At the lower doses, the severity and incidence of these changes were only minimal. The overall NOAEL was 300 mg/kg bw per day.

In dogs, the NOAEL in a 90-day feeding study was 10 000 ppm (equal to 323 mg/kg bw per day) on the basis of reduced body-weight gain, marginal reductions in albumin and calcium concentrations, and increased plasma alkaline phosphatase activities at 50 000 ppm. In a 1-year study in dogs given capsules containing glyphosate, the NOAEL was 30 mg/kg bw per day, on the basis of clinical signs (soft faeces, diarrhoea) and reduced body-weight gain at ≥ 300 mg/kg bw per day. In a 1-year feeding study, the NOAEL was 15 000 ppm (equal to 440 mg/kg bw per day) on the basis of reduced body-weight gain at 30 000 ppm.

Long-term studies of toxicity and carcinogenicity were conducted in mice and rats. In the study of carcinogenicity in mice, no toxic effects were observed at up to the highest dose tested (1000 mg/kg bw per day), and there was no evidence of carcinogenicity.

In a 1-year study of toxicity in rats, the NOAEL was 2000 ppm (equal to 141 mg/kg bw per day) on the basis of a reduction in body weight and clinical chemistry findings at 8000 ppm. Three new long-term studies in rats were evaluated. In the first study, the NOAEL was 8000 ppm (equal to 362 mg/kg bw per day) on the basis of a reduction in body weight in females and an increased incidence of cataracts and lens abnormalities in males at 20 000 ppm. In the second study, the NOAEL was 100 mg/kg bw per day on the basis of more pronounced alterations of the parotid and submaxillary salivary glands at ≥ 300 mg/kg bw per day. In the most recent 2-year study in rats, the NOAEL was 6000 ppm (equal to 361 mg/kg bw per day) on the basis of a reduction in body weight and food consumption and indications of kidney, prostate and liver toxicity at 20 000 ppm. There was no evidence of a carcinogenic response to treatment in rats.

The genotoxic potential of glyphosate has been extensively tested in a wide range of assays both in vitro and in vivo, including end-points for gene mutation, chromosomal damage and DNA repair. Negative results were obtained in studies performed in compliance with current test guidelines. The Meeting concluded that glyphosate is unlikely to be genotoxic.

In view of the absence of a carcinogenic potential in animals and the lack of genotoxicity in standard tests, the Meeting concluded that glyphosate is unlikely to pose a carcinogenic risk to humans.

Glyphosate had no effect on fertility in either two-generation study of reproductive toxicity in rats. The overall NOAEL for parental and offspring toxicity was 3000 ppm (equal to 197 mg/kg bw per day) on the basis of increased food and water consumption and reduced body-weight gain in F_1 animals, and increased incidences of alterations of parotid and submaxillary salivary glands in parental and F_1 animals at 10 000 ppm.

In studies of developmental toxicity in rats, the NOAEL for maternal and developmental toxicity was 300 mg/kg bw per day, on the basis of clinical signs and reduced body-weight gain in the dams and increased incidences of fetuses with delayed ossification and skeletal anomalies.

In studies of developmental toxicity in rabbits, the NOAEL for maternal toxicity was 100 mg/kg bw per day on the basis of clinical signs and reduced food consumption and body-weight gain. The NOAEL for developmental toxicity was 175 mg/kg bw per day on the basis of reduced fetal weight and delayed ossification and an increased incidence of postimplantation loss. The Meeting concluded that glyphosate is not teratogenic. The Meeting concluded that the existing database on glyphosate was adequate to characterize the potential hazard to fetuses, infants and children.

Hypertrophy and cytoplasmic alterations of the salivary glands (parotid and/or mandibular) was a common and sensitive end-point in six studies: in three 90-day studies (one in mice, two in rats), a 1-year study in rats, a 2-year study in rats and a two-generation study of reproductive toxicity in rats. Mechanistic studies available to the Meeting hypothesized that the mechanism was adrenergic. However, the inability of a β-blocker to significantly inhibit these effects indicates that glyphosate does not act as a β-agonist. Other proposed mechanisms for the salivary gland alterations include oral irritation caused by dietary administration of glyphosate, a strong organic acid. Although the mechanism of the cytoplasmic alterations in the salivary glands was unclear, the Meeting concluded that this treatment-related effect is of unknown toxicological significance.

In a study of acute neurotoxicity in rats, the NOAEL for neurotoxicity was 2000 mg/kg bw, the highest dose tested. In a short-term study of neurotoxicity in rats, the NOAEL for neurotoxicity was 20 000 ppm, equal to 1547 mg/kg bw per day, the highest dose tested. In a study of acute delayed peripheral neuropathy in hens, clinical and histopathological examination found no evidence for acute delayed peripheral neuropathy at a dose of 2000 mg/kg bw.

New toxicological data on AMPA (the primary degradation product of glyphosate in plants, soil and water, and the only metabolite of glyphosate found in animals) was submitted to the present Meeting for evaluation. AMPA was of low acute oral and dermal toxicity in rats (LD_{50}, > 5000 and > 2000 mg/kg bw, respectively), and was not a skin sensitizer in guinea-pigs. In a 90-day study of toxicity in rats, the NOAEL was 1000 mg/kg bw per day, the highest dose tested. AMPA had no genotoxic potential in vitro or in vivo. In a study of developmental toxicity in rats, no evidence for embryo- or fetotoxicity was found and the NOAEL for maternal and developmental toxicity was 1000 mg/kg bw per day, the highest dose tested.

On the basis of the new toxicological data, the present Meeting concluded that AMPA is of no greater toxicological concern than its parent compound, thus confirming the conclusion of the 1997 JMPR.

Routine medical surveillance of workers in production and formulation plants revealed no adverse health effects attributable to glyphosate. In operators applying glyphosate products, cases of eye, skin and/or respiratory tract irritation have been reported. Acute intoxication was reported in humans after accidental or intentional ingestion of concentrated glyphosate formulations, resulting in gastrointestinal, cardiovascular,

Glyphosate

pulmonary and renal effects and occasionally death. The acute toxicity of glyphosate formulations was likely to be caused by the surfactant in these products.

Toxicological evaluation

The Meeting established a group ADI for glyphosate and AMPA of 0–1.0 mg/kg bw on the basis of the NOAEL of 100 mg/kg bw per day for salivary gland alterations in a long-term study of toxicity and carcinogenicity in rats and a safety factor of 100. The ADI is supported by NOAELs of 141 and 197 mg/kg bw per day from the 1-year study and the two-generation study of reproductive toxicity in rats, respectively.

The Meeting concluded that it was not necessary to establish an ARfD for glyphosate in view of its low acute toxicity, the absence of relevant developmental toxicity in rats or rabbits as a consequence of acute exposure and the absence of any other toxicological effect that would be elicited by a single dose.

The NOAEL of 30 mg/kg bw per day in a 1-year study in dogs was not considered relevant for establishing either the ADI or ARfD, since the gastrointestinal effects seen in this study at 300 and 1000 mg/kg bw per day were related to high local concentrations of test substance resulting from the administration of glyphosate in capsules.

An addendum to the toxicological monograph was prepared.

Levels relevant to risk assessment

Species	Study	Effect	NOAEL	LOAEL
Mouse	3-month study of toxicity[a,e]	Toxicity	3125 ppm, equal to 507 mg/kg bw per day	6250 ppm, equal to 1065 mg/kg bw per day
	2-year study of carcinogenicity[a]	Toxicity	1000 mg/kg bw per day[d]	—
		Carcinogenicity	1000 mg/kg bw per day[d]	—
Rat	3-month study of toxicity[a,e]	Toxicity	300 mg/kg bw per day	12 500 ppm, equal to 811 mg/kg bw per day
	1-year study of toxicity[a]	Toxicity	2000 ppm, equal to 141 mg/kg bw per day	8000 ppm, equal to 560 mg/kg bw per day
	2-year study of toxicity and carcinogenicity[a,e]	Toxicity	100 mg/kg bw per day	300 mg/kg bw per day
		Carcinogenicity[d]	20 000 ppm, equal to 1214 mg/kg bw per day[d]	—
	Multigeneration reproductive toxicity[a,e]	Parental toxicity	3000 ppm, equal to 197 mg/kg bw per day	10 000 ppm, equal to 668 mg/kg bw per day
		Offspring toxicity	3000 ppm, equal to 197 mg/kg bw per day	10 000 ppm, equal to 668 mg/kg bw per day
	Developmental toxicity[b,e]	Maternal toxicity	300 mg/kg bw per day	1000 mg/kg bw per day
		Embryo- and fetotoxicity	300 mg/kg bw per day	1000 mg/kg bw per day
Rabbit	Developmental toxicity[b,e]	Maternal toxicity	100 mg/kg bw per day	150 mg/kg bw per day
		Embryo- and fetotoxicity	175 mg/kg bw per day	300 mg/kg bw per day
Dog	3-month study of toxicity[a]	Toxicity	10 000 ppm, equal to 323 mg/kg bw per day	50 000 ppm, equal to 1680 mg/kg bw per day
	1-year study of toxicity[a,c,e]	Toxicity	30 mg/kg bw per day[c,f]	300 mg/kg bw per day[c]

[a] Diet
[b] Gavage
[c] Capsules
[d] Highest dose tested
[e] Two or more studies combined
[f] Not used for establishing the ADI (or ARfD) since the NOAEL is based on an effect induced by high local concentrations

Estimate of acceptable daily intake for humans
    0–1.0 mg/kg bw

Estimate of acute reference dose
    Unnecessary

Studies that would provide information useful for continued evaluation of the compound
— Additional information on the mechanism of the changes in the salivary glands
— Further observations in humans

Critical end-points for setting guidance values for exposure to glyphosate

Absorption, distribution, excretion and metabolism in animals

Rate and extent of oral absorption	Rapid, approximately 30–36%
Dermal absorption	No information
Distribution	Widely distributed
Rate and extent of excretion	Largely complete within 48 h; approximately 30% in urine and 70% in faeces
Potential for accumulation	No evidence of accumulation (< 1% after 7 days)
Metabolism in mammals	Very limited (< 0.7%), hydrolysis leading to AMPA
Toxicologically significant compounds (animals, plants and the environment)	Parent compound, AMPA

Acute toxicity

Rat, LD_{50}, oral	> 5000 mg/kg bw
Rat, LD_{50}, dermal	> 2000 mg/kg bw
Rat, LC_{50}, inhalation	> 4.43 mg/l (4-h, nose-only exposure)
Rabbit, dermal irritation	Non-irritant
Rabbit, eye irritation	Moderately to severely irritant
Skin sensitization	Not sensitizing (Magnusson and Kligman test, Buehler test)

Short-term toxicity

Target/critical effect	Clinical signs (soft faeces, diarrhoea), reduced body-weight gain; liver (toxicity), salivary glands (hypertrophy)
Lowest relevant oral NOAEL	300 mg/kg bw per day (90-day study in rats)
Lowest relevant dermal NOAEL	—
Lowest relevant inhalation NOAEC	—

Genotoxicity

    No genotoxic potential

Long-term studies of toxicity and carcinogenicity

Target/critical effect	Reduced body-weight gain; liver (toxicity), salivary glands (hypertrophy), eye (cataract, lens fibre degeneration)
Lowest relevant NOAEL	100 mg/kg bw per day (2-year study in rats)

Glyphosate

Carcinogenicity	No evidence of carcinogenicity in rats or mice
Reproductive toxicity	
Reproductive target/critical effect	Reduced pup weight at parentally toxic doses
Lowest relevant reproductive NOAEL	197 mg/kg bw per day (two-generation study in rats)
Developmental target/critical effect	Embryo- and fetotoxicity at maternally toxic doses (rat, rabbit)
Lowest relevant developmental NOAEL	175 mg/kg bw per day (rabbit)
Neurotoxicity/delayed neurotoxicity	No evidence of neurotoxicty in any study conducted
Medical data	Medical surveillance of workers in plants producing and formulating glyphosate plants did not reveal any adverse health effects. In operators applying glyphosate products, cases of eye, skin and/or respiratory irritation have been reported. Cases of acute intoxication have been observed after accidental or intentional ingestion of glyphosate formulations.

Summary	Value	Study	Safety factor
ADI[a]	0–1.0 mg/kg bw	2-year study in rats (salivary gland effects)	100
ARfD	Unnecessary	—	—

[a] For the sum of glyphosate and AMPA

DIETARY RISK ASSESSMENT

Long-term intake

Estimated theoretical maximum daily intakes for the five GEMS/Food regional diets, based on recommended MRLs, were 1% of the ADI (Annex 3). The Meeting concluded that the long-term intake of residues of glyphosate resulting from uses that have been considered by the JMPR is unlikely to present a public health concern.

Short-term intake

The 2004 JMPR decided that an ARfD for glyphosate is unnecessary. The Meeting therefore concluded that the short-term intake of glyphosate residues is unlikely to present a public health concern.

4.14 MALATHION (049)

RESIDUE AND ANALYTICAL ASPECTS

Malathion has been evaluated many times since 1965. The company asked the CCPR at its Thirty-third Session in 2001 to reconsider withdrawal of the existing Codex MRLs recommended during the periodic review of the compound by the 1999 JMPR. The CCPR at its Thirty-sixth Session decided to retain the current CXL for apple, broccoli, cabbage (head), cereal grains, citrus fruit and grape, awaiting the review of the new residue data by the 2004 JMPR. The company submitted data on mandarin, orange, apple, peach, grape, strawberry, tomato, alfalfa fodder and forage (green) to the present Meeting.

Results of supervised trials on crops

Citrus fruit

Sixteen trials were conducted with mandarin and orange in Greece (GAP for citrus, 0.044–0.088 kg ai/hl, 7-day PHI), Italy (GAP, 0.053 kg ai/hl, 20-day PHI) and Spain (GAP for foliar application, 0.11–0.15 kg ai/hl, 7-day PHI) between 2000 and 2002. Malaoxon was analysed in all trials, and the maximum residue level found in fruit was 0.09 mg/kg on the day of the last application.

Ten trials conducted at 0.18 kg ai/hl and five trials performed at 0.30 kg ai/hl were evaluated against Spanish and Portuguese GAP for citrus (0.05–0.30 kg ai/hl, 7-day PHI), respectively. Decline studies conducted in both crops indicated that residue levels decreased slowly within 3 days of the last application, and samples harvested from 7 up to 10–11 days could be considered at GAP.

The residue levels in *mandarin* fruit were, in ranked order: 0.75, 1.8 (three), 2.4 (two), 2.9, 3.1 and 4.7 mg/kg. In mandarin pulp, the levels were: < 0.01, 0.01, 0.02, 0.03, 0.12, 0.15, 0.21 and 0.22 mg/kg.

The residue levels in *orange* fruit were, in ranked order: 0.58, 0.75, 0.89, 1.1, 1.4, 1.6, 1.7 and 2.1 mg/kg. The levels in orange pulp were: < 0.01 (three), 0.01 (two), 0.03 and 0.07 (two) mg/kg. One trial conducted at 0.4 kg ai/hl gave residue levels within the same range.

The Meeting agreed that the levels of residues of malathion in mandarin and orange from trials conducted at GAP could be combined to represent a residue population for citrus. In fruit, the levels were, in ranked order: 0.58, 0.75 (two), 0.89, 1.1, 1.4, 1.6, 1.7, 1.8 (three), 2.1, 2.4 (two), 2.9, 3.1 and 4.7 mg/kg. In citrus pulp, the levels were: < 0.01 (four), 0.01 (three), <u>0.02</u>, 0.03, 0.07 (two), 0.12, 0.15, 0.21 and 0.22 mg/kg.

The Meeting recommended a maximum residue level of 7 mg/kg, a STMR of 0.02 mg/kg and a highest residue levl of 0.22 mg/kg for malathion in citrus.

Apple

Twelve trials were conducted on apples in northern France in 2000 and 2001, with three applications at 0.18 kg ai/hl. There is no GAP for malathion in France; however, the results can be evaluated against Spanish or Italian GAP (up to 0.16 kg ai/hl). The PHI is 7 days in Spain and 20 days in Italy. The residue levels in apple fruit at the most critical PHI (7 days) were, in ranked order: 0.02, 0.05 (two), 0.07, 0.08, <u>0.09</u>, <u>0.13</u>, 0.14, 0.19, 0.24, 0.25 and 0.37 mg/kg. The levels of malaoxon were < 0.01–0.02 mg/kg.

The Meeting recommended a maximum residue level of 0.5 mg/kg, an STMR of 0.11 mg/kg and an HR of 0.37 mg/kg for malathion in apple.

Peach

Four trials were conducted in Italy in 1997 on peaches, with three applications of 0.16 kg ai/hl, corresponding to GAP. The residue levels of malathion in the fruit at 20 days' PHI were < 0.01 (three) and 0.01 mg/kg. No malaoxon was found in the fruit after the last application (0–20 days).

The Meeting agreed that four trials is not enough to recommend a maximum residue level for malathion in peaches. The Meeting confirmed the previous recommendation of 1999 JMPR to withdraw the CXs for malathion in peaches

Grape

Six trials were conducted in grapes in southern and northern France (no GAP) in 2000–01, with three applications at 0.27 kg ai/hl. The residue levels of malathion 7 days after the last application were 0.26–1.3 mg/kg.

In four trials conducted in northern Spain in 2000–01 within maximum GAP (0.11–0.15 kg ai/hl, 7-day PHI), the residue levels of malathion at 7 days' PHI were: 1.1, 1.5 (two) and 2.6 mg/kg. The levels of

Malathion 105

malaoxon were 0.02–0.13 mg/kg. Four trials conducted in Italy at GAP (0.05–0.16 kg ai/hl, 20-day PHI), evaluated against Spanish GAP, showed levels of 0.01, 0.03 and 0.21 (two) mg/kg.

The residue levels in trials conducted in Italy and Spain according to GAP were 0.01, 0.03, <u>0.21</u> (two), <u>1.1</u>, 1.5 (two) and 2.6 mg/kg.

The Meeting recommended a maximum residue level of 5 mg/kg, a STMR of 0.16 mg/kg and a highest residue level of 2.6 mg/kg for malathion grapes.

Strawberry

GAPs for malathion on strawberry in Europe are: Denmark, three times 1.1 kg ai/ha, 7-day PHI; Italy, up to 0.16 kg ai/hl, 20-day PHI; Portugal, 0.10 kg ai/hl, 1-day PHI; and Spain, up to 0.15 kg ai/hl, 7-day PHI. In three trials conducted in France, two in Italy and three in Spain with four or six applications at 0.15–0.18 kg ai/hl, the residue levels at 7 days' PHI were < 0.01–0.14 mg/kg.

The Meeting confirmed the currently recommended maximum residue level of 1 mg/kg of malathion in strawberry, which was set by the 1999 JMPR on the basis of trials conducted in the USA according to GAP with a 1-day PHI.

Tomato

GAP for malathion in tomato is up to 0.088 kg ai/hl with a 7-day PHI in Greece, up to 0.16 kg ai/hl with a 20-day PHI in Italy and up to 0.10 kg ai/hl with a 1-day PHI in Portugal. In five trials conducted in Italy and one in Spain in 1997–2001 according to Greek GAP, the residue levels were < 0.01 (three), 0.01 and 0.02 (two) mg/kg. Four other trials were conducted in these countries at higher or lower rates than GAP.

The Meeting confirmed the currently recommended maximum residue level of 0.5 mg/kg for malathion in tomato, which was set by the 1999 JMPR on the basis of trials conducted in the USA according to GAP with a 1-day PHI.

Alfalfa

Two trials were conducted in Italy in 2000–01 (no GAP) and two in Spain (GAP, 0.12–0.25 kg ai/hl or 1.5–2.8 kg ai/ha) with one application at 0.15 kg ai/hl (1.5 kg ai/ha). The Italian trials, evaluated against Spanish GAP, gave residue levels of malathion of 0.67 and 0.41 mg/kg on a dry weight basis in alfalfa forage (green) at 7 days' PHI, and 0.12 mg/kg in hay harvested at 7 days' PHI and allowed to dry on the field for 3 days. In the Spanish trials, the residue levels were 1.2 and 3.5 mg/kg in forage and 3.3 mg/kg in hay.

The Meeting confirmed the currently recommended maximum residue levels for malathion on alfalfa forage (green) and alfalfa hay of 500 and 200 mg/kg, respectively, which were set by the 1999 JMPR on the basis of trials conducted in USA according to GAP at 1 day PHI.

DIETARY RISK ASSESSMENT

Long-term intake

The current ADI for malathion is 0.3 mg/kg bw. IEDIs were calculated for commodities for human consumption for which STMRs were estimated by the 1999 JMPR and by the present Meeting (Annex 3). The IEDI for the five GEMS/Food regional diets was 0% of the maximum ADI. The Meeting concluded that long-term intake of residues of malathion resulting from uses considered by the JMPR is unlikely to present a public heath concern.

Short-term intake

An ARfD for malathion of 2 mg/kg bw was established by the 2003 JMPR. The IESTIs of malathion by the general population and by children were calculated for commodities for which STMR and highest

4.15 METALAXYL M (212)

RESIDUE AND ANALYTICAL ASPECTS

The toxicology of metalaxyl-M was evaluated by the 2002 JMPR, which established a group ADI of 0–0.08 mg/kg bw for metalaxyl and metalaxyl-M. Residue and analytical aspects were considered for the first time by the present Meeting. Metalaxyl-M is the biologically active enantiomer (R-enantiomer) of the racemic compound metalaxyl. Metalaxyl was first evaluated by the JMPR in 1982, and Codex MRLs for metalaxyl were established.

Metalaxyl-M is registered for use on fruit, nut and vegetable crops for the control of various fungal diseases such as those caused by *Phytophthora* and *Pythium* spp. It is applied to foliage, soil or seed and also as a post-harvest fruit treatment.

IUPAC name	(R)-2-[(2,6-dimethylphenyl)-methoxyacetylamino]-propionic acid methyl ester
Chemical Abstracts name	*N*-(2,6-dimethylphenyl)-*N*-(methoxyacetyl)-D-alanine methyl ester

The Meeting received information on the metabolism and environmental fate of metalaxyl-M and on methods of residue analysis, stability in freezer storage, national registered use patterns, the results of supervised trials and farm animal feeding studies, the fate of residues in processing and national MRLs.

As metalaxyl-M constitutes 50% of metalaxyl, investigations into the metabolism and fate of metalaxyl can legitimately be accepted as supporting the metabolism and fate of metalaxyl-M. When the metabolism of metalaxyl and metalaxyl-M was compared directly, it was found to be similar.

In the studies of animal and plant metabolism and environmental fate, metalaxyl or metalaxyl-M uniformly ^{14}C labelled in the aromatic ring was used.

Metabolism

In the list below, the numbering is preserved from the 2002 JMPR toxicology evaluation.

Metabolite 1: *N*-(2,6-dimethylphenyl)-*N*-(methoxyacetyl)alanine
Metabolite 3: *N*-(2,6-dimethylphenyl)-*N*-(hydroxyacetyl)alanine methyl ester
Metabolite 6: *N*-(2,6-dimethylphenyl)-*N*-(hydroxyacetyl)alanine
Metabolite 7: *N*-(2,6-dimethyl-5-hydroxyphenyl)-*N*-(methoxyacetyl)alanine methyl ester
Metabolite 8: *N*-(2-hydroxymethyl-6-methylphenyl)-*N*-(methoxyacetyl)alanine methyl ester (occurs as two isomers)
Metabolite P: *N*-[(2-hydroxymethyl)-6-methylphenyl]-*N*-(hydroxyacetyl)alanine (occurs as two isomers)

Animals

The Meeting received the results of studies on metabolism in rats, lactating goats and laying hens. When animals were dosed orally with radiolabelled metalaxyl, most of the radiolabel was excreted in the urine within a short time, with a small amount in the faeces. Numerous metabolites resulting from hydrolysis, oxidation and demethylation of metalaxyl and subsequent conjugate formation were identified. In a study in goats, metalaxyl itself was not detected as a component of the residue in tissues or milk. In a study in laying hens, low levels of metalaxyl were present in liver and eggs. The metabolic pathway for metalaxyl was similar in rats, goats and hens.

The absorption, distribution, metabolism and excretion of radiolabel were similar in *rats* dosed orally with ^{14}C-metalaxyl or ^{14}C-metalaxyl-M. Detailed information on metabolism in this species is reported in the 2002 JMPR toxicological evaluations.

Very little radiolabel was found in milk (0.003 mg/kg) or tissues (0.057 mg/kg in liver) from a *goat* dosed with ^{14}C-metalaxyl at the equivalent of 7 ppm in the feed for 10 days.

When two lactating dairy goats were dosed orally once daily for 4 consecutive days by gelatin capsule with ^{14}C-metalaxyl, equivalent to 77 ppm in the diet, the radiolabel was excreted rapidly: within 24 h of administration, 67% of the daily dose appeared in urine, 9% in faeces and 0.1% in milk. Metalaxyl was not detected as a component of the residue. Metabolite 6 was the main component of the residue in liver (0.19 mg/kg), leg muscle (0.014 mg/kg) and perirenal fat (0.065 mg/kg); metabolite 8 was the main residue component in kidney. The main metabolites in milk were C-10 and C-8 fatty acid conjugates of metabolite 3 (0.058 mg/kg). These fatty acids are conjugated through the hydroxyacetyl group of metabolite 3.

When five laying *hens* were dosed orally once daily for 4 consecutive days by gelatin capsule with ^{14}C-metalaxyl, equivalent to approximately 100 ppm metalaxyl in the diet, radiolabel recovered in edible tissues and eggs represented 0.97% of the administered dose; the remainder was recovered in excreta. Metabolite P (consisting of P1 and P2, steric isomers) was the main metabolite in egg white (0.056 mg/kg), egg yolk (0.072 mg/kg) and thigh muscle (0.31 mg/kg). Metalaxyl parent was identified in egg white (0.013 mg/kg), egg yolk (0.010 mg/kg) and liver (0.018 mg/kg), but not in thigh muscle or fat (< 0.001 mg/kg).

Plants

The Meeting received the results of studies on the metabolism of metalaxyl in grape, lettuce and potato and of metalaxyl-M in lettuce. No metabolites were identified in plants which had not already been identified in animals. Parent metalaxyl was the main component of the residue in grapes and in juice produced from the grapes when metalaxyl was used on grape vines. In treated lettuce, parent metalaxyl and metabolite 8 were each present at approximately 20% of the total residue. Metabolite 8 was the main residue component in lettuce in both cases in which metalaxyl and metalaxyl-M were compared. When metalaxyl was used on potato plants, some residue reached the tubers, where parent metalaxyl was the main residue component.

Grapevines in Switzerland were sprayed to runoff seven times at 14-day intervals with a ^{14}C-metalaxyl spray at a concentration of 0.050 kg ai/hl and were harvested 52 days after the final application. Parent metalaxyl (2.0 mg/kg) constituted 64% of the total residues in *grapes*.

When two grapevines in Switzerland were sprayed to runoff six times at approximately 14-day intervals with a ^{14}C-metalaxyl spray at a concentration of 0.030 kg ai/hl and harvested 68 days after the final application, parent metalaxyl (0.83 mg/kg) comprised 64% of total residues in the grapes. Metabolite 8 accounted for 20% of the residue, and metabolites 1, 6 and 7 were minor components (1.8–4.3%). When the grapes were separated into juice and presscake, metalaxyl was still the main part of the residue (62% and 57%, respectively).

Metalaxyl was the main identified component of the residue in *lettuce* (18.6% of the total ^{14}C residue) after seedlings in a greenhouse were sprayed twice, 2 weeks apart, with ^{14}C-metalaxyl at a rate equivalent to 0.25 kg ai/ha and harvested 2 weeks later. The identified metabolites (including glucose conjugates) were metabolite 8 (22% of the total ^{14}C residue), metabolite 6 (10%), metabolite 3 (8.9%), metabolite 7 (6.2%) and metabolite 1 (6.0%).

The metabolic pathways of metalaxyl-M and metalaxyl were compared in lettuce in a field in Switzerland treated three times at 10-day intervals with labelled compounds. The levels of total applied residue and parent compounds in the residue were generally comparable. Metabolite 8 (free and conjugated) was the main identified component of the residue in samples taken 14 and 21 days after treatment. Enantiomeric ratio measurements suggested similar disappearance rates for the two enantiomers and little interconversion.

Potato plants in the field in Switzerland received five foliar treatments of ^{14}C-metalaxyl at 0.2 kg ai/ha at 10-day intervals and were harvested at maturity 5 weeks after the final treatment. Little residue reached the tubers (0.02 mg/kg ^{14}C as metalaxyl). No parent metalaxyl was detected in tubers. In a second experiment, the level of ^{14}C as metalaxyl in tubers was < 0.0001 mg/kg after ^{14}C-metalaxyl was applied to the soil (residues in soil, approximately 0.5 mg/kg), indicating that metalaxyl is not taken up by the tubers directly from the soil.

Potato plants in the field in the USA received six foliar treatments, about 2 weeks apart, of ^{14}C-metalaxyl at 1.3 kg ai/ha. Tubers harvested at maturity, 1 week after the final treatment, contained 0.5 mg/kg ^{14}C as metalaxyl, of which 50–60% was parent metalaxyl. A number of metabolites were identified, but only the concentration of metabolite 8 exceeded 5% of the residue.

Environmental fate

Soil

The Meeting received information on the behaviour and fate of metalaxyl and metalaxyl-M during aerobic metabolism in a number of soils. The rate of degradation is strongly influenced by the properties of the soil, including its biological activity and the conditions of temperature, moisture and concentration of the residue, with recorded half-lives in the range of 5–180 days. In direct comparisons of metalaxyl and metalaxyl-M under aerobic conditions, metalaxyl-M was the more persistent in one case and less persistent in two others. The main soil metabolite is metabolite 1 or, in the case of metalaxyl-M, the specific enantiomer of metabolite 1.

Field dissipation studies for metalaxyl-M were provided from France, Italy, Spain and Switzerland. Metalaxyl-M residues disappeared from the soil with half-lives ranging from 5 to 35 days. The residues occurred mostly in the top 10 cm of soil, but some reached lower levels. The enantiomer of metabolite 1 was produced in all cases, and its level sometimes exceeded that of the parent metalaxyl-M. A comparison of enantiomeric ratios in metalaxyl residues in soil suggested that the R-enantiomer (i.e. the metalaxyl-M enantiomer) disappeared more quickly than the S-enantiomer. This resulted in a preponderance of S-enantiomer in the metalaxyl residue and a preponderance of R-enantiomer in metabolite 1.

The studies of dissipation in the field suggest that, after use of metalaxyl-M for seed treatment or at the time of sowing, little or none will remain as a soil residue when the crop is harvested.

Rotational crops

Information on the fate of radiolabelled metalaxyl in confined crop rotational studies and of unlabelled metalaxyl-M in field rotational crops was made available to the Meeting. The studies with radiolabel showed that parent metalaxyl was usually a minor part of the residue that reached the rotational crop. The identifiable metabolites were also usually minor, but metabolite 8 as glucose conjugates was detected in spring wheat stalks at 2.3 mg/kg. Metalaxyl-M residues were not detected in unconfined field rotational crops in Switzerland or the United Kingdom, but levels of 0.11 mg/kg were present in broccoli and 0.03 mg/kg in lettuce leaves from crops sown 29 days after treatment of the first crop in a study in Italy. The short interval was used in order to simulate the ploughing-in of a failed crop and the sowing of a new one.

Methods of analysis

The Meeting received descriptions and validation data for analytical methods for residues of metalaxyl in plant material, animal tissues, milk and eggs.

Common moiety methods rely on the 2,6-dimethylaniline moiety of metalaxyl and many of its metabolites, and these methods have been used to identify metalaxyl residues in animal commodities. The typical LOQs are 0.05 mg/kg for tissues and 0.01 mg/kg for milk. Metabolite 8, containing the 2-hydroxymethyl-6-methylaniline moiety, is apparently partially converted to 2,6-dimethylaniline, resulting in low and variable recoveries.

With gas–liquid chromatography and nitrogen–phosphorus detection and HPLC with mass spectrometry detection procedures for identifying metalaxyl or metalaxyl-M after a simple extraction and limited clean-up, the LOQs are 0.02–0.04 mg/kg for many crop substrates. A modification to the method (method REM 181.06), with the introduction of an HPLC chiral separation step before determination, allows for the analysis of specific enantiomers.

A multi-residue regulatory method (DFG S19) is available for metalaxyl.

Method REM 181.06 (gas–liquid chromatography with mass spectrometry detection) is not a multi-residue method, but it is enantioselective and suitable as a regulatory method for metalaxyl-M.

Stability of residues in stored analytical samples

The Meeting received information on the stability of residues of metalaxyl-M in crops (orange, potato, rape-seed, tomato, wheat) and animal commodities (beef muscle, beef liver, milk, eggs) during storage of analytical samples. Metalaxyl-M residues were stable in these substrates and under the conditions and intervals of storage (2 years). There was no evidence of epimerization during freezer storage. As a common moiety method was used for the animal commodity samples, storage stability refers to the total residue rather than to parent metalaxyl-M. As the common moiety method, which relies on the 2,6-dimethylaniline moiety, is less suitable for metabolite 8, the freezer stability of this metabolite during storage is not demonstrated.

Definition of the residue

Parent metalaxyl is the main identifiable component of the residue in crops resulting from use of metalaxyl, although metabolite 8 can occur at approximately the same levels. Metabolite 8 was not considered to be toxicologically significant.

The current residue definition of metalaxyl is metalaxyl. As metalaxyl-M is one enantiomer of metalaxyl, it is covered by the current residue definition. Non-enantioselective methods cannot distinguish metalaxyl-M from metalaxyl, but an enantioselective method is available. While metalaxyl-M and metalaxyl are both registered for crop uses, it is preferable, for enforcement purposes, to maintain a single residue definition. As the 2002 JMPR recommended a group ADI for metalaxyl and metalaxyl-M, the inclusive residue definition is also suitable for risk assessment purposes. The Meeting recommended that metalaxyl-M be contained within the metalaxyl residue definition and recommended amendment of the metalaxyl residue definition to provide definitions for enforcement and risk assessment purposes.

For plant commodities: Metalaxyl, including metalaxyl-M. Definition of the residue (for compliance with MRL and for estimation of dietary intake): metalaxyl. Note: Metalaxyl is a racemic mixture of an R-enantiomer and an S-enantiomer. Metalaxyl-M is the R-enantiomer.

In animals dosed with metalaxyl, parent metalaxyl was either a minor part of the residue or was not detected. Analytical methods for metalaxyl are based on a common moiety method, and residues in the farm animal feeding studies were measured by this method. Common moiety residues are acceptable for estimation of dietary intake when the parent compound is a minor part of the residue. The log P_{ow} for metalaxyl-M is 1.7, and the studies of animal metabolism confirm that metalaxyl is not fat-soluble.

For animal commodities: Metalaxyl including metalaxyl-M. Definition of the residue (for compliance with MRL and for estimation of dietary intake): metalaxyl and metabolites containing the 2,6-dimethylaniline moiety, expressed as metalaxyl.

Results of supervised trials on crops

The Meeting received data from supervised trials with metalaxyl-M used on citrus fruit, apple, grape, onion, tomato, pepper, lettuce, spinach, potato, sunflower and cacao. In some trials, residues were measured on samples taken just before and just after ('zero-day' residue) the final application. The residue level measured just before the final application, expressed as a percentage of zero-day residue, provides a measure of the contribution of previous applications to the final residue in the use pattern used in the trial. For grapes, the average carryover of residue was 32% in 12 trials in Australia, 57% in three trials in Germany and 46% in three trials in Italy. In lettuce, the average carryover was 1.7% in six trials in France, Germany and Italy. In spinach, the average carryover was 1.1% in eight trials in France.

Residue data were evaluated only when labels (or translations of labels) describing the relevant GAP were available to the Meeting.

Citrus fruit

Metalaxyl-M is registered for use as a post-harvest treatment on citrus in Israel. It is applied as a 0.1 kg ai/hl spray.

In the trials, the formulation of metalaxyl-M was mixed with a commercial wax to produce a spray solution, which was applied at a rate of 200 l /90 t of fruit (theoretical concentration of residue, 2.1 g/t). The residue levels in three trials on *oranges* were 1.2, 1.3 and 1.6 mg/kg in whole fruit and < 0.02 mg/kg in pulp. This method of post-harvest application includes control of the application rate in terms of the amount of metalaxyl-M per unit weight of fruit. The residue levels agreed substantially with expectations.

The Meeting noted that three supervised trials is generally an insufficient number for a major commodity such as oranges.

The residue levels of metalaxyl-M in the trials conducted in line with Israeli GAP did not exceed the current metalaxyl MRL of 5 mg/kg for citrus fruit.

Apple

Metalaxyl-M is registered in Spain for soil treatment around apple trees at 0.5–1.0 g ai/tree and in Italy at 0.5–4 g ai/tree. In two trials each in France, Italy and Spain at application rates of 0.78–10 kg ai/ha, no residues were detected in apples (< 0.02 mg/kg). For an assumed 500–1000 trees per ha, the rate of 10 kg ai/ha appears to be exaggerated.

The Meeting estimated a maximum residue level for metalaxyl-M in apples of 0.02* and an STMR value of 0 mg/kg.

Metalaxyl-M residue levels complying with the estimated maximum residue level of 0.02* mg/kg would not exceed the current metalaxyl MRL of 1 mg/kg for pome fruits.

Grape

In Australia, metalaxyl-M is registered for a maximum four applications on grapes at 0.11 kg ai/ha, with a PHI of 7 days. The residue levels in grapes in five Australian trials matching GAP, but with six applications instead of four, were: < 0.02, 0.03, 0.06, 0.14 and 0.52 mg/kg. As the final residue level should not be influenced by earlier applications, residue levels after six applications are acceptable as equivalent to residues levels in GAP trials.

No GAP was available to evaluate the data for grapes treated in Germany and Switzerland.

In Greece, grapes may be treated four times with metalaxyl-M at 0.1 kg ai/ha, with harvest 15 days after the final application. The residue levels in grapes in six trials in Italy and Portugal, conducted substantially according to Greek GAP, were: 0.04, 0.06, 0.18, 0.19, 0.21 and 0.55 mg/kg.

The residue levels in the Australian and European trials appear to be similar and can be combined. In summary, the residue levels in the 11 trials, in ranked order, were: < 0.02, 0.03, 0.04, 0.06, 0.06, 0.14, 0.18, 0.19, 0.21, 0.52 and 0.55 mg/kg

The Meeting estimated a maximum residue level for metalaxyl-M in grapes of 1 mg/kg and an STMR value of 0.14 mg/kg.

Metalaxyl-M residue levels complying with the estimated maximum residue level of 1 mg/kg would not exceed the current metalaxyl MRL of 1 mg/kg for grapes.

Onion

In Ecuador and Uruguay, metalaxyl-M is registered for a maximum of three applications on onions at 0.1 and 0.12 kg ai/ha, with a PHI of 7 days. Metalaxyl-M residue levels in bulb onions in three Brazilian trials matching Uruguayan GAP, but with four applications instead of three, were < 0.02 (two) and 0.02 mg/kg.

Metalaxyl-M is registered in Germany for a maximum of three applications on onions at 0.097 kg ai/ha, with a PHI of 21 days. In four trials in Switzerland with conditions matching German GAP, the residue levels were all below the LOQ (0.02 mg/kg).

Data on residues in trials in onions in Italy and Spain could not be evaluated because no relevant GAP was available.

In summary, the residue levels in the seven trials, in ranked order, were ≤ 0.02 (six) and 0.02 mg/kg.

The Meeting estimated a maximum residue level for metalaxyl-M in onions of 0.03 mg/kg and an STMR value of 0.02 mg/kg.

Metalaxyl-M residue levels complying with the estimated maximum residue level of 0.02 mg/kg would not exceed the current metalaxyl MRL of 2 mg/kg for bulb onions.

Tomato

Metalaxyl-M is registered for foliar application on tomatoes in Algeria, Chile, Ecuador, Greece, Israel and Morocco at 0.10–0.14 kg ai/ha, with a PHI of 3 days and a maximum of three or four treatments.

Residue levels in tomatoes in six greenhouse trials in France, two in Spain and four in Switzerland at 0.15 kg ai/ha, with harvest 3 days after treatment (equivalent to the stated GAP) were: 0.02 (two), 0.02, 0.03, 0.04, 0.04, 0.05, 0.05, 0.08, 0.09, 0.12 and 0.18 mg/kg.

The Meeting estimated a maximum residue level for metalaxyl-M in tomatoes of 0.2 mg/kg and an STMR value of 0.045 mg/kg.

Metalaxyl-M residue levels complying with the estimated maximum residue level of 0.2 mg/kg would not exceed the current metalaxyl MRL of 0.5 mg/kg for tomatoes.

Pepper

GAP for use of metalaxyl-M in Italy allows three soil applications of 1 kg ai/ha with a 15-day PHI. Data on residues from Italian and Spanish trials approximating Italian GAP were provided. In some of the trials, residues were measured 10 and 20 days after the final application instead of 15 days, but these trials were considered valid because the residue levels were relatively unchanged. The residue levels in the seven greenhouse trials were: < 0.02 (two), 0.02, 0.03, 0.08, 0.10 and 0.36 mg/kg; and those in the three outdoor trials in Italy were: < 0.02 (two) and 0.02 mg/kg.

The Meeting decided to use the data from the greenhouse trials: < 0.02 (two), 0.02, <u>0.03</u>, 0.08, 0.10 and 0.36 mg/kg.

The Meeting estimated a maximum residue level for metalaxyl-M in sweet peppers of 0.5 mg/kg and an STMR value of 0.03 mg/kg.

Metalaxyl-M residue levels complying with the estimated maximum residue level of 0.5 mg/kg would not exceed the current metalaxyl MRL of 1 mg/kg for peppers.

Lettuce

Metalaxyl-M is registered in Spain for use on lettuce at 0.10 kg ai/ha with a PHI of 14 days. The residue levels in lettuce were < 0.02 mg/kg in an Italian trial matching Spanish GAP. The residue levels in lettuce in four French trials matching Spanish GAP were: < 0.02 (two), 0.02 and 0.03 mg/kg.

Metalaxyl-M is registered in Germany for a maximum of two applications on lettuce at 0.097 kg ai/ha, with a PHI of 21 days. In six trials on head lettuce in Germany under conditions matching GAP, but with three applications instead of two, the residue levels were: < 0.02 (four), 0.02 and 0.03 mg/kg. In two German greenhouse trials at 0.10 kg ai/ha with harvest 21 days after the second application, the residue levels were < 0.02 and 0.41 mg/kg.

In two trials in The Netherlands matching German GAP, the residue levels were < 0.02 mg/kg.

Trials in Spain and Switzerland could not be evaluated because there was no matching GAP.

In summary, the residue levels in lettuce in the 15 trials, in ranked order, were: < <u>0.02</u> (10), 0.02 (two), 0.03 (two) and 0.41 mg/kg.

The Meeting estimated a maximum residue level for metalaxyl-M in head lettuce of 0.5 mg/kg and an STMR value of 0.02 mg/kg.

Metalaxyl-M residue levels complying with the estimated maximum residue level of 0.5 mg/kg would not exceed the current metalaxyl MRL of 2 mg/kg for head lettuce.

Spinach

Metalaxyl-M is registered in Switzerland for a maximum of six applications on spinach at 0.10 kg ai/ha with a PHI of 14 days. In three trials on spinach in Switzerland at 0.10 kg ai/ha and three at 0.14 kg ai/ha, with intervals before harvest of 10 or 14 days, the residue levels were all < 0.02 kg/ha. The levels were also < 0.02 mg/kg in two trials in Germany matching the conditions of GAP in Switzerland.

In a number of trials in France in which the application rate was 0.10 or 0.14 kg ai/ha, spinach was sampled for analysis 10 and 20 days after treatment. The Meeting noted that the residue levels generally changed slowly between 10 and 20 days post-treatment and decided to accept the residue levels at 10 days as sufficiently close to those expected at 14 days. The residue levels in the 10 French trials were: < 0.02 (two), 0.02 (four), 0.03, 0.04 (two) and 0.05 mg/kg.

In summary, metalaxyl-M residue levels in the 18 trials were < <u>0.02</u> (10), 0.02 (four), 0.03, 0.04 (two) and 0.05 mg/kg.

The Meeting estimated a maximum residue level for metalaxyl-M in spinach of 0.1 mg/kg and an STMR value of 0.02 mg/kg.

Metalaxyl-M residue levels complying with the estimated maximum residue level of 0.1 mg/kg would not exceed the current metalaxyl MRL of 2 mg/kg for spinach.

Potato

Labels were available from Algeria, Australia, Austria, Chile, Ecuador, Greece, Israel and Morocco from formulations for foliar application of metalaxyl-M to potatoes at 0.1–0.12 kg ai/ha. The information on GAP suggests that the recommended foliar application rate on potatoes is 0.1 kg ai/ha in many situations.

The results of supervised trials were available from Brazil (three at 0.1 kg ai/ha and three at 0.2 kg ai/ha), Germany (six at 0.1 kg ai/ha), Switzerland (three at 0.075 kg ai/ha) and the United Kingdom (four at 0.1 kg ai/ha). The residue levels in all 19 trials, measured at intervals of 0–28 days after the final treatment, were below the LOQ (0.02 mg/kg).

As residues were found in potato tubers in the metabolism studies after high application rates, the median residue values cannot be assumed to be nil. The Meeting estimated a maximum residue level for metalaxyl-M in potato of 0.02* mg/kg and an STMR value of 0.02 mg/kg. Metalaxyl-M residue levels complying with the estimated maximum residue level of 0.02* mg/kg would not exceed the current metalaxyl MRL of 0.05* mg/kg for potato.

Sunflower seed

Metalaxyl-M is registered in China and Serbia and Montenegro for use as a seed treatment at 0.105 kg ai/100 kg sunflower seed. The Meeting agreed that the results of trials from other countries could be evaluated with respect to this seed treatment GAP.

In six trials in France and two in Spain, metalaxyl-M was used as seed treatment at a nominal rate of 105 g ai/100 kg seed (measured, 61–80 g ai/100 kg seed). The residue levels in harvested sunflower seed 125–151 days after sowing were all below the LOQ (0.01 and 0.02 mg/kg). Because of the long interval between sowing and harvest and the solubility of metalaxyl-M, residues would not be expected in harvested sunflower seed.

The Meeting estimated a maximum residue level for metalaxyl-M in sunflower seed of 0.02* mg/kg and an STMR value of 0 mg/kg. Metalaxyl-M residue levels complying with the estimated maximum residue level of 0.02* mg/kg would not exceed the current metalaxyl MRL of 0.05* mg/kg for sunflower seed.

Cacao beans

Metalaxyl-M is registered in Côte d'Ivoire for use on cacao at 0.012 kg ai/ha. In eight trials in Côte d'Ivoire in which metalaxyl-M was applied as foliar treatment four times at 0.09 kg ai/ha (an exaggerated rate), with harvest 29–30 days after the final treatment, the residue levels in the cacao beans were: < 0.02 (four) and 0.02 (four) mg/kg. The cacao beans were fermented and dried before analysis. The Meeting agreed that the residue levels after application at the label rate would not exceed 0.02 mg/kg.

The Meeting estimated a maximum residue level for metalaxyl-M in cacao beans of 0.02 mg/kg and an STMR value of 0.02 mg/kg.

Metalaxyl-M residue levels complying with the estimated maximum residue level of 0.02 mg/kg would not exceed the current metalaxyl MRL of 0.2 mg/kg for cacao beans.

Fate of residues during processing

The Meeting received information on the fate of metalaxyl-M residues during the production of fruit juices and vinification. The Meeting also received information that metalaxyl-M is hydrolytically stable under hydrolysis conditions that simulate those occurring during food processing.

The following processing factors were calculated from the data from the trials. The factors are mean values, excluding those calculated in cases of undetectable residues.

Commodity	Processed product	Processing factor	No. of trials
Orange	Washed fruit	0.97	2
	Juice, pasteurized	0.060	4
	Oil	9.0	4
	Pomace, wet	1.1	4
	Pomace, dry	4.1	4
	Peel	2.5	3
	Pulp	0.091	1
	Marmalade	0.39	4
Grapes	Juice	0.36	6
	Young wine	0.87	8
	Wine	0.66	13

The Meeting used the processing factors to estimate STMR-Ps for processed commodities. The processing factor for wine (0.66) was applied to the grape STMR (0.14 mg/kg) to calculate an STMR-P of 0.092 mg/kg for wine. The processing factor for grape juice (0.36) was applied to the grape STMR (0.14 mg/kg) to calculate an STMR-P of 0.050 mg/kg for grape juice

Residues in animal commodities

Feeding studies

The Meeting received the results of studies of feeding metalaxyl to lactating dairy cows and laying hens, which provided information on probable residue levels in tissues, milk and eggs from residues in animal feeds.

A group of three lactating dairy cows were dosed daily with metalaxyl, equivalent to 75 ppm in their diet, and were slaughtered for tissue collection on days 14, 21 and 28. Liver, kidney, fat and muscle were analysed by a dimethylaniline common moiety method. The residues were transitory and did not accumulate, and the interval between final dose and slaughter (4 and 23.5 h) influenced the residue levels more than the duration of dosing. The level of residue in milk was 0.02 mg/kg. The residue levels in the tissues collected on day 28 from the animal slaughtered 23.5 h after the final dose were 0.11 mg/kg in kidney, 0.12 mg/kg in liver, < 0.05 mg/kg in fat and 0.06–0.08 mg/kg in muscle.

Groups of 15 laying hens were dosed daily for 28 days with metalaxyl at levels equivalent to 10, 30 and 100 ppm in the feed. Tissue and egg samples were analysed by a dimethylaniline common moiety method. No residues appeared in the eggs (< 0.05 mg/kg) at any dose. The residue levels in the tissues of hens fed 10 ppm were generally below the LOQ (< 0.05 mg/kg) or, in a few cases, just above the LOQ.

Maximum residue levels

The farm animal feeding studies suggest that residues would generally be undetected or transitory in meat, milk and eggs if metalaxyl was present in animal feeds.

Farm animals are therefore not exposed to residues in their feed from commodities in this evaluation, and no MRLs have been established for metalaxyl in animal commodities. Consequently, the Meeting agreed not to recommend animal commodity maximum residue levels.

DIETARY RISK ASSESSMENT

Long-term intake

Estimated Theoretical Maximum Daily Intakes for the five GEMS/Food regional diets, based on recommended MRLs for metalaxyl, were in the range of 2-10% of the ADI (Annex 3). The Meeting concluded that the long-term intake of residues of metalaxyl and metalaxyl-M resulting from their uses that have been considered by JMPR is unlikely to present a public health concern.

Short-term intake

The 2002 JMPR decided that an ARfD is unnecessary. The Meeting therefore concluded that the short-term intake of metalaxyl and metalaxyl-M residues is unlikely to present a public health concern.

4.16 METHAMIDOPHOS

RESIDUE AND ANALYTICAL ASPECTS

The 2003 JMPR evaluated supporting information on cucumbers and concluded that the residue data were insufficient to estimate maximum residue limits for acephate or for methamidophos arising from use of acephate. The Meeting noted that the report of the 2003 JMPR, while stating that the number of trials was inadequate for the purposes of estimating a maximum residue level for acephate on cucumber, made no recommendation with respect to the existing recommended maximum residue level for methamidophos arising from the use of acephate.

Recommendation

The Meeting recommended that the MRL of 1 mg/kg for cucumber (CCN: VC0424) be withdrawn.

4.17 METHOMYL (094)

RESIDUE AND ANALYTICAL ASPECTS

Data on methomyl residues were reviewed in 1975, and data from supervised field trials with various crops and related data were considered in 1976, 1978, 1986, 1991 and 2001. At its Thirty-sixth Session, the CCPR noted that new data on mint hay and pepper had been reported and decided to maintain the CXLs for these commodities for 4 years under the periodic review procedure. The evaluation was scheduled for 2004 (ALINORM 04/27/24, para. 124, p 15).

The 2004 Meeting received data on residues from supervised field trials on pepper and mint hay from the manufacturer. Information on labels and current GAP was also provided.

Methods of analysis

The gas chromatographic method for measuring residues of methomyl in many plant commodities, evaluated by the 2001 JMPR, was validated for pepper and mint hay. This method consists of extraction with an organic solvent, liquid–liquid partition and hydrolysis with sodium hydroxide. The latter converts methomyl to methomyl oxime. The final extract is analysed by gas chromatography, usually with a flame photometric detector in the sulfur mode.

A more recent method is based on HPLC. The plant matrix is extracted with solvent, cleaned up on a Florisil column and analysed by HPLC with post-column reaction to convert methomyl to methylamine. Methylamine is derivatized (on-line) and detected by fluorescence.

The gas chromatographic method has been validated for numerous plant commodities at an LOQ of 0.02 mg/kg. The HPLC method and its modifications have been validated at an LOQ of 0.02 mg/kg for methomyl.

The Meeting concluded that adequate methods exist for the determination of methomyl in pepper and mint hay.

Stability of residues in stored analytical samples

As described by the 2001 JMPR, the stability of methomyl under frozen conditions has been demonstrated in a number crop samples, including broccoli, orange, apple and grape, for up to 24 months.

Data were presented on the stability of methomyl under frozen storage (–10 °C) in mint hay. Adequate stability (> 90% remaining) was demonstrated after 6 months' storage.

The Meeting concluded that methomyl is stable under frozen conditions on mint hay and pepper.

Results of supervised trials on crops

Peppers

Supervised trials were conducted on peppers in Canada (no GAP) and the USA (GAP: 1.0 kg ai/ha, 3-day PHI). Fifteen trials (one in Canada, 14 in the USA) were conducted at US GAP, with residue concentrations of 0.02, 0.03, 0.04 (two), 0.08 (two), 0.10 (two), 0.11 (two), 0.12, 0.18, 0.24, 0.26, 0.39 and 0.44 mg/kg.

Supervised trials on peppers were conducted in France (no GAP), Greece (GAP: 0.45 kg ai/ha, 15-day PHI), Italy (GAP: 0.04 kg ai/hl, 10-day PHI), Portugal (no GAP) and Spain (no GAP). In nine trials (two in France, three in Italy, one in Portugal and three in Spain) conducted at about Italian GAP, the ranked order of concentrations was: < 0.02 (five), 0.02 (two), 0.03 and 0.04 mg/kg. The data from southern Europe and the USA were considered to represent different populations. Using only the data from the USA (higher values), the Meeting estimated an STMR value of 0.105 mg/kg, a highest residue of 0.44 mg/kg and a maximum residue level of 0.7 mg/kg, which replaces the previous estimate (1 mg/kg).

Mint hay

Supervised trials were conducted on fresh mint hay in the USA (GAP: 1.0 kg ai/ha, 14-day PHI, maximum of four applications). In the six trials conducted at GAP, the ranked order of concentrations of residues in fresh and spent mint hay was: 0.02 (three), 0.07 (two), 0.13, 0.16, 0.17, and 0.28 mg/kg. The dry matter in fresh and spent mint is 88%. The Meeting estimated an STMR value of 0.08 mg/kg, a highest residue value of 0.32 mg/kg and a maximum residue level of 0.5 mg/kg for mint hay on a dry weight basis. The Meeting agreed to withdraw the previous recommendation (2 mg/kg) and to replace it with the recommendation for mint hay (0.5 mg/kg, dry weight).

Residues in animal commodities

As mint hay is not considered to be a significant feed item, establishment of an MRL for mint hay would not significantly change the dietary exposure of animals. The Meeting therefore decided not to revise the previous MRL recommendations for animal commodities (edible offal, meat, milk) on the basis of the addition of mint hay (fresh and spent).

4.18 OXYDEMETON METHYL (166)

RESIDUE AND ANALYTICAL ASPECTS

Oxydemeton methyl was evaluated for residues by the 1998 JMPR within the CCPR periodic review programme and then for residues and toxicology by the JMPR in 1999 and 2002, respectively.

At its Thirty-first Session, the CCPR asked the JMPR to clarify whether demeton-*S*-methyl and demeton-*S*-methylsulfon should remain in the residue definition of oxydemeton methyl, as it was believed that registration of these compounds would not be retained. At its Thirty-second Session, the CCPR withdrew the draft MRLs for several commodities, as there was no existing GAP for them. The Committee advanced the proposed draft MRLs to Step 5 and returned the draft MRLs to Step 6 because of intake concerns, which would be considered at its next session. The Committee requested detailed information on oxydemeton methyl. The Committee discussed the residue definition that had been confirmed by the 1999 JMPR. It stated that, as demeton-*S*-methyl was no longer supported and there was no GAP, its use should be prevented by removing this compound from the residue definition. It was pointed out, however, that demeton-*S*-methyl could not be distinguished from oxydemeton methyl in analysis and that it could be generated from oxydemeton methyl during analysis. As no agreement was reached, the Committee agreed as a compromise to maintain the present residue definition but to specify that the residue definition and MRLs apply only to residues resulting from use of oxydemeton methyl. Those conditions would be met by adding a note to the residue definition, reading: "The residue definition and MRLs are based on the use of oxydemeton methyl only."

At its Thirty-third Session, the CCPR noted a written comment from the European Commission stating a general reservation (lack of an acute RfD) and a specific reservation on the MRLs for grape, lemon and oranges, sweet, sour (acute risk) and decided to return the draft MRLs to Step 6. The CCPR decided to return all the MRLs to Step 6 until calculations of short-term intake had been obtained from the JMPR. The Committee was informed by the manufacturer that data would be submitted to the 2004 JMPR for a review of the residue definition.

The Meeting received new data on physical and chemical properties (partially updated), analytical methods, fate of residues in processing, plant metabolism (apple), residue data (apples, pears, grapes, cabbage, Brussels sprouts, cauliflower, field peas, potatoes, sugar-beet, fodder beet, wheat, barley, rape and sunflower), GAP and national MRLs.

Metabolism

Plants

The metabolism of [ethylene-1-^{14}C]oxydemeton methyl in *apple* was investigated in the field. Apple trees were sprayed on two separate occasions about 4 months before harvest (pink bud stage, BBCH 57) and then about 3 months before harvest (flowers fading, BBCH 67). The formulated material was prepared at a nominal concentration of 1.4 g ai/l and applied at a nominal rate of 350 g ai/ha. Samples of fruit and leaves were taken for analysis 2 h after the first application (leaves only), 2 h after the second application (leaves only), about 60 days before harvest (first intermediate sample; fruit and leaves), about 30 days before harvest (second intermediate sample; fruit and leaves) and at harvest (fruit and leaves).

Analysis of fruit samples (flesh and peel) from the two intermediate samples showed the presence of oxydemeton methyl, but none was found at harvest. The two intermediate samples did, however, contain desmethyl-oxydemeton methyl sulfone (metabolite 7) and two polar materials (P2 and P3). In the first intermediate sample, polar radioactivity in the flesh accounted for 17% of the radioactivity in the fruit (0.2 mg/kg). In the second intermediate sample, polar radioactivity in the flesh had decreased to 10% of that in fruit (0.05 mg/kg), and by harvest no components were detected.

In the peel extracts, polar metabolites accounted for 3% of the radioactivity in the fruit (0.023 mg/kg) at the first intermediate sampling, and this had increased to 4% (0.026 mg/kg) by the second intermediate sampling. No components were detected at harvest.

The main metabolites detected at the first intermediate sampling time in both flesh and peel were oxydemeton methyl (26.6% fruit radioactivity, 0.245 mg/kg) and demeton-*S*-methylsulfone (2.9% fruit radioactivity, 0.026 mg/kg). In the second intermediate sample, oxydemeton methyl (1.7% fruit radioactivity, 0.012 mg/kg) and demeton-*S*-methylsulfone (0.2% fruit radioactivity, 0.001 mg/kg) were detected only in peel extracts. The main component detected in flesh extracts was an unidentified polar compound (14.1% radioactivity, 0.072 mg/kg). No components were detected at harvest in either flesh or peel extracts owing to the low levels of radioactivity in the extract samples (< 0.001 mg/kg). The results of this study do not change the conclusions reached in the 1998 evaluation.

Environmental fate

Soil

The aerobic degradation of oxydemeton methyl was studied in three soils for a maximum of 11 days under aerobic conditions in the dark at 20 °C. [ethylene-1-^{14}C]Oxydemeton methyl was applied at a nominal rate of 0.67 mg/kg dry soil, equivalent to the proposed single maximum annual use rate of 250 g ai/ha calculated for 2.5 cm depth of soil.

During the study, the total recovery of radioactivity in individual test vessels ranged from 90.8% to 100%, and the times to 50% and 90% decomposition (DT_{50} and DT_{90}) in the three soils ranged from 0.17 to 0.22 and from 0.58 to 0.74 days, respectively. The results also indicate that the main metabolites were continuously degraded, that no metabolite accumulated towards the end of the study and that the bound residues participated in the natural carbon cycle of soil.

Analysis of soil extracts showed two major and one semi-major degradation product, representing 10% or more of the applied radioactivity at any time during the study. The concentration of the 2-ethylsulfinyl ethane sulfonic acid metabolite reached a maximum on day 1 and then declined gradually until day 11. Its concentration was below the LOQ towards the end of the study in soils with higher microbial activity. The 2-ethylsulfonyl ethane sulfonic acid metabolite is an oxidation product of 2-ethylsulfinyl ethane sulfonic acid, and its concentration reached a maximum on day 3 in all soils; it ranked highest, at 16.8% of the applied radioactivity. The level declined towards the end of the study in all soils, and in the most active soil to below the LOQ by day 11. Significant formation of bound residues occurred during overall metabolism of parent compound. The concentration of bound residues reached a maximum on day 11 at about 50% of the applied radioactivity. Soils showed high mineralization capacity, yielding values for $^{14}CO_2$ of > 30% by day 11. The results of study demonstrate that oxydemeton methyl is quickly degraded in aerobic soils.

Water–sediment systems

The hydrolysis of oxydemeton methyl was studied in sterile 0.01 mol/l buffer solutions, which were adjusted to pH4, 7 or 9, for a maximum of 31 days in the dark at two temperatures. The experiment was carried out in compliance with good laboratory practice (GLP) and in accordance with guidelines of the US Environmental Protection Agency, the Society of Environmental Toxicology and Chemistry, the OECD and the European Commission. The test solutions were prepared with [ethylene-1-^{14}C]oxydemeton methyl at a concentration of about 5 mg/l. The pre-test solutions were incubated for 7 days under sterile conditions in the dark at 50 °C. The solutions at pH 4 and pH 7 in the main test were incubated for a maximum of 31 days under sterile conditions in the dark at 25 °C.

In the pre-test at 50 °C and in the main test at 25 °C, oxydemeton methyl was not stable at pH 4, 7 or 9, and considerable degradation occurred. Especially at higher pH values, the compound was thoroughly hydrolysed to desmethyl-oxydemeton methyl and 2-ethylsulfinyl-ethyl mercaptan by cleavage of the P–S bound. Furthermore, 2-ethylsulfonyl ethane sulfonic acid was observed as an oxidized P–S cleavage product at low percentages (maximum of 2.2% of the applied radioactivity), although it was identified only tentatively.

By calculation from the data obtained at 50 °C, orienting DT_{50} values (first order) for the hydrolysis of oxydemeton methyl were estimated to be 4.9, 3.5 and 0.2 days at pH 4, 7 and 9, respectively. Using the data obtained at 25 °C, the DT_{50} values (first order) were estimated to be 91, 42 and 2.5 days at pH 4, 7 and 9, respectively. At 20 °C, the DT_{50} values calculated from Arrhenius plots (1/T versus ln(k)) were 174, 73 and 4.5 days at pH 4, 7 and 9, respectively. The results indicate that hydrolytic processes contribute to the degradation of oxydemeton methyl in the environment.

The quantum yield from direct photodegradation of oxydemeton methyl in water was determined according to the European Centre for Ecotoxicilogy and Toxicology of Chemicals method in polychromatic light. The quantum yield calculated from the ultraviolet absorption data and the kinetics of photodegradation was 0.00078. The resulting quantum yield and data on ultraviolet absorption in aqueous solution were used to estimate the environmental half-life of oxydemeton methyl after direct photodegradation in water in two simulation models. The calculated half-lives were 112 days in summer and 274 days in winter at 30° latitude and 200 days in May) and 790 days in October at 50° latitude.

Methods of analysis

A number of methods have been developed for the analysis of residues of oxydemeton methyl in various matrices, many of which were reviewed by the 1998 Meeting as part of the periodic review of this compound. The Meeting was provided with additional methods based on the same principle as those evaluated earlier, i.e. use of the oxidation process to produce demeton-*S*-methylsulfone as the analyte. The LOQs were 0.005 mg/kg for potato; 0.005–0.01 mg/kg for apples and grapes; 0.01 mg/kg for pear, Brussels sprouts, cauliflower, cabbage, corn, sunflower, rape-seed and rape (green plant material); 0.02 mg/kg for wheat grain and 0.05 mg/kg for wheat straw.

Stability of residues in stored analytical samples

The stability of oxydemeton methyl in stored cabbage, maize, lettuce and papaya was evaluated by the 1998 JMPR, which concluded that data on the stability of stored analytical samples of raw agricultural commodities containing quantifiable residues of oxydemeton methyl were highly desirable. The available information was not representative of the various crop groups, did not cover extended storage intervals and suggested variable storage stability. The manufacture submitted new study data on the storage stability of oxydemeton methyl in several crops.

A study was conducted to determine the stability of oxydemeton methyl in spiked samples of stored apple, dried peas, potato and oil-seed rape (meal and oil at − 20 °C). Samples for the study were obtained from a commercial source. The samples of apple and potato were prepared for use by fine chopping in an industrial food processor; the preparation of samples of dried peas was not recorded in the report. For each crop, 20 g of prepared sample were spiked with formulated oxydemeton methyl at a nominal concentration of 0.1, 1.0 or 10 mg/kg and placed in storage at − 20 °C. Two samples of each were removed for analysis after 0, 3, 6, 12 and 24 months of storage. The analytical method was validated for each crop at each sampling time.

There was no substantial loss of residue during 24 months' storage from apple, potato or oil; however, the residues in dried peas and rape meal apparently decreased to half the initial values after 3 or 6 months of storage. The control samples (un-spiked samples) also showed residues of oxydemeton methyl, with < 0.005–0.022 mg/kg in apple, < 0.005–0.016 mg/kg in dried peas, < 0.01–0.016 mg/kg in potatoes, 0.013–0.03 mg/kg in rape meal and < 0.005–0.011 mg/kg in rape oil. The Meeting concluded that the data submitted on storage stability were insufficient or inadequate, and the former requirement was maintained.

Oxydemeton methyl

Definition of the residue

The Meeting received the results of new studies of plant metabolism and supplementary information on the analytical method. The new data did not, however, provide a basis for changing the current residue definition. The Meeting confirmed its previous recommendation.

Results of supervised trials on crops

The results of supervised field trials on apples, pears, grapes, cabbage, Brussels sprouts, cauliflower, field peas, potatoes, sugar-beet, fodder beet, wheat, barley, rape and sunflower were submitted to the Meeting. The new data were evaluated against current GAPs, and highest residue levels were estimated for commodities evaluated by the 1998 JMPR, as the 1998 JMPR did not do so. When no residues were found in any sample in older trials in which the analytical methods used had higher LOQs than current methods, the Meeting decided to use only data from the newly submitted trials in order to avoid unnecessarily high maximum residue levels.

Citrus fruit

No data from new supervised trials were submitted. From 11 trials on *orange* and *lemon*, the 1998 JMPR estimated a maximum residue level of 0.2 mg/kg and an STMR of 0.01 mg/kg.

In the 1998 evaluations, the reported residues in pulp, in ranked order, were < 0.01 (seven), 0.01, 0.02 (two) and 0.04 mg/kg. The Meeting estimated a highest residue level of 0.04 mg/kg for orange and lemon.

Pome fruit

The results of 15 new supervised trials on apples and two on pears were submitted to the Meeting

One supervised trial on *apples* in northern France in 1998 involved higher doses than used in German GAP (0.13 kg ai/ha once, immediately after flowering). Nevertheless, the residue data could be used, since the residue levels were below the LOQ. Three supervised trials on apples in southern France in 1997 and 1998, three on apples in Italy in 1997 and 1998 and four on apples in Spain in 1997 and 1998 were conducted according to Italian GAP (0.023–0.028 kg ai/hl, 90-day PHI). As two of the four Spanish trials (conducted in 1997) were carried out in the same location under almost identical trial conditions, one residue level was taken from each. The residue levels of oxydemeton methyl in apples were < 0.01 (nine) and 0.01 mg/kg.

One supervised trial on *pears* in southern France in 1998 was conducted according to Italian GAP (0.023–0.028kg ai/hl, 90-day PHI), and one trial on pears in Germany in 1998 was conducted according to Austrian GAP (0.024 kg ai/hl once, up to 2 weeks after flowering). The residue level of oxydemeton methyl in pears was < 0.01 (two) mg/kg.

The 1998 JMPR evaluated 10 supervised trials on apples and pears and estimated a maximum residue level of 0.05 mg/kg and an STMR of 0.01 mg/kg for apple and pear, based on residue levels, in ranked order, of < 0.01 (seven), 0.03 and < 0.04 (two) mg/kg.

The Meeting confirmed that the residue levels found in the newly submitted trials did not exceed the formerly estimated maximum and highest residue values of 0.04 mg/kg for apple and pear in the 1998 JMPR evaluations.

Grape

Three new supervised trials from Germany were conducted within German GAP (0.027 kg ai/hl, up to fully developed inflorescence). The residue levels were < 0.01 (three).

The 1998 JMPR evaluated the results of five supervised trials and estimated a maximum residue level of 0.1 mg/kg and an STMR of 0.04 mg/kg.

The Meeting confirmed that the residue levels found in the newly submitted trials did not exceed the formerly estimated maximum and highest residue values of 0.06 mg/kg for grapes in the 1998 JMPR evaluations. The residue levels, in ranked order, were < 0.04 (four) and 0.06 mg/kg.

Brassica vegetables

Cabbage (head)

The results of four new supervised trials were submitted to the Meeting.

Two supervised trials on red and white cabbage in northern France in 1996 and two supervised trials on Savoy cabbage in Germany were conducted according to German GAP (0.16 kg ai/ha (< 50 cm) once, 21-day PHI). The residue levels were < 0.01 (four) mg/kg.

The 1998 JMPR evaluated 16 supervised trials and estimated a maximum residue level of 0.05* mg/kg and an STMR of 0.03 mg/kg. The highest value of < 0.06 mg/kg was disregarded because of the high LOQ in the older trials (1976).

The Meeting confirmed that the residue levels in the submitted trials did not exceed the formerly estimated maximum and highest residue values of 0.05 mg/kg in the 1998 JMPR evaluations. The residue levels, in ranked order, were < 0.01 (six), 0.02, < 0.03 (three), < 0.04 and < 0.05 (four) mg/kg.

Kale

No data from new supervised trials were submitted to the Meeting. Four supervised trials were evaluated by the 1998 JMPR, which estimated a maximum residue level of 0.01* mg/kg and an STMR of 0.01 mg/kg.

On the basis of the reported residue level, < 0.01 (four) mg/kg, the Meeting estimated a highest residue level of 0.01mg/kg.

Kohlrabi

No data from new supervised trials were submitted to the Meeting. Four supervised trials were evaluated by the 1998 JMPR, which recommended a maximum residue level of 0.05 mg/kg and an STMR of 0.02 mg/kg. The reported residue levels, in ranked order, were < 0.01 (two), 0.03 and < 0.06 mg/kg. The highest value was disregarded because of the high LOQ associated with the older trials (1979).

The Meeting estimated a highest residue level of 0.05 mg/kg, at the same level as the maximum residue level recommended by the 1998 JMPR.

Brussels sprouts

Five supervised trials conducted in Germany in 1997 and 1998, two conducted in Belgium in 1997 and 1998, one conducted in the United Kingdom in 1998 and one conducted in northern France in 1998 were submitted to the Meeting; however, no comparable GAP was submitted. The Meeting could not therefore estimate a maximum residue level, an STMR or a highest residue level.

Cauliflower

The results of eight new supervised trials were submitted to the Meeting. Two trials in Germany in 1996 were conducted according to German GAP (0.16 kg ai/ha (< 50 cm) once, 21-day PHI) and also according to Belgian GAP (0.15 kg ai/ ha once, 28-day PHI).

Four supervised trials in southern France in 1996 and 1997 and one in the United Kingdom in 1996 were conducted according to German GAP. One of the trials in France in 1996 involved higher doses than in German GAP, but the residue data could be used for evaluation as the level was below the LOQ.

In all the trials the residue level was ≤ 0.01 (eight) mg/kg. The Meeting estimated a maximum residue level of 0.01* mg/kg and STMR and highest residue values of 0.01 mg/kg for cauliflower.

Field peas (dry)

Data from two supervised trials on field peas were submitted to the Meeting, but no information on GAP was provided. The Meeting could therefore not estimate a maximum residue level, an STMR or a highest residue level.

Potatoes

The results of 20 supervised trials were submitted to the Meeting.

Two supervised trials in Germany in 1996 and one in the United Kingdom in 1996 were not matched by comparable GAP. Nine supervised trials in France in 1996 and 1998, one in Greece in 1998, three in Italy in 1997 and 1998 and four in Spain in 1997 and 1998 were conducted according to Greek GAP (0.05kg ai/hl three times, 28-day PHI). As two of four Spanish trials conducted in 1997 were carried out in the same location under similar trial conditions, one residue level was taken from each. The residue levels, in ranked order, were < 0.005 (five) and ≤ 0.01 (11) mg/kg.

The 1998 JMPR evaluated the results of 16 supervised trials and estimated a maximum residue level of 0.05* mg/kg and an STMR of 0.02 mg/kg. The reported residue levels, in ranked order, were < 0.01 (seven), < 0.02 (nine) and < 0.05 (two) mg/kg.

The Meeting decided to use the data from the newly submitted trials and estimated a maximum residue level of 0.01*mg/kg and STMR and highest residue values of 0.01 mg/kg to replace the former recommendations.

Sugar-beet (root)

The results of seven supervised trials on sugar-beet and two on fodder beet were submitted to the Meeting.

Three supervised trials on sugar-beet in Spain in 1997 and 1998 and four in Italy in 1997 and 1998 involved higher doses than in Spanish GAP (0.025 kg ai/hl, 30-day PHI) or Italian GAP (0.023–0.028 kg ai/hl, 30-day PHI); however, the residue data could be used for evaluation as all the levels in leaf and root were below the LOQ. As two of the three Spanish trials were conducted at the same location under the same trial conditions, one residue level was taken from each.

Two supervised trials on fodder beet in southern France in 1998 involved higher doses than in Spanish GAP (0.025 kg ai/hl, 30-day PHI), but the residue data could be used for evaluation as all the levels in leaf and root were below the LOQ. Six trials on sugar-beet and two on fodder beet could be evaluated together, as all the residue levels were ≤ 0.01 (eight) mg/kg.

The 1998 JMPR evaluated seven supervised trials on sugar-beet and two on fodder beet in Germany, and estimated a maximum residue level of 0.05* mg/kg and an STMR of 0.04 mg/kg. The reported residue levels, in ranked order, were < 0.01 (four) and < 0.04 (five) mg/kg.

The Meeting decided to use the data from the newly submitted trials and estimated a maximum residue level of 0.01*mg/kg and an STMR of 0.01 mg/kg, to replace the former recommendations.

Wheat, barley and rye

Eleven supervised trials on wheat and two on barley were submitted to the Meeting.

Four supervised trials on wheat in Italy in 1997 and 1998, six in southern France in 1997 and 1998 and one in Spain in 1998 involved higher doses than in Italian GAP (0.023–0.028 kg ai/hl, 30-day PHI), but the data on residues on wheat grain could be used for evaluation as all the levels were below the LOQ. The residue levels, in ranked order, were < 0.01 (six) and < 0.02 (five) mg/kg.

Oxydemeton methyl

Two supervised trials on barley in southern France in 1997 and 1998 involved slightly higher doses than in the Italian GAP for wheat. The Meeting concluded that this GAP could be applied to trials on barley, as the two crops are cultivated similarly. The residue levels were < 0.01 (two) mg/kg.

The 1998 JMPR evaluated seven supervised trials on wheat and three on barley and estimated a maximum residue level of 0.05* mg/kg and an STMR of 0.04 mg/kg. The reported residue levels, in ranked order, were < 0.04 (seven) and < 0.05 (three) mg/kg.

The Meeting decided to use the data from the newly submitted trials to estimate the maximum residue level. The combined residue levels, in ranked order, were ≤ 0.01 (eight) and < 0.02 (five) mg/kg. The Meeting estimated a maximum residue level of 0.02*mg/kg and an STMR of 0.01 mg/kg, to replace the former recommendations.

Rape-seed

One supervised trial on rape was submitted to the Meeting; however, no information on GAP was provided. The Meeting could therefore not estimate a maximum residue level, an STMR or a highest residue level.

Sunflower seed

Six supervised trial data on sunflower were submitted to the Meeting; however, no information on GAP was provided. The Meeting could therefore not estimate a maximum residue level, an STMR or a highest residue level.

Sugar-beet (tops)

The residue levels in the leaves of sugar-beets and fodder beets treated according to GAP, in ranked order, were ≤ 0.01 (five) and < 0.04 (three) mg/kg. The Meeting estimated a maximum residue level of 0.05 mg/kg, an STMR of 0.01 mg/kg and a highest residue level of 0.04 mg/kg, to replace the former recommendations.

Wheat, barley and rye straw and fodder

The residue levels in straw and fodder from wheat and barley, in ranked order, were ≤ 0.04 (eight), < 0.05 (three) and 0.06 mg/kg. The Meeting estimated a maximum residue level of 0.1 mg/kg, an STMR of 0.04 mg/kg and a highest residue level of 0.06 mg/kg, to replace the former recommendations.

Fate of residues during processing

The results of studies on residues in processed peas were provided to the Meeting. The samples were fortified with diluted oxydemeton methyl formulation by soaking because the levels of residue in samples produced under GAP conditions were expected to be too low. The reported processing factors were 0.034 for processed marrowfat peas (canning), 0.024 for vining peas (canning), 0.43 for vining peas (freezing) and 0.43 for vining peas (freezing and domestic cooking).

Residues in animal commodities

The 1998 JMPR, estimated the dietary burden of farm animals and concluded that quantifiable residues of demeton-*S*-methyl, oxydemeton methyl or demeton-*S*-methylsulfone are unlikely to occur in commodities of animal origin (meat, milk, poultry and egg). Therefore, MRLs could be set at the practical LOQ of 0.05* mg/kg for all commodities except milk and at 0.01* mg/kg for milk. The current Meeting did not recommend the addition of further feed items or an increase in the recommended residue levels. It therefore confirmed the previous maximum residue levels and STMRs for commodities of animal origin and estimated a highest residue level of 0 mg/kg for cattle fat, eggs, meat of cattle, pigs and sheep, pig fat, poultry fats, poultry meat and sheep fat.

Further work or information
Desirable

Data on the stability of stored analytical samples of raw agricultural commodities containing quantifiable residues of oxydemeton methyl are highly desirable, as the information provided was not representative of the various crop groups, did not cover extended storage and suggested variable storage stability.

DIETARY RISK ASSESSMENT

Long-term intake

STMR or STMR-P values were estimated by the 1998 JMPR and by the present Meeting for 27 commodities. When data on consumption were available, these values were used in the estimates of dietary intake.

The dietary intake from the five GEMS/Food regional diets, on the basis of the STMR values, represented 3–30% of ADI (Annex 3). The Meeting concluded that the intake of residues of oxydemeton methyl resulting from uses that have been considered by the JMPR is unlikely to present a public health concern.

Short-term intake

The IESTI for oxydemeton methyl was calculated for the commodities for which maximum residue levels, STMR values and highest residue levels were established and for which data on consumption (of large portions and unit weight) were available. The results are shown in Annex 4.

The ARfD for oxydemeton methyl is 0.002 mg/kg bw. The IESTI represented 0–220% of the ARfD for children and 0–90% of that for the general population. For children, 100% of the ARfD was exceeded in apple (130%), cabbage (120%), grape (220%) and orange (120%).

The Meeting concluded that the short-term intake of residues of oxydemeton methyl from uses on commodities other than apples, cabbages, grapes and oranges that have been considered by the JMPR is unlikely to present a public health concern.

4.19   PARAQUAT (057)

RESIDUE AND ANALYTICAL ASPECTS

Paraquat, a non-selective contact herbicide, was first evaluated by the JMPR for toxicology and residues in 1970. Subsequently, it was reviewed for toxicology in 1972, 1976, 1982, 1985 and 1986 and for residues in 1972, 1976, 1978 and 1981. The Meeting reviewed paraquat toxicologically within the periodic review programme in 2003 and established an ADI of 0–0.005 mg/kg bw and an ARfD of 0.006 mg/kg bw as paraquat cation. Currently, there are 22 Codex MRLs for plant commodities, their derived products and animal commodities.

The CCPR at its Thirty-second Session identified paraquat as a priority for periodic review by the 2002 JMPR, but residue evaluation was postponed to the present Meeting.

Paraquat is usually available in the form of paraquat dichloride or paraquat bis(methylsulfate). The Meeting received data on metabolism, environmental fate, analytical methods, storage stability, supervised field trials, processing and use patterns.

Metabolism

Animals

The WHO Expert Group of the 2003 JMPR reviewed studies on the the excretion balance of paraquat in *rats* given a single dose of 1 or 50 mg/kg bw [1,1'-^{14}C-dimethyl]paraquat dichloride or 14 daily doses of 1 mg/kg bw unlabelled paraquat dichloride followed by 1 mg/kg bw of the labelled compound. They also evaluated studies of the biotransformation of paraquat in rats given the same doses of radiolabelled paraquat and other studies of metabolism and toxicity in rats. They concluded that orally administered paraquat is not well absorbed. Excretion was rapid, with 60–70% in faeces and 10–20% in urine; 90% was excreted within 72 h. Paraquat was eliminated largely unchanged: 90–95% of radiolabelled paraquat in urine was identified as the parent compound.

When 23 mg/kg [1,1'-^{14}C-dimethyl]paraquat dichloride were administered through a rumen fistula to one *sheep*, all the administered radiolabel was excreted within 10 days in urine (4%) and faeces (96%), indicating that residues of orally administered paraquat would not remain or accumulate in sheep tissues. Most of the radiolabel in urine and faeces was attributed to unchanged paraquat and 2–3% to paraquat monopyridone. Less than 1% 4-carboxy-1-methylpyridinium ion, paraquat dipyridone and monoquat were found.

When 0.92 mg/kg [1,1'-^{14}C-dimethyl]paraquat dichloride was administered subcutaneously to a sheep, paraquat was again excreted rapidly. Over 80% of the administered radioactivity was excreted in urine, 69% 1 day after treatment. Unchanged paraquat accounted for most of the radiolabel. The monopyridone was present at 2–3% and monoquat as a trace metabolite. The excretion patterns in the two sheep were virtually identical, regardless of the route of administration.

A *pig* weighing about 40 kg was fed twice daily with a diet containing [1,1'-^{14}C-dimethyl]paraquat ion at a rate equivalent to 50 mg/kg for 7 days. At sacrifice, 69% of the administered radiolabel had been excreted in faeces and 3.4% in urine; 13% was present in the stomach contents and viscera. All the radiolabel found in tissues, except in liver, was attributed to paraquat. About 70% of the radiolabel in the liver was identified as paraquat, with 7% as monoquat ion and about 0.6% as monopyridone ion. This result indicates that there is no significant metabolism of paraquat in pigs.

In a similar study, a pig was fed a diet containing [2,2',6,6'-^{14}C]paraquat ion at a rate equivalent to 50 mg/kg for 7 days. At sacrifice, 72.5% of the administered radiolabel had been excreted in faeces and 2.8% in urine. In the liver, about 70% of the radiolabel was identified as paraquat and 4% as monoquat ion.

A Fresian *cow* weighing 475 kg given a single dose of about 8 mg/kg [1,1'-^{14}C-dimethyl]paraquat dichloride from a balling gun excreted 95.6% of the administered radioactivity in faeces within 9 days; 89% was excreted within the first 3 days. Analysis indicated that 97–99% of the radioactivity in 1–4-day faeces and 100% of that in 5–6-day faeces co-chromatographed with paraquat. A total of 0.7% of the administered dose was excreted in urine, 80% of which was excreted within the first 2 days. Paraquat accounted for 90% of the radiolabel in urine on day 1, 70% on day 3 and 62% on day 5. The remaining activity was attributed to paraquat monopyridone and monoquat. Only 0.0032% of the administered radiolabel was recovered from milk within 9 days. The traces of radioactivity in milk (a maximum of 0.005 mg/l as paraquat ion equivalent milk taken in the morning of day 2) were attributed mainly to paraquat and its monopyridone and to a naturally occurring compound which appeared to be lactose. The residue level of any one compound in milk was ≤ 0.002 mg/kg.

When a lactating *goat* was dosed with [2,2′,6,6′-^{14}C]paraquat dichloride twice daily at each milking for 7 days at a total daily rate equivalent to approximately 100 mg/kg in the diet, 50.3% of the administered radioactivity was excreted in faeces, 2.4% in urine and 33.2% in stomach contents by the time of sacrifice. The total radioactivity, expressed in paraquat ion equivalents, in milk increased during the experimental period, reaching a maximum of 0.0092 mg/kg (equivalent to 0.003% of the daily dose) 4 h before slaughter. Of this radioactivity, 75.7% was attributed to paraquat, and 15.8% did not show a cationic character. There appeared to be no significant metabolism of paraquat in any tissue, except liver and peritoneal fat, where about half the radiolabel was attributed to paraquat, < 5% as monopyridone ion and 5% as monoquat ion.

Warren laying *hens* given [2,2′,6,6′-^{14}C]paraquat ion in gelatin capsules at a rate equivalent to 30 mg/kg normal diet for 10 days had excreted 99% of the administered radiolabel in faeces at the time of sacrifice; 96.6% of the radiolabel was attributed to unchanged paraquat. The amount of radiolabel in egg albumen did not exceed 0.0014 mg/kg in paraquat ion equivalents throughout the experimental period, while that in the yolk was < 0.001 mg/kg on day 1 and increased gradually to 0.18 mg/kg (in one bird) on day 8. All the radiolabel in yolk was identified as paraquat.

The studies on the fate of orally administered paraquat show that most is excreted unchanged, mainly in faeces and to a much smaller extent in urine. Excretion of paraquat was rapid in all the species studied, hens showing the most efficient excretion. Little paraquat was absorbed from the gastrointestinal tract, and the small amount absorbed was not significantly metabolized. Less than 0.05 mg/kg of paraquat was found in muscle, milk and eggs, even at the high dose rates used in these studies. These findings indicate that no significant bioaccumulation of paraquat is expected to occur in these species.

The metabolism of paraquat in these species was similar. Four metabolites were identified: monoquat, paraquat monopyridone, 4-carboxy-1-methylpyridinium ion and paraquat dipyridone. In all tissues except liver of all the species tested and in goat peritoneal fat, 80–100% of the total radiolabel was attributable to the parent compound, paraquat. In liver and goat peritoneal fat, 50–80% of the radiolabel was associated with paraquat, and absorbed paraquat was metabolized to monoquat and paraquat monopyridone and to a much smaller extent to 4-carboxy-1-methylpyridinium ion. The metabolism of paraquat involves oxygenation of one pyridine ring to form paraquat monopyridone and desmethylation of one pyridine ring to form monoquat. Cleavage of the pyridine–pyridine linkage produces 4-carboxy-1-methylpyridinium ion. The other *N*-methylpyridine moiety would produce carbon dioxide and methylamine.

Plants

When paraquat is used as a directed spray before sowing, before planting, before emergence and after emergence, it is present in soil as residues, but no direct contact occurs with crops. Sandy loam soil in pots in which *lettuce* and *carrots* were sown was sprayed with [U-^{14}C-bipyridyl]paraquat ion immediately after sowing at rates equivalent to 14.3 kg ai/ha for lettuce and 14.7 kg ai/ha for carrots, which are 13 times the highest current application rates for those crops, and maintained in a greenhouse. The radiolabel in mature lettuce and carrots harvested 65 and 96 days after treatment represented 0.0034 and 0.0048 mg/kg in paraquat ion equivalents, respectively. This result confirms the lack of significant translocation of residues of paraquat from treated soil to lettuce leaves or carrot roots.

Paraquat is also used as a crop desiccant and harvest aid, when it is in direct contact with crops. The foliage of *potatoes* and *soya beans* growing in pots in a greenhouse was treated with ^{14}C-paraquat at rates equivalent to 8.7 or 8.8 kg ai/ha (potato) and 8.2 kg ai/ha (soya beans), 14–16 times the highest current use for desiccation on potato and soya beans. The average TRR, expressed in paraquat ion equivalents, in soya and potato plants harvested 4 days after treatment were 638 mg/kg in soya foliage, 0.747 mg/kg in soya beans and 0.082 mg/kg in potato tuber. In all the samples, 89–94% of the TRR was identified as paraquat. The rest of the radioactive residue consisted of two or three fractions, none of which exceeded 10% of the respective

TRR. In soya foliage extracts, small amounts of 4-carboxy-1-methylpyridinium ion (0.3% TRR) and monoquat (0.3 % TRR) were found. The latter is a known photodegradation product of paraquat.

As paraquat is strongly adsorbed by soil (see above), its uptake by plants after pre-emergence or post-emergence directed use is insignificant, even at exaggerated application rates. When paraquat was applied as a desiccant to potato and soya bean at a rate > 10 times the highest recommended application rate, with a 4-day PHI, the main component in potato tuber, soya beans and soya foliage was paraquat. In soya foliage, monoquat and 4-carboxy-1-methylpyridinium ion were also found. Although the latter is a known photodegradation product and was not found in soya beans or potato tuber, biotransformation cannot be excluded because the TRR was too low for reliable identification. As the fate of paraquat in soya foliage appears to involve photodegradation, its fate is considered to be common among plants.

The metabolism of paraquat involves desmethylation of one pyridine ring to form monoquat. 4-Carboxy-1-methylpyridinium ion appears to be produced by photolysis of monoquat, with breakdown of the pyridine–pyridine linkage, but involvement of biotransformation cannot be excluded. Paraquat monopyridone and dipyridone, which are found in animals, were not found in plants even at much higher than normal application rates. The transformation of paraquat in plants is similar to its metabolism in animals.

Environmental fate

Soil

Paraquat was applied to slurries of loam, loamy sand, silty clay loam or coarse sand in 0.01 mol/l aqueous calcium chloride at rates higher than normal, to give 0.01 mg/l in the equilibrium solution after a 16-h equilibration. The calculated adsorption coefficients ranged from 480 in the coarse sand to 50 000 in the loam. At normal application rates, the concentration of paraquat in the equilibrium solution could not be determined (< 0.0075 mg/l). No significant desorption was observed.

A field survey of 242 agricultural soils in Denmark, Germany, Greece, Italy, The Netherlands and the United Kingdom showed that paraquat was strongly adsorbed to all the soil types studied. The adsorption coefficients calculated at application rates much higher than normal ranged from 980 to 400 000, and those adjusted for the organic carbon content of soil were 8400–40 000 000. Adsorption coefficients could not be calculated at normal application rates because the concentration in equilibrium solution was below the limit of determination (0.01 mg/l). On the McCall scale, paraquat was classified as 'immobile' in all these soils, without leaching.

[2,6-^{14}C]Paraquat was applied to sandy loam soil in pots at a nominal rate of 1.05 kg/ha and incubated in the dark at 20 ± 2 °C under aerobic conditions in order to study the aerobic degradation of paraquat. After 180 days of incubation, paraquat accounted for > 93% of the applied radiocarbon, with no detected degradation products. Less than 0.1% of the applied radioactivity evolved as $^{14}CO_2$ over the 180-day incubation period. The half-life of paraquat in soil under aerobic conditions could not be estimated, although a long half-life in soil was implied by the results of the study.

In long-term field dissipation studies conducted on cropped plots in Australia, Malaysia, The Netherlands, Thailand, the United Kingdom and the USA, the location had no major effect on the field dissipation rate. Generally, paraquat residue levels had declined to about 50% 10–20 years after the start of the studies. This implies a DT_{50} of 10–20 years after application of single, large doses of paraquat to soil. The DT_{90} could not be estimated in these studies, however, as the experimental periods were too short.

Conventional laboratory studies could not provide useful information on the route or rate of degradation of paraquat in soil because of its strong adsorption to soil minerals and organic matter. In order to obtain information, microbiological degradation studies were conducted with microorganisms isolated from soil. The most effective soil organism for decomposing paraquat was a yeast species, *Lipomyces starkeyi*. When incubated with radiolabelled paraquat, the yeast culture or cultures originating from two sandy loam soils decomposed most of the paraquat, released CO_2 and formed oxalic acid at 24–25 °C.

An unidentified bacterium isolated from soil metabolized [1,1′-^{14}C]paraquat to monoquat and 4-carboxy-1-methylpyridinium ion. Extracts of *Achromobacter* D were found to produce CO_2, methylamine, succinate and formate as metabolites of 4-carboxy-1-methylpyridinium ion. The results showed that the CO_2 originated from a carboxyl group, methylamine from the *N*-methyl group and the carbon skeletons of formate and succinate from the C-2 and C-3–C-6 atoms of the pyridine ring, respectively. These results indicate that the pyridine ring is split between C-2 and C-3.

The degradation rate of paraquat in soil was determined by cultivating 10 mg/kg [U-^{14}C-dipyridyl]paraquat with *Lipomyces* and mixed cultures derived from two soils. The degradation of paraquat was rapid, with a DT_{50} between 0.02 and 1.3 days after a lag phase of about 2 days, accompanied by rapid mineralization to CO_2 and the formation of several unidentified minor polar metabolites.

The photolysis of [2,2′,6,6′-^{14}C]-paraquat was studied by applying it to the surface of a highly sandy soil which was exposed to natural sunlight. The proportion of paraquat in samples declined during 85 weeks, at which time paraquat represented 86.6–89.5% of the total radiolabel found in unmixed and mixed soil samples. Thin-layer chromatographic analysis of the 6 mol/l HCl extracts of mixed and unmixed soils contained monoquat ion, paraquat monopyridone ion and an uncharacterized compound, which accounted for 1.4–2.4%, 1.2–1.3% and 1.8–2.4%, respectively, of the total radioactivity after 85 weeks. Photodegradation on the soil surface is not considered to be a major environmental degradation process for paraquat.

Water–sediment systems

Aqueous photolysis of paraquat was examined by maintaining ring-labelled paraquat in sterilized 0.01 mol/l phosphate buffer solution (28 mg/l) at 25 °C under light. After 36 days of irradiation simulating summer sunlight in Florida (USA), most of the recovered radioactivity was attributed to paraquat, with 0.13% as CO_2 and no photodegradation products. When solutions of radiolabelled paraquat were exposed to unfiltered ultraviolet light, no paraquat remained after 3 days, with formation of CO_2, methylamine and 4-carboxy-1-methylpyridinium ion; the last metabolite further degraded to CO_2 and methylamine. These results indicate that, while paraquat appears to be stable to photolysis at pH 7, it readily degrades into CO_2 and methylamine when exposed to unfiltered ultraviolet light.

[U-^{14}C-dipyridyl]Paraquat in deionized water was applied to the water surface of two continuously aerated sediment–water systems at a rate equivalent to 1.1 kg ai/ha. Paraquat was strongly adsorbed to the sediment in both systems, even immediately after treatment. After 100 days of incubation, 0.1–0.2% of the applied radioactivity was found in the aqueous phase, 92.9–94.9% in extracts from sediment fractions and 4.2–4.5% in unextracted sediment fractions. Most of the radiolabel recovered from the aqueous phase and sediment extract was attributed to paraquat, while no degradation products were detected. The DT_{50} or the DT_{90} could not be estimated as no significant degradation of paraquat was observed during the experimental period.

Residues in succeeding crops

Seeds of wheat, lettuce and carrot were sown into individual pots containing a sandy loam soil 0, 30, 120 and 360 days after treatment of the soil with [2,2′,6,6′-^{14}C]paraquat at an application rate equivalent to 1.05 kg ai/ha, and were maintained in a glasshouse until maturity. Over the course of the study, the TRR in soil represented an average of 99.2% of the applied radioactivity. ^{14}C-Paraquat accounted for 72.7–99.3% of the TRR in soil extracts and no other radioactive compounds were detected in any soil sample. Radioactive residues, expressed in paraquat ion equivalents per kilogram, were below the LOQ in most crop samples sown 0, 30 and 120 days after treatment. The highest radioactive residue level, 0.009 mg/kg in paraquat ion equivalents, was found in wheat straw sown 30 days after treatment.

Seeds of lettuce and carrot were sown in pots containing sandy loam soil, and the soil was treated immediately afterwards with [U-^{14}C-dipyridyl]paraquat at exaggerated rates of 14.3 and 14.7 kg/ha respectively, corresponding to approximately 13 times the highest current application rate. The lettuce was harvested 65 days after treatment and the carrots 96 days after treatment. The levels of radioactive residues in lettuce leaf and carrot root at harvest were 0.0034 and 0.0048 mg/kg in paraquat ion equivalents, respectively. There is therefore no significant uptake of paraquat into rotational crops, even when the soil is treated at exaggerated rates.

Methods of analysis

With the long history of registration of paraquat in many countries, many analytical methods have been developed and used for measuring residues in plant and animal commodities. All the methods provided to the Meeting were for analysis of paraquat only. Some analytical methods allow separate determination of paraquat and diquat in a sample.

Samples of plant origin

Six analytical methods for the determination of paraquat in plant commodities and oil and oil cake were submitted.

Three of the methods involve extraction of paraquat by refluxing homogenized or comminuted samples in 0.5 mol/l sulfuric acid for 5 h; filtration, cation-exchange chromatography from which paraquat is eluted with saturated ammonium chloride, conversion of paraquat to its coloured free radical with 0.2% (w/v) sodium dithionite in 0.3 mol/l NaOH and spectrophotometric measurement. The methods differ only in the spectrophotometric measures used: absorption of the free radical in the range 360–430 nm measured against a control solution or absorption in the range of 380–430 nm measured in second derivative mode against a paraquat standard.

In the most recent method, the eluate from cation-exchange chromatography is further cleaned up on a C18 SepPak solid phase extraction cartridge, and the second 5-ml eluate is analysed by reverse-phase ion-pair HPLC with ultraviolet detection at 258 nm.

Two other methods developed for the determination of paraquat in liquid samples, such as oil, also involve second derivative spectrophotometry (360–430 nm), but they do not involve extraction with sulfuric acid. Reverse-phase ion-pair HPLC is also used as the confirmatory method.

All these methods were validated in one or several laboratories for vegetables and fruits, cereal grains and seed, grass and straw, sugar-cane juice, oil seeds, oil and oil cake. The LOQ of these methods ranged from 0.01 to 0.05 mg/kg, except for oil cake, for which the LOQ was 0.5 mg/kg. The mean procedural recoveries were 61–107% at fortification rates reflecting both the LOQ and the actual levels of incurred residues. In general, lower recoveries were made from oil and oil cake. The mean recovery from rape-seed oil cake and olive oil was 67% and that from coffee beans was 61%; those from other commodities were > 70%. The relative standard deviation of recoveries ranged from 2% to 19%.

Samples of animal origin

Three analytical methods for the determination of paraquat in animal products were submitted.

Two methods, including the most recent, for determining paraquat in milk, eggs and animal tissues involve extraction of paraquat by homogenizing samples in 10% trichloroacetic acid, centrifugation, dilution with water, application to a cation-exchange column, sequential washing, elution of paraquat with saturated ammonium chloride, determination by reverse-phase ion-pair HPLC with ultraviolet detection at 258 nm. Fat in milk, skin with subcutaneous fat and fat samples must be removed by hexane extraction before cation exchange.

A method for analysing liquid samples, including milk, does not involve acid extraction or defatting, and milk is mixed directly with cation exchange resin before packing. Otherwise, this method is the same as those described above.

The LOQs were reported to be 0.005 mg/kg for milk, eggs and bovine, ovine and chicken tissues. The mean procedural recoveries were 75–105%, with a relative standard deviation of 2–13%.

The currently used methods for plant and animal samples were found to be suitable for quantification of paraquat in plant and animal commodities for enforcement purposes. The methods are fully validated and include confirmatory techniques. The earlier methods for quantification of paraquat in plant and animal samples were also found to be suitable in validation; however, a mean recovery < 70% was seen for rape-seed cake, olive oil and coffee beans analysed by one of the methods.

Stability of residues in stored analytical samples

Investigations were reported of the stability of residues in ground samples of prunes, banana, cabbage, potato, carrot, tomato, maize (grain, forage, fodder and silage), wheat grain, coffee beans, birdsfoot trefoil (forage and hay), meat, milk and eggs stored in a deep freezer at a temperature < –15 °C for 1–4 years.

No decrease in residue levels of paraquat, whether fortified or incurred, was observed in any of the crop matrices during the test period, the longest being 46 months. The exception was a slight decrease in birdsfoot trefoil forage that had been treated at a rate equivalent to 0.54 kg ai/ha and contained incurred residues at 57 mg/kg.

No decrease in the levels of residues of paraquat in animal commodity matrices over time was observed under storage for up to 28 months. The test matrices represented a diverse selection of animal tissues, and the studies demonstrate the stability of paraquat under various storage conditions.

Definition of the residue

Paraquat is usually available as the dichloride salt or the bis(methylsulfate) salt but is determined as paraquat ion in analysis. Paraquat is known to adsorb strongly to soil, and most of the small amount incorporated into plant remains as paraquat (90%). Its metabolites were not found when paraquat was applied at normal rates. When it was applied post-emergence, most of the applied compound remained, with minimal amounts of photodegradation products, indicating the involvement of photolysis in the transformation of paraquat. The residue of concern in plants is paraquat ion.

In studies of metabolism in rats, cattle, goats, pigs and hens, the metabolic pathway was similar, producing minor levels of oxidized metabolites. The metabolic pathways in animals and plants are similar. In animals, the residue of concern is also paraquat ion.

The definition of the residue in all countries that provided national MRLs to the Meeting was paraquat ion.

All the identified metabolites have been covered by toxicological evaluations, owing either to their occurrence in rats or in independent studies. The ADI recommended by the JMPR is for paraquat cation.

The Meeting therefore agreed that the definition of residues for plant and animal commodities should be: Paraquat cation (for both compliance with MRLs and estimation of dietary intake).

Results of supervised trials on crops

When used for weed control, paraquat is not sprayed directly onto crops and is strongly adsorbed to soil. Therefore, little paraquat is expected to be found in harvested crops. After pre-emergence application, no residues were expected to be detected in the harvested crops, although some samples contained residues. After use as a harvest aid desiccant, however, paraquat is in direct contact with crops, and the residue levels tend to be much higher than when it is used for weed control.

The Meeting agreed that data from trials of pre-plant and pre-emergence application should be evaluated against any GAP available to the Meeting, regardless of the country or region; while data on trials of post-emergence application and harvest aid desiccation should be evaluated against GAP of the country in which the trials were conducted or of a neighbouring country.

As degradation of paraquat on the surface of crops appears to involve photolysis, residue levels are expected to be similar in all crops, justifying estimation of group MRLs for paraquat.

For estimating STMR from the results of two or more sets of trials with different LOQs in which no residues exceeding the LOQs are reported, the lowest LOQ should be used, as stated in the 2002 *FAO Manual*, unless the residue level can be assumed to be essentially zero. The size of the trial database supporting the lowest LOQ was taken into account in making decisions in these cases.

Since maximum residue levels were estimated for a number of vegetable groups in which the levels were below the LOQ, the Meeting decided to withdraw the previous recommendation for vegetables (except as otherwise listed) of 0.05 * mg/kg.

In Germany, information is required on the possible contamination of fruits that have fallen onto ground treated with pesticides. Therefore, tests were carried out on apples, stone fruits, grapes and olives to simulate the residue situation in fruit used for juice and other processed products. Nevertheless, direct consumption of fruit picked up from the ground is regarded as inappropriate.

Citrus fruit

Numerous supervised residue trials have been carried out over several seasons and in several locations on orange in Italy and in California and Florida, USA, and on lime, lemon and grapefruit in Florida.

Paraquat is registered for the control of weeds around the base of citrus fruit trees at a maximum rate of 1 kg ai/ha as an inter-row spray, with no PHI, in Italy and at a maximum rate of 1.14 kg ai/ha as a directed spray, with no PHI, in the USA. The residue levels of paraquat in whole mature *oranges* in trials in Italy and the USA were below the LOQs of 0.01, 0.02 or 0.05 mg/kg, even when paraquat was applied at twice or 30 times the maximum application rate, except in two trials. In one trial with an application rate of 2.44 kg ai/ha, mature fruit from one plot contained paraquat residues at a level of 0.01 mg/kg. In a trial with an application rate of 1.12 kg ai/ha, residue levels of 0.06 and 0.08 mg/kg were found in whole fruit. In this trial, however, the lower fruit-bearing branches were deliberately sprayed, the fruit fell onto sprayed weeds, and they were picked up from the ground within 3 days of spraying for analysis. Even though this represents the worst-case scenario, it does not reflect GAP in any country and is therefore inappropriate for use in estimating a maximum residue level. The residue levels in whole mature oranges in valid trials were, in ranked order: < 0.01 (15), 0.01, < 0.02 (two) and < 0.05 mg/kg (one).

In one trial in the USA, both juice and pulp were analysed for paraquat residues. Although the levels were below the LOQ of 0.01 mg/kg, the procedural recovery was too low for the results to be regarded as reliable.

In trials in the USA on *grapefruit*, *lemon* and *lime* in 1970 and 1972, with application rates reflecting GAP in the USA, the paraquat residue levels were < 0.01 (one) and < 0.05 mg/kg (three).

As the residue situation in oranges and other citrus fruits is similar and GAP is recommended for citrus fruits as a group in Italy and the USA, the Meeting considered it appropriate to establish a group maximum residue level for citrus fruits. The combined residue levels, in ranked order, were: < <u>0.01</u> (16), 0.01, < 0.02 (two) and < 0.05 (four) mg/kg. The Meeting estimated a maximum residue level of 0.02 mg/kg, an STMR of 0.01 mg/kg and a highest residue level of 0.02 mg/kg for paraquat in citrus fruits. The value of 0.02 mg/kg covers only the finite residue level found at 0.01 mg/kg.

Pome fruit

Trials were carried out on apples in Canada, Germany and the United Kingdom and on pears in Canada and Germany.

Paraquat is registered for use to control weeds around the base of pome fruit trees at a maximum rate of 0.66 kg ai/ha with one application and no PHI in the United Kingdom and at a maximum rate of 1.14 kg ai/ha with no PHI in the USA. No information on GAP was available for Canada or Germany, but the results of trials conducted in those countries were reviewed against the GAP of the USA and United Kingdom, respectively.

Trials on *apple* were conducted at rates of 1.12–4.48 kg ai/ha, and in one trial in the United Kingdom at a highly exaggerated rate of 12.3 kg ai/ha, about 20 times the maximum rate permitted in that country. In the latter trial, paraquat was applied directly to the bark of the trees to simulate worst-case conditions. In some cases, two applications were made, in the same or subsequent years. Apples were harvested 0–780 days after the last application. In trials on *pear*, paraquat was applied at rates of 1.0–4.48 kg ai/ha once or twice, and pears were harvested 0–77 days after the last application. Paraquat residue levels were below the LOQ of 0.01 mg/kg in all apples and pears taken from trees, even after treatment at rates as high as 20 times the maximum GAP rate.

In the trials in Germany, apples and pears taken from the trees were placed on the ground 6–7 days after application and collected about 7 days later for analysis. Residue levels of paraquat of 0.02–0.19 mg/kg were found in the apples, which could be attributed to the transfer of paraquat from the sprayed weed. The Meeting concluded that these data are not appropriate for use in estimating a maximum residue level.

As the residue situations in apples and pears are similar, and GAP is recommended for pome fruits or orchard fruits as a whole in all the countries that provided information on GAP, the Meeting considered it appropriate to establish a group maximum residue level for pome fruits. As the paraquat residue levels in all the valid trials were below the LOQ, even after application at exaggerated rates, the Meeting estimated a maximum residue level for pome fruits of 0.01* mg/kg, an STMR of 0 mg/kg and a highest residue level of 0 mg/kg.

Stone fruit

Trials were carried out on peaches, plums, apricots and cherries in Canada, Germany, the United Kingdom and the USA.

Paraquat is registered for use to control weeds around the base of stone fruit trees at a maximum rate of 0.66 kg ai/ha, with one application and no PHI for stone fruits in the United Kingdom and at a maximum rate of 1.14 kg ai/ha, with three applications and a 28-day PHI for stone fruits other than peaches in the USA; the PHI for use on peach trees in the USA is 14 days. No information on GAP was available from Canada or Germany, and the results of trials conducted in those countries were reviewed against the GAP of the USA and the United Kingdom, respectively.

The application rates in the supervised trials ranged from 0.22 to 4.48 kg ai/ha, applied to the base of the fruit trees up to three times in a season; the fruit was harvested from the trees 0–103 days after the last application. No residues of paraquat above the LOQ of 0.01 or 0.05 mg/kg were found in fruit harvested directly from the trees in any trial, even after spraying three times at a rate four times the maximum permitted rate. In most of the US trials, paraquat was applied one or two times instead of the maximum of three, but because of the higher application rates, the total amount applied was higher than the maximum allowed by GAP.

In trials on plums in the United Kingdom, paraquat was applied directly to suckers at rates of 0.22–1.34 kg ai/ha. No residues were found above the LOQ of 0.01 mg/kg in fruit harvested 21 or 55 days later.

In the trials in Germany, fruit were placed on sprayed weeds and collected for analysis about 1 week later. Small amounts of paraquat residues were found (0.02 and 0.04 mg/kg on peach, < 0.01 mg/kg on plum and 0.07 mg/kg on cherry) in the fruit samples, due to transfer from the sprayed weeds. As stone fruit intended for juice production is usually grown in orchards in which herbicides are rarely used, these data were not used for estimating a maximum residue level.

As the residue situations in stone fruits are similar and GAP is recommended for stone fruits or similar GAPs are established for peach and stone fruits excluding peach, the Meeting considered it appropriate to establish a group maximum residue level for stone fruits. As the paraquat residue levels were below the LOQ, even when applied at exaggerated rates and the methods of analysis in most of the trials had a LOQ of 0.01 mg/kg, the Meeting estimated a maximum residue level for stone fruits of 0.01* mg/kg and STMR and highest residue values of 0 mg/kg.

Berries and small fruit

Grape

Trials on residues in grapes have been conducted in Canada, Japan, Switzerland and the USA at rates of 0.3–4.4 kg ai/ha applied one to five times. Grapes were harvested from the vines at maturity 0–196 days after the last application. Four trials were conducted in Germany in which paraquat was applied between the rows of established vines at a rate of 1.0 kg ai/ha and grapes were sampled from the vines 0–14 days after application.

Paraquat is registered for weed control around grape vines at a maximum rate of 0.72 kg ai/ha, with five applications and a 30-day PHI in Japan and a maximum rate of 1.14 kg ai/ha, with the number of applications and the PHI unspecified in the USA. No information on GAP was available from Canada, Germany or Switzerland, but the results of trials in Canada were reviewed against US GAP.

In all trials in Canada, Japan and the USA reviewed against respective GAP, grapes obtained directly from the vine did not contain paraquat residues at levels above the LOQ of 0.01 or 0.02 mg/kg, even when applied at five times the recommended rate or with a shorter PHI.

In the German trials, bunches of grapes were also placed on the sprayed weed a few days after application and collected 7 days later for analysis. Small amounts of paraquat residues (0.04, 0.07, 0.09, 0.10, 0.13 and 0.17 mg/kg) were found in the grapes due to transfer from the sprayed weeds. When the fruits were sampled directly from the vine, the levels of residues were always below the LOQ of 0.01 mg/kg (six trials), which supports the results of the trials conducted in Canada, Japan and the USA.

The residue levels of paraquat in grapes in the trials that met the respective GAP or were conducted at higher rates were: < 0.01 (16), < 0.02 (three) and < 0.05 (two) mg/kg.

Cane fruit

Trials on residues were conducted in Canada on red and blackcurrants, blueberries, loganberries, gooseberries and raspberries at rates of application of paraquat of 0.56–2.24 kg ai/ha. Paraquat was applied once and the fruit was harvested 20–111 days after application.

GAP for cane fruit in the USA is a maximum rate of 1.14 kg ai/ha, with the number of applications and PHI unspecified.

Even at double the application rate, cane fruit did not contain paraquat residues at levels above the LOQ of 0.01 mg/kg. The residue levels in 25 trials following GAP or conducted at higher rates were < 0.01 mg/kg.

Strawberry

Supervised trials were conducted in France, Germany and the United Kingdom in which paraquat was used to control runners of strawberry plants at rates of 0.42–1.32 kg ai/ha once or twice. Berries were harvested 47–226 days after the last application. Three trials in Germany were conducted in plastic greenhouses.

GAP in the United Kingdom for strawberries is a maximum rate of 0.66 kg ai/ha, with one application and PHI unspecified.

The residue levels of paraquat in strawberries in trials following GAP or conducted at higher application rates were < 0.01 (six) and < 0.05 mg/kg.

As the samples analysed in all the trials except that in which grapes were kept and taken from the ground did not contain paraquat residues at levels above the LOQs and the application rate in the respective GAP is similar, the Meeting decided to propose a group maximum residue level for small fruits and berries. The residue levels in these fruits, in ranked order, were: ≤ 0.01 (47), < 0.02 (three) and < 0.05 mg/kg (three). The Meeting, considering that use of modern analytical methods would enable lower LOQs, agreed to disregard residue levels of < 0.05 mg/kg and < 0.02 mg/kg and estimated a maximum residue level of 0.01* mg/kg and STMR and highest residue values of 0 mg/kg.

Olive

Trials on residues in olives have been carried out in Greece, Italy, Spain and the USA (California).

Paraquat is registered for controlling weeds around the base of olive trees at a maximum rate of 1 kg ai/ha, with the number of applications unspecified and a 40-day PHI in Italy and at a maximum rate of 1.14 kg ai/ha, with four applications and a 13-day PHI in the USA. The results of trials conducted in Greece and Spain were reviewed against GAP in Italy.

In trials in Italy, paraquat was applied at rates of 0.54–1.8 kg ai/ha to the base of trees, and olives were harvested from the ground or trees 7–21 days after application. Although the delay was shorter than the recommended PHI of 40 days, the residue levels in the olives were < 0.05 and < 0.1 (two) mg/kg, indicating that at a PHI of 40 days the levels are likely to be < 0.1 mg/kg. No residues (< 0.05 mg/kg) of paraquat were detected in the oil from these fruits.

In one trial in the USA, paraquat was applied four times at an exaggerated rate (5.6 kg ai/ha; 22.4 kg/ha total) and the fruit was harvested from the trees 13 days later for analysis. The residue levels of paraquat were below the LOQ of 0.05 mg/kg, as were the levels in oil and cake prepared from the olives.

In six trials in Spain, olives were harvested from the ground 0, 1 and 7 days after application of paraquat at 0.60 kg ai/ha, simulating the worse-case scenario of collecting olives intended for oil production. In these trials, the application rate was 60% of the maximum allowed in Italy, but the olive fruit were harvested much earlier than the PHI of 40 days. The residue levels in whole fruit were 0.64–10 mg/kg, indicating that there had been transfer of paraquat from the sprayed weeds to the olives. In all the oil produced from these samples, however, the maximum residue levels of paraquat were 0.06 mg/kg, indicating that paraquat is not extracted into oil, as might be expected from its chemical nature.

In other trials in Spain, mature olives were sprayed directly on the ground with paraquat at rates of 0.36–1.3 kg/ha, and the fruit was analysed 3–17 days after application. The residue levels of paraquat in the olives were 0.08–4.4 mg/kg. Residues of paraquat did not transfer to extracted oil, and washing appeared to reduce the levels on the fruit.

In one trial in Greece, mature olives were sprayed directly with paraquat at a rate of 1.0 kg ai/ha to simulate direct spraying on fallen fruit in collection nets during weed control. No residues were found at levels above the LOQ (0.05 mg/kg) in oil extracted from treated fruit harvested 5 days after application.

Olives for oil production are often harvested from the ground and paraquat used for weed control may occasionally be applied directly to the fallen fruit on the ground. The whole fruit will contain some paraquat residue, either through transfer from treated vegetation or through direct spraying. Although the olives may contain relatively high levels of paraquat, no transfer of paraquat to oil occurs. This practice is not in compliance with GAP for olives.

The residue levels in olives taken directly from trees were: < 0.05 and < 0.10 mg/kg (two). In another trial, the level was < 0.05 mg/kg in olives taken from ground that had not been directly sprayed. The residue levels in one US trial conducted at five times the usual rate were below the LOQ of 0.05 mg/kg, indicating that when paraquat is applied in accordance with GAP no residues are expected to occur in olive fruit. The Meeting estimated a maximum residue level of 0.1 mg/kg to replace the previous recommendation for olive at 1 mg/kg. The Meeting also estimated an STMR of 0.05 mg/kg and a highest residue level of 0.1 mg/kg.

Assorted tropical fruits minus inedible peel

Trials on residues were carried out on *passion fruit* in Hawaii, USA, at an application rate of 1.12–4.48 kg ai/ha, to control weeds. Fruit was harvested 1–28 days after application. GAP in the USA for use on passion fruit is a maximum rate of 1.05 kg ai/ha, with an unspecified number of applications and PHI. The residue level in whole fruit in a trial complying with the maximum GAP was 0.13 mg/kg. After application at a rate higher than the maximum GAP, residue levels of up to 0.19 mg/kg were found in whole fruit. The levels in the edible pulp of all passion fruits analysed in the trials, regardless of PHI, ranged from < 0.01 to 0.02 mg/kg at 1.12 kg ai/ha and from < 0.01 to 0.06 mg/kg at higher rates. Higher levels were found in peel than in the edible portion.

Trials on residues were carried out on *kiwifruit* in California, USA, at an application rate of 0.56–2.24 kg ai/ha, three times, to control weeds. Fruit was harvested 7–14 days after the last application. The US GAP for kiwifruit is a maximum rate of 1.14 kg ai/ha, with the number of applications unspecified and a 14-day PHI. The residue level in kiwifruit in one trial conducted in accordance with the maximum US GAP was < 0.01 mg/kg. Even at a higher application rate or a shorter PHI, the levels were below the LOQ of 0.01 mg/kg.

Trials on *guava* were carried out in two locations in Hawaii, USA, with three different application rates of 1.12–4.48 kg ai/ha at each location. Fruit was harvested 1–28 days after application. The US GAP for guava is identical to that for passion fruit. The residue levels of paraquat in all edible pulp and peel analysed were below the LOQ of 0.01 mg/kg at the maximum GAP rate and at rates up to four times the maximum GAP. No residue was found at levels above the LOQ of 0.01 or 0.02 mg/kg in juice, discarded skin or seed obtained from guava treated at 1.12 or 4.48 kg/ha with a 6-day PHI. Although no information was available on residues in whole fruit, levels above the LOQ were not expected in whole fruit in view of the residue situation in pulp, peel and other fractions.

Trials were carried out on *banana* in Honduras, with three applications of paraquat at 1.4 kg ai/ha or a single application at double this rate, to control weeds in established plantations. Fruit was harvested 0–90 days after the last application. As no information was available on GAP in Honduras, the data were reviewed against GAP of the USA (maximum rate of 1.14 kg ai/ha). The residue levels of paraquat in flesh (0- and 3-day PHI) and whole fruit (≥ 7-day PHI) were below the LOQ (0.01 mg/kg) in three trials, except in skin from fruit harvested immediately after application.

Except in the trials on passion fruit, the residue levels in tropical fruits in 10 trials conducted according to the respective GAP were all below the LOQ (< 0.01 mg/kg). The Meeting estimated a maximum residue level for paraquat in assorted tropical fruits with inedible peel, excluding passion fruit, of 0.01* mg/kg. The Meeting decided to withdraw the previous recommendation for passion fruit.

The residue levels in edible portions of these fruit were below the LOQ: <u>≤ 0.01</u> (11) mg/kg. The Meeting estimated STMR and highest residue values for paraquat in assorted tropical fruits minus inedible peel, excluding passion fruit, of 0.01mg/kg.

Bulb vegetables

Trials on residues were conducted on *onion* in Canada, Germany and the United Kingdom in the 1960s. Paraquat is registered in the USA for pre-plant or pre-emergence application to onion in a limited number of states at a maximum rate of 1.14 kg ai/ha, with one application and a 60-day PHI (200 days in California). Uses on bulb vegetables are not included in the label in the United Kingdom.

In one Canadian trial at twice the GAP rate and with a shorter PHI (36 days), the residue levels were below the LOQ of 0.01 mg/kg. In another Canadian trial at an application rate of 1.12 mg/kg, the levels were also < 0.01 mg/kg, but the PHI was 143 days.

Trials were conducted in Germany for post-emergence directed application and for harvest aid uses, but there was no related GAP.

In one trial conducted in the United Kingdom of pre-emergence application on spring onion, the residue level was 0.02 mg/kg, but the application rate was > 30% higher than the maximum rate allowed in the USA. A further trial on spring onion involved directed post-emergence application, for which no information on GAP was available.

The Meeting concluded that there were insufficient data to recommend a maximum residue level for paraquat in onion bulb or bulb vegetables.

Brassica vegetables

Residue trials were carried out on *broccoli* in Canada; *Brussels sprouts* in The Netherlands (harvest aid); *cabbage* in Canada, Japan, Spain and the USA; and *cauliflower* in Canada. Paraquat was applied once or twice at 0.67–2.2 kg ai/ha for inter-row weed control, and the crop was harvested 5–52 days after the last application.

Paraquat is registered for use in the cultivation of *Brassica* vegetables during seed-bed preparation as a pre-plant or pre-emergence treatment, or applied as a post-emergence directed or guarded spray for inter-row weed control. GAP in Japan is a maximum rate of 0.36 kg ai/ha, with three applications and a 30-day PHI, for broccoli, cabbage, cauliflower and Chinese cabbage as pre-plant inter-row applications. GAP in the USA is a maximum rate of 1.14 kg ai/ha, with the number of applications and PHI unspecified, for *Brassica* vegetables as pre-plant, pre-emergence treatment.

In trials conducted on broccoli, cabbage and cauliflower in Canada, the residue levels were below the LOQ of 0.01 mg/kg, even when applied at double the rate. The exception was one trial in Canada in which cabbage harvested 51 days after treatment at twice the rate contained a residue level of 0.06 mg/kg. The residue levels were < 0.01 (two) and 0.06 mg/kg.

In two trials conducted on cabbage in Japan, the residue levels were below the LOQ of 0.03 mg/kg even after application at a higher rate of 0.96 kg ai/ha and a shorter PHI of 5 days. At a highly exaggerated rate of 19.2 kg ai/ha but with only one application and a longer PHI of 52 days, the residue levels were also < 0.03 mg/kg.

No information was available on GAP that would allow evaluation of trials conducted in Spain.

Trials on Chinese cabbage were conducted in the USA in which paraquat was applied once as pre-emergence treatment at 1.05 kg ai/ha, followed by three post-emergence directed applications at 0.56 kg ai/ha. The residue levels were < 0.05 and 0.07 mg/kg. The US label allows only pre-plant and pre-emergence applications.

Trials on Brussels sprouts in The Netherlands involved a direct harvest aid application to the vegetable. In these trials, the unwashed vegetable contained a residue level of 7.3 mg/kg after 31 days, while washed vegetable had a reduced level of 1.6 after 31 days. Harvest aid desiccation was not, however, included in the labels provided to the Meeting.

The residue levels in these crops in trials that followed GAP and in trials that showed residue levels below the LOQ were, in ranked order: < 0.01 (two), <u>< 0.03</u> (two) and 0.06 mg/kg. The Meeting concluded that there were insufficient data for estimating a maximum residue level for *Brassica* vegetables.

Fruiting vegetables

Numerous residue trials were carried out on tomatoes in Canada and the USA, on cucumbers, melons and summer squash in the USA and on peppers in Canada and the USA.

Paraquat is registered in the USA for use on tomatoes for pre-plant or pre-emergence application at a maximum rate of 1.14 kg ai/ha, with an unspecified number of applications and a 30-day PHI; on tomatoes for post-emergence directed spray at a maximum rate of 0.55 kg ai/ha, with an unspecified number of applications and a 30-day PHI; on peppers by directed spray application at a maximum rate of 0.55 kg ai/ha, with three applications and no PHI; and on other fruiting vegetables for pre-plant or pre-emergence application at a maximum rate of 1.14 kg ai/ha, with unspecified number of applications and PHI.

The trials in Canada on *tomatoes* were for pre-emergence or pre-planting weed control, in which paraquat was used at a low rate of 0.11 kg ai/ha. Trials on tomatoes in the USA involved post-emergence directed application at 0.56–2.24 kg/ha and an exaggerated single high pre-emergence application at a rate of 11.2 kg ai/ha or pre-emergence application of 1.12 kg ai/ha followed by three inter-row directed applications at 2.8 kg ai/ha. Although samples were harvested 21 days after treatment, 30% shorter than the PHI in US GAP of 30 days, the residue levels in tomatoes were below the LOQ of 0.01 mg/kg after application at 0.56 kg ai/ha for post-emergence directed application, except in one trial in which levels up to 0.04 mg/kg were found. After application at exaggerated rates, the residue levels were still below the LOQ of 0.005 or 0.01 mg/kg or at a maximum of 0.02 mg/kg.

The residue levels in trials following GAP or conducted at higher application rates were, in ranked order: < 0.005 (two), ≤ 0.01 (seven) and 0.04 mg/kg.

The trials on *sweet peppers* were for use of paraquat in inter-row weed control at 0.56–2.2 kg ai/ha. The residue levels in trials at maximum GAP were < 0.01 and 0.01 mg/kg. The levels after exaggerated application rates were either below the LOQ of 0.01 mg/kg, 0.03 mg/kg (once at 1.12 kg ai/ha pre- emergence and four times at 1.12 or 2.24 kg ai/ha post-emergence applications) or 0.02 mg/kg (one trial).

The Meeting considered it appropriate to evaluate residues in tomato and peppers together for estimating the maximum residue level for fruiting vegetables, other than cucurbits. The combined levels were: < 0.005 (two), ≤ 0.01 (eight), 0.01 and 0.04 mg/kg. The Meeting estimated a maximum residue level for fruiting vegetables, other than cucurbits, of 0.05 mg/kg, an STMR of 0.01 mg/kg and a highest residue level of 0.04 mg/kg.

In trials on *cucumbers*, *melons* and *summer squash* in California (USA), paraquat was applied at 1.12 kg ai/ha pre-emergence, followed by three inter-row applications at 0.56 kg ai/ha. While US GAP allows pre-emergence application at a maximum of 1.12 kg ai/ha, the residue levels of paraquat in all 12 trials were below the LOQ of 0.025 mg/kg. The Meeting estimated a maximum residue level for cucurbits of 0.02 mg/kg and STMR and highest residue values of 0 mg/kg.

Leafy vegetables

Trials for residues were conducted on lettuce in Canada, Germany, Spain, the United Kingdom and the USA, on kale in France, Italy and the United Kingdom and on turnip greens in the USA.

Paraquat is registered for pre-emergence application on collard and lettuce in the USA at a maximum rate of 1.14 kg ai/ha, with the number of applications and PHI unspecified. Uses on leafy vegetables are not included on labels in Italy or the United Kingdom.

Trials on residues on *lettuce* were conducted in Canada, Germany, Spain, the United Kingdom and the USA at application rates of 0.42–2.24 kg/ha; lettuce was sampled 0–147 days after application. In trials conducted in Canada and the USA following US GAP, the residue levels in untrimmed head or bunch were 0.01, 0.04 and 0.05 mg/kg.

The results of trials in the United Kingdom were evaluated against US GAP, as the uses were similar in trials in the two countries. The residue levels in unwashed lettuce head in trials following US GAP were < 0.01, 0.01 and 0.02 mg/kg.

Residue levels up to 1.4 mg/kg were found in German trials on lettuce harvested immediately after one or two applications of paraquat for post-emergence inter-row weed control. The residues were believed to have derived from spray drift onto the outer leaves. In most of these trials, the whole lettuce head was analysed without removal of outer wrapper leaves that were yellow and withered. The residue levels had declined to close to the LOQ (< 0.01 mg/kg) by 21 days after harvest. The results of trials in Germany and Spain could not be evaluated as no information on GAP in Europe was available.

Residue trials on *kale* were carried out in France, Italy and the United Kingdom at rates of 1.0–2.24 kg/ha, and kale was sampled 0–147 days after application. As no information was available on GAP in Europe, these data were not evaluated.

Six trials on *turnip greens* were carried out in the USA at a rate of 1.12 kg/ha, with sampling 55–128 days after application. The levels of paraquat residue were < 0.025 (three), 0.03, 0.04 and 0.05 mg/kg.

As the US GAPs for collard and lettuce are identical and the residue situations for these crops were similar, the Meeting considered it appropriate to combine the results for estimating a maximum residue level for leafy vegetables. The combined residue results, in ranked order were: < 0.01, 0.01 (two), 0.02, ≤ 0.025 (three), 0.03, 0.04 (two) and 0.05 (two) mg/kg. The Meeting estimated a maximum residue level for paraquat in leafy vegetables of 0.07 mg/kg, an STMR of 0.025 mg/kg and a highest residue level of 0.05 mg/kg.

Legume vegetables and pulses

Residue trials were conducted on beans (with pod and dry) in Canada, Germany, Italy, The Netherlands and Spain, on broad beans in Spain, on peas in Australia, Canada and the USA, and on soya beans in Brazil and the USA.

Paraquat is registered for weed control and harvest aid on legume vegetables and pulses in Australia, Brazil and the USA as follows:

Country	Maximum rate (kg ai/ha)	No. of applications	PHI (days)	Crop	Type of application
Australia	0.2		14	Chickpea	Over-the-top spray
	0.2		14	Field pea	Over-the-top spray
	0.43			Soya bean	Pre-plant
Brazil	0.6	1	7	Soya bean	Pre-plant
	0.5	1	7	Soya bean	Desiccation
USA	1.14		–	Beans (lima, snap)	Pre-plant, pre-emergence
	1.14		–	Pea	Pre-plant, pre-emergence
	0.55	2	7	Pulses	Harvest aid
	1.14		–	Soya bean	Pre-plant or pre-emergence Should not exceed 1.9 l per season
	0.14	2	–	Soya bean	Post-emgence directed spray Second and final application 7–14 days later if needed
	0.28		15	Soya bean	Harvest aid

Uses on legumes and pulses were not included in the European labels provided to the current Meeting.

Residue trials were carried out on *dry beans* (genus *Phaseolus*) in Germany, Italy, The Netherlands and Spain, in which paraquat was used for pre-emergence weed control at single application of 0.56 or 2.24 kg ai/ha or post-emergence directed inter-row weeding at rates of 0.28–1.12 kg ai/ha. In trials in Europe, young pods were harvested 0–7 days after treatment and analysed. The residue levels in beans in pods were < 0.05–0.10 mg/kg (five trials). As no related GAP was available, these results were not used in estimating a maximum residue level. The Meeting concluded that there were insufficient data to estimate a maximum residue level for legume vegetables.

The residue levels of paraquat in dry beans in Canadian trials after pre-emergence application following GAP were < 0.01 (two), < 0.05 and 0.07 mg/kg.

Residue trials were conducted on *broad beans* in Spain after post-emergence directed spray. The residue levels in seeds harvested on the day of application were < 0.05 mg/kg (two); however, no information was available on related GAP.

Residue trials were carried out on *peas* in Canada and the United Kingdom with paraquat used for pre-emergence weed control at single applications or post-emergence directed inter-row weeding at rates of 0.14–1.68 kg ai/ha and harvesting 55–152 days after application. The residue levels of paraquat in seeds were below the LOQ of 0.01 or 0.05 mg/kg in trials with post-emergence application; however, no GAP was available for post-emergence application on peas.

Paraquat was applied at 0.20 or 1.12 kg ai/ha to field peas and chick peas as a harvest aid desiccant in Australia and the USA, with samples taken 1–38 days after application. The resulting residues of paraquat in seed in trials following GAP were found at levels of: 0.05, 0.15, 0.23, 0.25, 0.31 and 0.41 mg/kg.

A number of trials were conducted on *soya beans* in Brazil between 1981 and 1983 with a harvest aid desiccation application of paraquat at 0.25–0.80 kg/ha and sampling 2–21 days after application. The residue levels of paraquat in seed in trials following GAP in Brazil were: < 0.02, 0.03 (two), < 0.05 (two), 0.07, 0.08, 0.09, 0.10, 0.11 (two), 0.13, 0.16 (two) and 0.28 (three) mg/kg.

In trials conducted in the USA with pre-emergence application with or without a post-emergence directed application at 0.14–1.4 kg/ha, the residue levels of paraquat in soya beans harvested 3–147 days after the last application in trials following GAP were < 0.025 (nine) and 0.03 mg/kg.

Other trials were conduced in the USA on harvest aid desiccation application at 0.28 or 0.56 kg/ha and sampling 6–36 days after application. The residue levels of paraquat in seeds in trials following GAP were: < 0.01, 0.02 (four), 0.03 (two), 0.04 (two), 0.05, 0.06, 0.07, 0.08 (two), 0.09, 0.12 and 0.13 mg/kg. The hulls of treated soya beans contained higher residues than seeds.

The results of these trials clearly indicate that the levels of residues arising from harvest desiccant uses are higher than those from pre-emergence or post-emergence application.

The Meeting considered it appropriate to combine the results of trials on field peas and chick peas in Australia and on soya beans in Brazil and the USA in which paraquat was used as a harvest aid desiccant to estimate a group maximum residue level for pulses. The combined residue levels in seeds were, in ranked order: < 0.01 (two), < 0.02, 0.02 (four), 0.03 (four), 0.04 (two), < 0.05 (two), 0.05 (two), 0.06, 0.07 (two), 0.08 (three), 0.09 (two), 0.10, 0.11 (two), 0.12, 0.13 (two), 0.15, 0.16 (two), 0.23, 0.25, 0.28 (three), 0.31 and 0.41 mg/kg. The Meeting estimated a maximum residue level of 0.5 mg/kg to replace the previous recommendation for soya bean and an STMR of 0.08 mg/kg and a highest residue level for pulses of 0.41 mg/kg.

Root and tuber vegetables

Paraquat is registered for use at a maximum rate of 0.36 kg ai/ha with three applications and a 30-day PHI in Japan for pre-plant, inter-row application on carrot and in the USA at a maximum rate of 1.14 kg ai/ha for pre-emergence treatment of root and tuber vegetables excluding potatoes.

Two residue trials carried out on *beetroot* in Canada and the United Kingdom for pre-emergence application in compliance with US GAP resulted in residue levels of < 0.01 and 0.03 mg/kg.

Residue trials were conducted in the United Kingdom on beetroot and *sugar-beet* in which paraquat was used pre-sowing or pre-emergence at 1.68 kg ai/ha, followed by two directed inter-row applications at

2.24 kg ai/ha after crop emergence. No information was available, however, on GAP for post-emergence application from Europe.

In trials conducted in four states of the USA with pre-emergence application at 1.12 kg ai/ha, the residue levels in sugar-beet roots harvested 136–178 days after application were < 0.05 mg/kg (six) after a single pre-emergence application at 1.12 kg ai/ha. After application at an exaggerated rate of 5.6 kg ai/ha, the residue levels in unwashed root were < 0.05 mg/kg.

Residue trials on *carrots* with use of paraquat for pre-emergence or inter-row weed control have been carried out in Canada, Japan, Germany and the United Kingdom. The residue levels of paraquat in carrot in the Japanese trials after both pre-emergence and inter-row applications were all below the LOQ of 0.03 mg/kg, despite a shorter PHI or use of a highly exaggerated rate of 19.2 kg ai/ha. The residue levels in carrot in four trials following GAP or conducted at higher rates or shorter PHI were < 0.03 mg/kg. In Canadian trials, the residue levels were below the LOQ of 0.01 mg/kg, even in one trial in which the rate was doubled and the PHI shorter.

As no information was available on GAP in Europe, the data from German trials with post-emergence application were not considered in estimating the maximum residue level.

Residue trials were carried out on *parsnips* and *swedes* in the United Kingdom and on *turnips* in Canada and United Kingdom with use of paraquat for pre-emergence weed control (Canada) or pre-emergence followed by inter-row weed control (United Kingdom). The rates of application were 0.56–2.24 kg ai/ha. Turnip, swede and parsnip roots were harvested 49–122 days after application. The residue levels of paraquat in turnips in two Canadian trials that followed US GAP were < 0.01 mg/kg. No information on GAP was available for post-emergence application in Europe.

One trial was conducted in France on *black salsify*, in which paraquat was applied as an inter-row treatment at 0.5 and 0.8 kg ai/ha. There were no residues (< 0.02 mg/kg) in salsify roots harvested 8 and 80 days after treatment; however, no information on GAP was available.

The combined residue levels in beetroot, sugar-beet, carrots and turnips were, in ranked order: < 0.01 (four), < 0.03 (four), 0.03 (two) and < 0.05 (six) mg/kg.

Potato

Trials were carried out on potatoes in Canada, Germany, the United Kingdom and the USA for pre-emergence, post-emergence and harvest aid applications of paraquat.

Paraquat is registered in the United Kingdom for pre-emergence use at a maximum rate of 0.66 kg ai/ha with one application. It is registered in the USA for pre-plant and pre-emergence broadcast application at a maximum rate of 0.55 kg ai/ha and for broadcast application for pre-harvest vine killing and weed dessication at a maximum rate of 0.42 kg ai/ha with a 3-day PHI. The latter application is restricted to fresh market produce, with a restriction of 2.3 l/ha per season; split applications must be applied a minimum of 5 days apart.

Trials were carried out in Germany with post-emergence directed application. The residue levels were below the LOQ of 0.01 mg/kg.

Several residue trials were carried out in Canada and the USA in which paraquat was applied for weed control by pre-emergence or post-crop emergence application at a rate of 0.20–1.12 kg ai/ha. The residue levels in the tubers in trials following US GAP were < 0.01 (eight) and 0.02 mg/kg. At double the application rate, the residue levels were below the LOQ of 0.01 mg/kg.

Trials were also carried out on harvest aid desiccant use in Canada, the United Kingdom and the USA. The US label allows use of paraquat for vine killing and weed desiccation at a maximum of 0.42 kg ai/ha, with a PHI of 3 days, but in these trials rates equivalent to or higher than twice the maximum rate or a much longer PHI were used. Harvest aid use is not included in the United Kingdom label.

The residue levels in trials of pre- and post-emergence application were < 0.01 (eight) and 0.02 mg/kg. The levels in trials with double the application rate in the USA and in trials conducted in Germany were all below the LOQ.

The Meeting decided to combine the results from trials on beetroot, sugar-beet, carrot, turnip and potato. The combined residue levels, in ranked order, were: < 0.01 (12), <u>0.02</u>, < 0.03 (four), 0.03 (two) and < 0.05 (six) mg/kg. The Meeting estimated a maximum residue level of 0.05 mg/kg, an STMR of 0.02 mg/kg and a highest residue level of 0.05 mg/kg for root and tuber vegetables. The maximum residue level replaces the previous recommendation for potato.

Stem vegetables

Residue trials have been carried out on asparagus, celery and globe artichokes in Canada and the USA with use of paraquat for post-emergence directed inter-row weeding at rates of 1.12–3.25 kg ai/ha in a single application. Three applications of 1.12 or 1.35 kg/ha on artichokes were also tested.

Paraquat is registered in the USA for *asparagus* at a maximum rate of 1.14 kg ai/ha for pre-plant and pre-emergence broadcast or banded over-row application and at the same maximum rate with a 6-day PHI for asparagus more than 2 years old by broadcast or banded over-row application. The residue levels were < 0.02 (two) and < 0.05 mg/kg.

Although trials were conducted on *celery* in Canada and on *artichoke* in the USA, no information on GAP for these crops was available. The Meeting concluded that the data were insufficient for estimating a maximum residue level for asparagus.

Cereal grains

Maize

Residue trials were conducted on maize in Canada, Italy, the United Kingdom and the USA with pre- and post-emergence applications and harvest aid uses.

Paraquat is registered for use in the USA at a maximum rate of 1.14 kg ai/ha for pre-plant or pre-emergence broadcast of banded over-row applications and at a maximum rate of 0.55 kg ai/ha for post-emergence directed spray. Residue trials were conducted with use of paraquat for pre-emergence weed control or for post-emergence directed spray in Canada and the USA at rates of 0.28–1.12 kg ai/ha.

In a series of trials in the USA in 1987, one pre-emergence application at 1.12 kg ai/ha and two post-emergence applications at 0.31 kg ai/ha were made. Although the post-emergence application rate was not as high as the maximum rate, the pre-emergence application rate was the maximum allowed for pre-emergence application. The Meeting considered that these trials were conducted in accordance with US GAP. The residue levels in trials in Canada and the USA conducted in accordance with US GAP were: < 0.01 (eight) and < 0.025 mg/kg (16). In trials with higher application rates (up to four times), the residue levels were below the LOQ. The levels in maize cobs were also below the LOQ of 0.01 mg/kg (two trials).

In two residue trials in Italy, paraquat was applied pre-emergence at 0.92 kg ai/ha. The residue levels in cob were < 0.05 mg/kg; however, no analysis of kernels or grain was reported.

Trials were conducted in South Africa and the United Kingdom with post-emergence application; however, owing to the lack of relevant GAP for South Africa and the fact that post-emergence application is not included on the label in the United Kingdom, the results of these trials could not be evaluated by the Meeting.

Several trials were conducted in the USA on use of paraquat as a harvest aid desiccator at rates of 0.56–1.12 kg/ha. This use is not included in US GAP, although it is allowed in Argentina, Brazil and Uruguay.

On the basis of the residue levels in maize grain in trials with paraquat applied pre- or post-emergence in Canada and the USA, < 0.01 (eight) and < <u>0.025</u> mg/kg (16), the Meeting estimated a maximum residue level of 0.03 mg/kg to replace the previous recommendation for maize and STMR and highest residue values of 0.025 mg/kg.

Sorghum

A number of residue trials were conducted in the USA, where paraquat is registered for use on sorghum at a maximum rate of 1.14 kg ai/ha, with a PHI of 48 days for grain and 20 days for forage, for pre-plant or pre-emergence broadcast application, and at a maximum rate of 0.55 kg ai/ha in two applications with the same PHIs for post-emergence directed spray. In the latter application, the applications must not exceed 2.5 l per season.

Several residue trials were carried out in the USA in several years and locations, in which paraquat was applied for weed control, either pre-emergence, post-crop emergence directed or as a harvest aid, at rates of 0.21–7.8 kg ai/ha. Samples were taken 20–131 days after pre-emergence or post-emergence directed application. The residue levels in grain in 12 trials conducted in accordance with maximum GAP for pre-emergence or post-emergence applications were all <u>< 0.025</u> mg/kg. When both pre- and post-emergence applications were made, if the post-application rate was in compliance with GAP, the residue results were taken into consideration in estimating the maximum residue level. In one trial with one pre-emergence application at 0.56 kg ai/ha followed by a post-emergence application at 0.56 kg ai/ha, a residue level of 0.01 mg/kg was found.

In harvest aid desiccation applications, paraquat was applied at a rate of 0.21–2.8 kg/ha, and sorghum was sampled 7–49 days after application. Harvest aid desiccant use is not included on the US label.

The Meeting estimated a maximum residue level of 0.03 mg/kg to replace the previous recommendation and STMR and highest residue values of 0.025 mg/kg for sorghum.

Rice

Trials on residues of paraquat on rice were conducted in Guatemala, Italy and the USA. Paraquat is registered for use on rice in the USA by pre-plant or pre-emergence broadcast at a maximum rate of 1.14 kg ai/ha, with no PHI specified.

Two trials were conducted in Italy in 1993, in which paraquat was applied at a rate of 0.92 kg ai/ha to the seed bed 5 days before rice was sown. Rice grain and straw samples taken at harvest did not contain residues of paraquat at levels above the LOQ of 0.05 mg/kg.

Three residue trials were conducted in Guatemala in 1983 in which paraquat was applied as a pre-emergence treatment at rates of 0.30 and 1.0 kg ai/ha to rice. Rice grain and straw samples were taken at harvest. The residues in de-husked rice in one trial conducted in compliance with the maximum rate in US GAP were < 0.05 mg/kg, but residues in rice grain were not analysed.

Residue trials were conducted in the USA in 1978 and 1982 in which paraquat was applied as a pre-emergence treatment at rates of 0.56 and 1.12 kg ai/ha to rice. In trials conducted at the maximum GAP, the residue levels in rice grain were below the LOQ of 0.01 (two) or 0.02 mg/kg. No trials were conducted at rates higher than the maximum allowed in US GAP for rice.

The Meeting concluded that there were insufficient data to estimate a maximum residue level and withdrew the previous recommendation for rice and rice, polished.

Tree nuts

It is common practice to harvest nuts from the ground, and this may result in residues of paraquat in the nuts.

Paraquat

Supervised residue trials were carried out over a number of years in Italy on *hazelnuts* and in the USA on *almonds* (California), *macadamia nuts* (Hawaii), *pecans* (Alabama and Texas), *pistachio nuts* (California) and *walnuts* (California).

Paraquat is registered for use on hazelnuts in Italy at a maximum rate of 1 kg ai/ha with a 40-day PHI and on walnuts at the same maximum rate but with no PHI specified. In the USA, paraquat is registered for use on pistachio nuts at a maximum rate of 1.14 kg ai/ha with a 7-day PHI, with the proviso that no more than two applications should be made after the nuts have split. It is registered for use in the USA on other tree nuts at the same maximum rate with no specification of the number of applications or PHI.

Two trials were conducted in Italy in which hazelnuts were harvested from the ground 1–10 days after treatment around the base of the trees at rates of 0.54–1.8 kg ai/ha. Although the PHI was shorter than 40 days, the residue levels in shelled nuts were below the LOQ of 0.05 mg/kg in one trial. At almost twice the maximum application rate and with a shorter PHI of 10 days, the levels were still below the LOQ.

In a trial in the USA, paraquat was applied at rates of 0.56–4.5 kg ai/ha one to eight times, to control weeds under mature nut trees. In some cases, applications were made over 2 years. Nuts were harvested, in some cases immature, 1–171 days after the last application. The residue levels in shelled nuts in trials following GAP were: < 0.01 (seven), 0.01, 0.02 and < 0.05 (three) mg/kg.

The combined results of all the trials, in ranked order, were: ≤ 0.01 (seven), 0.01, 0.02 and < 0.05 (four) mg/kg. The Meeting estimated a maximum residue level for paraquat in tree nuts of 0.05 mg/kg, an STMR of 0.01 mg/kg and a highest residue level of 0.05 mg/kg.

Oil seeds

Cotton-seed

Paraquat is registered for use on cotton in the USA at a maximum rate of 1.14 kg ai/ha, with no specification of the number of applications of PHI, for pre-plant or pre-emergence treatment, and at a maximum rate of 0.55 kg ai/ha, with repeated application if necessary and a 3-day PHI as a harvest aid, with the proviso that a total of 1.5 l should not be exceeded in this use.

Residue trials were conducted in the USA over several years and locations, involving pre-emergence applications at 1.12 kg/ha and harvesting 4–176 days after application. The residue levels in fuzzy seed in trials at the maximum GAP were < 0.01 (four) and 0.04 mg/kg.

In numerous trials with pre-emergence application followed by harvest aid desiccation application or a single application as harvest aid desiccant, the residue levels of paraquat in fuzzy seed in trials following maximum GAP were: 0.07, 0.09, 0.15, 0.16 (two), 0.18, 0.21, 0.23, 0.30, 0.34, 0.35, 0.38, 0.44, 0.46, 0.49, 0.50, 0.58 and 2.0 mg/kg. On the basis of residue levels arising from harvest aid uses, the Meeting estimated a maximum residue level for cotton-seed of 2 mg/kg, to replace the previous recommendation, an STMR of 0.34 mg/kg and a highest residue level of 2 mg/kg.

Sunflower seed

In the USA, paraquat is registered for use on sunflower at a maximum rate of 1.14 kg ai/ha with no PHI specified for pre-plant or pre-emergence broadcast or banded over-row application and at a maximum rate of 0.55 kg ai/ha with a 7-day PHI for dessication use.

Trials were conducted with pre-emergence application to sunflowers at 1.12 or 5.6 kg/ha and sampling 41–131 days after application. The residue levels in seeds in four trials conducted in compliance with maximum GAP were < 0.05 mg/kg. When paraquat was applied at five times the maximum recommended rate, the levels were still below the LOQ of 0.05 mg/kg.

In further trials, paraquat was applied as a harvest aid desiccator at 0.28–1.12 kg/ha, and sunflower seeds were harvested 7–21 days after application. The residue levels of paraquat in seeds in trials conducted at maximum GAP were: 0.09, 0.14, 0.15, 0.16 (three), 0.19, 0.22, 0.24, 0.32, 0.35, 0.51, 0.60, 0.74, 0.81

(two) and 0.93 mg/kg. The Meeting used the residue levels arising from harvest aid uses to estimate a maximum residue level for sunflower seed of 2 mg/kg, an STMR of 0.22 mg/kg and a highest residue level of 0.81 mg/kg.

Hops

Residue trials were conducted in Canada and the USA. Paraquat was registered in the USA for use as a directed spray or for suckering and stripping on hops at a maximum rate of 0.55 kg ai/ha in three applications with a 14-day PHI; no more than two applications or applications at no more than 1.5 l/ha were recommended.

In a trial in Canada, a single post-emergence directed application of 1.12 kg ai/ha, which is double the maximum recommended dose, resulted in residue levels of < 0.01 mg/kg in green hops harvested 53 days after application.

In the USA, trials were conducted in the states of Idaho, Oregon and Washington with three post-emergence directed applications of paraquat at 2.8 kg ai/ha. The residue levels of paraquat in dried hops prepared from hops harvested 14 days after the last of three directed application at the maximum GAP rate were 0.05 mg/kg in two trials. At double this rate, the levels in dried hops prepared from green hops harvested 13 or 14 days after the last treatment were below the LOQ of 0.1 mg/kg (0.01 and 0.07 mg/kg). Two applications at higher rates than that of maximum GAP resulted in 0.02 and 0.03 mg/kg in dried hops.

The residue levels in dried hops were 0.05 mg/kg (two). In view of the low levels of residues in the other trials, the Meeting estimated a maximum residue level of 0.1 mg/kg, to replace the previous recommendation, and STMR and highest residue values of 0.05 mg/kg for hops, dry.

Tea, green, black

Residue trials on tea were conducted in India, where paraquat is registered for use for pre-emergence or post-emergence directed application between rows at a maximum rate of 0.75 kg ai/ha in one application, with no PHI specified.

Six trials were conducted at a total application rate of 0.57–2.0 kg ai/ha over 5–6 months. Green tea leaves were harvested 7 or 21 days after blanket application (after the first or last spot application) and processed into black tea, which was analysed. The residue levels of paraquat in black tea from tea plants treated in accordance with GAP in India or at higher rates were almost always below the LOQ of 0.05 mg/kg. In trials conducted in accordance with GAP, the levels in black tea were: ≤ 0.05 (three), 0.07, 0.09 and 0.12 mg/kg.

In other trials in India, with application rates of 0.05–0.06 kg ai/ha, black tea samples from green tea leaves harvested 5 or 7 days after application contained 0.05 mg/kg (one) or < 0.05 mg/kg. As the application rate was much lower than the maximum, these results were not considered in estimating the maximum residue level.

The Meeting estimated a maximum residue level for teas, green, black of 0.2 mg/kg and an STMR of 0.06 mg/kg.

Animal feedstuffs

Soya forage and hay or fodder

Paraquat is registered for use in Australia, Brazil and the USA for weed control andas a harvest aid on soya beans. In the USA, it is registered for use at a maximum rate of 1.14 kg ai/ha for pre-plant or pre-emergence treatment, not to exceed 1.9 l per season, at a maximum rate of 0.14 kg ai/ha as a post-emergence directed spray with a second and final application 7–14 days later; it can also be used at a maximum rate of 0.28 kg ai/ha with a 15-day PHI as a harvest aid.

The residue levels in forage in trials conducted in the USA in accordance with US GAP were: < 0.025 (12), ≤ 0.05 (13), 0.05, 0.06 (four), 0.07, 0.08, 0.15, 0.28 and 1.8 mg/kg, expressed on a dry weight basis.

The Meeting estimated a maximum residue level for soya bean forage (green) of 2 mg/kg, an STMR of 0.05 mg/kg and a highest residue level of 1.8 mg/kg.

The residue levels in hay or fodder in trials conducted in accordance with US GAP were: < 0.025 (five), 0.04, ≤ 0.05 (four), 0.05, 0.1, 0.2 and 0.3 mg/kg, on a dry weight basis. The Meeting estimated a maximum residue level for soya bean fodder of 0.5 mg/kg, an STMR of 0.05 mg/kg and a highest residue level of 0.3 mg/kg.

Sugar-beet tops

Trials were conducted on beet and sugar-beet in the United Kingdom and the USA. The residue levels in sugar-beet tops in six trials conducted in accordance with US GAP were < 0.025 mg/kg, on a fresh weight basis. The Meeting estimated a maximum residue level of 0.2 mg/kg and an STMR of 0.11 mg/kg. On the basis of 23% dry matter and a highest residue level on a fresh weight basis of 0.025 mg/kg, the Meeting calculated the highest residue level on a dry weight basis to be 0.11 mg/kg. As there is no code for sugar-beet tops, the maximum residue level was recommended for fodder beet leaves and tops.

Maize forage and fodder

Trials were conducted in Italy and the USA. The residue levels in maize forage in trials in the USA conducted in accordance with US GAP were ≤ 0.025 (eight), 0.09, 0.6, 2 (two) and 3 (two) mg/kg on a dry weight basis. The Meeting estimated a maximum residue level for maize forage of 5 mg/kg, an STMR of 0.025 mg/kg and a highest residue level of 3 mg/kg.

The levels of residues in silage were mostly below the LOQ of 0.025 or 0.05 mg/kg, except in one trial in which levels up to 0.04 mg/kg were found.

The residue levels in maize fodder in trials in the USA conducted in accordance with US GAP were: ≤ 0.025 (eight), 0.03, 0.05, 0.06, 0.2, 1, 2 and 6 mg/kg on a dry weight basis. The Meeting estimated a maximum residue level for maize fodder of 10 mg/kg, an STMR of 0.025 mg/kg and a highest residue level of 6 mg/kg.

Sorghum forage (green) and straw and fodder, dry

In trials conducted in the USA in accordance with GAP, the residue levels in sorghum forage were: ≤ 0.025 (six), 0.025 (three), 0.04, 0.06 and 0.2 mg/kg. The Meeting estimated a maximum residue level for sorghum forage (green) of 0.3 mg/kg, an STMR of 0.025 mg/kg and a highest residue level of 0.2 mg/kg.

The residue levels in sorghum fodder or hay (whichever gave higher levels) in trials conducted in accordance with GAP were: < 0.025 (four), 0.03, 0.04, 0.05, 0.06 (two), 0.09, 0.1 and 0.2 mg/kg. The Meeting estimated a maximum residue level for sorghum straw and fodder, dry, of 0.3 mg/kg, an STMR of 0.035 mg/kg and a highest residue level of 0.2 mg/kg.

Rice straw and fodder, dry

The Meeting concluded that there were insufficient data for estimating a maximum residue level for rice straw and fodder, dry.

Almond hulls

In three trials conducted in the USA in accordance with GAP, the residue levels in almond hulls were < 0.01 mg/kg. The Meeting estimated maximum residue, STMR and highest residue values of 0.01 mg/kg.

Cotton fodder

The Meeting concluded that there were insufficient data for estimating a maximum residue level for cotton fodder.

Fate of residues during processing

Numerous studies of residue levels after processing conducted in conjunction with supervised trials were submitted. Residue levels found after processing of raw agricultural commodities into animal feedstuffs are described in the section above. Some processed commodities for which maximum residue levels and STMR-Ps were estimated are also described in that section.

In this section, processing factors from raw commodities to processed food products and by-products are discussed. Information on processing was provided for orange, plum, grape, olive, tomato, sugar-beet, maize, sorghum, cotton-seed, sunflower seed and hop. Processing factors could not be reliably calculated for the processing of orange, plum, grape, tomato and sugar-beet because the paraquat residue levels in both raw commodities and processed products were all below the respective LOQs.

Processing factors were calculated for olive (oil), potato (crisps and granules), maize (milling fractions and oil), sorghum (milling fractions), cotton-seed (trash, gin products and oil), sunflower seed (oil) and hop (dried hop and beer) and are shown below.

Commodity	Processing factor	STMR-P (mg/kg)
Olive		0.05
Unwashed olives before processing	0.57	
Washed olives before processing	< 0.43	
Virgin oil	< 0.35	0.018
Refined oil	< 0.35	0.018
Potato		0.02
Wet peel	> 1.9	0.04
Dry peel	> 11	0.22
Peeled potato	0.27[a]	0.01
Crisps	> 0.95	0.02
Granules	> 2.7	0.05
Maize		0.025
Wet milling		
Coarse starch	< 0.25[a]	0.006
Starch	< 0.25[a]	0.006
Crude oil	< 0.25[a]	0.006
Refined oil	< 0.25[a]	0.006
Dry milling		
Germ	0.3[a]	0.0075
Grits	0.25–0.5[a]	0.0006–0.013
Coarse meal	1[a]	0.025
Meal	0.5[a]	0.013
Flour	1.5[a]	0.038
Crude oil	< 0.25[a]	0.006
Refined oil	< 0.05[a]	0.001
Sorghum		0.025

Commodity	Processing factor	STMR-P (mg/kg)
Hulled grain	0.07[a]	0.002
Dry milled bran	3.9	0.097
Coarse grits	0.17	0.004
Flour	0.14	0.004
Wet milled bran	2.3	0.058
Starch	0.07	0.002
Shorts	2.6	0.065
Germ	0.52[a]	0.013
Cotton (from cotton including trash and bolls)		
Fuzzy seed	0.08	0.34
Crude oil	< 0.006	0.01[b]
Meal	< 0.009	0.04
Sunflower seed		0.3
Hulls	2.8[a]	0.64
Meal	0.05[a]	0.01
Oil	< 0.05[a]	0[b]
Hop		
Dry cones	1.2	0.05[b]
Beer	< 0.28	0.0001[c]

[a] Based on only one trial.
[b] Estimated from supervised trials
[c] Calculated from a factor of 0.0001

The STMR values for processed products from raw commodities with no residues or for which the results of many supervised trials were available were estimated on the basis of supervised trials.

In four trials in the USA, orange fruit was processed into juice, and the paraquat residues were measured; in all cases, the levels were below the LOQ of 0.01 mg/kg. The residue levels in *orange juice*, including those in trials conducted at rates higher than the maximum application rate, were all below the LOQ of 0.01 mg/kg. The Meeting estimated an STMR-P for orange juice of 0 mg/kg.

No residues of paraquat were found at levels above the LOQ of 0.05 mg/kg in *dried prunes* prepared from plums in two trials. The STMR-P for dried prunes was estimated to be 0 mg/kg.

In a number of trials, olives were processed into oil for analysis of residues. *Olive oil* prepared from olive fruits harvested directly from trees did not contain levels above the LOQ of 0.05 mg/kg. Most samples of olive oil prepared from olive fruits picked up from ground or sprayed directly did not contain paraquat residues at levels above the LOQ; however, in some samples, paraquat residues were found at levels up to 0.06 mg/kg, and fruit harvested at the same time contained 6.8 mg/kg of paraquat residues. As paraquat is unlikely to be transferred into oil owing to its chemical and physical characteristics, its STMR-P is calculated from the processing factor to be 0.018 mg/kg.

Tomato juice and *ketchup* prepared from tomato in trials conducted at an exaggerated rate did not contain paraquat residues at levels above the respective LOQ (0.005 mg/kg for juice and 0.025 mg/kg for ketchup). The STMR values for these products were estimated to be 0 mg/kg.

The residue levels in oil prepared from soya bean treated with paraquat as a harvest aid desiccant in accordance with GAP were below the LOQ of 0.01 mg/kg in five trials. The Meeting estimated an STMR-P for *soya bean oil* of 0.01 mg/kg.

The residue levels in cotton-seed oil, crude, were below the LOQ of 0.01 mg/kg in two trials. The Meeting estimated an STMR-P for *cotton-seed oil* of 0.01 mg/kg and decided to withdraw the previous recommendation for cotton-seed oil, edible.

The residue levels in *sunflower seed oil* obtained from sunflower seed in eight trials conducted at the maximum GAP were < 0.01 mg/kg. Oil obtained from sunflower seed in a trial at double the rate did not contain residues at levels above the LOQ of 0.01 mg/kg. The Meeting estimated an STMR-P for sunflower seed oil of 0 mg/kg and decided to withdraw the previous recommendation for sunflower seed oil, crude and edible.

The residue levels of paraquat in *cotton gin by-product* in trials for harvest aid uses were (including results for cotton harvested 13–17 days after treatment): 5.2, 5.3, 5.9, 6.2, 7.3, 8.0, 9.4, 11, 12 (two), 18, 23, 32, 34 and 69 mg/kg. The Meeting estimated an STMR-P of 10.2 mg/kg for cotton gin by-products.

As *maize flour* contained a higher concentration of paraquat residues than maize grain in one trial, the Meeting estimated a maximum residue level of 0.05 mg/kg.

Residues in animal commodities

Dietary burden of farm animals

The Meeting estimated the dietary burden of paraquat residues for farm animals on the basis of the diets described in Appendix IX to the *FAO Manual* (FAO, 2002), by summing the contribution of each feed to the residue.

Estimated maximum dietary burden of farm animals

Crop	Residue (mg/kg)	Basis	Group	Dry matter (%)	Residue/ Dry matter (mg/kg)	Dietary content (mg/kg) Beef cattle	Dairy cows	Poultry	Residue contribution (mg/kg) Beef cattle	Dairy cows	Poultry
Sugar-beet tops	0.025	HR	AV	23	0.11						
Cotton-seed	2	HR	SO	88	2.27	25	25		0.57	0.57	
Cotton gin by-product	10.2	STMR-P		90	11.3	20	20		2.27	2.27	
Maize grain	0.025	HR	GC	88	0.03			80			0.023
Maize forage	3	HR	AF		3	40	50		1.2	1.5	–
Potato, wet peel	0.04	STMR-P	VR	15	0.27						
Sorghum grain	0.025	HR	GC	86	0.03				–	–	
Sorgum forage	0.2	HR	AF	–	0.20				–	–	
Soya bean	0.41	HR	VD	89	0.46			20			0.092
Soya bean, forage	1.8	HR	AL	–	1.8	15	5		0.27	0.09	–
Soya bean, hay	0.3	HR	AL	–	0.3				–	–	
Sunflower meal	0.011	STMR-P	AL	92	0.01	–	–	–	–	–	–
Turnip tops	0.05	HR	VL	30	0.17						
Total									4.30	4.43	0.11

Estimated maximum dietary burden of farm animals

Crop	Residue (mg/kg)	Basis	Group	Dry matter (%)	Residue/ Dry matter (mg/kg)	Dietary content (mg/kg) Beef cattle	Dairy cows	Poultry	Residue contribution (mg/kg) Beef cattle	Dairy cows	Poultry
Sugar-beet tops	0.025	STMR	AV	23	0.11						–
Cotton-seed	0.34	STMR	SO	88	0.39	25	25		0.098	0.098	
Cotton gin by-product	10.2	STMR-P		90	11.3	20	20		2.27	2.27	
Maize grain	0.025	STMR	GC	88	0.028			80			0.02
Maize forage	0.025	STMR	AF		0.03	40	50		0.010	0.013	–
Potato wet peel	0.55	STMR-P	VR	15	0.27						
Sorghum grain	0.025	STMR	GC	86	0.03				–		–
Sorgum forage	0.025	STMR	AF		0.03				–		–
Soya bean	0.08	STMR	VD	89	0.09			20			0.02
Soya bean, forage	0.05	STMR	AL		0.05	15	5		0.008	0.003	–
Soya bean, hay	0.05	STMR	AL		0.05				–		–
Sunflower meal	0.011	STMR-P	AL	92	0.01	–	–	–	–	–	–
Turnip tops	0.025	STMR	VL	30	0.08						
Total									2.39	2.38	0.04

The dietary burdens of paraquat for estimation of MRL and STMR values for animal commodities are: beef cattle, 4.30 and 2.39 ppm; dairy cattle, 4.43 and 2.38 ppm; and poultry, 0.11 and 0.04 ppm.

Feeding studies

In a study of metabolism in goats (see above), one goat was dosed at a rate equivalent to 100 mg/kg of total diet. This is considerably higher than the estimated maximum dietary burden for cattle of 4.30 or 4.43 mg/kg. At 100 mg/kg of diet, the maximum TRRs, expressed in paraquat ion equivalents, found in milk and edible goat tissues were 0.009 mg/kg in milk, 0.12 mg/kg in meat, 0.03 mg/kg in fat, 0.56 mg/kg in liver and 0.74 mg/kg in kidney. In milk, 75.9% of the radiolabel was identified with paraquat.

At the estimated maximum animal burden of 4.30 or 4.43 mg/kg, the levels of paraquat residues were calculated to be < 0.005 mg/kg in milk, 0.005 mg/kg in meat, 0.025 mg/kg in liver and 0.033 mg/kg in kidney. The Meeting estimated maximum residue levels of 0.005* mg/kg for milks, 0.005 mg/kg for mammalian meat and 0.05 mg/kg for edible mammalian offal. These levels replace the previous recommendations for related animal commodities. The STMR values were estimated to be 0.00002 mg/kg for milk, 0.0003 mg/kg for meat and 0.0018 mg/kg for edible offal; and the highest residue level values were estimated to be 0.005 mg/kg for meat and 0.033 mg/kg for edible offal.

In the study of metabolism in hens (see above), birds were dosed at a rate equivalent to 30 mg/kg of total diet, which is considerably higher than the estimated maximum dietary burden for poultry of 0.11 mg/kg. At 30 mg/kg diet, the maximum TRRs, expressed in paraquat ion equivalents, found in eggs and edible chicken tissues were 0.18 mg/kg in egg yolk, 0.001 mg/kg in egg albumen, 0.05 mg/kg in meat, 0.05 mg/kg in fat and 0.09 mg/kg in liver.

At the estimated maximum animal burden of 0.11 mg/kg, the maximum residue levels were calculated to be far below the LOQ of 0.005 mg/kg in eggs and other tissues. The Meeting estimated the maximum residue levels to be 0.005* mg/kg for eggs, poultry meat and edible poultry offal. The STMR and highest residue level values were estimated to be 0 for these commodities.

DIETARY RISK ASSESSMENT

Long-term intake

The IEDIs were calculated for the five GEMS/Food regional diets from the STMR values for fruit, vegetables, maize, sorghum, cotton-seed, sunflower, hops, tea and animal commodities and the STMR-P values for their processed products, as estimated by the current Meeting (Annex 3). The ADI is 0–0.005 mg/kg bw, and the calculated IEDIs were 2–5% of the ADI. The Meeting concluded that the intake of residues of paraquat resulting from uses considered by the current JMPR was unlikely to present a public health concern.

Short-term intake

The IESTIs of paraquat by the general population and by children were calculated for commodities for which STMR or STMR-P values had been estimated by the current Meeting when information on consumption was available (Annex 4). The ARfD is 0.006 mg/kg; the calculated IESTIs for children up to 6 years range from 0 to 50% and those for the general population from 0 to 20% of the ARfD. The Meeting concluded that the short-term intake of residues of paraquat from uses considered by the current Meeting was unlikely to present a public health concern.

4.20 PHORATE (112)

TOXICOLOGY

Phorate is the ISO approved name for phosphorothioic acid, *O*-diethyl *S*-(ethyl thio)methyl ester, which is an organophosphate insecticide that inhibits acetylcholinesterase activity and is a systemic and contact insecticide and acaricide. Phorate was first evaluated by the JMPR in 1977. In 1985, an ADI of 0–0.0002 mg/kg bw was established. Phorate was re-evaluated in 1994 when an ADI of 0–0.0005 mg/kg bw was established. In 1994, because it was reported in a limited study of metabolism in rats that < 40% of the administered dose was excreted, the Meeting requested adequate studies on absorption, for review in 1996. Such studies were received and the ADI established previously was confirmed.

Since the 1994 JMPR, a study of acute neurotoxicity and a 13-week study of neurotoxicity in rats have been submitted. The present Meeting re-evaluated phorate within the periodic review programme of the CCPR, using new data that had not been reviewed previously and relevant data from previous evaluations.

After oral administration of radiolabelled phorate to rats, 77% of the administered dose was recovered in the urine within 24 h after dosing. Faecal excretion accounted for approximately 12% of the administered dose. Over the total duration of the study (192 h), essentially the entire administered dose was eliminated by excretion.

Phorate was highly toxic when administered orally, dermally or by inhalation. The oral LD_{50} values for rats were 3.7 mg/kg bw in males and 1.4 mg/kg bw in females. The dermal LD_{50} values for rats were 9.3 mg/kg bw in males and 3.9 mg/kg bw in females. The LC_{50}s for rats after exposure for 1 h were 0.06 and 0.011 mg/l of air in males and females respectively. Studies of dermal and eye irritation and of dermal sensitization were not performed owing to the high acute toxicity of phorate by skin contact.

Phorate

The toxicological effects of phorate are associated with inhibition of acetylcholinesterase activity. Inhibition of acetylcholinesterase activity and clinical signs occurred at similar doses in rats, rabbits and dogs, while mice appeared to be somewhat less sensitive. The NOAELs for toxicologically significant inhibition of brain acetylcholinesterase activity were 0.05–0.07 mg/kg bw per day in 13-week and 2-year studies in rats and in 1-year studies in dogs. The NOAELs for clinical signs were generally higher. The Meeting noted that the dose–response curve for acetylcholinesterase inhibition is steep.

In an 18-month study in mice and in a 24-month study in rats, phorate did not increase the incidence of tumours or cause any non-neoplastic effects other than clinical signs secondary to inhibition of acetylcholinesterase activity.

Phorate was tested for genotoxicity in vitro and in vivo in an adequate battery of assays. In view of the lack of genotoxicity in vitro and in vivo and on the basis of the results of studies of carcinogenicity in rodents, the Meeting concluded that phorate is not likely to pose a carcinogenic risk to humans.

In a multigeneration study of reproductive toxicity in mice, the NOAEL was 1.5 ppm (equal to 0.30 mg/kg bw per day) on the basis of slightly reduced lactation indices in four out of the six litters at 3 ppm (equal to 0.60 mg/kg bw per day).

In a two-generation study of reproductive toxicity in rats, phorate showed effects on pup growth and mortality at maternally toxic doses. The NOAEL was 2 ppm (equal to 0.17 mg/kg bw per day) on the basis of decreased brain acetylcholinesterase activity, decreased parental and pup body weights and decreased pup survival at 4 ppm (equal to 0.35 mg/kg bw per day).

In a study of developmental toxicity in rats, the NOAELs for maternal and developmental toxicity with phorate were 0.3 mg/kg bw per day on the basis of mortality, cholinergic signs of toxicity, significantly decreased body weights and food consumption in the dams, decreased fetal body weights and delays in skeletal ossification at 0.4 mg/kg bw per day. No fetal malformations were produced, even at the lethal dose (0.4 mg/kg bw per day), the highest dose tested. The Meeting concluded that phorate is not teratogenic in rats.

Phorate was not embryotoxic, fetotoxic or teratogenic in rabbits at doses of up to and including 1.2 mg/kg bw per day, a dose that produced severe maternal toxicity. The NOAEL for maternal toxicity with phorate was 0.15 mg/kg bw per day on the basis of mortality observed at 0.5 mg/kg bw per day. The NOAEL for developmental toxicity was 1.2 mg/kg bw per day, the highest dose tested.

The Meeting concluded that the existing database on phorate was adequate to characterize the potential hazards to fetuses, infants and children.

In a study of acute neurotoxicity in rats treated by gavage, phorate at a dose of 1 mg/kg bw caused miosis in 2 out of 20 males and 5 out of 20 females, tremors in 2 out of 20 females, fasciculations, slightly impaired locomotion and splayed or dragging hindlimbs in one female and significant inhibition of brain and erythrocyte acetylcholinesterase activity in females (65%), but not in males (14–21%). No histopathological signs were observed. At 0.5 mg/kg bw, miosis was observed in 2 out of 20 males and 2 out of 20 females. Although miosis was observed in a small number of animals (and in 1 out of 20 controls) in the absence of inhibition of erythrocyte and brain acetylcholinesterase activity, it could not be dismissed as a compound-related effect. The NOAEL was 0.25 mg/kg bw on the basis of miosis.

Phorate did not cause acute delayed neurotoxicity in hens. Although measurements of neuropathy target esterase were not carried out, the Meeting noted that the dose used (approximately equal to the LD$_{50}$) was sufficiently high to indicate that dietary exposure to phorate would not cause delayed polyneuropathy.

The mammalian and plant metabolites of phorate, phorate sulfone and phorate sulfoxide, had similar toxicity to the parent compound. In rats, the oral LD$_{50}$ values for these metabolites were 1.2–3.5 and 2.2–2.6 mg/kg bw, respectively. The NOAELs for inhibition of brain acetylcholinesterase activity were 0.80 ppm (equal to 0.08 and 0.06 mg/kg bw per day) for phorate sulfone and sulfoxide, respectively, in 90-day studies in rats.

Several cases of occupational and non-occupational poisoning in humans have been reported. The subjects showed typical cholinergic symptoms, including gastrointestinal effects, bradycardia and neurological effects (headache, giddiness, fatigue). Skin and eye irritation were also observed.

Toxicological evaluation

An ADI of 0–0.0007 mg/kg bw was established on the basis of a overall NOAEL of 0.07 mg/kg bw per day for inhibition of brain acetylcholinesterase activity in rats and dogs and a safety factor of 100. This ADI includes the phorate metabolites, phorate sulfone and phorate sulfoxide.

An ARfD of 0.003 mg/kg bw was also established based on the NOAEL of 0.25 mg/kg bw for miosis in the study with single doses in rats. Although inhibition of acetylcholinesterase activity is a C_{max}-dependent phenomenon, a safety factor of 100 was used in view of the steep dose–response curve and the slow recovery of brain acetylcholinesterase activity because of irreversibility of its inhibition. This ARfD includes the metabolites of phorate, phorate sulfone and phorate sulfoxide.

A toxicological monograph was prepared.

Levels relevant to risk assessment

Species	Study	Effect	NOAEL	LOAEL
Mouse	18-month study of toxicity and carcinogenicity[a]	Toxicity	3 ppm, equivalent to 0.45 mg/kg bw per day	6 ppm, equivalent to 0.90 mg/kg bw per day
		Carcinogenicity	6 ppm, equal to 0.90 mg/kg bw per day[d]	—
	Multigeneration study of reproductive toxicity[a]	Parental and offspring toxicity	1.5 ppm, equal to 0.30 mg/kg bw per day	3 ppm, equal to 0.60 mg/kg bw per day
Rat	2-year study of toxicity and carcinogenicity[a]	Toxicity	1 ppm, equal to 0.05 mg/kg bw per day	3 ppm, equal to 0.16 mg/kg bw per day
		Carcinogenicity	6 ppm, equal to 0.32 mg/kg bw per day[c, d]	—
	Multigeneration reproductive toxicity[a]	Parental and offspring toxicity	2 ppm, equal to 0.17 mg/kg bw per day	4 ppm, equal to 0.35 mg/kg bw per day
	Developmental toxicity[a]	Embryo- and fetotoxicity and maternal toxicity	0.3 mg/kg bw per day	0.40 mg/kg bw per day
	Single-dose study[c]	Miosis	0.25 mg/kg bw	0.50 mg/kg bw per day
	13-week study of neurotoxicity[a]	Neurotoxicity	0.07 mg/kg bw per day	0.3 mg/kg bw per day
Rabbit	Developmental toxicity[a]	Maternal toxicity	0.15 mg/kg bw per day	0.50 mg/kg bw per day
		Embryo- and fetotoxicity[a]	1.2 mg/kg bw per day[d]	—
Dog	1-year study of toxicity[b]	Toxicity	0.05 mg/kg bw per day	0.25 mg/kg bw per day

[a] Diet
[b] Capsules
[c] Gavage
[d] Highest dose tested

Estimate of acceptable daily intake for humans

0–0.0007 mg/kg bw

Estimate of acute reference dose

0.003 mg/kg bw

Studies that would provide information useful for the continued evaluation of the compound

Further observation in humans

Phorate

Critical end-points for setting guidance values for exposure to phorate

Absorption, distribution, excretion and metabolism in animals	
Rate and extent of oral absorption	Rapid, approximately 90% within 24 h
Dermal absorption	Extensive based on acute toxicity
Distribution	Rapid and extensive
Potential for accumulation	None
Rate and extent of excretion	89% within 24 h; urinary excretion predominated (77%); faecal excretion (12%)
Metabolism in animals	Major pathway: cleavage of phosphorus–sulfur bond, methylation of the liberated thiol group and oxidation of the resulting divalent moiety to the sulfoxide and sulfone
Toxicologically significant compounds (plants, animals and the environment)	Parent, phorate sulfoxide and phorate sulfone
Acute toxicity	
Rat, LD_{50}, oral	3.7 mg/kg bw in males, 1.4 mg/kg bw in females
Rat, LD_{50}, dermal	9.3 mg/kg bw in males, 3.9 mg/kg bw in females
Rat, LC_{50}, inhalation	0.06 mg/l of air in males (1-h), 0.011 mg/l of air (1-h) in females
Rabbit, skin irritation	Highly toxic by skin contact — could not be tested
Rabbit, eye irritation	Highly toxic by eye contact — could not be tested
Skin sensitization	Highly toxic by skin contact — could not be tested
Short-term studies of toxicity	
Target/critical effect	Brain and erythrocyte acetylcholinesterase activity and miosis (rats)
Lowest relevant oral NOAEL	0.07 mg/kg bw per day
Lowest relevant dermal NOAEL	No data
Lowest relevant inhalation NOAEC	No data
Genotoxicity	Negative results in vivo and in vitro
Long-term studies of toxicity and carcinogenicity	
Target/critical effect	Inhibition of erythrocyte and brain cholinesterase activity
Lowest relevant NOAEL	0.07 mg/kg per day (rat)
Carcinogenicity	Not carcinogenic in mice and rats
Reproductive toxicity	
Reproduction target/critical effect	Reduced pup growth at maternally toxic dose
Lowest relevant reproductive NOAEL	2 ppm, equivalent to 0.17 mg/kg bw per day
Developmental target/critical effect	Decreased pup weights and delayed ossification at maternally toxic doses (rats)
Lowest relevant developmental NOAEL	0.3 mg/kg bw per day (rats)
Neurotoxicity/delayed neurotoxicity	
Single dose study of neurotoxicity	
Target/critical effect	Signs consistent with acetylcholinesterase inhibition; no neuropathological effects
Relevant NOAEL	0.25 mg/kg bw
Delayed neuropathy	No delayed neurotoxicity in hens

Medical data		Findings consistent with inhibition of acetylcholinesterase activity; no record of permanent sequelae	
Summary	**Value**	**Study**	**Safety factor**
ADI	0–0.0007 mg/kg bw	Rats and dogs, short- and long-term studies, inhibition of brain acetylcholinesterase activity	100
ARfD	0–0.003 mg/kg bw	Rats, single-dose study, miosis	100

DIETARY RISK ASSESSMENT

Long-term intake

The estimated theoretical maximum daily intakes in the five GEMS/Food regional diets, based on recommended MRLs, were in the range of 40–200% of the ADI (Annex 3). Further refinements of dietary intake estimates will be undertaken during the periodic review of phorate residues scheduled for 2005.

Short-term intake

The Meeting established an ARfD for phorate of 0.003 mg/kg bw but was unable to finalize the risk assessment before the residue evaluation, scheduled for 2005, had been completed.

4.21 PIRIMICARB (101)

TOXICOLOGY

Pirimicarb is the ISO approved common name for 2-dimethylamino-5,6-dimethylpyrimidin-4-yl dimethylcarbamate. It is a selective aphicide that is used extensively on a broad range of crops, including vegetable, cereal and orchard crops. The mode of action of pirimicarb is by inhibition of acetylcholinesterase activity.

Pirimicarb was evaluated by the JMPR in 1976, 1978 and 1982; an ADI of 0–0.02 mg.kg bw was established in 1983. Pirimicarb was reviewed by the present Meeting within the periodic review programme of CCPR, using new data not previously reviewed and relevant data from previous evaluations.

Kinetic studies in rats have demonstrated that pirimicarb administered orally to male and female rats is rapidly and extensively absorbed (> 70% of the administered dose) and widely distributed. Radioactivity from [^{14}C]pyrimidinyl-labelled pirimicarb was excreted predominantly in the urine, while radioactivity from [^{14}C]carbamoyl-labelled pirimicarb was excreted predominantly in expired air. Tissue retention of radioactivity was low. There were no pronounced sex differences in the routes or rates of excretion. Pirimicarb was extensively metabolized, giving rise to 24 metabolites, 17 of which were identified. The main metabolic pathway involves the loss of the carbamate moiety to produce a range of substituted hydroxypyrimidines, some of which are glucuronide conjugates.

The acute oral median LD$_{50}$ for pirimicarb was 152 mg/kg bw in male rats and 142 mg/kg bw in female rats, while the acute dermal median LD$_{50}$ of pirimicarb was > 2000 mg/kg in both male and female rats. The 4-h inhalation median LC$_{50}$ of pirimicarb in rats was 0.948 and 0.858 mg/l of air in males and females respectively. Pirimicarb is not irritating to the rabbit eye or skin. It does, however, have skin sensitizing potential under the conditions of the Magnusson & Kligman maximization test.

In a 21-day study of dermal toxicity in rats, there were no signs of irritation and no clinical signs of systemic toxicity, but there was a reduction in brain cholinesterase activity at 1000 mg/kg bw per day. The NOAEL was 200 mg/kg bw per day.

Acetylcholinesterase that has been inhibited by pirimicarb is rapidly reactivated (both in vivo and in vitro). This property hampers the reliable determination of acetylcholinesterase inhibition in erythrocytes and brain in treated animals, and special attention must be given to crucial methodological features (e.g. time between sampling and measurement, sample temperature and dilution). Consequently, the most reliable indicators of an effect are clinical signs, which usually occur when acetylcholinesterase inhibition is > 50% at nerve terminals.

In experiments with multiple doses, common toxicological targets are blood and acetylcholinesterase.

Three dietary studies of up to 90 days in duration have been conducted in rats. In the first study (in which animals were given diets containing pirimicarb at a concentration of 250 or 750 ppm for 90 days) there were no adverse clinical, haematological or other pathological effects. A reduction in plasma cholinesterase activity was seen at 750 ppm, providing evidence for the absorption of the compound from the intestinal tract. In the second study (in which animals were given diets containing pirimicarb at a concentration of 250 or 750 ppm for 8 weeks) a clear reduction in body-weight gain was seen at 750 ppm, with a slight reduction at 250 ppm. These growth reductions were completely reversible after an 8-week recovery period. In the final study (in which animals were given diets containing pirimicarb at a concentration of 100, 175, 250 or 750 ppm for 8 weeks) there were no adverse clinical effects. A reduction in body-weight gain and food consumption was seen at 750 ppm. Owing to the effect on body-weight gain in the second study at 250 ppm, the Meeting concluded that the overall NOAEL in short-term studies in rats was 175 ppm, equivalent to 17.5 mg/kg bw per day.

Reports of three studies of 13–16 weeks in duration in dogs were available. In the first study, beagle dogs were given diets delivering pirimicarb at a dose of 0, 4, 10 or 25 mg/kg bw per day for at least 90 days. Body weight was reduced at the highest dose and plasma cholinesterase activity was reduced at 10 and 25 mg/kg bw per day. Bone-marrow changes, indicative of increased erythropoiesis, were observed in the terminal blood films in all treatment groups. Three out of 32 animals (two receiving 25 mg/kg bw per day and one receiving 10 mg/kg bw per day) developed anaemia. Two dogs of each sex per group were killed after 90 days (one male only in the group receiving 25 mg/kg bw per day, as one had been killed after 10–11 weeks) and the remainder were allowed to recover untreated for 28 days. Partial recovery from the bone-marrow changes was evident at 28 days after cessation of treatment. A NOAEL for this study was not identified, therefore a second study with diets delivering pirimicarb at a dose of 0.4, 1.8 and 4 mg/kg bw per day for at least 90 days was conducted. An additional group received pirimicarb at 4 mg/kg bw per day for 180 days. There were no adverse clinical or pathological effects, but dogs at 4 mg/kg bw per day showed evidence of increased erythropoietic activity in the bone marrow. The NOAEL was 1.8 mg/kg bw per day. In the third study, foxhounds were given diets delivering pirimicarb at a dose of 0, 2 or 25/50 mg/kg bw per day for 16 weeks, followed by a 7-week recovery period. The dose of 25 mg/kg bw per day was increased to 50 mg/kg bw per day from week 8. Anaemia and reticulocytosis developed in dogs receiving a dose of 50 mg/kg bw per day and bone-marrow changes, characterized by an increase in normoblasts and hypoplasia, were observed. Both the anaemia and bone-marrow changes were reversible when the dose was reduced to 25 mg/kg bw per day, or upon cessation of treatment. The NOAEL in this study in dogs was 2 mg/kg bw per day.

In addition, there were two 2-year dietary studies in dogs and a more recent, guideline-compliant, 1-year study in which pirimicarb was administered in capsules. In the first 2-year dietary study, designed to reproduce and characterize the anaemia, two out of four unrelated beagles had an immune haemolytic anaemia when exposed to pirimicarb at a dose of 25 or 50 mg/kg bw per day for at least 3 months. Dogs at ≤ 2 mg/kg bw per day did not show such effects. The anaemia was completely reversible after withdrawal of the compound. Other dogs in the same study showed no haematological changes when exposed to primicarb at a dose of up to 50 mg/kg bw per day for 2 years. In the second dietary study, pirimicarb was administered at a dose of 0.4, 1.8 or 4 mg/kg bw per day for 2 years. There were no adverse changes in growth rate, blood and urine clinical chemistry, organ weights or histopathology. At 4 mg/kg bw per day, there were reductions in haemoglobin concentration and erythrocyte volume fraction in males and a slight increase in the erythroid to myeloid ratio in two females. No adverse changes were detected in the bone marrow. None of the dogs developed overt anaemia. In the most recent study, groups of beagles were dosed orally with gelatine capsules containing pirimicarb delivering a dose of 0, 3.5, 10 or 25/35 mg/kg bw per day for 1 year. The highest dose of 35 mg/kg bw per day could not be sustained owing to adverse clinical signs in week 1, so from week 4 onwards, the dose was reduced to 25 mg/kg bw per day. One female dog receiving 25 mg/kg bw per day was killed humanely in week 36 after significant body-weight loss and the development of anaemia. The haematological changes in this dog were characterized by increased erythropoietic activity in the bone marrow and by histological changes consistent with increased erythrocyte breakdown. No other dog showed any treatment-related haematological changes; however, increased haemosiderin deposition was observed in the liver and spleen of dogs at 25 mg/kg bw per day. The NOAEL in this study was 3.5 mg/kg bw per day; this value is very close to the LOAEL of 4 mg/kg bw identified in three other experiments. The overall NOAEL was 2 mg/kg bw per day in dogs.

The carcinogenic potential of pirimicarb has been assessed in feeding studies of 80 and 96 weeks' duration in mice and 104 weeks in rats. In both species, the highest dose tested induced moderate levels of toxicity. Two studies of carcinogenicity were conducted in Alderley Park Swiss-derived mice and one in C57 black Alderley Park mice; however, one of the studies in Swiss-derived mice, a study that pre-dated the establishment of good laboratory practice (GLP), was not considered adequate for assessing the carcinogenic potential of pirimicarb, owing to a high incidence of respiratory disease. Similarly, three pre-GLP studies in rats were not adequate for carcinogenicity assessment owing to high incidences of respiratory disease.

In Alderley Park Swiss-derived mice given diets containing pirimicarb at a concentration of 0, 0, 200, 400 or 1600 ppm for up to 96 weeks, there was a significant increase in the incidence of liver tumours [classified as type A (hyperplastic nodules and benign neoplasms) and type B nodules (which showed characteristics of malignancy)] at the highest dose (equivalent to approximately 240 mg/kg bw per day), with no evidence of nodule induction at lower doses. The incidence of liver nodules was above the historical control range for the test laboratory. This finding was not confirmed in C57 black Alderley Park mice given diets containing pirimicarb at a concentration of 0, 50, 200 or 700 ppm for at least 80 weeks, where there was no evidence that liver tumours induced by pirimicarb at doses of up to 700 ppm (equal to approximately 94 and 130 mg/kg bw per day in males and females, respectively). Thus, the significant response was only seen at a very high dose and after a prolonged exposure time in a mouse strain with a high and variable background incidence of liver tumours.

Also in Alderley Park Swiss-derived mice, the incidence of pulmonary adenoma was significantly increased in both sexes at the highest dose, but with no significant response at lower doses. Concurrent and historical control data indicate a high and variable spontaneous background incidence of pulmonary adenomas in this strain of mouse. Given the overall variability in the incidence of pulmonary adenoma in these mice, the observation of an increased tumour incidence at the highest dose does not give cause for concern in terms of cancer risk. A small, statistically significant, increase in the incidence of pulmonary adenoma was also observed in female (but not male) C57 black Alderley Park mice at the highest dose tested, with no evidence of pulmonary adenoma induction at lower doses. In contrast to Swiss-derived mice, C57 black mice have a low spontaneous background incidence of pulmonary adenoma. Therefore, the occurrence of these tumours is considered to be treatment-related. The absence of a significant response in male mice could be a chance difference in the incidence of an uncommon tumour type. The Meeting concluded that oral administration of pirimicarb at a dose of up to 700 ppm, equal to 94 and 130 mg/kg bw per day for males and females, respectively, for at least 80 weeks produced a small increase in the incidence of benign lung tumours in females, but not in males. The NOAEL for non-neoplastic effects was 50 ppm, equal to 6.7 mg/kg bw per day, on the basis of slight haematological changes at 200 ppm, equal to 26.6 mg/kg bw per day in the 80-week study.

A 2-year study in rats showed that dietary administration of pirimicarb at 0, 75, 250 or 750 ppm resulted in reduced body-weight gains and food consumption in both sexes at 750 ppm, indicating that a maximum tolerated dose had been achieved. There was also a slight reduction in body-weight gain at 250 ppm in females. There were increases in plasma cholesterol at all observation times at 750 ppm, at weeks 13 and 26 at 250 ppm and (in females only) at week 13 at 75 ppm. Plasma concentrations of triglycerides were increased at 750 ppm in males at weeks 52 and 78 and in females at weeks 13 and 26. Males fed diets containing pirimicarb at 750 ppm showed a small increase in incidence and severity of necrosis in the brain. The significance of this is equivocal, but could not be dismissed as being incidental to treatment with pirimicarb. Females fed diets containing pirimicarb at 750 ppm showed an increased severity of sciatic nerve demyelination and an increased severity and incidence of voluntary muscle degeneration that were considered to be an exacerbation of a spontaneous age-related change. Overall, the findings in the brain, sciatic nerve and voluntary muscle were minor, confined to the highest dose and did not elicit any clinical signs of increased neurological dysfunction. Plasma cholinesterase activity was slightly reduced in females at 250 and 750 ppm, demonstrating the absorption of the test substance, but brain and erythrocyte cholinesterase activities were not affected at any dose. The NOAEL for non-neoplastic effects was 75 ppm, equal to 3.7 mg/kg bw per day, on the basis of reductions in body weights and increases in plasma cholesterol and triglycerides at 250 ppm, equal to 12.3 mg/kg bw per day. There was an overall higher, but non-significant incidence in the number of male rats with tumours at 250 and 750 ppm. This reflected an increased incidence of males with multiple tumours at 250 ppm and of males with single tumours at 750 ppm. There were also small increased incidences of astrocytoma of the brain in all treated groups and in females at 750 ppm, but these were not statistically significant and no dose–response relationship was evident. As there was a decreased incidence of males with multiple tumours at 750 ppm and there were no consistent effects across doses, the Meeting concluded that pirimicarb did not induce a carcinogenic response in any tissue.

The Meeting concluded that pirimicarb had no clear carcinogenic potential in mice or rats. Liver tumours were not consistently found in the two studies in mice, while the benign lung tumours were found only at the highest dose and with clear evidence for a threshold. There were no compound-related increases in the incidences of any tumour type in rats.

Pirimicarb was tested for genotoxicity in an adequate range of studies, both in vitro and in vivo. The results observed were largely negative. A small increase in mutant frequency in the assay for mutation in L5178Y mammalian cells, in the presence of metabolic activation, was considered not to be a significant alert for genotoxicity. Pirimicarb has shown no evidence of genotoxic potential in several test systems in vivo. The Meeting concluded that pirimicarb is unlikely to pose a genotoxic risk to humans.

Because the results of the studies of carcinogenicity in rodents were judged not to provide evidence of carcinogenic potential, an evaluation supported by the lack of genotoxic potential, the Meeting concluded that pirimicarb is unlikely to pose a carcinogenic risk to humans.

In a two-generation study of reproductive toxicity in rats, the NOAEL for adult rats and for their offspring was 200 ppm, equal to 23 mg/kg bw per day in adults, on the basis of systemic toxicity in the parental rats and reduced body-weight gain in the parental rats and the offspring at a dose of 750 ppm, equal to 88 mg/kg bw per day; no other signs of reproductive toxicity were observed at this dose, the highest tested. In studies of developmental toxicity in rats, the NOAEL for foetal toxicity and maternal toxicity was 25 mg/kg bw per day on the basis of reduced fetal weight and maternal body-weight gains at 75 mg/kg bw per day. In studies of developmental toxicity in rabbits, the NOAEL for fetal and developmental toxicity was 60 mg/kg bw per day, the highest dose tested, and the NOAEL for maternal toxicity for 10 mg/kg bw per day, on the basis of reduced food consumption and body-weight gains at 60 mg/kg bw per day. The results from the two studies of developmental toxicity and the study of reproductive toxicity demonstrated that fetuses and pups were not more susceptible than adults to toxicity caused by pirimicarb.

In a study of acute neurotoxicity in rats, a single oral administration of pirimicarb at 110 mg/kg bw per day by gavage resulted in early mortalities, adverse clinical signs and reductions in brain, erythrocyte and plasma cholinesterase activities. These clinical and enzyme activity changes were transient and were not associated with histopathological changes in the nervous system. At a dose of 40 mg/kg bw, there was evidence of toxicity seen as a single mortality, transient adverse clinical signs in a few rats and reduced motor activity on day 1. Plasma cholinesterase activity was reduced, but this observation was not accompanied by biologically significant reductions in brain or erythrocyte cholinesterase activity at this dose. The Meeting concluded that the NOAEL for acute neurotoxic potential was 10 mg/kg bw per day and this value formed the basis for the ARfD. The acute toxic effects of pirimicarb are due to inhibition of acetylcholinesterase activity at nerve terminals. Inhibition of acetylcholinesterase by carbamates (such as pirimicarb) and organophosphates involves the carbamoylation or phosphorylation of the active site on the enzyme. Plasma cholinesterase is inhibited by a similar mechanism; therefore, although this is a toxicologically irrelevant target, its inhibition serves as an indicator of exposure and a surrogate for the response of acetylcholinesterase. The degree of enzyme inhibition is dependent on the concentration of inhibitor, a property that is particularly significant for carbamates because of the short occupation half-life at the active site of the enzyme (a few minutes, both in vitro and in vivo). Consequently, plasma cholinesterase activity was inhibited after a single dose of pirimicarb at 25 mg/kg bw by gavage, but not after dietary exposure corresponding to a daily dose of about 40 mg/kg bw, when the C_{max} would have been lower.

In a 90-day study of neurotoxicity, rats fed diets containing pirimicarb at a concentration of 250 or 1000 ppm resulted in toxicity evident as reduced growth and food consumption or utilization. There were no treatment-related effects on the functional observational battery, motor activity, cholinesterase and neurotoxic esterase activities or neuropathology. The NOAEL for neurotoxicity in this study was 1000 ppm, equal to 77 mg/kg bw per day, the highest dose tested.

Studies of toxicity have been conducted on a number of metabolites of pirimicarb: three carbamate metabolites, three hydroxypyrimidine metabolites and three guanidine metabolites. The acute toxicities of two carbamates (the desmethyl pirimicarb and the desmethylformamido pirimicarb metabolites) were of the same order as that of pirimicarb itself, whereas the LD_{50} values of all the other seven metabolites were less or considerably less than that of pirimicarb itself. In addition, some of these metabolites tested in studies of toxicity with repeated doses and some to assays for genotoxicity. Desmethyl pirimicarb and desmethylformamido pirimicarb had effects on cholinesterase that were similar to those caused by pirimicarb itself and are included in the residue definition, since they occur in plants. In 28- and 90-day studies of toxicity in rats, the hydroxypirimidine metabolite, 2-dimethylamino-5,6-dimethylpyrimidin-4-ol (and, by implication, its mammalian metabolite, 5,6-dimethyl-2-(methylamino)pyrimidin-4-ol) was of low toxicity; the NOAEL was 240 ppm, equal to 19.5 mg/kg bw per day, on the basis of blood chemistry changes at 800 ppm, equal to 65.6 mg/kg bw per day. NOAELs could not be identified because testing was restricted to single doses in the cases of desmethyl pirimicarb (100 mg/kg bw per day for 2 weeks) and desmethylformamido pirimicarb (25 mg/kg bw per day for 2 weeks). Each of these metabolites caused slight hypochromia (reduced haemoglobin concentrations per cell). Genotoxicity tests were conducted with the hydroxypirimidines, 2-dimethylamino-5,6-dimethylpyrimidin-4-ol and 5,6-dimethyl-2-(methylamino)pyrimidin-4-ol. Both metabolites, like pirimicarb itself, produced some weak evidence of mutagenic effects in the assay in mouse lymphoma cells, but not in other assays. The Meeting concluded that, within the limitations of studies conducted (short-term and only in rats), desmethyl pirimicarb and desmethylformamido pirimicarb have toxicological profiles similar to that of pirimicarb itself.

The Meeting concluded that the existing database on pirimicarb was adequate to characterize the potential hazards to fetuses, infants and children.

Production workers have been reported to show inhibition of plasma and erythrocyte cholinesterase activity of sufficient severity to result in their movement to other work areas.

Pirimicarb

Toxicological evaluation

An ADI of 0–0.02 mg/kg bw was established for pirimicarb and its dimethyl carbamate metabolites on the basis of the overall NOAEL of 2 mg/kg bw per day in 90-day and 2-year studies in dogs treated by dietary administration and with a safety factor of 100.

The Meeting established an ARfD of 0.1 mg/kg bw for pirimicarb on the basis of a NOAEL of 10 mg/kg bw in a study of acute neurotoxicity in rats. Although a reduced safety factor would be supported by the reversibility of clinical signs and the C_{max}-dependency of the effects, a safety factor of 100 was used in consideration of the steep dose–response curve (indicated by a mortality at the LOAEL) and the lack of reliable measurement of acetylcholinesterase inhibition. Haematotoxicity in dogs was also considered as a possible end-point for an ARfD; however, in one study in dogs haematological parameters were measured during treatment before the onset of anaemia, indicating that this condition did not occur after a single dose.

A toxicological monograph was prepared.

Levels relevant to risk assessment

Species	Study	Effect	NOAEL	LOAEL
Mouse	18-month and 21-month studies of toxicity and carcinogenicity[a]	Toxicity	50 ppm, equal to 6.7 mg/kg bw per day	200 ppm, equal to 27 mg/kg bw per day
		Carcinogenicity	200 ppm, equal to 37mg/kg bw per day	700 ppm, equal to 94 mg/kg bw per day
Rat	24-month study of toxicity and carcinogenicity[a]	Toxicity	75 ppm, equal to 3.7mg/kg bw per day	250ppm, equal to 12.3 mg/kg bw per day
		Carcinogenicity	750 ppm, equal to 37 mg/kg bw per day[c]	—
	Two-generation study of reproductive toxicity[a]	Parental toxicity	200 ppm, equal to 22 mg/kg bw per day	750 ppm, equal to 88 mg/kg bw per day
		Offspring toxicity	200 ppm, equal to 23 mg/kg bw per day	750 ppm, equal to 88 mg/kg bw per day
	Developmental toxicity[b]	Maternal toxicity	25 mg/kg bw per day	75 mg/kg bw per day
		Embryo- and fetotoxicity	25 mg/kg bw per day	75 mg/kg bw per day
	Single-dose neurotoxicity[b]	Neurotoxicity	10 mg/kg bw	40 mg/kg bw per day
	3-month study of neurotoxicity[a]	Neurotoxicity	1000 ppm, equal to 81 mg/kg per day[c]	—
Rabbit	Developmental toxicity[b]	Maternal toxicity	10 mg/kg bw per day	60 mg/kg bw per day
		Embryo and foetal toxicity	60 mg/kg bw per day[c]	—
Dog	90-day and 2-year studies of toxicity[a]	Toxicity	2 mg/kg bw per day	4 mg/kg bw per day

[a] Dietary administration
[b] Gavage administration
[c] Highest dose tested

Estimate of acceptable daily intake for humans

    0–0.02 mg/kg bw

Estimate of acute reference dose

    0.1 mg/kg bw

Studies that would provide information useful to the continued evaluation of the compound

Further observations in humans

Critical end-points for setting guidance values for exposure to pirimicarb

Absorption, distribution, excretion and metabolism in animals	
Rate and extent of oral absorption	Rapid; > 80% absorbed
Dermal absorption	No study of direct dermal absorption available, but brain cholinesterase activity was inhibited after application of pirimicarb to rat skin, indicating absorption by this route
Distribution	Distributed throughout the body; highest concentrations in liver and fat
Potential for accumulation	Low, owing to rapid excretion
Rate and extent of excretion	Rapid, > 80 % excretion within 24 h
Metabolism in animals	Extensive
Toxicologically significant compounds (animals, plants and environment)	Parent and the metabolites desmethyl pirimicarb and desmethylformamido pirimicarb
Acute toxicity	
Rat, LD_{50}, oral	142 mg/kg bw
Rat, LC_{50}, inhalation	0.858 mg/l (4 h)
Rabbit, LD_{50}, dermal	> 2000 mg/kg bw
Rabbit, skin irritation	Not irritating
Rabbit, eye irritation	Not irritating
Skin sensitization	Sensitizing (Magnusson and Kligman test)
Short-term studies of toxicity	
Target/critical effect	Body-weight gain decrement, haemolytic anaemia or cholinesterase inhibition
Lowest relevant oral NOAEL	1.8 mg/kg bw per day: (3-month study in dogs)
Lowest relevant dermal NOAEL	2000 mg/kg bw per day (21-day study in rats)
Lowest relevant inhalation NOAEC	No data available
Genotoxicity	No genotoxic potential: negative in vivo, positive results in one study in vitro
Long-term studies of toxicity and carcinogenicity	
Target/critical effect	Blood/anaemia, increased plasma lipids
Lowest relevant NOAEL	2 mg/kg bw per day (24-month study in dogs)
	3.7 mg/kg bw per day (24-month study in rats)
Carcinogenicity	Benign lung tumours in mice induced by a non-genotoxic mode of action; a clear NOAEL was identified; therefore pirimicarb is unlikely to pose a carcinogenic risk to humans
Reproductive toxicity	
Reproductive target/critical effect	Reduced parental and offspring body weight, clinical signs
Lowest relevant reproductive NOAEL	23 mg/kg bw per day
Developmental target/critical effect	Not teratogenic; reduced fetal body weight at maternally toxic doses
Lowest relevant developmental NOAEL	25 mg/kg bw per day (rat)
Neurotoxicity/delayed neurotoxicity	

Pirimicarb

Target/critical effect	Nervous system/cholinergic signs
Lowest relevant NOAEL	10 mg/kg bw

90-day neurotoxicity

Target/critical effect	Nervous system/cholinergic signs
Lowest relevant NOAEL	77 mg/kg bw per day
Other toxicological studies	Desmethyl pirimicarb and desmethylformamido pirimicarb inhibited acetylcholinesterase activity in rats (no studies in dogs)
Medical data	There have been a few reports of cholinesterase inhibition in workers exposed during manufacture.

Summary	Value	Study	Safety factor
ADI	0–0.02 mg/kg bw	Dog; haematological changes in short- and long-term studies	100
ARfD	0.1 mg/kg bw	Rat; mortality and clinical signs of neurotoxicity in a study of acute neurotoxicity	100

DIETARY RISK ASSESSMENT

Long-term intake

The estimated theoretical maximum daily intakes in the five GEMS/Food regional diets, based on recommended MRLs, were in the range of 3–20% of the ADI (Annex 3). The Meeting concluded that the long-term intake of residues of pirimicarb resulting from uses that have been considered by the JMPR is unlikely to present a public health concern.

Short-term intake

The Meeting established an ARfD of 0.1 mg/kg bw for pirimicarb but was unable to finalize the risk assessment before the residue evaluation, scheduled for 2006, had been completed.

4.22 PIRIMIPHOS-METHYL (086)

RESIDUE AND ANALYTICAL ASPECTS

Pirimiphos-methyl was evaluated for residues by the 2003 JMPR within the CCPR periodic review programme. As no data were reported on the storage stability in animal tissues or eggs, the JMPR recommended withdrawal of the CXLs for eggs and meat (from mammals other than marine mammals). Data on the storage stability of pirimiphos-methyl in animal tissues, milk and eggs were provided to the present Meeting.

Stability of residues in stored analytical samples

Pirimiphos-methyl residues at 0.5 mg/kg were shown to be stable for up to 12 months in beef muscle, liver, kidney, fat and milk and hens' eggs stored below – 18 °C.

Residues in animal commodities

The 2003 JMPR calculated the dietary burden of pirimiphos-methyl for estimated MRLs and STMR values for animal commodities as 6.4 and 2.1 mg/kg for beef cattle, 5.6 and 2.9 mg/kg for dairy cattle and 6.3 and 2.1 mg/kg for poultry, respectively. In one study submitted to the 2003 Meeting, the residue levels were below the LOQ (0.01 mg/kg) in heart, liver, kidney, fat and pectoral muscle from cows fed diets containing 0, 5, 15 or 50 ppm (dry weight basis) of pirimiphos-methyl for 30 days. Adductor muscle from one of three cows at the highest feeding level contained detectable residues (0.02 mg/kg). In another study submitted to the previous Meeting, eggs from hens receiving 32 ppm pirimiphos-methyl for 7 days contained residues at a maximum of 0.01 mg/kg. No detectable residues were found in leg or breast muscle.

Maximum residue levels

The Meeting agreed that it is unlikely that pirimiphos-methyl residues would remain in tissues of cows fed commodities treated with the insecticide, and recommended a maximum residue level of 0.01* mg/kg and STMR and highest residue values of 0 mg/kg for pirimiphos-methyl in edible offal (mammalian) and meat (fat) from mammals, other than marine mammals. For calculating dietary intake, the Meeting also recommended an STMR and a highest residue level of 0 mg/kg for pirimiphos-methyl in muscle and fat from mammals other than marine mammals.

The Meeting agreed that it is unlikely that pirimiphos-methyl residues would remain in poultry tissues after birds have been fed commodities treated with the insecticide, and recommended a maximum residue level of 0.01* mg/kg and STMR and highest residue values of 0 mg/kg for pirimiphos-methyl in poultry meat and poultry edible offal. For estimating dietary intake, the Meeting also recommended STMR and highest residue values of 0 for pirimiphos-methyl in poultry muscle and fat.

Pirimiphos-methyl was detected in eggs of hens at the highest feeding level, which corresponded to five times the dietary burden of poultry for MRL estimation. The Meeting recommended maximum and highest residue levels of 0.01 mg/kg for pirimiphos-methyl in eggs. As the dietary burden for STMR corresponds to 15 times the feeding level, the Meeting agreed to recommend an STMR of 0 mg/kg for eggs.

DIETARY RISK ASSESSMENT

Long-term intake

The Meeting agreed that the STMR of 0 mg/kg estimated by the present Meeting for pirimiphos methyl in animal commodities will not affect the IEDIs calculated by the 2003 JMPR, which concluded that the long-term intake of residues of pirimiphos methyl is unlikely to present a public heath concern.

Short-term intake

The IESTIs of pirimiphos-methyl by the general population and by children were calculated for commodities for which highest residue levels were estimated by the current Meeting (Annex 4). For all the commodities, the IESTI was 0–0.08 µg/kg bw. Although it might be necessary, no ARfD has yet been established for pirimiphos methyl, and the short-term risk assessment could not be finalized.

4.23 PROCHLORAZ (142)

RESIDUE AND ANALYTICAL ASPECTS

Prochloraz is a broad-spectrum imidazole fungicide that is active against a range of diseases in field crops, fruit and vegetables and is also used on mushrooms, as a post-harvest treatment of fruit and as a seed treatment on cereals. It was evaluated initially in 1983 for residues and toxicology; six additional reviews of residues were carried out between 1985 and 1992, and a periodic toxicological review in 2001. The CCPR at its Twenty-ninth Session scheduled prochloraz for periodic review with respect to residues, and it was included on the 2004 JMPR agenda. The Meeting received information on the metabolism and environmental fate of prochloraz, methods of residue analysis, freezer storage stability, national registered use patterns, the results of supervised residue trials, farm animal feeding studies, fate of residues in processing and national MRLs. Information on GAP and national MRLs was submitted by Australia and Japan.

The formulations that are available include emulsifiable concentrates, suspo-emulsions and wettable powders. A number of formulations with other fungicides are also available, mainly for use on cereal crops. Wettable powder formulations of a 4:1 complex of prochloraz and manganese chloride are available for use on crops susceptible to phytotoxicity.

In this evaluation, the term 'total prochloraz' refers to the parent compound and metabolites containing the common 2,4,6-trichlorphenol moiety, expressed as prochloraz equivalents (using a correction factor of 1.9). The term 'free prochloraz' refers to the parent compound only.

The following abbreviations are used for the metabolites:

BTS 44595	N-propyl-N'-2-(2,4,6-trichlorophenoxy)ethylurea
BTS 44596	N'-formyl-N-propyl-N-[2-(2,4,6-trichlorophenoxy)ethyl]urea
BTS 44770	N-2-(2,4,6-trichlorophenoxy)ethylurea
BTS 9608	2,4,6-trichlorophenoxyacetic acid
BTS 45186	2,4,6-trichlorophenol
BTS 54906	2-(2,4,6-trichloro-3-hydroxyphenoxy)ethanol
BTS 54908	N-2-(2,4,6-trichloro-3-hydroxyphenoxy)ethyl-N-propylurea

Metabolism

Animals

The Meeting received information on the metabolism of prochloraz in rats, lactating goats lactating cows and laying hens.

Prochloraz was extensively metabolized in rats, no unchanged parent compound being detected in urine; it was, however, detected in faeces and was the most abundant component on day 1. Faeces contained significant quantities of the plant metabolites BTS 44595 and BTS 44596, formed by opening of the imidazole ring. The most abundant metabolite in urine was BTS 9608, comprising around 35% of the excreted radioactivity. A more recent study in rats (2003) generally confirmed the results of the earlier studies, although a more complex pattern of metabolism was reported, additional metabolites being detected in urine and faeces. The metabolism of prochloraz in the rat proceeds via cleavage of the imidazole ring, oxidation of the side-chain, phenyl-ring hydroxylation and substitution of chlorine by a hydroxyl group. Other processes were revealed in the latest study in rats, including N-dealkylation, N-deacetylation and sulfate conjugation of hydroxy groups.

Straw from field plots treated 11 weeks before harvest with ^{14}C-prochloraz and containing the equivalent of 19 mg/kg was fed to a lactating goat daily for 4 days. Milk and blood samples were taken twice daily, and the animal was killed on the fifth day. The highest residue levels were found in liver (0.05 mg/kg), kidney fat and rumen wall (0.04 mg/kg), expressed as equivalents. All other tissues contained ≤ 0.03 mg/kg, milk contained ≤ 0.006 mg/l; the maximum level in plasma was 0.08 mg/l.

A lactating cow was given gelatin capsules containing ^{14}C-prochloraz at a rate providing 1.5 mg/kg bw per day twice a day for 3 days, equivalent to 37.5 mg/kg of diet. The radioactivity in plasma reached a plateau at 72 h, and the levels in milk rose to a plateau of 0.14 mg/l after 24 h. Most of the radioactivity was found in the liver (10 mg/kg) and kidney (1.7 mg/kg), with lower levels in other tissues. Parent prochloraz was not found in the gut contents, plasma, milk or tissues. Analysis of the gut contents indicated that prochloraz was rapidly degraded to the imidazole ring-opened metabolites BTS 44596, BTS 44595 and BTS 44770, and these metabolites were also detected in tissues and in early plasma samples. The phenolic metabolites BTS 54906 and BTS 54908 were prevalent in milk, in addition to BTS 4496 (23% TRR).

In laying hens given ^{14}C-prochloraz in gelatin capsules daily at a rate of 1.5 mg/day for 14 days, equivalent to 10 mg/kg of diet, 85% of the TRR had been excreted within 24 h, and the levels of radioactivity in eggs, mostly in the yolk, reached a plateau of 1.7 mg/kg by day 8. The highest residue levels were found in the liver (0.9 mg/kg) and gastrointestinal tract (0.8 mg/kg), and levels ≤ 0.19 mg/kg were found in skin, ≤ 0.09 mg/kg in fat, 0.05 mg/kg in breast muscle and 0.07 mg/kg in thigh muscle. Parent prochloraz was not found in excreta, eggs or tissues; BTS 9608 and BTS 44596 were the main metabolites in liver, muscle, fat and eggs.

Generally, prochloraz is rapidly absorbed, metabolized and excreted, and it is not detected in milk, eggs or tissues. BTS 44596, the formyl urea metabolite, is the residue component found predominantly in egg yolk, and BTS 54906 and BTS 44596 are the main components in milk. Except in liver and to a lesser extent in kidney, the residue levels in tissues are generally low and consist mainly of BTS 44596, BTS 44595 and BTS 44770.

Plants

The Meeting received the results of studies of the metabolism of prochloraz in wheat and oil-seed rape after foliar application, in mushrooms after treatment of the casing and in wheat after seed treatment.

In two studies in which young wheat plants received foliar treatment with radiolabelled prochloraz at 0.25–0.39 kg ai/ha, the residue levels of parent compound were 0.6–1% TRR after 19–20 days. The main metabolites detected were BTS 44596 (32–38%), free and conjugated BTS 44595 (31%) and BTS 45186 (8%). In one of the studies, mature grain harvested 14 weeks after treatment contained < 0.05 mg/kg, representing < 0.2% of the TRR present in straw; the residues in grain were mostly bound in fibre, while the residues in straw consisted mainly of BTS 44596 (26% TRR) and BTS 44595 (8% TRR), with less than 0.1% parent compound.

In two studies on mature wheat harvested 13 weeks after foliar treatment with radiolabelled prochloraz at 1 kg ai/ha, the residues in grain represented 3–8% TRR as free BTS 45186 and 40–54% TRR as stable polar conjugates containing the 2,4,6-trichlorophenol moiety. Of the residues in straw, about 5% TRR was free BTS 45186 and 38–58% was conjugates containing the 2,4,6-trichlorophenol moiety.

In wheat plants grown from seed treated with radiolabelled prochloraz (0.4 g ai/kg of seed), 5.6% TRR was measured in aerial plant portions during the first 6 weeks of growth; no further translocation was seen. At maturity, 58% of the applied radiolabel was found in soil and 15% in the root system; no radioactivity was observed in grain.

The metabolism of prochloraz in oil-seed rape was studied after foliar treatment of young plants. Leaves sampled 19 days after treatment contained < 3% of the TRR, and the main metabolites were BTS 44596 (20%), BTS 44595 (29%) and material of polar origin (30%). A similar distribution was reported in mature plants; about 3% of the TRR was detected in plant parts that had not been treated directly, and the residues in mature seeds accounted for about 0.1% TRR.

In mushrooms treated with the prochloraz–manganese chloride complex at 3 g ai/m^2 and analysed 8 and 30 days later, unchanged parent compound accounted for 75% and 83% of the extracted radioactivity, respectively, and BTS 9608 accounted for a further 9–10%.

In summary, prochloraz is metabolized to BTS 44596 via cleavage of the imidazole ring, followed by 'deformylation' to generate BTS 44595. Low levels of conjugates of both these metabolites are formed, which are resistant to the initial extraction solvents, being released only under more exhaustive conditions (e.g. microwave acetonitrile:water extraction). The fact that polar materials could be converted to BTS 45186 by pyridinium hydrochloride hydrolysis indicates that the trichlorophenol moiety is present in this material. The metabolic fate of prochloraz in oil-seed rape is similar to that in wheat. In mushrooms, prochloraz–manganese complex underwent dissociation to free prochloraz and subsequent metabolism to BTS 9608 and conjugates containing the BTS 45186 moiety.

Rotational crops

The Meeting received information on the behaviour and fate of prochloraz in soil and in rotational crops.

Under normal agricultural conditions, prochloraz is moderately persistent in soil, with a DT$_{50}$ of < 40 days. The products of biotic and photolytic degradation, BTS 44596 and BTS 44595, have been detected occasionally in soil samples collected in the field but at levels close to the LOQ.

When prochloraz is applied to bare soil, it is metabolized in rotational crops to BTS 44596, BTS 44595, BTS 45186 and BTS 9608. The levels of total residues declined sharply between 30- and 120-day crops and declined further with soil ageing. The metabolites in rotational crops were essentially the same as those reported in the studies of plant metabolism, except that the levels of BTS 9608 were lower. The concentrations of metabolites were low in all crops (< 0.01 mg/kg), exceeding 0.05 mg/kg only in wheat forage and straw.

Methods of analysis

The Meeting received information on methods for analysis for free prochloraz and for total prochloraz (prochloraz plus metabolites containing the common 2,4,6-trichlorophenol moiety), in plant material, animal tissues and soils. Analytical methods for specific metabolites in plant and animal tissues were also provided.

The 'common moiety' method, involving hydrolysis of prochloraz and its metabolites to 2,4,6-trichlorophenol, was used in most of the supervised residue trials and for enforcement purposes. In this method (RESID/88/72) and in an earlier, related method (RESID/82/88), samples are Soxhlet-extracted with acetone, concentrated and hydrolysed with pyridine hydrochloride to break down all components to 2,4,6-trichlorophenol. This hydrolysate is then extracted into petroleum ether by steam distillation, with further clean-up by extraction into the aqueous layer with alkali and re-extraction into toluene after acidification. Total 2,4,6-trichlorophenol residues are determined by gas chromatography (with electron capture detection for plant material and soil and MS detection or mass spectrometry for milk and animal tissues), and the results are expressed as prochloraz equivalents, with a correction factor of 1.9.

Methods for measuring free prochloraz in plant materials are based on acetone extraction, acidification with hydrochloric acid, evaporation and extraction under acid conditions with petroleum ether. The aqueous extract is neutralized, further extracted with petroleum ether and evaporated, and the residue is dissolved in ethyl acetate before analysis by gas chromatography with electron capture detection.

An HPLC method for measuring prochloraz and the major metabolites BTS 44596 and BTS 44595 and the hydroxyamide metabolite BTS 54908 has also been reported for milk (liquid chromatography with tandem mass spectrometry), with an LOQ of 0.01 mg/kg for prochloraz and 0.005 mg/kg for each metabolite.

The LOQs of 'common moiety' methods for prochloraz equivalent in most substrates are 0.01–0.05 mg/kg, although a higher LOQ of 0.1 mg/kg may be required for some materials (immature cereal plants and straw, sub-tropical citrus fruit peel) in which there are high background levels of residues. Recovery efficiencies of about 90% are common, usually ranging from 75% to 110%.

Stability of residues in stored analytical samples

The Meeting received information on the stability of prochloraz in various commodities under freezer storage (–18 to –20 °C). Less than 30% of the residues had degraded during storage in wheat stored for 24 months, in barley at 23 months, in sugar-beet roots and tops at 14 months, in maize plants at 24 months, in rape-seed at 36 months, in muscle at 12 months, in milk at 12 months and in eggs at 12 months.

Definition of the residue

Studies of metabolism in lactating goats and cows and in hens indicate that the parent compound is not found in tissues, milk or eggs; however, a number of metabolites containing the 2,4,6-trichlorophenol moiety occur, BTS 44596, BTS 44595, BTS 44770, BTS 54906, BTS 54908 and BTS 9608 being found in one or more of the above substrates.

In plants, the metabolic pathway is consistent, involving cleavage of the imidazole ring to form the aldehyde BTS 44596, oxidation of the side-chain to form the urea (BTS 44595) and the carboxylic acid (BTS 9608, generally only in conjugated form), with eventual formation of the phenol BTS 45186 and its polar conjugates.

The Meeting therefore considered that the residues of toxicological concern would be those of unchanged parent compound, the non-polar metabolites BTS 44595 and BTS 44596 and low levels of free BTS 45186, which are readily extracted from plant material with acetone. The Meeting noted that these compounds are also found in rats and are therefore covered by the toxicological assessment.

The Meeting confirmed the current prochloraz residue definition, 'Sum of prochloraz and its metabolites containing the 2,4,6-trichlorphenol moiety, expressed as prochloraz', for compliance with MRLs and for estimation of dietary intake from both animal and plant commodities.

Taking into account the log P_{ow} of prochloraz of 3.5 and the results of the animal feeding studies, the Meeting decided that residues of prochloraz should be classified as fat-soluble.

Results of supervised trials on crops

The results of supervised trials were available for use of prochloraz on citrus fruit (lemon, mandarin, orange), avocado, banana, mango, papaya, pineapple, onion, melon, mushroom, tomato, lettuce, bean, pea, sugar-beet, rape-seed, sunflower seed, linseed, soya bean, barley, oat, rice, rye, wheat and pepper, black.

The results of trials or relevant GAP were not submitted for coffee beans or stone fruits, for which maximum residue levels are currently recommended. The Meeting agreed to withdraw the previously recommended maximum residue levels for these commodities.

The Meeting noted that a post-treatment interval has not been defined for most approved uses on citrus and sub-tropical fruit (inedible peel), and that in the relevant studies on orange, mandarin, avocado and banana, the residue levels, although variable, were often high in fruit sampled after day 0. As the residues in post-harvest-treated fruit did not appear to degrade appreciably during storage, the Meeting agreed to use the results of analyses up to 21 days after treatment to reflect the residue levels expected in fruit immediately after treatment.

Prochloraz

Citrus fruit

The results of trials of post-harvest dipping and spraying on lemon, mandarin and orange were available from Argentina, Australia, Greece, Italy, Morocco, Spain and the United Kingdom, and the results of post-harvest brushing trials on oranges were available from South Africa.

Lemon

GAP in Argentina for citrus includes a post-harvest spray application of 0.2–0.29 kg ai/hl (no post-treatment interval specified). In trials in Spain reflecting this GAP, the residue levels in lemons sampled 12–16 days after treatment were 3.8 and 4.5 mg/kg. The corresponding residue levels in lemon pulp were 0.16 mg/kg and 0.23 mg/kg

Orange

GAP for citrus in Argentina includes a post-harvest spray application of 0.2–0.29 kg ai/hl (no post-treatment interval specified). While no trials in Argentina matched this GAP, one trial in Spain that did showed a residue level of 1.7 mg/kg in oranges (0.07, 0.1, 0.13, 0.14 and 0.17 mg/kg in pulp) sampled 14 days after treatment.

In South Africa, GAP is for a post-harvest brush treatment with 0.15 kg ai/hl (no post-treatment interval specified). In one trial in Argentina reflecting this GAP (± 25–30%), the residue level was 5.3 mg/kg in oranges (0.02 mg/kg in pulp).

In Greece, GAP is for use of prochloraz as a post-harvest dip or spray at up to 0.09 kg ai/hl. Trials in Australia, Morocco and Spain were evaluated against the GAP of Greece. The residue levels in oranges were: 1.3, 1.4, 1.5, 1.7, 2.0, 3.7, 5.9 and 6.8 mg/kg, and those in pulp (edible portion) were: 0.02, 0.06 (two), < 0.1 (two), 0.26, 0.33, 0.56 and 0.92 mg/kg.

The residue levels in oranges, in ranked order, were: 1.3, 1.4, 1.5, 1.7 (two), 2.0, 3.7, 5.3, 5.9 and 6.8 mg/kg, and those in pulp were: 0.02 (two), 0.06 (two), 0.07, < 0.1 (two), 0.1, 0.13, 0.14, 0.17, 0.26, 0.33, 0.56 and 0.92 mg/kg.

Mandarin

Two post-harvest trials in Spain matching GAP in Argentina (up to 0.29 kg ai/hl) showed residue levels of 2.1 and 3.5 mg/kg, with corresponding levels of 0.1 and 0.35 mg/kg in pulp.

In Greece, GAP is for use of prochloraz as a dip or spray at up to 0.09 kg ai/hl, while in Spain GAP is for use at up to 0.08 kg ai/hl. The trials in Australia (no GAP), Morocco (no GAP) and Spain were evaluated against the GAP of Greece. The residue levels in mandarins were 2.0, 2.1 (three), 2.3, 3.2, 3.4, 3.5, 3.9, 4.3, 4.6, 5.4 and 5.9 mg/kg, and the corresponding levels in pulp (edible portion) were: 0.07 (two), 0.09 (three), 0.1, 0.12, 0.26 and 0.31 mg/kg.

The residue levels in mandarins, in ranked order, were: 2.0, 2.1 (four), 2.3, 3.2, 3.4, 3.5 (two), 3.9, 4.3, 4.6, 5.4 and 5.9 mg/kg, and the corresponding levels in pulp (edible portion) were: 0.07 (two), 0.09 (three), 0.1 (two), 0.12, 0.26, 0.31 and 0.35 mg/kg.

The Meeting agreed to combine the data for lemon, orange and mandarin, to give a data set for citrus from 27 trials of: 1.3, 1.4, 1.5, 1.7 (two), 2.0 (two), 2.1 (four), 2.3, 3.2, <u>3.4</u>, 3.5 (two), 3.7, 3.8, 3.9, 4.3, 4.5, 4.6, 5.3, 5.4, 5.9 (two) and 6.8 mg/kg, and residue levels in pulp of: 0.02 (two), 0.06 (two), 0.07 (three), 0.09 (three), < 0.1 (two), <u>0.1</u> (three), 0.12, 0.13, 0.14, 0.16, 0.17, 0.23, 0.26 (two), 0.31, 0.33, 0.35, 0.56 and 0.92 mg/kg.

The Meeting estimated a maximum residue level of 10 mg/kg for prochloraz in citrus, replacing the previous recommendation of 5 mg/kg for oranges, sweet and sour. The Meeting also estimated an STMR of 0.1 mg/kg and a highest residue level of 0.92 mg/kg for prochloraz in the edible portion.

Assorted tropical and sub-tropical fruits minus inedible peel

Avocado

The results of trials on post-harvest dipping and spraying on avocado were made available to the Meeting from Australia, Colombia (no GAP) and South Africa.

GAP in South Africa is for post-harvest spray at 0.05 kg ai/hl (no post-treatment interval specified). The residue levels in trials in Australia and South Africa matching this GAP were: 0.42, 1, 1, 1.3, 2.3 and 3.5 mg/kg in whole fruit with stone and < 0.1 and 0.11 mg/kg in pulp.

The residue levels in post-harvest trials in Australia and South Africa matching Australian GAP (0.025 kg ai/hl, no post-treatment interval specified) were: 0.39, 0.83, 0.87, 0.92, 1, 1.2 (two) and 2.4 mg/kg in whole fruit with stone and < 0.1 (four) and 0.12 mg/kg in the edible portion.

The residue levels in 14 trials in avocados, in ranked order, were: 0.39, 0.42, 0.83, 0.87, 0.92, 1 (three), 1.2 (two), 1.3, 2.3, 2.4 and 3.5 mg/kg in whole fruit with stone and ≤ 0.1 (five), 0.11 and 0.12 mg/kg in the edible portion.

The Meeting noted that, while data for pre-harvest foliar application were available, no matching GAP was provided.

Banana

The results of trials of post-harvest dipping on banana were available from Australia, South Africa and the West Indies, and trials of spray or drench application were reported from the Canary Islands and the Philippines.

GAP in the Philippines is for a spray application at 0.09 kg ai/hl, but no trials matched this GAP. In China, GAP is for use of prochloraz as a dip at up to 0.05 kg ai/hl (no post-treatment interval specified). The residue levels in trials of dipping in Australia, the Canary Islands, South Africa and the West Indies matching this GAP were: 1.8, 2.5, 2.6, 2.7, 3.0, 3.3, 3.5 and 5.1 mg/kg in whole fruit and: 0.04, < 0.1, 0.1, 0.11, 0.12 and 0.17 (two) mg/kg in pulp.

In Australia, GAP (0.025 kg ai/hl) is for use of prochloraz as a dip. The residue levels in dipping trials in Australia, the Canary Islands, the Philippines, South Africa and the West Indies matching this GAP were: < 0.1, 0.69, 1.1, 1.3, 1.6, 1.7 (two), 2.3, 2.4, 2.9, 3.0 and 3.4 mg/kg in whole fruit and: 0.03, 0.06, 0.07, 0.08, < 0.1, 0.1 (two), 0.12 (two), 0.13 and 0.21 (two) mg/kg in pulp.

The residue levels in banana, in ranked order, were: < 0.1, 0.69, 1.1, 1.3, 1.6, 1.7, 1.8, 2.3, 2.4, 2.5, 2.6, 2.7, 2.9, 3.0 (two), 3.4, 3.5 and 5.1 mg/kg in whole fruit, and the residue levels in pulp were: 0.03, 0.04, 0.06, 0.07, 0.08, < 0.1 (two), 0.1 (three), 0.11, 0.12 (three), 0.13, 0.17 (two) and 0.21 (two) mg/kg.

The Meeting noted that, while data for pre-harvest foliar applications were available, no matching GAP was provided.

Mango

The Meeting was provided with the results of trials on pre-harvest foliar spray in Israel (no GAP), Malaysia (GAP: 0.056 kg ai/hl, 15-day PHI), South Africa (no GAP) and Taiwan (no GAP), and of trials on post-harvest dipping in Australia (post-harvest spray at 0.025 kg ai/hl, no post-treatment interval specified), Colombia (0.025 kg ai/hl, no post-treatment interval specified), Israel (no GAP) and South Africa (maximum of 0.08 kg ai/hl, no post-treatment interval specified).

None of the pre-harvest trials from Malaysia matched Malaysian GAP.

None of the post-harvest treatment trials matched GAP in China (0.1 kg ai/hl) or South Africa (maximum of 0.08 kg ai/hl), but four trials in Australia, Colombia, Israel and South Africa matched GAP in

Brazil (0.05 kg ai/hl) and Peru (0.045 kg ai/hl). In these trials, the residue levels in total fruit were: 1, 1.2, 1.3 and 1.4 mg/kg, while those in the edible portion were: 0.18 and 0.44 mg/kg.

In three trials in Australia and South Africa that matched GAP in Australia and Colombia (0.025 kg ai/hl), the residue levels were 0.48, 0.68 and 1.8 mg/kg in whole fruit without stone and 0.1 and 0.47 mg/kg in pulp.

The residue levels in the post-harvest trials on mango, in ranked order, were: 0.48, 0.68, 1, 1.2, 1.3 and 1.4 (two) mg/kg, and those in pulp were: 0.1, 0.18, 0.44 and 0.47 mg/kg.

Papaya

The results of post-harvest dipping trials were provided from Australia (GAP: 0.025 kg ai/hl, no post-treatment interval specified), Brazil (GAP: 0.034 kg ai/hl, 3-day post-treatment interval) and South Africa (no GAP).

In three post-harvest dipping trials in Australia and South Africa that matched the GAP of Australia (0.025 kg ai/hl), the residue levels were: 0.41, 0.61 and 1.4 mg/kg in whole fruit (including pips) and < 0.1 and 0.7 mg/kg in the edible portion.

Pineapple

The results of post-harvest dipping trials were made available from Australia (GAP: 0.025 kg ai/hl, no post-treatment interval specified) and Kenya (no GAP). In the one trial matching Australian GAP, the residue level was 1.1 mg/kg in whole fruit and 0.18 mg/kg in the edible portion.

The Meeting considered that the available data on residue levels in avocado, banana, mango, papaya and pineapple were sufficient to mutually support a group maximum residue level for assorted tropical and sub-tropical fruits minus inedible peel. The residue levels, in ranked order, were: < 0.1, 0.39, 0.41, 0.42, 0.48, 0.61, 0.68, 0.69, 0.83, 0.87, 0.92, 1 (four), 1.1 (two), 1.2 (three), 1.3 (three), 1.4 (three), 1.6, 1.7, 1.8, 2.3 (two), 2.4 (two), 2.5, 2.6, 2.7, 2.9, 3.0 (two), 3.4, 3.5 (two) and 5.1 mg/kg in whole fruit in 43 trials, and: 0.03, 0.04, 0.06, 0.07, 0.08, < 0.1 (eight), 0.1 (four), 0.11 (two), 0.12 (four), 0.13, 0.17 (two), 0.18 (two), 0.21 (two), 0.44, 0.47 and 0.7 mg/kg in the edible portion in 33 trials.

The Meeting estimated a maximum residue level of 7 mg/kg for prochloraz in assorted tropical and sub-tropical fruits minus inedible peel, replacing the previous recommendations of 5 mg/kg for avocado and banana, 2 mg/kg for mango and 1 mg/kg for papaya. The Meeting also estimated an STMR of 0.1 mg/kg and a highest residue level of 0.7 mg/kg for prochloraz in the edible portion.

Onion

The results of field trials on onions were made available to the Meeting from The Netherlands (no GAP) and from Thailand (no GAP). The results of post-harvest dipping trials were provided from Australia (no GAP).

Melon

The results of field trials on melons involving foliar drenching and flood irrigation were made available to the Meeting from Spain. None of the trials matched Spanish GAP (up to four applications at 0.9 kg ai/ha, 15-day PHI). The results of post-harvest dipping trials were provided from Australia and Colombia, but no matching GAP was available.

Mushroom

The Meeting noted two distinct patterns of use of prochloraz on mushrooms: one established in the United Kingdom, involving two to three casing sprays of 0.3–0.6 g ai/m^2, with a PHI of 2 days, and the other common in a number of other European countries, Australia and New Zealand, involving one or more treatments at 1.5 g ai/m^2 and a PHI of 10–14 days.

In seven trials in The Netherlands, Switzerland and the United Kingdom matching GAP in Denmark, Italy, The Netherlands, New Zealand and Poland (one or two treatments at 1.5 g ai/m^2, 10-day PHI), the residue levels were: 0.21, 0.25, 0.48, 0.71 and 0.74 mg/kg.

The maximum GAP of two sprays of 0.6 g ai/m^2 (2-day PHI) in the United Kingdom is supported by the results of trials in Germany and the United Kingdom, with residue levels of: 0.81, <u>3.6</u>, <u>6.2</u> and 37 mg/kg.

The Meeting noted that these two residue populations are different and, on the basis of the data supporting the United Kingdom GAP, estimated a maximum residue level of 40 mg/kg for prochloraz in mushrooms, an STMR of 4.9 mg/kg and a highest residue level of 37 mg/kg. The recommended maximum residue level of 40 mg/kg for mushrooms replaces the previous recommendation of 2.0 mg/kg.

Tomato

The results of field trials of both foliar application and soil drenching on tomatoes were made available to the Meeting from Israel (no GAP) and the USA (no GAP).

Lettuce

The results of field trials on lettuce, head, were made available to the Meeting from Australia (GAP: 0.18 kg ai/ha, 0.023 kg ai/hl, 7-day PHI) and from the United Kingdom (no GAP). Trials on protected lettuce crops were also provided by the United Kingdom (no GAP). Four trials in Australia matching Australian GAP showed residue levels of: 0.06, 0.16, 0.41 and 0.59 mg/kg.

The Meeting agreed that the available data were insufficient to estimate a maximum residue level for lettuce

Beans, dry

The results of field trials on beans were made available to the Meeting from Germany, and two trials of seed treatment were provided from Brazil. No matching GAP was available for either use pattern.

Peas, dry

The results of field trials on peas were made available to the Meeting from Germany and the United Kingdom; however, no matching GAP was available.

Sugar-beet

The results of field trials on sugar-beet were made available to the Meeting from Italy (GAP: one or two applications at 0.48–0.8 kg ai/ha, 20-day PHI). Three of the trials matched Italian GAP (± 30%); however, because the control samples apparently contained high residue levels, the Meeting agreed that the available data were insufficient to estimate a maximum residue level for sugar-beet.

Oil seeds

Rape-seed

The results of field trials from Canada (no GAP), Denmark (GAP: 0.45–0.7 kg ai/ha, 28-day PHI), France (GAP: 0.45–0.6 kg ai/ha), Germany (GAP: 0.6 kg ai/ha, 56-day PHI), Sweden (no GAP) and the United Kingdom (GAP: 0.2–0.5 kg ai/ha, 42-day PHI, maximum of 1 kg ai/ha per season) were made available to the Meeting.

In trials in Denmark, France and Germany that matched German GAP (0.6 kg ai/ha, 56-day PHI), the residue levels in mature seed were: 0.05, 0.07, 0.08, 0.09 (two), < 0.1 (four), 0.11, 0.12 and 0.15 mg/kg.

In field trials from the United Kingdom matching the corresponding GAP (up to 0.5 kg ai/ha, 42-day PHI), the residue levels in mature rape-seed were < 0.1 (five), 0.1 (three), 0.12, 0.14, 0.17, 0.18, 0.19, < 0.2 (two), 0.22, 0.24, 0.36, 0.39, 0.46 and 0.48 mg/kg.

Prochloraz

The residue levels in rape-seed, in ranked order, were: 0.05, 0.07, 0.08, 0.09 (two), < 0.1 (nine), <u>0.1</u> (three), 0.11, 0.12 (two), 0.14, 0.15, 0.17, 0.18, 0.19, < 0.2 (two), 0.22, 0.24, 0.36, 0.39, 0.46 and 0.48 mg/kg in 33 trials.

The Meeting estimated a maximum residue level of 0.7 mg/kg for prochloraz in rape-seed, replacing the previous recommendation of 0.5 mg/kg. The Meeting also established an STMR of 0.1 mg/kg and a highest residue level of 0.48 mg/kg.

Sunflower seed

The results of field trials in France (GAP: one or two applications at 0.32–0.6 kg ai/ha, no PHI specified) were made available to the Meeting. In view of the similarity between GAP in France and that in Croatia (GAP: up to two applications at 0.6 kg ai/ha, 63-day PHI), the Meeting evaluated the trials in France against GAP in Croatia.

In the 11 trials, the residue levels were: <u>< 0.1</u> (eight), 0.14, 0.27 and 0.32 mg/kg,

One trial on seed treatment was reported from France, but no matching GAP was available

The Meeting estimated a maximum residue level of 0.5 mg/kg, an STMR of 0.1 mg/kg and a highest residue level of 0.32 mg/kg for sunflower seed.

Linseed

The results of six trials of seed treatment were reported from the United Kingdom (GAP: 0.4 g ai/kg seed) in which seed treated with prochloraz according to GAP was grown to maturity and the daughter seeds analysed for residues 149–178 days after planting. In all the trials, the residue levels of total prochloraz were below the LOQ of 0.05 mg/kg. As a residue level of 0.04 mg/kg was reported in the control samples in one trial, the Meeting decided to evaluate the remaining five trials, with residue levels < 0.05 mg/kg.

The Meeting estimated a maximum residue level of 0.05* mg/kg, an STMR of 0.05 mg/kg and a highest residue level of 0.05 mg/kg for linseed.

Soya beans

The results of two field trials by foliar application on soya beans were made available to the Meeting from France, but no GAP was available.

Cereal grains

Barley

The results of field trials on barley were made available to the Meeting from Austria (GAP: one application at 0.45 kg ai/ha, 35-day PHI), Brazil (GAP: 0.45 kg ai/ha, 32-day PHI), Canada (no GAP), Denmark (maximum GAP: one application at 0.45 kg ai/ha, PHI up to Zadoks 39), France (maximum GAP: 0.6 kg ai/ha, PHI up to stem elongation), Germany (maximum GAP: one application at 0.48 kg ai/ha, 35-day PHI), Greece (maximum GAP: 0.19 kg ai/ha, 56-day PHI), The Netherlands (GAP: 0.45 kg ai/ha, 42-day PHI), Italy (maximum: GAP: 0.8 kg ai/ha, 40-day PHI), Portugal (GAP: 0.45 kg ai/ha, 35-day PHI), Spain (maximum GAP: one application at 0.72 kg ai/ha, 60-day PHI), Sweden (one application at 0.45 kg ai/ha, PHI not specified) and the United Kingdom (maximum GAP: 0.45 kg ai/ha, 42-day PHI).

In five trials in Denmark, The Netherlands and the United Kingdom, involving a single application of prochloraz, which matched German GAP (0.48 kg ai/ha, 35-day PHI), the residue levels in grain were: < 0.05, 0.06, 0.08, 0.13 and 0.21 mg/kg.

The residue levels in 34 trials in Belgium, Denmark, France, Germany, Sweden and the United Kingdom with two applications and matching GAP in the United Kingdom (0.45 kg ai/ha, 42-day PHI) were: < 0.02 (two), 0.03, 0.07, 0.08 (three), 0.1 (three), 0.11 (two), 0.12 (two), 0.14, 0.16 (three), 0.23 (two), 0.24, 0.26 (two), 0.3, 0.31, 0.35, 0.38, 0.45, 0.48, 0.5, 0.53, 0.59, 0.65 and 0.68 mg/kg.

In 15 trials in France, Greece, Italy, Portugal and Spain involving two applications of prochloraz according to GAP in Portugal (0.45 kg ai/ha, 35-day PHI), the residue levels were: 0.13, 0.21, 0.22, 0.23, 0.26, 0.3, 0.35, 0.36, 0.41, 0.43, 0.46, 0.47, 0.51, 0.87 and 0.88 mg/kg.

The residue levels in 54 trials in barley, in ranked order, were: < 0.02 (two), 0.03, < 0.05, 0.06, 0.07, 0.08 (four), 0.1 (three), 0.11 (two), 0.12 (two), 0.13 (two), 0.14, 0.16 (three), 0.21 (two), 0.22, 0.23 (three), 0.24, 0.26 (three), 0.3 (two), 0.31, 0.35 (two), 0.36, 0.38, 0.41, 0.43, 0.45, 0.46, 0.47, 0.48, 0.5, 0.51, 0.53, 0.59, 0.65, 0.68, 0.87 and 0.88 mg/kg.

The Meeting also received data from field trials in Denmark (GAP: 20 g ai/100 kg seed) and Germany (no GAP) on barley grown from seed treated with prochloraz. In the 17 trials matching GAP in Denmark, the residue levels of total prochloraz were < 0.01 (two) and < 0.05 (15) mg/kg..

Oats

Prochloraz is registered for use on cereals in Austria, Belgium, Croatia, Denmark and Germany, with a common GAP of one or two foliar applications at 0.45 kg ai/ha, and a 35-day PHI. None of the four trials provided from Denmark matched this GAP.

The Meeting was also provided with the results of trials of seed treatment in Germany (GAP: 20 g ai/kg seed), in which the residue levels in oats grown from seed treated with prochloraz were: < 0.05 (eight) and < 0.1 (two) mg/kg.

Rice

Field trials of foliar application on rice were made reported to the Meeting from Japan (no GAP), Spain (GAP: 0.45 kg ai/ha, 15-day PHI) and Taiwan China (no GAP). None of the trials in Spain matched Spanish GAP.

Rye

Field trials of foliar application on rye were reported to the Meeting from Denmark (maximum GAP: one application at 0.45 kg ai/ha, 28-day PHI) and Germany (maximum GAP: one to two applications at 0.48 kg ai/ha, 35-day PHI). In three trials in Germany that matched German GAP, the residue levels were: 0.06, 0.09 and < 0.1 mg/kg. The Meeting noted that GAP for rye in the United Kingdom is similar to that in Germany but with a PHI of 42 days. It therefore considered that the trials in Germany supported the United Kingdom GAP. The residue levels in trials matching GAP in the United Kingdom were: < 0.05, 0.05, 0.06 (three), 0.09 and < 0.1 mg/kg.

The Meeting also received the results of field trials of seed treatment in Germany (GAP: 20 g ai/100 kg seed) with regard to residues in rye grown from seed treated with prochloraz. In five trials matching German GAP, the residue levels of total prochloraz were < 0.02 (two) and < 0.05 (three) mg/kg.

Wheat

The results of field trials on wheat were made available to the Meeting from Austria (GAP: one application at 0.45 kg ai/ha, 35-day PHI), Brazil (GAP: 0.45 kg ai/ha, 40-day PHI), the former Czechoslovakia (no GAP), Denmark (maximum GAP: one application at 0.45 kg ai/ha, 28-day PHI), France (maximum GAP: 0.6 kg ai/ha, PHI up to stem elongation), Germany (maximum GAP: one application at 0.48 kg ai/ha, 35-day PHI) Italy (GAP: 0.45 kg ai/ha, 42-day PHI), The Netherlands (GAP: 0.45 kg ai/ha, 42-day PHI), Portugal (no GAP), Spain (maximum GAP: one application at 0.72 kg ai/ha, 60-day PHI), Sweden (one application at 0.45 kg ai/ha, PHI not specified), the United Kingdom (maximum GAP: 0.45 kg ai/ha, 42-day PHI) and the USA (no GAP).

None of the trials in Spain trials matched Spanish GAP. One trial in Italy matching Italian GAP (maximum GAP: 0.8 kg ai/ha, 40-day PHI) showed a residue level of 0.12 mg/kg. In six trials in the former Czechoslovakia, The Netherlands and the United Kingdom involving a single application of prochloraz, which matched German GAP (maximum GAP: one application at 0.48 kg ai/ha, 35-day PHI), the residue levels in grain were: < 0.05, 0.09, 0.12, 0.21, 0.23 and 0.24 mg/kg.

The residue levels in trials in southern France, Greece, Italy, Portugal and Spain involving two applications of prochloraz according to GAP in Portugal (0.45 kg ai/ha, 35-day PHI) were: < 0.05 (six), 0.07 (two), 0.09, 0.13, 0.14, 0.15, 0.52 and 1.2 mg/kg. The residue levels in trials involving two applications of prochloraz to wheat in northern France, Germany and the United Kingdom and matching GAP in Belgium and the United Kingdom (maximum GAP: 0.45 kg ai/ha, 42-day PHI) were: 0.03, < 0.05 (10), 0.05 (two), 0.06 (two), 0.07 (two), 0.08, 0.09, < 0.1 (four), 0.11, 0.12 (two), 0.13, 0.15, 0.16, 0.17, 0.2 and 0.31 (two) mg/kg.

The residue levels in 54 trials in wheat, in ranked order, were: 0.03, < 0.05 (17), 0.05 (two), 0.06 (two), 0.07 (four), 0.08, 0.09 (three), < 0.1 (four), 0.11, 0.12 (four), 0.13 (two), 0.14, 0.15 (two), 0.16, 0.17, 0.2, 0.21, 0.23, 0.24, 0.31 (two), 0.52 and 1.2 mg/kg.

The Meeting also received the results of field trials on residues in wheat grown from seed treated with prochloraz in Denmark (no GAP), Germany (GAP: 20 g ai/100 kg seed), Greece (no GAP) and the United Kingdom (GAP: 14 g ai/100 kg seed). In the 26 trials matching GAP in Germany, the residue levels of total prochloraz were: < 0.01 (three), < 0.02 (nine) and < 0.05 (14) mg/kg..

The Meeting considered that the available data on barley, rye and wheat treated by foliar application were sufficient to mutually support a group maximum residue level for cereal grains. The residue levels, in ranked order, in 118 trials were: < 0.02 (two), 0.03 (two), < 0.05 (19), 0.05 (three), 0.06 (seven), 0.07 (five), 0.08 (five), 0.09 (five), < 0.1 (six), 0.1 (three), 0.11 (three), 0.12 (six), 0.13 (four), 0.14 (two), 0.15 (two), 0.16 (four), 0.17, 0.2, 0.21 (three), 0.22, 0.23 (four), 0.24 (two), 0.26 (three), 0.3 (two), 0.31 (three), 0.35 (two), 0.36, 0.38, 0.41, 0.43, 0.45, 0.46, 0.47, 0.48, 0.5, 0.51, 0.52, 0.53, 0.59, 0.65, 0.68, 0.87, 0.88 and 1.2 mg/kg.

The Meeting estimated a maximum residue level of 2 mg/kg for prochloraz in cereal grains, replacing the previous recommendations of 0.5 mg/kg for barley, oats, rye and wheat. The Meeting also estimated an STMR of 0.11 mg/kg and a highest residue level of 1.2 mg/kg.

The Meeting agreed that the proposed maximum residue level, the STMR and the highest residue level for cereal grains based on foliar application would also accommodate seed treatment use of prochloraz.

Pepper, black

The results of trials on foliar application on black pepper were made available to the Meeting from Malaysia (GAP: 0.05 kg ai/hl, 30-day PHI). In trials matching this GAP, the residue levels were 5.0 and 5.1 mg/kg.

The Meeting estimated a maximum residue level of 10 mg/kg, an STMR of 5.1 mg/kg and a highest residue level of 5.1 mg/kg for prochloraz in pepper, black.

Animal feed commodities

Barley straw and fodder, dry

In four trials on barley in Denmark and The Netherlands involving a single application of prochloraz and which matched the GAP of Germany, the residue levels in barley straw were: 5.0, 6.8, 7.0 and 17 mg/kg.

The residue levels in straw in 34 trials in Belgium, Denmark, France, Germany, Sweden and the United Kingdom with two applications and matching GAP in the United Kingdom were: 0.68, 0.7, 1.1 (two), 1.4, 1.6, 2.1, 2.3, 2.4, 3.3, 3.5, 3.6, 3.7, 4.1 (two), 4.5, 4.8, 5.4, 5.7, 6.0, 6.5, 6.7, 7.6, 9.7 (two), 9.8, 12, 13 (two), 14 (two), 21, 24 and 30 mg/kg.

In 15 trials in France, Greece, Portugal and Spain, involving two applications of prochloraz according to GAP in Portugal, the residue levels in straw were: 4.0, 4.1, 4.6, 5.0, 6.4, 7.0, 7.1, 8.2, 8.4 (two), 8.8, 12 (two), 13 and 20 mg/kg.

The residue levels in barley straw in the 53 trials, in ranked order, were: 0.68, 0.7, 1.1 (two), 1.4, 1.6, 2.1, 2.3, 2.4, 3.3, 3.5, 3.6, 3.7, 4.0, 4.1 (three), 4.5, 4.6, 4.8, 5.0 (two), 5.4, 5.7, 6.0, 6.4, <u>6.5</u>, 6.7, 6.8, 7.0 (two), 7.1, 7.6, 8.2, 8.4 (two), 8.8, 9.7 (two), 9.8, 12 (three), 13 (three), 14 (two), 17, 20, 21, 24 and 30 mg/kg.

In field trials in Denmark (GAP: 20 g ai/100 kg seed) and Germany (no GAP) on residues in barley grown from seed treated with prochloraz, the residue levels of total prochloraz in straw in 17 trials matching Danish GAP were: < 0.05 (three), < 0.1 (13) and < 0.2 mg/kg.

Oat straw and fodder, dry

In 10 field trials of seed treatment in Germany that matched German GAP (20 g ai/kg seed), the residue levels of total prochloraz in oat straw grown from seed treated with prochloraz were: < 0.05 (two), < 0.1 (seven) and < 0.2 mg/kg.

Rye straw and fodder, dry

In three trials in Germany matching German GAP, the residue levels in straw were: 1.1, 1.5 and 1.7 mg/kg. The residue levels in additional trials in Germany matching GAP in the United Kingdom were: 0.09, 3.1, 3.4 and 4.7 mg/kg. The residue levels in rye straw, in ranked order, were: 0.09, 1.1, 1.5, <u>1.7</u>, 3.1, 3.4 and 4.7 mg/kg.

In five trials in Germany matching German GAP on residues in straw from rye grown from seed treated with prochloraz, the residue levels of total prochloraz were: < 0.1 (four) and < 0.2 mg/kg.

Wheat straw and fodder, dry

In six trials on wheat conducted in the former Czechoslovakia, The Netherlands and the United Kingdom involving a single application of prochloraz and matching German GAP, the residue levels in straw were: 1.5, 4.6, 5.2, 6.4, 6.8 and 11 mg/kg.

The residue levels in straw in 14 trials on wheat in southern France, Greece, Italy, Portugal and Spain involving two applications of prochloraz according to GAP in Portugal (0.45 kg ai/ha, 35-day PHI), the residue levels were: 3.5, 4.3, 8.0, 8.2 (two), 8.3, 9.6 (two), 10 (two), 11, 13 (two) and 22 mg/kg.

The residue levels in 32 trials involving two applications of prochloraz to wheat in northern France, Germany and the United Kingdom matching GAP in Belgium and the United Kingdom were: 1.7, 1.8, 2.3, 2.4, 2.5, 2.6, 2.7 (two), 2.8, 3.0, 3.7, 4.3, 5.1, 5.2, 5.3, 5.6, 5.8, 6.5, 6.6, 6.9, 8.0, 8.8, 10, 11 (three), 13, 15, 16, 19, 20 and 22 mg/kg.

The residue levels in wheat straw in all 52 trials, in ranked order, were: 1.5, 1.7, 1.8, 2.3, 2.4, 2.5, 2.6, 2.7 (two), 2.8, 3.0, 3.5, 3.7, 4.3 (two), 4.6, 5.1, 5.2 (two), 5.3, 5.6, 5.8, 6.4, 6.5, 6.6, <u>6.8</u>, <u>6.9</u>, 8.0 (two), 8.2 (two), 8.3, 8.8, 9.6 (two), 10 (three), 11 (five), 13 (three), 15, 16, 19, 20 and 22 (two) mg/kg.

In 24 trials of seed treatment matching German GAP, the residue levels of total prochloraz in wheat straw were: < 0.05 (four), < 0.06, < 0.1 (17) and < 0.2 (two) mg/kg.

The Meeting considered that the available data on barley, rye and wheat straw and fodder treated by foliar application were sufficient to mutually support a group maximum residue level for straw and fodder, dry, of cereal grains. The residue levels in the 112 trials, in ranked order, were: 0.09, 0.68, 0.7, 1.1 (three), 1.4, 1.5 (two), 1.7 (two), 1.6, 1.8, 2.1, 2.3 (two), 2.4 (two), 2.5, 2.6, 2.7 (two), 2.8, 3.0, 3.1, 3.3, 3.4, 3.5 (two), 3.6, 3.7 (two), 4.0, 4.1 (three), 4.3 (two), 4.5, 4.6 (two), 4.7, 4.8, 5.0 (two), 5.1, 5.2 (two), 5.3, 5.4, 5.6, 5.7, 5.8, 6.0, <u>6.4</u> (two), <u>6.5</u> (two), 6.6, 6.7, 6.8 (two), 6.9, 7.0 (two), 7.1, 7.6, 8.0 (two), 8.2 (three), 8.3, 8.4 (two), 8.8 (two), 9.6 (two), 9.7 (two), 9.8, 10 (three), 11 (five), 12 (three), 13 (five), 14 (two), 15, 16, 17, 19, 20 (two), 21, 22 (two), 24 and 30 mg/kg.

Prochloraz

Allowing for a dry matter content of 90% (*FAO Manual*), the Meeting estimated a maximum residue level of 40 mg/kg for prochloraz in straw and fodder, dry, of cereal grains, replacing the previous recommendations of 15 mg/kg for barley straw and fodder, dry; oats straw and fodder, dry; rye straw and fodder, dry, and wheat straw and fodder, dry. The Meeting also estimated an STMR of 7.2 mg/kg and a highest residue level of 33 mg/kg.

The Meeting agreed that the available data indicated that the proposed maximum residue level, STMR and highest residue level for straw and fodder, dry, after foliar application would also accommodate seed treatment use of prochloraz.

Fate of residues during storage

The Meeting received the results of a study on the fate of prochloraz residues in oranges. Fruit dipped in prochloraz at 0.1–0.2 kg ai/hl and shipped under refrigeration for 44 days were stored at 4 °C or 20 °C for a further 7, 14 or 21 days. No significant degradation of residue was observed during the post-shipping 21-day storage period, as > 87% of the residue remained in the stored fruit. The median retention value was 99% for ambient-stored fruit and 117% for cool-stored fruit.

Fate of residues during processing

The effect of processing on levels of residues of prochloraz was studied in barley, rape-seed and wheat, and the residue levels in oil and press cake were reported in several field trials on sunflower seed; residue levels in green black pepper and in processed white pepper were reported in one trial on pepper. A study on residues in dehydrated and preserved mushrooms was also provided to the Meeting. The processing factors of relevance to estimation of maximum residue levels, the dietary burden of farm animals and dietary risk assessment, shown below, were derived from these studies.

Raw agricultural commodity	Processed product	No. of samples	Mean processing factor
Barley	Beer	4	0.09
Wheat	Bran (total)	1	4.3
	Flour (unspecified)	3	0.23
	Bread (whole grain)	1	1.3
Rape-seed	Seed cake (meal)	18	0.79
	Refined oil	4	< 0.6
Sunflower seed	Seed cake (meal)	6	0.49
Pepper	Black peppercorns	4	0.96
	White peppercorns	4	0.35
Mushrooms	Dehydrated	3	3.7
	Preserved	2	0.4
	Preservation liquor	2	0.65

Wheat was processed into milled by-products (bran), flour and whole-grain bread, with processing factors of 4.3, 0.23 and 1.3, respectively. On the basis of the STMR value of 0.11 mg/kg for cereal grains, the STMR-Ps were 0.025 mg/kg for wheat flour and 0.14 mg/kg for wholemeal bread.

Wheat milled by-products (bran) is listed as animal feed in the *FAO Manual* (Appendix IX). Allowing for the standard 88% dry matter, the Meeting estimated an STMR-P of 0.54 mg/kg for wheat bran (dry weight).

On the basis of the highest residue level of 1.2 mg/kg, the processing factor of 4.3 and the standard dry matter content of 88%, the Meeting recommended a maximum residue level of 7 mg/kg for wheat bran, unprocessed (dry weight basis).

Barley was processed into beer, with a processing factor of 0.09. On the basis of the STMR value of 0.11 mg/kg for cereal grains, the STMR-P for beer was 0.01 mg/kg.

The residue levels in seed cake in trials on rape in Denmark and Germany, matching the GAP of the United Kingdom and Germany, respectively, and used in estimating the maximum residue levels, were: < 0.05, 0.05 (three), 0.07, 0.08 (two) and 0.1 mg/kg. The Meeting established an STMR-P of 0.06 mg/kg for rape-seed meal.

In four processing studies, the residue levels in refined oil from rape-seed containing 0.07–0.12 mg/kg were below the LOQ (< 0.05 mg/kg). Using a processing factor of < 0.6 and an STMR of 0.1 mg/kg for rape-seed, the Meeting established an STMR-P of 0.06 mg/kg for rape-seed oil, edible.

The residue levels in seed cake in three trials on *sunflower seed* in France that were used in estimating the maximum residue level were ≤ 0.1 (two) and 0.15 mg/kg. Taking into account the STMR for sunflower seed (0.1 mg/kg) and the processing factor of 0.49, the Meeting established an STMR-P of 0.05 mg/kg for sunflower seed meal.

The Meeting agreed to use the STMR of 0.1 mg/kg for sunflower seed and the processing factor of < 0.6 derived for refined rape-seed oil to estimate an STMR-P of 0.06 mg/kg for sunflower seed oil (refined).

In a field study in Malaysia on green (fresh) peppercorns, the residues were not concentrated during the sun-drying process used to produce black peppercorns (mean processing factor, 0.96), and the residue levels decreased during husking to produce white peppercorns (mean processing factor, 0.35).

Residues in animal commodities

Dietary burden of farm animals

The Meeting estimated the dietary burden of total prochloraz in cows and poultry on the basis of the diets listed in Appendix IX of the *FAO Manua*. Calculations from MRLs and highest residue levels provide the levels in feed suitable for estimating MRLs for animal commodities, while calculations from STMR values for feed are suitable for estimating STMR values for animal commodities. The percentage of dry matter is taken as 100% when MRLs and STMR values are already expressed as dry weight.

Estimated maximum dietary burden of farm animals

Commodity	Group	Residue (mg/kg)	Basis	Dry matter (%)	Residue/Dry matter (mg/kg)	Dietary content (%)			Residue contribution (mg/kg)		
						Beef cattle	Dairy cattle	Poultry	Beef cattle	Dairy cattle	Poultry
Rape meal	–	0.06	STMR-P	88	0.07						
Sunflower meal	–	0.05	STMR-P	92	0.05						
Barley straw	AS	33	HR	100	33	**10**	**60**		3.3	19.8	
Wheat straw	AS	33	HR	100	33						
Rye straw	AS	33	HR	100	33						
Oat straw	AS	33	HR	100	33						
Wheat milled by-products	CF	0.54	STMR-P	100	0.54	**10**		**20**	0.05		0.11
Barley grain	GC	1.2	HR	88	1.36						

Commodity	Group	Residue (mg/kg)	Basis	Dry matter (%)	Residue/Dry matter (mg/kg)	Dietary content (%) Beef cattle	Dietary content (%) Dairy cattle	Dietary content (%) Poultry	Residue contribution (mg/kg) Beef cattle	Residue contribution (mg/kg) Dairy cattle	Residue contribution (mg/kg) Poultry
Corn grain	GC	1.2	HR	88	1.36	**80**	40	80	1.09	0.55	1.09
Rye grain	GC	1.2	HR	88	1.36						
Wheat grain	GC	1.2	HR	89	1.35						
Oat grain	GC	1.2	HR	89	1.35						
Total						100	100	100	**4.4**	**20**	**1.2**

Estimated median dietary burden of farm animals

Commodity	Group	Residue (mg/kg)	Basis	Dry matter (%)	Residue/Dry matter (mg/kg)	Dietary content (%) Beef cattle	Dietary content (%) Dairy cattle	Dietary content (%) Poultry	Residue contribution (mg/kg) Beef cattle	Residue contribution (mg/kg) Dairy cattle	Residue contribution (mg/kg) Poultry
Rape meal	–	0.06	STMR-P	88	0.07						
Sunflower meal	–	0.05	STMR-P	92	0.05						
Barley straw	AS	7.2	STMR	100	7.2	**10**	60		0.72	4.32	
Wheat straw	AS	7.2	STMR	100	7.2						
Rye straw	AS	7.2	STMR	100	7.2						
Oat straw	AS	7.2	STMR	100	7.2						
Wheat milled by-products	CF	0.54	STMR-P	100	0.54	10		20	0.05		0.11
Barley grain	GC	0.11	STMR	88	0.13						
Corn grain		0.11	STMR	88	0.13	**80**	40	80	0.1	0.05	0.1
Rye grain	GC	0.11	STMR	88	0.13						
Wheat grain	GC	0.11	STMR	89	0.12						
Oat grain	GC	0.11	STMR	89	0.12						
Total						100	100	100	**0.87**	**4.4**	**0.21**

The total dietary burdens of prochloraz for estimating MRLs for animal commodities (residue levels in animal feeds expressed as dry weight) are 4.4 ppm for beef cattle, 20 ppm for dairy cattle and 1.2 ppm for poultry. The associated median dietary burdens for estimating STMR are 0.87 ppm for beef cattle, 4.4 ppm for dairy cattle and 0.21 ppm for poultry.

Feeding studies

The Meeting received information from two studies on the residue levels in tissues and milk from dairy cows dosed with prochloraz for 28 days at an equivalent of 10, 30 and 100 ppm in the diet and from a feeding study in which calves were fed a diet containing prochloraz twice daily, resulting in a rate of 0.263 mg/kg bw (dietary concentration could not be estimated).

In one of the studies in dairy cows, tissues were analysed for total prochloraz residues by gas chromatography with mass spectrometry detection after conversion of the metabolites to 2,4,6-trichlorphenol. The mean recovery efficiency was 89%, and the LOQ was 0.05 mg/kg. At the end of the 28-day treatment period, the mean residue levels in muscle ranged from < 0.05 mg/kg to 0.37 mg/kg in cows at the highest dose. In subcutaneous fat, the mean residue levels ranged from 0.09 mg/kg at the lowest dose to 1.2 mg/kg at the highest dose, and those in peritoneal fat ranged from 0.16 mg/kg to 1 mg/kg for the three groups. The mean residue levels in kidney were 0.52 mg/kg at the lowest dose to 3.2 mg/kg at the highest, and those in liver were 2.8 mg/kg at the lowest dose, 6.4 mg/kg at 30 ppm and 23 mg/kg at 100 ppm.

The results of the study in which calves were dosed at 0.26 mg/kg bw for 28 days were similar to those for cows receiving the lowest dose, with mean residue levels of 2.2 mg/kg in liver, 0.55 mg/kg in kidney, 0.09 mg/kg in fat and 0.06–0.09 mg/kg in muscle.

Residues of free prochloraz and three metabolites were measured in milk from cows dosed twice daily after milking with prochloraz for 28 days at an equivalent of 10, 30 and 100 ppm in the diet. Traces of prochloraz were detected in the group at 100 ppm from day 4 and in the group at 30 ppm after day 28, but all the residue levels were below the reported LOQ of 0.01 mg/kg. No residues of the metabolites BTS 54906 and BTS 54908 were detected in milk from cows at any dose. Trace levels (< 0.01 mg/kg) of BTS 44596 were reported in milk from cows at 30 ppm from day 22, and the average levels in milk from cows at 100 ppm reached a plateau of 0.01 ± 0.003 mg/kg from day 4. Milk sampled and separated on day 24 contained average levels of BTS 44596 of 0.005 mg/kg in skim milk and 0.032 mg/kg in cream, suggesting preferential partitioning (six times) into milk fat.

Maximum residue levels

As the total dietary burdens of *beef and dairy cattle* are 4.4 and 20 ppm, respectively, the maximum residue levels to be expected in tissues can be obtained by interpolating the results of feeding at a level of 10 or 30 ppm. The maximum residue levels reported were 3.3 mg/kg and 9 mg/kg in liver, 0.24 mg/kg and 0.51 mg/kg in fat, 0.05 mg/kg and 0.14 mg/kg in muscle and 0.59 mg/kg and 1.8 mg/kg in kidney.

The median dietary burdens were 0.87 ppm for beef cattle and 4.4 ppm for dairy cattle. STMR values can be extrapolated from the mean residue levels in tissues of animals at 10 ppm, i.e. 2.8 mg/kg in liver, 0.13 mg/kg in fat, < 0.05 mg/kg in muscle and 0.52 mg/kg in kidney. The mean residue level in milk at both feeding levels was < 0.01 mg/kg,

Dietary burden (mg/kg)[a] Feeding level [ppm][b]		Prochloraz residue level (mg/kg)[c]							
		Milk (mean)	Fat		Muscle		Liver		Kidney
			High	Mean	High	Mean	High	Mean	High	Mean
MRL beef cattle	(20) [10:30]		*(0.38)* 0.24:0.51		*(0.1)* 0.05:0.14		*(6.2)* 3.3:9		*(1.2)* 0.59:1.8	
MRL dairy cattle	(20) [10:30]	*(< 0.01)* < 0.01:< 0.01								
STMR beef cattle	(4.4) [5]			*(0.057)* 0.13		*(< 0.022)* < 0.05		*(1.23)* 2.8		*(0.229)* 0.52
STMR dairy cattle	(4.4) [5]	*(< 0.0044)* < 0.01								

[a] In parentheses, estimated dietary burden
[b] In square brackets, actual feeding levels in transfer studies
[c] Values in parentheses in italics are derived from the dietary burden, feeding levels and residue levels found in the transfer studies. 'High' is the highest residue level in an individual tissue in the relevant feeding group. 'Mean' is the mean residue level in tissue (or milk) in the relevant feeding group.

On the basis of the above considerations, the Meeting estimated highest residue levels of 0.1 mg/kg in meat (muscle), 0.38 mg/kg in meat (fat), 6.2 mg/kg in edible offal (mammalian) and 0 mg/kg in milks.

The Meeting estimated maximum residue levels of 0.5 mg/kg (fat) in meat (from mammals other than marine mammals); 10 mg/kg in edible offal, mammalian and 0.05 (*) mg/kg in milks. These recommendations replace the previous recommendations of 0.5 mg/kg for cattle fat, 0.1 (*) mg/kg for cattle meat, 5.0 mg/kg for cattle, edible offal of, and 0.1 (*) mg/kg for milks. The Meeting estimated STMRs of 0.02 mg/kg for meat (muscle), 0.06 mg/kg for meat (fat), 1.2 mg/kg for edible offal (mammalian) and 0 mg/kg for milks.

Prochloraz

For poultry, the information provided by the study of metabolism in hens at feeding rates of 5 and 10 ppm in hens was considered by the Meeting to be sufficient for use in estimating maximum residue levels in eggs and poultry tissues. In tissues from birds at 5 ppm, the maximum residue levels were 0.41 mg/kg in liver, 0.029 mg/kg in fat and 0.02 mg/kg in muscle. The average residue level in eggs after a plateau had been reached at day 8 was 0.28 mg/kg. The average residue levels in hens at 5 ppm were 0.34 mg/kg in liver, 0.028 mg/kg in fat, 0.019 mg/kg in muscle and 0.28 mg/kg in eggs (after day 8).

The total dietary burden of poultry is 1.2 mg/kg, and the median dietary burden is 0.21 mg/kg. The Meeting agreed that extrapolation from the results for hens at the 5 ppm feeding level in the metabolism study was appropriate for estimating maximum residue levels, STMRs and highest residue levels

Dietary burden (mg/kg)[a] Feeding level [ppm][b]		Prochloraz residue levels (mg/kg)[c]						
		Eggs (mean)	Fat		Muscle		Liver	
			High	Mean	High	Mean	High	Mean
MRL poultry	(1.2) [5]	*(0.0672)* 0.28	*(0.007)* 0.029		*(0.0048)* 0.02		*(0.0984)* 0.41	
STMR poultry	(0.21) [5]	*(0.0118)* 0.28		*(0.0012)* 0.028		*(0.0008)* 0.019		*(0.0143)* 0.34

[a] In parentheses, estimated dietary burden
[b] In square brackets, actual feeding levels in transfer studies
[c] Values in parentheses in italics are derived from the dietary burden, feeding levels and residue levels found in the transfer studies. 'High' is the highest residue level in an individual tissue in the relevant feeding group. 'Mean' is the mean residue level in tissue in the relevant feeding group.

On the basis of this extrapolation, the Meeting estimated highest residue levels of 0.005 mg/kg for poultry meat, 0.007 mg/kg for poultry fats, 0.1 mg/kg for poultry, edible offal of, and 0.07 mg/kg for eggs.

The Meeting estimated maximum residue levels of 0.05 (*) mg/kg for poultry meat, 0.2 mg/kg for poultry, edible offal of, and 0.1 mg/kg for eggs; it also estimated STMRs of 0.001 mg/kg in poultry meat (muscle), 0.001 mg/kg in poultry meat (fat), 0.015 mg/kg in poultry, edible offal of, and 0.012 mg/kg in eggs.

DIETARY RISK ASSESSMENT

Long-term intake

The evaluation of prochloraz resulted in recommendations for MRLs and STMRs for raw and processed commodities. Data were available on the consumption of 35 food commodities, and these were used in calculating dietary intake. The results are shown in Annex 3.

The IEDIs in the five GEMS/Food regional diets, on the basis of the estimated STMRs, represented 7–10% of the ADI of 0–0.01 mg/kg bw (Annex 3). The Meeting concluded that the long-term intake of residues of prochloraz from uses that have been considered by the JMPR is unlikely to present a public health concern.

Short-term intake

The IESTI of prochloraz was calculated for the food commodities (and their processing fractions) for which maximum and highest residue levels had been estimated and for which data on consumption were available. The results are shown in Annex 4.

The IESTI varied from 0 to 130% of the ARfD (0.1 mg/kg bw) for the general population and from 0 to 150% of the ARfD for children ≤ 6 years. The short-term intake of mushrooms, for which the calculation was made, represented 150% of the ARfD for children ≤ 6 years and 130% of the ARfD for the general population. The information provided to the Meeting precluded a conclusion that the short-term dietary intake of mushrooms would result in residue levels below the ARfD.

4.24 PROPICONAZOLE (160)

TOXICOLOGY

Propiconazole is the ISO approved name for 1-[2-(2,4-dichlorophenyl)-4-propyl-1,3-dioxolan-2-yl-methyl]-1*H*-1,2,4-triazole, a systemic fungicide that acts by inhibition of ergosterol biosynthesis. Propiconazole was evaluated toxicologically by the JMPR in 1987, when an ADI of 0–0.04 mg/kg bw was established on the basis of the NOAEL of 4 mg/kg bw per day for effects on body weight, clinical chemistry and haematology in a 2-year study in rats, and this was supported by the NOAEL of 7 mg/kg bw per day in a 1-year study in dogs. Propiconazole was considered by the present Meeting within the periodic review programme of the CCPR. The Meeting reviewed new data on propiconazole that had not been reviewed previously and relevant data from the previous evaluation.

After oral administration of radiolabelled propiconazole to rats and mice, the radiolabel is rapidly (C_{max} at 1 h) and extensively (> 80% of the administered dose) absorbed and widely distributed, with the highest concentrations being found in the liver and kidney. Excretion of the radiolabel is rapid (80% in 24 h) with significant amounts being found in the urine (39–81%) and the faeces (20–50%), the proportions varying with dose, species and sex. There is a significant degree of biliary excretion and subsequent enterohepatic recirculation. There was no evidence for bioaccumulation with tissue or carcass residues being typically < 1% of the administered dose 6 days after dosing. Propiconazole is extensively metabolized and < 5% of the dose remains as parent compound; however, many metabolites have not been identified. The primary metabolic steps involve oxidation of the propyl side-chain on the dioxolane ring to give hydroxy or carboxylic acid derivatives. Hydroxylation of the chlorophenyl and triazole rings followed by conjugation with sulfate or glucuronide was also detected. There is evidence for only limited cleavage between the triazole and chlorophenyl rings. The extent of cleavage of the dioxolane ring was significantly different according to species and sex, representing about 60% of urinary radioactivity in male mice, 30% in female mice and 10–30% in male rats. In rats, about 30% propiconazole is absorbed within 10 h of dermal application.

Propiconazole has moderate acute oral toxicity in rats and mice (LD_{50} values, about 1500 mg/kg bw) and low acute dermal (LD_{50} values, > 4000 mg/kg bw) and inhalation toxicity (LC_{50}, > 5 mg/l of air). Propiconazole is not an eye irritant in rabbits, but is irritating to rabbit skin and is a skin sensitizer in guinea-pigs in the Magnusson and Kligman test.

Decreased body-weight gain was seen in short- and long-term studies of toxicity and studies of developmental and reproductive toxicity and was often linked with reduced food consumption. In studies of repeated doses, liver was the primary target organ for toxicity attributable to propiconazole. In rats, erythrocyte parameters were reduced and a range of clinical chemistry changes were seen, however, with the exception of reduced chloride and cholesterol concentrations, there was no consistent pattern between sexes and studies, and results were generally within the physiological range.

In two studies, mice given diets containing propiconazole at ≥ 850 ppm for up to 17 weeks had increases in liver weight, reduced concentrations of serum cholesterol and increased hepatocyte hypertrophy, vacuolation and necrosis. The findings were present after 4 weeks and did not progress with increased duration of dosing. The NOAEL was 500 ppm (equal to 65–85 mg/kg bw per day) in both studies.

Rats given propiconazole at 450 mg/kg bw per day for 28 days by gavage exhibited a range of effects. Males had reductions in body-weight gain, while females had clinical signs of toxicity and reductions in erythrocyte parameters. Both sexes had increased liver weights and hepatocyte hypertrophy, with hepatocyte necrosis also being seen in females. Increases in liver weight with hepatocyte hypertrophy were seen at 150 mg/kg bw per day, but these effects were not considered to be adverse and the NOAEL was this dose. In a 13-week dietary study in rats, reductions in body-weight gain, increased relative liver weight and increased γ-glutamyltranspeptidase activity was seen in both sexes at 6000 ppm. In females, erythrocyte parameters were reduced at this dose. The NOAEL was 1200 ppm (equal to 76 mg/kg bw per day).

Dogs appeared to be sensitive to the local effects of propiconazole as manifested by gastrointestinal tract irritation at ≥ 8.4 mg/kg bw per day; the NOAELs were 250 ppm (equal to 6.9 mg/kg bw per day) after 90 days and 1.9 mg/kg bw per day after 1 year. No systemic effects were seen in dogs receiving 8.4 mg/kg bw per day for 1 year or 1250 ppm (equal to 35 mg/kg bw per day) for 90 days, the highest doses tested.

In a 3-week (five applications per week) study of dermal toxicity in rabbits, tremors, dyspnoea and ataxia were increased at ≥ 1000 mg/kg bw per day. The NOAEL was 200 mg/kg bw per day. In a 13-week (5 days per week; 6 h per day) study in rats treated by inhalation, reduced body-weight gain was seen in females at 0.19 mg/l of air; the NOAEC was 0.085 mg/l of air.

The carcinogenic potential of propiconazole was studied in one study in rats and in two studies in mice. In a 2-year dietary study in male and female mice and an 18-month dietary study in male mice, the liver was the only target organ. At ≥ 500 ppm, there were decreases in body-weight gain and serum concentration of cholesterol and increases in liver weight, hepatocellular hypertrophy and hepatocellular vacuolation. The NOAEL for non-neoplastic effects in both studies was 100 ppm (equal to 11 mg/kg bw per day). Propiconazole was a hepatocarcinogen only in male mice, on the basis of significant increases in the incidence of liver tumours at ≥ 850 ppm (equal to 108 mg/kg bw per day), with a NOAEL of 500 ppm (equal to 59 mg/kg bw per day). Assays for hepatocyte proliferation (measured by bromodeoxyuridine incorporation) in mice showed qualitative similarities between propiconazole and phenobarbital. The doses that produced increases in tumour incidences (≥ 850 ppm) also produced cell proliferation, increased liver weight and hepatocyte hypertrophy. Studies of liver enzyme induction in mice showed that propiconazole increased the activity of a number of cytochrome P450s, particularly Cyp2b and exhibited similar characteristics to a phenobarbital-type inducer of xenobiotic metabolizing enzymes. The progression from cytochrome P450 (Cyp2b) induction, initial mitogenic response, hepatocyte hypertrophy and increased liver weight to tumours is consistent with a mode of action similar to that of phenobarbital.

At 2500 ppm (96 mg/kg bw per day) in a 2-year dietary study in rats, there were reductions in body-weight gain in both sexes. Increased incidences of enlarged hepatocytes were present in males and increases in atrophy of the exocrine pancreas and dilatation of the uterine lumen in females. Slight (< 10%), transient reductions in body-weight gain, variations in clinical chemistry and haematology parameters that fell within physiological ranges at 500 ppm (equal to 18 mg/kg bw per day) were not considered to be adverse. Propiconazole was not carcinogenic in rats at doses of up to 2500 ppm (equal to 96 mg/kg bw per day). The NOAEL in the 2-year study in rats was 500 ppm (equal to 18 mg/kg bw per day).

Propiconazole gave negative results in an adequate battery of studies of genotoxicity in vitro and in vivo.

The Meeting concluded that propiconazole was unlikely to be genotoxic.

On the basis of the above consideration of the male mouse liver tumours, the high doses required to induce tumours, the likely mechanism of action, the absence of tumorigenicity in rats and the negative results in studies of genotoxicity, the Meeting concluded that propiconazole was unlikely to pose a carcinogenic risk to humans.

In a two-generation study of reproductive toxicity in rats, reproductive parameters were not affected by treatment with propiconazole. At 500 ppm (equivalent to 35 mg/kg bw per day) dams had reduced body-weight gains ($p < 0.01$) and both sexes exhibited hepatoxicity, thus the NOAEL for parental toxicity was 100 ppm (equivalent to 7 mg/kg bw per day). The NOAEL for offspring toxicity was 100 ppm (equivalent to

7 mg/kg bw per day) on the basis of decreased pup body weights in the F_{2b} litters ($p < 0.01$). The NOAEL for reproductive effects was 500 ppm (equivalent to 35 mg/kg bw per day) on the basis of reduced pup survival at 2500 ppm (equivalent to 175 mg/kg bw per day).

Three studies of developmental toxicity were conducted in rats and one in rabbits. In the first study in rats, at the highest dose of 300 mg/kg bw per day there was evidence of maternal toxicity and retarded development, but no malformations. In the second study, propiconazole caused developmental delay (incomplete ossification of sternebrae and rudimentary cervical ribs) at a dose of 90 mg/kg bw per day, which also produced a slight, transient reduction in food consumption and body-weight gain at the initiation of dosing. The NOAEL was 90 mg/kg bw per day for maternal effects and 30 mg/kg bw per day for developmental effects. A low incidence of cleft palate was observed at 90 mg/kg bw per day (one fetus, 0.3%) and at 360/300 mg/kg bw per day (two fetuses, 0.7%) in the presence of severe maternal toxicity. The maternal toxicity included lethargy, ataxia, salivation and reductions in food consumption and body-weight gain at the start of the dosing period. The cleft palate finding was also seen at a low incidence in rats in an extensive study that specifically investigated the palate and jaw at a single dose of 300 mg/kg bw per day. Cleft palates were detected in two out of 2064 fetuses of treated animals versus none in the 2122 fetuses of controls, in the presence of severe maternal toxicity. Marked maternal toxicity was observed throughout the treatment period, included reductions in food consumption and body-weight gain, ataxia, coma, lethargy and prostration and three treatment-related deaths among 189 dams. Cleft palate is a very rare but occasional finding in control rats and there were published data that indicated testing compounds at maternally toxic doses is associated in some way with the induction of a number of malformations, including cleft palate.

Propiconazole was not teratogenic in rabbits. The NOAEL for fetal effects was 250 mg/kg bw per day on the basis of an increased incidence of the formation of thirteenth ribs at 400 mg/kg bw per day in the presence of maternal body-weight loss, signs of toxicity and abortions. The NOAEL for maternal toxicity was 100 mg/kg bw per day on the basis of reduced food consumption and body-weight loss at 250 mg/kg bw per day.

No studies of neurotoxicity with propiconazole were available; however, no evidence of neurotoxicity was apparent in any of the available studies.

Humans exposed to formulated products containing propiconazole have shown local irritant reactions. No evidence of sensitization was seen in an epicutaneous test in 20 volunteers.

The Meeting concluded that the existing database on propiconazole was adequate to characterize the potential hazards to fetuses, infants and children.

Toxicological evaluation

The Meeting established an ADI of 0–0.07 mg/kg bw based on the NOAEL of 7 mg/kg bw per day in a multigeneration study of reproductive toxicity in rats and a 100-fold safety factor. This value covers all other end-points and is supported by NOAELs of 11 mg/kg bw per day in a 24-month study in mice and 18 mg/kg bw per day in a 2-year study in rats. This ADI is protective against the local effects seen in the gastrointestinal tract in dogs (NOAEL, 1.9 mg/kg bw per day), which were considered to be concentration-dependent and hence would merit a safety factor of 25.

An ARfD of 0.3 mg/kg bw was established based on the NOAEL of 30 mg/kg bw per day in the study of developmental toxicity in rats and a 100-fold safety factor. The NOAEL was identified on the basis of slight increases in rudimentary ribs and unossified sternebrae at 90 mg/kg bw per day, which could not be discounted. This provides an adequate margin over the maternal toxicity and cleft palate seen at 300 mg/kg bw per day. The Meeting noted that the highest dose tested in dogs was 35 mg/kg bw per day and that the proposed ARfD would be protective for any potentially acute effects observed in dogs.

A toxicological monograph was prepared.

Propiconazole

Levels relevant to risk assessment

Species	Study	Effect	NOAEL	LOAEL
Mouse	24-month study of toxicity and carcinogenicity[a, e]	Toxicity	100 ppm, equal to 11 mg/kg bw per day	500 ppm, equal to 59 mg/kg bw per day
		Carcinogenicity	500 ppm, equal to 59 mg/kg bw per day	850 ppm, equal to 108 mg/kg bw per day
Rat	2-year study of toxicity and carcinogenicity[a]	Toxicity	500 ppm, equal to 18 mg/kg bw per day	2500 ppm, equal to 96 mg/kg bw per day
		Carcinogenicity	2500 ppm, equal to 96 mg/kg bw per day[c]	—
	Two-generation study of reproductive toxicity[a]	Parental toxicity	100 ppm, equivalent to 7 mg/kg bw per day	500 ppm, equivalent to 35 mg/kg bw per day
		Offspring toxicity	100 ppm, equivalent to 7 mg/kg bw per day	500 ppm, equivalent to 35 mg/kg bw per day
	Developmental toxicity[b]	Maternal toxicity,	90 mg/kg bw per day	300 mg/kg bw per day
		Embryo- or fetotoxicity	30 mg/kg bw per day	90 mg/kg bw per day
Rabbit	Developmental toxicity[b]	Maternal toxicity	100 mg/kg bw per day	250 mg/kg bw per day
		Embryo- or fetotoxicity	250 mg/kg bw per day	400 mg/kg bw per day
Dog	3-month study of toxicity[a]	Systemic effects	1250 ppm, equal to 35 mg/kg bw per day[c]	—
		Local effects on gastrointestinal tract	250 ppm, equal to 6.9 mg/kg bw per day	1250 ppm, equal to 35 mg/kg bw per day[c]
	12-month study of toxicity[d]	Systemic effects	8.4 mg/kg bw per day[c]	—
		Local effects on gastrointestinal tract	1.9 mg/kg bw per day	8.4 mg/kg bw per day[c]

[a] Diet
[b] Gavage
[c] Highest dose tested
[d] Capsules
[e] Two studies

Estimate of acceptable daily intake for humans

    0–0.07 mg/kg bw

Estimate acute reference dose

    0.3 mg/kg bw

Studies that would provide information useful for continued evaluation of the compound

    Further observations in humans

Critical end-points for setting guidance values for exposure to propiconazole

Absorption, distribution, excretion and metabolism in animals	
Rate and extent of oral absorption:	> 80% in 48 h
Dermal absorption	About 30% in 10 h (rat)
Distribution:	Widely distributed; highest concentrations in the liver and kidney
Potential for accumulation:	Limited
Rate and extent of excretion:	> 95% in the faeces and urine in 48 h; extensive enterohepatic recirculation (68% of administered dose in bile)

Propiconazole

Metabolism in animals	Extensive; oxidation of propyl side-chain; hydroxylation of phenyl and triazole rings, plus conjugation; cleavage of dioxolane ring
Toxicologically significant compounds (animals, plants and the environment)	Propiconazole; triazolyl alanine and triazolyl acetic acid are produced in plants but not animals

Acute toxicity

Rat, LD$_{50}$, oral	1517 mg/kg bw
Rat, LD$_{50}$, dermal	> 4000 mg/kg bw
Rat, LC$_{50}$, inhalation	> 5 mg/l of air (4-h; nose only)
Rabbit, dermal irritation	Irritating
Rabbit, eye irritation	Not irritating
Skin sensitization	Sensitizing (Magnusson and Kligman test)

Short-term studies of toxicity

Target/critical effect	Body weight, liver (mice, rats); erythrocytes (rat); stomach (dog)
Lowest relevant oral NOAEL	50 ppm, equal to 1.9 mg/kg bw per day (1-year study in dogs)
Lowest relevant dermal NOAEL	200 mg/kg bw per day (5 days/week)
Lowest relevant inhalation NOAEL	0.085 mg/l (6 h/day; 5 days/week)

Genotoxicity

Not genotoxic in vitro or in vivo

Long-term studies of toxicity and carcinogenicity

Target/critical effect	Liver hypertrophy and tumours (mice)
	Liver, body weight, uterine lumen dilatation (rats)
Lowest relevant NOAEL	100 ppm, equal to 11 mg/kg bw per day (mice)
Carcinogenicity	Hepatocellular tumours in male mice (≥ 850 ppm, equal to 108 mg/kg bw per day). Phenobarbital-type mechanism. The NOAEL was 500 ppm (equal to 59 mg/kg bw per day).
	Unlikely to pose a carcinogenic risk to humans

Reproductive toxicity

Reproduction target/critical effect	Reduced pup weight at parentally toxic dose
Lowest relevant reproductive NOAEL	100 ppm, equivalent to 7 mg/kg bw per day (rat)
Developmental target/critical effect	Skeletal variations
Lowest relevant developmental NOAEL	30 mg/kg bw per day (rat)

Neurotoxicity/delayed neurotoxicity

No specific studies; no findings in other studies

Other toxicological studies

Mechanism of induction of liver tumours	Phenobarbital-type mode of action indicated by cell proliferation, liver weight and microsomal enzyme induction patterns

Medical data

Local irritation associated with exposure to the formulated product

Summary	Value	Study	Safety factor
ADI	0–0.07 mg/kg bw	Reproductive toxicity in rats: pup and parental body weight	100
ARfD	0.3 mg/kg bw	Developmental toxicity in rats: embryo- or fetoxicity	100

DIETARY RISK ASSESSMENT

Long-term intake

Theoretical maximum daily intake were estimated for the commodities of human consumption for which Codex (CX) MRLs existed (Annex 3). The intakes in the five GEMS/Food regional diets ranged from 0 to 1% of the maximum ADI. The Meeting concluded that the long-term intake of residues of propiconazole resulting from uses considered by the JMPR is unlikely to present a public health concern.

Short-term intake

An ARfD of 0.3 mg/kg bw was established for propiconazole at this Meeting, but IESTIs could not be calculated, as the residues of the compound were evaluated before procedures for estimating STMRs and highest residue levels in the edible portion of a commodity found in trials used to estimate an MRL had been established. Propiconazole was scheduled for periodic evaluation of residues in 2007, when the risk assessment would be finalized.

4.25 PROPINEB (105)

TOXICOLOGY

At the present Meeting, the FAO Panel of Experts asked the WHO Core Assessment Group to establish an ARfD for propineb on the basis of the data available to the 1993 JMPR. The results of this evaluation are given in section 2.3.1.

RESIDUE AND ANALYTICAL ASPECTS

Propineb is a broad-spectrum dithiocarbamate fungicide used on many crops. It has been evaluated several times, the initial evaluation being in 1977 and the latest in 1993. It was listed in the periodic review programme of the CCPR at its Thirty-third Session for residue review by the 2003 JMPR (ALINORM 99/24) but was re-scheduled for evaluation in 2004. The Meeting received information on the metabolism and environmental fate of propineb, methods of residue analysis, freezer storage stability, national registered use patterns, the results of supervised residue trials and national MRLs. Information on GAP, national MRLs and residue data were submitted by Australia and Japan.

The 1993 JMPR established an ADI for propineb of 0–0.007 mg/kg bw, and the 1999 JMPR established an ADI of 0–0.0003 and an ARfD of 0.003 mg/kg bw for the metabolite propylenethiourea.

Metabolism

Animals

The Meeting received the results of studies of the metabolism of propineb in rats and a lactating goat. The biotransformation and degradation pathways in the goat were similar to those established in studies of rat metabolism. The metabolism of ^{14}C-propineb proceeds mainly via propylenethiourea and propylene diamine.

Once formed, propylenethiourea undergoes further reactions, leading to propylene urea, which can in turn be transformed by methylation to 2-methoxy-4-methylimidazoline. Other metabolites of propylenethiourea include 2-methylthio-4-methylimidazoline and 2-sulfonyl-4-methylimidazoline; the latter can undergo further metabolism to 4-methylimidazoline and N-formylpropylene diamine. In the lactating goat, the main metabolites detected were 2-methylthio-4-methylimidazoline in milk (48% TRR), kidney (25% TRR) and muscle (17% TRR), a sulfonyl conjugate of propylenethiourea in liver (23% TRR) and kidney (18% TRR) and propylenethiourea in fat and muscle (23% TRR).

Plants

The Meeting received the results of studies on the metabolism of propineb in apples, grapes, potato vines and tomato. The metabolism of ^{14}C-propineb was similar. It proceeds mainly via propylenethiourea (apple, 15% TRR; grape, 5.3% TRR; tomato, 30% TRR; potato vine, 3.5% TRR), which is itself further metabolized to propylene urea (apple, 5% TRR; tomato, 6.7% TRR; potato vine, 9.7% TRR). Propylenethiourea is also transformed to 4-methylimidazoline (apple, 10% TRR; tomato, 5% TRR; potato vine, 9.4% TRR), which on ring opening and oxidation gives N-formyl-propylene diamine (tomato, 6.7% TRR). The main metabolites identified in potato tubers after foliar spray were propylene urea (21% TRR) and a conjugate of its oxidation product 5-methylhydantoin (11% TRR). In a study on grapes harvested 0, 21 and 43 days after the last of one or three foliar applications of [1-propane-^{14}C]propineb, 83% of the ^{14}C was located on the surface of the fruit 43 days after three foliar sprays. Propineb was the main component of the radiolabelled residue at all times sampled (about 42% TRR at 43 days), metabolites each accounting for < 6% of the residues.

In contrast, when two applications of ^{14}C-propineb were made at the pre-blossom growth stage and grapes harvested about 100 days after the last application, most of the ^{14}C was associated with small molecules arising from incorporation of ^{14}C into natural plant products. Only low levels of propineb, propylene urea and N-formylpropylene diamine were detected, all at < 2% TRR.

After one or three applications of ^{14}C-propineb to individual fruit on an apple tree, most (55–59%) of the ^{14}C residue 14 days after application was located on the surface of the fruit. After 14 days, propineb accounted for 15–22% of the TRR, and no individual metabolite was present at > 10% TRR.

The metabolism of ^{14}C-propylenethiourea was also studied after application to apples. Propylenethiourea (metabolite 1) undergoes rapid degradation on apples, only 0.7% of the applied ^{14}C remaining on or in the peel 3 days after application. The main metabolite of propylenethiourea is the main metabolite of propineb, 4-methyl-imidazoline.

In greenhouse tomatoes harvested 7 days after four foliar applications of ^{14}C-propineb, most of the TRR was located on the surface of the fruit (about 70%), propineb accounting for 11% of the TRR. With the exception of propylenethiourea, which accounted for 30% of the TRR, all other metabolites were present at < 10% of the TRR.

In potato tubers and vines harvested 14 days after four foliar applications of ^{14}C-propineb, the ^{14}C residues in vines were mainly propineb (29%), with smaller amounts of propylene urea (10%) and 4-methylimidazoline (6.4%). Propylenethiourea was only a minor metabolite (3.5% TRR). In contrast, propineb and propylenethiourea were not detected in tubers. Propylene urea was the main metabolite (21%), with smaller amounts of a derivative of 5-methylhydantoin (11%). Most of the ^{14}C in tubers was incorporated into natural products (33%).

Environmental fate

The Meeting received information on the behaviour and fate of propineb during solution photolysis in aerobic soil metabolism. Information was also provided on the soil adsorption properties of propineb and on its behaviour and fate during anaerobic soil metabolism and column leaching of aged residues. Consistent with the policy outlined by the 2003 JMPR, only data on environmental fate relevant to residues of propineb in crops were evaluated.

Crop rotation studies were not provided; however, the aerobic soil metabolism of propineb was rapid, with inferred degradation half-lives of < 1 day. The main degradate formed was propylene urea. In aqueous solution, propineb is readily hydrolysed, the rate of hydrolysis increasing with pH; the DT_{50} values were 1–5 days. The rate of degradation in the field and in aquatic environments is fast, and propineb is not expected to persist in the environment.

Methods of analysis

Propineb residues are measured as CS_2 or propylene diamine formed by a common acid hydrolysis step. Samples in the field trials were analysed for propineb as CS_2 (spectrophotometry) or propylene diamine (gas chromatography with electron capture or mass spectrometry detection) and for propylenethiourea (HPLC with ultraviolet detection, gas chromatography with flame photometricf detection). LOQs of 0.05–0.1 mg/kg for propineb and 0.01 mg/kg for propylenethiourea were reported to be achievable in numerous commodities.

Stability of residues in stored analytical samples

The Meeting received information on the stability of propineb residues during storage of analytical samples at freezer temperatures. The available data indicate that the combined residues of propineb and propylenethiourea are stable under frozen storage conditions (−20 °C) in and on the following commodities (storage interval in parentheses): tomatoes (2 years); tomato juice (2 years); tomato marc (2 years) and potatoes (2 years for propineb, 2 weeks for propylenethiourea).

Definition of the residue

The studies of metabolism in grapes, apples and tomatoes after spraying with propineb demonstrated rapid degradation of the residues on the surface of plant parts. The patterns of metabolites found were similar in different species of plants. The main metabolites found in plants—propylenethiourea, propylene urea, 4-methylimidazoline, 2-sulfonyl-4-methylimidazoline and *N*-formylpropylene diamine—were also detected in animals. The Meeting agreed that propineb and propylenethiourea should be regarded as the residues of toxicological concern.

For estimating dietary intake and to enable comparison of the calculated intakes with the ADI, the residues should be expressed in terms of propineb (propineb = $1.9 \times CS_2$).

Currently, the residue definition for dithiocarbamates including propineb is 'total dithiocarbamates, determined as CS_2, evolved during acid digestion and expressed as mg CS_2/kg'. Propineb can be determined by a specific method that measures both CS_2 and the amine (propylene diamine) released on acid hydrolysis. Therefore, separate MRLs could be established for propineb. Until specific methods are developed for all dithiocarbamates, however, the listing of one compound under two different residue definitions would be confusing for analysts and enforcement agencies. The *FAO Manual* (page 51) states that no compound, metabolite or analyte should be listed in more than one residue definition. In national systems, the residue definition for propineb is generally in terms of CS_2.

The Meeting agreed that the residue definition applicable to propineb should continue to be that for dithiocarbamates in general. For estimation of dietary intake and for the risk assessment component relating to exposure, the metabolite propylenethiourea is considered to be toxicologically relevant and must be accounted for. For an overall risk assessment of 'thyroid-active' dithiocarbamates such as propineb, the 1997 JMPR "agreed that it is necessary to combine not only the intake of different parent pesticides but also the intake of [ethylene thiourea] or propylenethiourea" and recommended that an ADI adjustment approach be used. Therefore, in estimating dietary intake, residues of both propineb and propylenethiourea must be accounted for and their relative toxicity taken into account. A conservative approach is to sum the residues after scaling the propylenethiourea residues for 'potency' on the basis of the ratio of the ADIs for propineb and propylenethiourea (2.3), in order to estimate STMRs, and the ratio to ARfDs (3.3) for estimating the highest residue levels. This approach has been used for dimethoate–omethoate and acephate–methamidophos. The ratios are based on mass and do not require correction for relative molecular mass.

For estimation of the STMR for propineb, residue = propineb + (2.3 × propylenethiourea)

For estimation of the highest reside level for propineb, residue = propineb + (3.3 × propylenethiourea)

Definition of propineb residue for compliance with MRLs: Total dithiocarbamates, determined as CS_2, evolved during acid digestion and expressed as mg CS_2/kg

Definition of propineb residue for estimation of dietary intake: propineb and propylenethiourea

These definitions apply to plant and animal commodities.

Results of supervised trials on crops

The results of supervised trials were available on the use of propineb on apple, asparagus, cabbage, cherry, Chinese cabbage, celery, citrus (orange), cucumber, garlic, grape, leek, lettuce, melon, onion, olive, pear, pepper, potato, tomato and watermelon.

The Meeting decided to use only data from trials in which propineb was determined as CS_2 for estimation of maximum residue, STMR and highest reside levels. In some cases, untreated control samples also contained residues of CS_2. Trials were considered acceptable if the residue levels in untreated control samples were < 10% of the residue in the treated crop or, when propineb was also determined as propylene diamine, there was satisfactory agreement between the results for propineb determined as CS_2 and propylene diamine.

The following relation is useful when considering the data: CS_2 residue (mg/kg) = 0.52 × propineb residues (mg/kg).

Citrus fruit

Trials on citrus were conducted in Brazil and Japan but were provided only in summary form, which was unsuitable for the purpose of estimating maximum residue levels.

Pome fruit

Trials on apple and pear were conducted in Belgium (GAP, 0.49–0.71 kg ai/ha fruit tree leaf wall, equivalent to 0.84–1.6 kg ai/ha for a standard orchard, applied just after flowering), Germany (GAP, 1.58 kg ai/ha, 0.105 kg ai/hl, 28-day PHI), Italy (GAP, 0.105–0.14 kg ai/hl, 28-day PHI) and Spain (GAP for pome fruit, 0.14–0.21 kg ai/hl, 28-day PHI). The trials conducted in Germany, Italy and Spain did not match GAP in the respective countries and were evaluated against the GAP of Belgium.

In two trials in Belgium and one in Germany on *apple*, the residue levels of propineb (measured as CS_2) in untreated controls were unacceptable. One trial in Belgium approximated Belgian GAP, with levels of propineb residues < 0.10 mg/kg (propylenethiourea, < 0.01 mg/kg). In a further trial in Germany and one in Spain that approximated Belgian GAP, the residue levels were < 0.10 and < 0.10 mg/kg (propylenethiourea, < 0.01 (two) mg/kg).

Trials on *pear* were conducted in Belgium (GAP, 0.49–0.71 kg ai/ha fruit tree leaf wall, equivalent to 0.84–1.6 kg ai/ha for a standard orchard, applied just after flowering), Germany (GAP, 1.58 kg ai/ha, 0.105 kg ai/hl, 28-day PHI) and Italy (GAP, 0.105–0.14 kg ai/hl, 28-day PHI). One trial in Belgium and one in Germany matched GAP in Belgium, with levels of propineb residues of < 0.10 and 0.10 mg/kg, respectively. The levels of propylenethiourea residues were: < 0.01 mg/kg.

The Meeting considered that the number of trials on apples and pears was inadequate for the purpose of estimating maximum residue levels and agreed to withdraw its previous recommendation for propineb of 2 mg/kg as CS$_2$ for apples and pears.

Cherry

Six trials on cherry were conducted in Germany (GAP, 0.105 kg ai/hl, 28-day PHI) which approximated German GAP. Two trials were conducted at two locations, which differed only in the formulation used; one trial at each location was selected for estimating maximum residue levels. The residue levels in the six trials were < 0.05 (two), 0.05, 0.06, 0.13 and 0.15 mg/kg for CS$_2$ and < 0.01 (four) and 0.02 (two) mg/kg for propylenethiourea.

The residue levels of propineb (1.9 × CS$_2$) and propylenethiourea, combined as explained above (residue = propineb + 2.3 × propylenethiourea), used for estimating the STMR were < 0.12 (three), 0.14, 0.29 and 0.33 mg/kg. The highest reside level for dietary intake was estimated to be 0.35 mg/kg (residue = propineb + 3.3 × propylenethiourea). The Meeting estimated a maximum residue level for propineb in cherries of 0.2 mg/kg as CS$_2$, an STMR of 0.13 mg/kg as propineb and a highest reside level of 0.35 mg/kg as propineb.

Grape

Trials on wine grapes were conducted in France (GAP, 0.68 kg ai/ha, 21-day PHI) and Germany (GAP, 2.8 kg ai/ha, 0.14 kg ai/hl, 56-day PHI) after pre-blossom application. None of the trials approximated GAP in the respective countries. Trials were also conducted on table and wine grapes after pre- and post-blossom applications in France (GAP, 0.68 kg ai/ha, 21-day PHI), Greece (GAP, 0.14 kg ai/hl, 7-day PHI for table grapes, 21-day PHI for wine grapes), Italy (GAP, 0.14 kg ai/hl, 28-day PHI) and Spain (GAP, 0.28 kg ai/hl, 15-day PHI). None of the trials matched GAP. The Meeting agreed to withdraw the previous recommendations for propineb in grapes of 2 mg/kg as CS$_2$.

Olive

Trials on olives were conducted in Spain (GAP, 0.21 kg ai/hl, 15-day PHI), but none matched GAP.

Onion

Trials on onion were conducted in Australia (GAP, 1.4 kg ai/ha, 0.14 kg ai/hl, 14-day PHI) and Brazil (GAP 2.1 kg ai/ha, 7-day PHI), but the latter was available only in the form of a summary. The residue levels of propineb (measured as propylene diamine and not CS$_2$) in the Australian trials approximating GAP were < 0.2 and 1.2 mg/kg. The number of trials was considered by the Meeting to be inadequate for estimating a maximum residue level, and the Meeting agreed to withdraw its previous recommendation for propineb in onion, bulb, of 0.2 (*) mg/kg as CS$_2$.

Garlic

Trials on garlic were conducted in Brazil; however, the data were supplied only in summary form and were therefore not suitable for estimating a maximum residue level.

Lettuce

Trials on lettuce were conducted in Australia (GAP, 1.4 kg ai/ha, 0.14 kg ai/hl, 3-day PHI) and Brazil (no information on GAP). The latter was available only in the form of a summary. The residue levels of propineb (measured as propylene diamine and not CS$_2$) in the Australian trials approximating GAP were 0.3 and 2.5 mg/kg. The number of trials was considered by the Meeting to be inadequate for the purposes of estimating a maximum residue level.

Brassica vegetables

Trials on head cabbage were available from Brazil (no GAP) and on Chinese cabbage from Thailand (no GAP). As no relevant GAP was available and as the data were provided only in summary form, the Meeting was unable to estimate a maximum residue level for these vegetables.

Cucumber

Trials on cucumbers grown in greenhouses in Greece (GAP for vegetables, 0.18 kg ai/ha, 3-day PHI), Italy (no GAP) and Spain (GAP, 0.21 kg ai/ha, 3-day PHI) were made available to the Meeting. The trials in Italy and Spain did not match GAP for those countries and were assessed against the GAP of Greece. The levels of propineb residues (measured as CS_2) in three trials in Greece approximating GAP in Greece were 0.60, 0.90 and 1.1 mg/kg (propylenethiourea, 0.01, < 0.01 and 0.02 mg/kg). The levels of propineb residues in one trial in Italy and three in Spain matching GAP ± 25% in Greece were 0.20, 0.20, 0.43 and 0.47 mg/kg (propylenethiourea, < 0.01 (four) mg/kg). Conversion of the residue levels expressed in terms of propineb to CS_2 gives values of 0.10 (two), 0.22, 0.24, 0.31, 0.47 and 0.57 mg/kg. The Meeting estimated a maximum residue level for propineb in cucumbers of 1 mg/kg as CS_2.

The appropriately scaled and totalled residue levels of propineb and propylenethiourea for estimating the STMR were: 0.22 (two), 0.45, <u>0.49</u>, 0.62, 0.92 and 1.1 mg/kg. The highest reside level was estimated to be 1.1 mg/kg. For estimation of dietary intake, the Meeting estimated STMR and highest reside levels for propineb in cucumbers of 0.49 and 1.1 mg/kg, respectively.

Melon (except watermelon)

Trials on melons (except watermelon) were reported from Greece (GAP for vegetables, 0.18 kg ai/ha, 3-day PHI) and Spain (GAP, 0.21 kg ai/hl, 15-day PHI; GAP for cucurbits, 0.21 kg ai/hl, 3-day PHI). The levels of propineb residues (measured as propylene diamine) in two trials in Spain matching Spanish GAP ± 25% were 0.52 and 1.5 mg/kg (propylenethiourea, 0.05 and 0.06 mg/kg). One field trial in Greece, in which 0.43 mg/kg were found (propylenethiourea, < 0.01 mg/kg) also matched GAP in that country. Data were not available for propineb measured as CS_2 in any of the trials at the relevant PHI. The Meeting considered three trials inadequate for the purposes of estimating a maximum residue level for melon (except watermelon) and agreed to withdraw its previous recommendation of 0.1 (*) mg/kg as CS_2.

Watermelon

Trials on watermelon were reported from Greece (GAP for vegetables, 0.18 kg ai/hl, 3-day PHI) and Italy (no GAP). The levels of propineb residues in two trials in Greece matching Greek GAP ± 25% were 0.17 and 0.31 mg/kg (propylenethiourea, < 0.01 and 0.02 mg/kg). Two field trials in Italy approximating GAP in Greece showed residue levels of 0.17 and 0.29 mg/kg (propylenethiourea, 0.01 and 0.02 mg/kg). Data were not available for propineb determined as CS_2 at the relevant PHI in any of the trials. The Meeting considered the number of trials inadequate for the purposes of estimating a maximum residue level for watermelon.

Tomato

Trials on field tomatoes were reported from France (GAP, 0.21 kg ai/hl, 7-day PHI), Germany (GAP, 0.84 kg ai/ha at crop height < 0.5 m; 1.26 kg ai/ha at crop height 0.5–1.25 m; 1.68 kg ai/ha at crop height > 1.25 m; 7-day PHI) and Spain (GAP, 0.21 kg ai/hl, 3-day PHI). The CS_2 residue levels in four trials in Germany matching GAP were 0.11, 0.14, 0.40 and 0.55 mg/kg, equivalent to 0.21, 0.27, 0.76 and 1.0 mg/kg as propineb (propylenethiourea, < 0.02 (three) and 0.02 mg/kg).

Four trials were available from France and four from Spain which were conducted according to GAP in the respective countries. As GAP in France and Spain differs only with respect to the PHI, the Meeting decided to evaluate the French and Spanish trials against the GAP of Spain to obtain a representative data set. The residue levels of propineb in these trials were 0.14, 0.22, 0.26, 0.35, 0.49, 0.94, 1.0 and 1.1 mg/kg (propylenethiourea, < 0.01, 0.02 (two), 0.04, 0.05 (two) and 0.06 (two) mg/kg).

Propineb

Additional trials on tomatoes grown under protected cover (greenhouse) were reported from France, Germany and Spain and evaluated against the GAP of Spain, which is the same for tomatoes grown in the field and protected under cover. The residue levels of CS_2 reported in terms of propineb ≥ 3 days after the last application were 0.82, 1.1, 1.3, 1.5, 2.3 and 2.4 mg/kg. The levels of propylenethiourea residues were 0.04, 0.05, 0.06, 0.08, 0.09 and 0.16 mg/kg.

The Meeting considered that the residue levels in field trials conducted in Germany according to German GAP and the trials under cover and in the field conducted according to GAP in Spain represent similar residue populations and could be combined for the purposes of estimating a maximum residue level. The residue levels expressed in terms of CS_2, were: 0.07, 0.11 (two), 0.14 (two), 0.18, 0.25, 0.40, 0.42, 0.49, 0.52, 0.54, 0.57 (two), 0.68, 0.78 and 1.2 (two) mg/kg.

The Meeting estimated a maximum residue level for propineb in tomatoes of 2 mg/kg as CS_2 to replace the previous recommendation for tomatoes of 1 mg/kg as CS_2.

The appropriately scaled and totalled residue levels of propineb and propylenethiourea in the 18 trials used for estimating the STMR were: 0.16, 0.26, 0.27, 0.31 (two), 0.44, 0.61, 0.81, 0.89, 1.1 (three), 1.2 (two), 1.3, 1.7, 2.5 and 2.8 mg/kg. The Meeting estimated the STMR for propineb in tomatoes at 1.0 mg/kg and the highest reside level at 2.9 mg/kg.

Peppers (sweet)

Trials on field-grown peppers in France (no GAP) and Spain (GAP, 0.21 kg ai/hl, 3-day PHI) were made available to the Meeting. The French trials were evaluated against GAP of Spain. In two trials in France matching GAP in Spain, the residue levels of propineb were 0.22 and 0.83 mg/kg (propylenethiourea, 0.02 and 0.07 mg/kg). Four trials in Spain that matched GAP for peppers showed propineb residue levels of 0.60, 1.4 (two) and 1.7 mg/kg (propylenethiourea, 0.09, 0.12, 0.17 and 0.18 mg/kg). The levels of propineb residues in field-grown peppers were thus: 0.22, 0.60, 0.83, 1.4 (two) and 1.7 mg/kg. The corresponding levels of propylenethiourea residues were: 0.02, 0.07, 0.09, 0.12, 0.17 and 0.18 mg/kg.

Trials on peppers grown in greenhouses in France (no GAP), Germany (no GAP) and Spain (GAP, 0.21 kg ai/hl, 3-day PHI) were made available to the Meeting. The trials in France and Germany were evaluated against GAP in Spain. Residues of propineb in sweet peppers grown indoors were 1.3, 2.1 and 11 mg/kg (propylenethiourea, 0.05, 0.23 and 0.71 mg/kg) in three trials in Spain; and 0.75, 1.4, 1.5 and 1.7 mg/kg (propylenethiourea, 0.06, 0.07, 0.10 and 0.11 mg/kg) in four trials in France. Thus, the levels of propineb in sweet peppers grown in greenhouses were: 0.75, 1.3, 1.4, 1.5, 1.7, 2.1 and 11 mg/kg (propylenethiourea: 0.05, 0.06, 0.07, 0.10, 0.11, 0.23 and 0.71 mg/kg). Conversion of the levels of CS_2 residues reported in terms of propineb back to CS_2 gave levels of 0.11, 0.31, 0.39, 0.43, 0.68, 0.73 (three), 0.78, 0.88 (two), 1.1 and 5.7 mg/kg. The Meeting estimated a maximum residue level for propineb in peppers, sweet, of 7 mg/kg as CS_2.

The appropriately scaled and totalled residue levels of propineb and propylenethiourea in the 13 trials used for estimating the STMR were: 0.27, 0.89, 0.99, 1.0, 1.4, 1.6 (two), 1.7 (two), 2.0, 2.1, 2.6 and 13 mg/kg as propineb. The STMR was 1.6 mg/kg and the highest reside level was estimated to be 13 mg/kg.

Potato

Field trials on potatoes were made available to the Meeting from France (GAP, 0.21 kg ai/hl, PHI not specified), Germany (GAP, 1.3 kg ai/ha, 7-day PHI), Spain (GAP, 0.21 kg ai/hl, 15-day PHI) and the United Kingdom (no GAP). The trials in Germany and the United Kingdom did not comply with the relevant GAP. The trials in France were evaluated against GAP in Spain.

In three trials in France approximating Spanish GAP, the levels of propineb residues on potatoes were < 0.10 (two) and 0.14 mg/kg (propylenethiourea, < 0.01 (three) mg/kg). Three trials in Spain approximating GAP in that country showed propineb residue levels of < 0.10 (three) mg/kg (propylenethiourea, < 0.01 (three) mg/kg). Conversion of the residue levels determined as CS_2 but reported in terms of propineb to CS_2 gave levels of < 0.05 (five) and 0.073 mg/kg. The Meeting estimated a maximum residue level for propineb in potatoes of 0.1 mg/kg as CS_2, which replaces the previous recommendation of 0.1 (*) mg/kg.

The appropriately scaled and totalled residue levels of propineb and propylenethiourea in six trials used for estimating the STMR were: 0.12 (five) and 0.16 mg/kg. The Meeting estimated an STMR for propineb in potatoes of < 0.12 mg/kg and a highest reside level of 0.16 mg/kg.

Celery

Two trials on celery were reported from Australia (GAP, 1.4 kg ai/ha, 0.14 kg ai/hl, 7-day PHI), which showed propineb residue levels of < 0.2 and 0.4 mg/kg (propylenethiourea not analysed). The Meeting considered the number of trials inadequate for the purpose of estimating a maximum residue level for celery.

Asparagus

In a single trial on asparagus in Peru (GAP, 2.1 kg ai/ha, 0.21 kg ai/hl, 30-day PHI) that matched GAP in that country, the residue levels of propineb were < 0.01 mg/kg (propylenethiourea not measured).

The Meeting considered the number of trials inadequate for the purpose of estimating a maximum residue level for asparagus.

Fate of residues during processing

The Meeting received the results of studies on incurred residues of propineb and propylenethiourea in apples, pears, cherries, tomatoes, grapes and olives after washing and further processing in a range of fractions. Only the studies relevant to commodities for which maximum residue levels have been estimated are reported below.

It would not usually be appropriate to derive processing factors for propylenethiourea, as these would reflect both the effect of processing and also the formation of propylenethiourea from propineb, especially after boiling steps. In the present case, the use of processing factors would result in overestimates of the residue levels of propylenethiourea in processed commodities, and the Meeting decided to continue to use this approach. Nevertheless, if concern about dietary intake were identified, the Meeting would consider refining the approach to estimate propylenethiourea residues in processed commodities.

In trials in Germany, cherries were processed according to simulated household and commercial practices into washed fruit, juice, jam and preserves. The processing factors for juice and jam prepared by household procedures in two trials each were 0.5–0.6 (mean, 0.55) for juice and 0.3–0.4 (mean, 0.35) for jam. Propylenethiourea residues did not concentrate in juice or jam, with mean processing factors of < 0.68 for juice and < 0.78 for jam. After simulated commercial preparation, the mean processing factors for propineb in three trials each were 0.63 (range, 0.6–0.7) for washed fruit and 0.15 (range, 0.13–0.16) for preserves. The corresponding mean values for propylenethiourea were 1 for washed fruit and < 0.5 for preserves.

The Meeting considered that it would be appropriate to use the mean processing factors from the various studies, to reflect different commercial practices. For cherries, it estimated processing factors for propineb of 0.63 in washed fruit, 0.55 in juice, 0.15 in preserves and 0.35 in jam. The processing factors for propylenethiourea were 1 in washed fruit, < 0.68 in juice, < 0.5 in preserves and < 0.78 in jam.

Processing studies for tomatoes with respect to washed fruit, juice, ketchup, paste and preserves were reported. For washed fruit, the mean processing factors in four studies were 0.45 (range, 0.3–0.6) for propineb and 0.4 (range, 0.3–0.5) for propylenethiourea. In the case of juice, the mean processing factor for propineb in 10 studies was < 0.12 (range, < 0.06–0.2), while that for propylenethiourea in nine studies was 0.91 (range, 0.3–2.3). The levels of residues of propineb were significantly reduced during the preparation of preserves and ketchup, with mean processing factors of 0.15 in four studies on preserves (range, 0.1–0.2) and < 0.12 in six studies on ketchup (range, < 0.06–< 0.25). Residues were concentrated during preparation of paste, with a mean processing factor in four studies of 1.1 (range, 0.4–2.0). The mean processing factors for propylenethiourea were 0.75 (*n* = 4; range, 0.5–1) for preserves, 0.54 (*n* = 5; range, 0.3–0.7) for ketchup and 11 (*n* = 4; range, 6.8–17) for paste.

The Meeting considered that it would be appropriate to use the mean processing factors from the various studies to reflect different commercial practices. For tomato, it estimated processing factors for propineb of 0.45 in washed fruit, < 0.12 in juice, 0.15 in preserves, < 0.12 in ketchup and 1.1 in paste. For propylenethiourea, processing factors of 0.4 in washed fruit, 0.91 in tomato juice, 0.75 in preserves, 0.54 in ketchup and 11 in paste were established.

Commodity	Processing factor$_{propineb}$	Propineb residues (mg/kg) For STMR/ STMR-P	For HR/ HR-P	Processing factor$_{propylenethiourea}$	Propylenethiourea residues (mg/kg) For STMR/ STMR-P	For HR/ HR-P	Adjusted values (mg/kg) STMR[1]	HR[2]
Cherry		0.128	0.351		0.01	0.02		
Washed	0.63	0.0803	0.221	1	0.01	0.02	0.103	0.287
Juice	0.55	0.0701		0.68	0.0068		0.0858	
Preserves	0.15	0.0191		0.5	0.005		0.0306	
Jam	0.35	0.0446		0.78	0.0078		0.0626	
Tomato		1.0	2.93		0.03	0.16		
Washed	0.45	0.45	1.32	0.4	0.012	0.064	0.478	1.53
Juice	0.12	0.12		0.91	0.0273		0.183	
Preserves	0.15	0.15		0.75	0.0225		0.202	
Ketchup	0.12	0.12		0.54	0.0162		0.157	
Paste	1.1	1.1		11	0.33		1.86	

[1] Adjusted STMR-P = STMR-P$_{propineb}$ + 2.3 × STMR-P$_{propylenethiourea}$
[2] adjusted HR-P = HR-P$_{propineb}$ + 3.3 × HR-P$_{propylenethiourea}$

Residues in animals commodities

Dietary burden of farm animals

The Meeting estimated the dietary burden of propineb residues of farm animals on the basis of the diets described in Appendix IX of the *FAO Manual*. As no relevant items were identified, the dietary burdens for estimating MRLs and STMRs for animal commodities (residue levels in animal feeds expressed in dry weight) are zero for all the relevant animal diets.

Maximum residue levels

The Meeting estimated maximum residue levels of 0.05 (*) mg/kg for meat (from mammals other than marine mammals), 0.05 (*) mg/kg for edible offal (mammalian) and 0.01 (*) mg/kg for milks.

The Meeting estimated maximum residue levels of 0.05 (*) mg/kg for poultry meat, 0.05 (*) for poultry offal and 0.01 (*) mg/kg for eggs. The STMRs for animal commodities are zero.

DIETARY RISK ASSESSMENT

The Meeting considered how best to approach the dietary risk assessment of mixed residues of propineb and propylenethiourea and decided that an appropriately conservative approach would be to calculate the sum of the residues after scaling the propylenethiourea residues to account for the difference in toxicity. The relevant factors for long-term and short-term intake were derived from the ratios of the ADI and ARfD values for propineb and propylenethiourea, which are 2.3 and 3.3, respectively. Dietary intake estimates for the residues, adjusted for potency and combined, were compared with the ADI and interim ARfD for propineb (See general item 2.2, Interim acute reference dose).

Long-term intake

The evaluation of propineb resulted in recommendations for MRLs and STMRs for raw and processed commodities. Data were available on the consumption of 15 food commodities and were used in the dietary intake calculation. The results are shown in Annex 3.

The IEDIs in the five GEMS/Food regional diets, based on estimated STMRs, were 4–30% of the ADI of 0–0.007 mg/kg bw for propineb (Annex 3). The Meeting concluded that the long-term intake of residues of propineb and propylenethiourea from uses that have been considered by the JMPR is unlikely to present a public health concern.

Short-term intake

The IESTI for propineb was calculated for the food commodities (and their processing fractions) for which maximum and highest reside levels had been estimated and for which data on consumption were available. The results are shown in Annex 4.

The IESTI was 0–110 % of the interim ARfD (0.1 mg/kg bw) for the general population and 0–120% of the interim ARfD for children ≤ 6 years. The values 110% and 120% represent the estimated short-term intake of sweet peppers by the general population and children, respectively.

The Meeting concluded that the short-term intake of residues of propineb from uses other than on sweet peppers that have been considered by the JMPR is unlikely to present a public health concern.

4.26 PYRACLOSTROBIN (210)

RESIDUE AND ANALYTICAL ASPECTS

Residue and analytical aspects of pyraclostrobin were considered for the first time by the present Meeting.

Pyraclostrobin, chemical name (IUPAC) methyl N-(2-{[1-(4-chlorophenyl)-1H-pyrazol-3-yl]-oxymethyl}phenyl)-N-methoxycarbamate, is a new fungicidal active ingredient. It represents a modification of the structural pattern of natural fungicides called strobilurins.

The Meeting received information on the metabolism and environmental fate of pyraclostrobin, methods of residue analysis, freezer storage stability, national registered use patterns, the results of supervised residue trials, farm animal feeding studies, fate of residues in processing and national MRLs.

Metabolism

Animals

The Meeting received the results of metabolism studies in rats, lactating goats and laying hens. The metabolism and distribution of pyraclostrobin in plants and livestock was investigated with [chlorophenyl-^{14}C]pyraclostrobin and [tolyl-^{14}C]pyraclostrobin

The main metabolite are methyl-*N*-(2{[1-(4-chlorophenyl)-1*H*-pyrazol-3-yl]oxymethyl}phenyl) carbamate (500M07) and 1-(4-chlorophenyl)-1*H*-pyrazol-3-yl hydrogen sulfate.

Metabolism in laboratory rats was evaluated by the WHO panel of the 2003 JMPR, which concluded that the metabolism proceeds through three main pathways. The methoxy group on the tolyl-methoxycarbamate moiety is readily lost, with few main metabolites retaining this group. Hydroxylation of the benzene or pyrazole ring is followed by conjugation with glucuronide. Many metabolites are derived from the chlorophenol pyrrazole or tolyl-methoxycarbamate moieties of pyraclostrobine. The metabolites were similar in both sexes and at all doses. No unchanged parent compound was found in bile or urine, and only small amounts were found in faeces.

Studies of the metabolism of pyraclostrobin in *goats* showed that residues in products of animal origin derive from the parent compound as well as from its *N*-desmethoxylation product. The metabolism and distribution of pyraclostrobin were investigated in lactating goats given material labelled in the chlorophenyl or in the tolyl ring. After five consecutive daily oral administrations of ^{14}C-pyraclostrobin at a nominal dosage of 12 or 50 mg/kg of feed, there was rapid absorption from the gastrointestinal tract. Radioactivity was excreted mainly via the faeces. The radiolabel in milk accounted for only 0.1–0.5% of the total applied radioactivity. There was no indication of accumulation of ^{14}C-pyraclostrobin in tissues. The parent compound was found in fat, muscle and, at lower amounts, in liver. Metabolites are formed in liver and kidney by hydroxylation of the chlorophenyl and tolyl rings and by cleavage of the molecule. Little extraction was seen in liver.

^{14}C-Pyraclostrobin is thus metabolized in goats by three key steps: (1) desmethoxylation at the oxime ether bond, (2) hydroxylation of the chlorophenyl, the pyrazole or the tolyl ring system and (3) cleavage of the two ring systems with subsequent oxidation of the two resulting molecules.

Pyraclostrobin was present in all tissues and in milk and was the main residue component in muscle and in fat (log P_{ow} = 3.99).

Tissues and eggs from *hens* that received an exaggerated dose of 12 mg/kg feed of [chlorophenyl-^{14}C]pyraclostrobin or 13 mg/kg [tolyl-^{14}C]pyraclostrobin contained low residue levels consisting of three main metabolites. The parent compound was found in fat and eggs but not in liver. The main metabolite in liver was the glucuronic acid conjugate, which was bound to the tolyl ring of the demethoxylated parent structure. The desmethoxy metabolite 500M07 was also present in fat and eggs,

Five routes of biotransformation were detected. The predominant transformation was the demethoxylation step. Second, the demethoxylated metabolite was oxygenated at the tolyl ring, followed by conjugation with glucuronic acid. Third, the demethoxylated metabolite was hydroxylated at the chlorophenyl or the pyrazole ring, again followed by a conjugation reaction with glucuronic acid. Fourth, the parent compound was hydroxylated at the chlorophenyl ring in the *para* position, whereby the C-l was shifted to the *meta* position (NIH shift). Fifth, the parent compound was cleaved at the methylene ether bridge. A specific variation was substitution of C-l by glucuronic acid.

The main metabolite in fat and eggs was 500M07, and that in liver was the glucuronic acid conjugate. The metabolism in rats, goats and hens were comparable.

Plants

The Meeting received the results of studies of the metabolism of pyraclostrobin in grapes, potatoes and wheat.

The metabolism of pyraclostrobin in *grapes* was investigated with material labelled in the tolyl or the chlorophenyl ring. Applications were made six times at a rate of 0.25 kg ai/ha, and the grapes were harvested

40 days after the last application. The relevant residue in grapes consists of the parent compound and its desmethoxy metabolite 500M07. Some other compounds were identified as products formed by cleavage of the molecule. *O*-Glucosylation and methoxylation were of minor importance, representing much less than 10% of the TRR.

Studies of metabolism in *potato* were conducted with material labelled in the tolyl or the chlorophenyl ring. Six post-emergence applications were made at the intended use rate of 300 g/ha. The relevant residue in potato green matter and tuber consisted of the parent compound (65% and 2.5% of the TRR, respectively) and its desmethoxy metabolite 500M07 (6.2% and 0.6% of the TRR, respectively) at growth stage 70.

Some other compounds were identified as products formed by cleavage of the molecule. *O*-Glucosylation and methoxylation were of minor importance, representing far less than 10% of the TRR. The total residue levels in the edible portion (potato tubers) were low. One derivative found in larger amounts in tubers was identified as the naturally occurring amino acid L-tryptophan. This compound represented 10% of the TRR in tuber at growth stage 70, but its contribution increased to 29.2% of the TRR at growth stage 85–89. It should not therefore be regarded as a relevant residue that must be covered by the analytical method.

Wheat received two application at 0.3 kg ai/ha, and samples were collected 0, 31 and 41 days after the last treatment. The relevant residue of ^{14}C-pyraclostrobin in wheat consists of unchanged parent compound and its desmethoxy metabolite 500M07. Tryptophan, which is formed in considerable amounts from pyraclostrobin in grain, is a natural ingredient and is therefore of no toxicological concern. All the other metabolites identified represented < 10% TRR and are thus of minor importance. The low levels of unextractable residues in forage and straw indicate that pyraclostrobin and its metabolites are not firmly associated with cell wall polymers. Somewhat larger amounts of unextractable were found in grain, as some of the radioactivity was incorporated into or associated with grain protein and starch.

The metabolic pathways in grapes, potatoes and wheat were qualitatively similar. Pyraclostrobin and its desmethoxy metabolite 500M07 constituted the main part of the residue. In addition, hydroxylation in the tolyl and the chlorophenyl rings and cleavage reactions between the two ring systems were observed. The hydroxylation reaction is followed by glucosylation or methylation, whereas the intermediates of the cleavage reaction are further transformed by conjugation or the shikimate pathway. Transformation via the shikimate pathway resulted in the formation of the natural amino acid L-tryptophan in potato tubers and wheat grain.

Environmental fate

Soil

The Meeting received the results of studies on the fate and behaviour in soil of [tolyl-U-^{14}C]pyraclostrobin and [chlorophenyl-U-^{14}C]pyraclostrobin.

Pyraclostrobin was investigated for aerobic metabolism in a number of soils. The degradation of pyraclostrobin in aerobic soil studies is characterized by a relatively low mineralization rate (about 5% of the total applied radioactivity within 100 days) and formation of large amounts of bound residues (about 55% of the total). The same metabolites, the *trans*-azooxy and the *trans*-azo dimers (or *N,N'*-bis-[2-(1*H*-pyrazol-3-yloxymethyl)phenyl]diazene) of pyraclostrobin, were found in all soil types. The amount of the *trans*-azooxy dimer generally exceeded 10% the total applied radioactivity (maximum, 31%), whereas that of the *trans*-azo dimer slightly exceeded 10% of the total applied in only one of the investigated soils. The amount of bound residue increased with time, and the most of the radiolabel was associated with insoluble humins and high-molecular-mass humic acids. No release of pyraclostrobin or its metabolites was observed, even with harsh extraction methods (NaOH) or with intensive activity of soil-eating animals (earthworms). Photolytic degradation leads to the same degradation products; however, all the metabolites were formed in amounts less than 10% of the total applied radioactivity.

Pyraclostrobin

Pyraclostrobin is degraded in soil under laboratory conditions, with DT_{50} values ranging from 12 to 101 days in five microbially active soils. Higher soil moisture contents generally accelerated the degradation. Photolysis did not significantly influence the degradation rate; however, it reduced the amounts of the *trans*-azooxy and the *trans*-azo dimers of pyraclostrobin. In field studies, the DT_{50} values of pyraclostrobin were much lower, ranging from 2 to 37 days. The DT_{90} values in the field were 83–230 days. The DT_{50} values of the soil metabolites in the laboratory were 60–166 days for the *trans*-azooxy dimer and 38–159 days for the *trans*-azo dimer. (The high values for the latter were calculated for soils in which the metabolite was formed in amounts < 10% total applied radioactivity.) Under field conditions, however, the metabolites 500M07 and *trans*-azo dimer were not detected. Only the *trans*-azooxy dimer was found sporadically in trace amounts close to the LOQ.

With regard to mobility, no radiolabel was found in leaching studies, and pyraclostrobin remained in the first layer of soil (< 12 cm). Thus, pyraclostrobin and its metabolites are not mobile in soil.

The results indicate that pyraclostrobin and its metabolites are not stable in soil. They were degraded quickly and were not mobile.

Succeeding crops

The residue levels and the nature of the residues of pyraclostrobin were investigated in three succeeding crops, radish, lettuce and wheat, after application at a rate of 900 g ai/ha. The total residues in the edible parts of the succeeding crops were low at all plant-back intervals. There was no accumulation of pyraclostrobin or its degradation products in the parts of plants used for human or animal consumption.

Methods of analysis

Methods for the determination of pyraclostrobin in plant and animal matrices are based on HPLC with ultraviolet, mass spectrometry or tandem mass spectrometry detection. The LOQ is 0.02 mg/kg in plant matrices, 0.01 mg/kg in milk and 0.05 mg/kg in others animal matrices.

Plant matrices are extracted with methanol:water and purified on a micro-C_{18} column with a micro silica gel column step. Independent laboratory validation showed good performance of the methods.

Animal matrices can be extracted with acetone or acetonitrile and purified by liquid–liquid partition. Further clean-up is necessary before determination.

For enforcement, HPLC with ultraviolet detection was used, but some difficulties were found for crops like hops and oilseed crops.

Stability of residues in stored analytical samples

The stability of pyraclostrobin in plant matrices was shown to be 19 months. Untreated samples were fortified with 1.0 mg/kg pyraclostrobin and its metabolite 500M07. The residues in peanut meat, peanut oil, wheat grain, wheat straw, sugar beet tops, sugar beet roots, tomatoes and grape juice were stable during storage (range, 88–106% for the parent and 84–120% for the metabolite 500 M07).

Untreated samples of muscle, liver and milk from a cow were fortified with pyraclostrobin at 0.5 mg/kg (0.1 mg/kg in milk) or a mixture of 0.5 mg/kg (0.1 mg/kg for milk) pyraclostrobin and the same amount of a hydroxylated metabolite. Other potential metabolites also form these analytes on cleavage of the methylene ether bridge. After about 0, 30, 60, 90, 120 and 240 days, samples were analysed by BASF methods Nos 439 and 446. The results used to calculate stability were corrected for individual procedural recoveries. The average results of analysis for the parent compound with method 446 show degradation in muscle and milk. The model hydroxylated metabolite appeared to be less stable after 240 days' storage (68–86% in liver, milk and muscle). Nevertheless, this result does not affect the validity of the cow feeding study, as milk samples, in which degradation of the metabolite was fastest, were analysed within 91 days of sampling.

Definition of the residue

Three studies were performed on metabolism in three crop categories: grape for fruits, potato for root and tuber vegetables and wheat for cereals. Pyraclostrobin (grape fruits, potato green matter, wheat forage, wheat straw) and the desmethoxy metabolite 500M07 (grape fruits, potato green matter, wheat forage, wheat straw) accounted for most of the residue in most plant samples investigated. As the desmethoxy metabolite occurred in much smaller amounts than parent pyraclostrobin, the metabolite was not included in the definition of the relevant residue.

Studies of metabolism in goats and hens showed that the residues in products of animal origin derive from the parent compound and from its *N*-desmethoxylation product. Oxidation of the aromatic rings to several hydroxylated compounds and cleavage of the molecule led to further metabolites. As these transformations occur in matrices with small amounts of parent or little extractability, residue data obtained by this method represent reasonable worst-case estimates for risk assessment in all matrices. Furthermore, a method for parent only was developed to monitor residues of pyraclostrobin.

The Meeting agreed that the parent compound is suitable for enforcement in plant and animal commodities and is also the compound of interest for dietary risk assessment.

Definition of the residue for compliance with MRL and for estimation of dietary intake: pyraclostrobin.

The residue is fat-soluble.

Results of supervised trials on crops

The Meeting received data from supervised trials on citrus, nuts, apple, stone fruit, grape, strawberry, raspberry, blueberry, banana, mango, papaya, carrot, radish, sugar beet, garlic, onion, tomato, red pepper, summer squash, cucumber, lettuce, bean, lentil, pea dry, peanut, soya bean, oat, wheat, barley, maize and coffee. Most of the trials were carried out in the USA. All of the information from Europe and the USA was acceptable. The trials were conducted according to GLP.

Citrus fruit

GAP trials was reported from the Republic of Korea (citrus), South Africa (grapefruit and orange) and the USA (grapefruit, lemon, lime, orange, tangelo and tangerine).

Orange

Trials were conducted in Argentina (one) at 0.075 kg ai/ha, with four applications, including a study on the decline of residues, and in the USA (13) at GAP (0.274 kg ai/ha, four applications, 14-day PHI). The residue levels in orange were: 0.12, 0.13, 0.17 (two), 0.18, 0.19, 0.23, <u>0.24</u>, 0.25, 0.26, 0.34, 0.35, 0.37 and 0.51 mg/kg. The Meeting estimated a maximum residue level of 1 mg/kg, an STMR of 0.24 mg/kg and a highest reside level of 0.51 mg/kg for orange.

No residue was detected in pulp in five trials (< 0.02 mg/kg).

Grapefruit

Six trials were carried out in the USA at GAP (0.27 kg ai/ha, four applications,14-day PHI). The pyraclostrobin residue levels were: 0.07, 0.08, 0.11, <u>0.12</u>, 0.19 and 0.24 mg/kg. The Meeting estimated a maximum residue level of 0.5 mg/kg, an STMR of 0.12 mg/kg and a highest reside level of 0.24 mg/kg for grapefruit.

Pyraclostrobin

The residue in pulp was below the LOQ in one trial.

Lemon

Trials were conducted in Argentina (two trials at 0.075 kg ai/ha, four applications with two decay curves), Brazil (four trials with one decay curve) and the USA (five trials at GAP: 0.27 kg ai/ha, four applications, 14-day PHI). The trials in Brazil could not be evaluated (no GAP), and no results were available from the trial in Argentina at 14 days.

Pyraclostrobin residue levels in lemons in the five US trials were 0.15, 0.19, <u>0.20</u>, 0.28 and 0.32 mg/kg. The Meeting estimated a maximum residue level of 0.5 mg/kg, an STMR of 0.20 mg/kg and a highest reside level of 0.32 mg/kg for lemon.

The Meeting agreed to combine the above results in order to estimate a maximum residue level for citrus fruit. The combined results from the trials in Argentina and the USA, in ranked order, were: 0.07, 0.08, 0.11, 0.12 (two), 0.13, 0.15, 0.17 (two), 0.18, <u>0.19</u> (three), 0.20, 0.23, 0.24 (two), 0.25, 0.26, 0.28, 0.32, 0.34, 0.35, 0.37 and 0.51 mg/kg.

The Meeting estimated a maximum residue level of 1 mg/kg, with an STMR of 0.19 mg/kg and a highest reside level of 0.51 mg/kg for citrus.

Apple

GAP in Brazil was reported to be a rate of 0.1 kg ai/ha, with four applications and a 14-day PHI. Eight trials were conducted in Brazil at 0.15 kg ai/ha with two decay curves and six trials at 0.3 kg ai/ha. The Meeting agreed that no maximum residue level for apple could be established.

Stone fruit

GAP was reported from Canada (stone fruit) and the USA (peach, nectarine, apricot, plum, prune and cherry). The rate of application in the two countries is the same, 0.13 kg ai/ha with five applications. The waiting period is 10 days in Canada and 0 day in the USA.

Peach

Eighteen trials were carried out in the USA according to GAP. Two trials with decay curves were available. Pyraclostrobin residue levels in peaches were 0.07, 0.08 (two), 0.10 (two), 0.11, 0.13, 0.14, <u>0.15</u> (two), 0.16 (two), 0.20, 0.21, 0.23, 0.26, 0.28 and 0.31 mg/kg. The Meeting estimated a maximum residue level of 0.5 mg/kg, an STMR of 0.15 mg/kg and a highest reside level of 0.31 mg/kg for peaches.

Cherry

Twelve trials on sour cherries were conducted in the USA according to GAP. Pyraclostrobin residue levels were 0.25 (two), 0.27, 0.34, 0.38, 0.42, <u>0.43</u>, 0.48, 0.50 (two), 0.51 and 0.63 mg/kg. The Meeting estimated a maximum residue level of 1 mg/kg, an STMR of 0.43 mg/kg and a highest reside level of 0.63 mg/kg for cherry.

Plum: Twelve trials were carried out in the USA according to GAP, including two with decay curves. Pyraclostrobin residue levels were 0.02 (two), 0.03, 0.04 (two), 0.05, <u>0.06</u> (three), 0.12, 0.13 and 0.19 mg/kg. The Meeting estimated a maximum residue level of 0.3 mg/kg, an STMR of 0.06 mg/kg and a highest reside level of 0.19 mg/kg for plums.

Berries and small fruit

Grape

A total of 48 trials were performed in representative growing areas in Brazil, Europe and the USA. GAP was 0.1 kg ai/ha with two applications and a 7-day PHI in Brazil (four trials), 0.16 kg ai/ha with eight and three applications and a 35-day PHI in Europe (30 trials) and 0.168 kg ai/ha with three applications and a 14-day PHI in the USA (14 trials).

The pyraclostrobin residue levels in grapes in trials conducted according to GAP in Brazil were 0.36, 0.79, 1.1 and 1.4. The levels in trials conducted according to GAP in Europe (France, Germany, Italy and Spain) were 0.13, 0.16, 0.17 (two), 0.20, 0.23, 0.25, 0.26, 0.36, 0.40, 0.44 (two), 0.47, 0.56, 0.59 (two), 0.64, 0.67, 0.74, 0.75, 0.76, 0.78 (two), 1.2 and 1.3 mg/kg. The levels in trials conducted according to GAP in the USA were 0.09, 0.10 (two), 0.12 (two), 0.22, 0.24, 0.35, 0.43, 0.49 (two), 0.55, 0.67 and 1.2 mg/kg.

The Meeting agreed to combine the results in order to estimate a maximum residue level for grapes. The levels, in ranked order, were: 0.09, 0.10 (two), 0.12 (two), 0.13, 0.16, 0.17 (two), 0.20, 0.22, 0.23, 0.24, 0.25, 0.26, 0.35, 0.36 (two), 0.40, 0.43, 0.44 (two), 0.47, 0.49 (two), 0.55, 0.56, 0.59 (two), 0.64, 0.67 (two), 0.74, 0.75, 0.76, 0.78 (two), 0.79, 1.1, 1.2 (two), 1.3 and 1.4. The Meeting estimated a maximum residue level of 2 mg/kg, an STMR of 0.44 mg/kg and a highest reside level of 1.4 mg/kg for grapes.

Strawberry

GAP was reported for Canada and the USA. Eight trials were carried out in the USA at GAP (0.2 kg ai/ha, five applications, 0-day PHI), one with a decay curve.

The levels of pyraclostrobin residues in strawberries in trials conducted according to GAP in the USA were: 0.06, 0.10, 0.13, 0.15, 0.16, 0.19, 0.24 and 0.26 mg/kg. The Meeting estimated a maximum residue level of 0.5 mg/kg, an STMR of 0.16 mg/kg and a highest reside level of 0.26 mg/kg for strawberry on the basis outdoor uses of pyraclostrobin.

Raspberry

GAP is reported for the USA only, with four applications at 0.2 kg ai/ha and 0-day PHI. The results of three trials were provided.

The Meeting agreed that no maximum residue level for raspberries could be established.

Blueberry

GAP was available in Canada and the USA. The rate of application (0.2 kg ai/ha) and the number of applications (four) were the same, but the PHI in the USA is 0 days. Six trials were performed in the USA, but one included 50% ripe fruit.

The levels of pyraclostrobin residues in blueberries in trials conducted according to GAP in the USA were, in ranked order: 0.19, 0.30, 0.33, 0.35, 0.48 and 0.57 mg/kg. The Meeting estimated a maximum residue level of 1 mg/kg, an STMR of 0.34 mg/kg and a highest reside level of 0.57 mg/kg for blueberry.

Assorted tropical fruit minus inedible peel

Banana

Twelve trials were conducted in the main banana-growing regions of Central and South America. In all the trials, the formulation BAS 500 00F was applied eight times at a rate of 0.1 kg ai/ha. According to regional agricultural practice, the bananas were treated both bagged and unbagged and collected separately. Samples of whole bananas with peel were taken directly after the last application. No levels > 0.02 mg/kg were found in any sample.

The levels of pyraclostrobin residues in bananas in trials conducted according to GAP in Colombia (two), Costa Rica (three), Ecuador (three), France, Guatemala and Mexico were < 0.02.mg/kg. The Meeting estimated a maximum residue level of 0.02* mg/kg, an STMR of 0.02 mg/kg and a highest reside level of 0.02 mg/kg for bananas.

Mango

GAP in Brazil requires a maximum rate of 0.1 kg ai/ha, with two applications and a 7-day PHI. Four trials were conducted in Brazil at a rate of 0.225 kg ai/ha with three applications, and three trials were conducted at a rate of 0.45 kg ai/ha. No residues were detected at 0 or 7 days (< 0.05 mg/kg).

The Meeting agreed to propose a maximum residue level of 0.05* mg/kg and STMR and highest reside values of 0.05 mg/kg.

Papaya

GAP in Brazil requires a maximum rate of 0.1 kg ai/ha, four applications and a 7-day PHI. Four trials were reported from Brazil at a rate of 0.125 kg ai/ha and three trials at 0.25 kg ai/ha. No residues were detected at 7 days (< 0.05 mg/kg)

The Meeting estimated a maximum residue level of 0.05* mg/kg and STMR and highest reside values of 0.05 mg/kg for papaya.

Bulb vegetables

GAP was reported for Brazil (onions), Canada (bulb vegetable) and the USA (garlic and onions). The maximum rate of application is 0.1 kg ai/ha in Brazil and 0.17 kg ai/ha in Canada and the USA. The PHI is 3 days for onions and 7 days for garlic in Brazil and 7 days in Canada and the USA.

Seven trials were conducted on garlic in Brazil, but only four in accordance with GAP, and seven trials were conducted on onions, none of which conformed to GAP. Nine trials on onions were conducted in the USA.

Garlic

In the four trials in Brazil conforming to GAP, all the residue levels were < 0.05 mg/kg.

The Meeting estimated a maximum residue level of 0.05* mg/kg and STMR and highest reside values of 0.05 mg/kg for garlic.

Onion

Four trials in Brazil conducted at a rate of 0.15 kg ai/ha and a PHI of 7 days and three trials at a rate of 0.30 kg ai/ha and 7-day PHI could not be used to evaluate the residue levels.

Bulb onion

In six trials carried out in the USA according to GAP, the levels of pyraclostrobin residues in dry onions were: <u>0.02</u> (five) and 0.09 mg/kg. The Meeting estimated a maximum residue level of 0.20 mg/kg, an STMR of 0.02 mg/kg and a highest reside level of 0.09 mg/kg for onions, dry.

Spring onion

In three trials conducted according to GAP in the USA, the levels of pyraclostrobin residues in spring onions were 0.05, 0.42 and 0.53 mg/kg.

The Meeting agreed that no maximum residue level for spring onions could be established

Fruiting vegetables

Tomato

GAP was reported for Brazil, Canada (fruiting vegetables), Chile and the USA. The critical GAP was a maximum rate of 0.224 kg ai/ha, six applications and a 0-day PHI. Three outdoor trials were conducted in Brazil and 21 in the USA, which included two with decay curves.

The levels of pyraclostrobin residues in tomatoes in trials conducted according to GAP in Brazil were, in ranked order: 0.02, 0.03 and 0.12 mg/kg. The levels in the trials in the USA were, in ranked order: 0.06, 0.07 (two), 0.08, 0.10, 0.11 (four), 0.12 (three), 0.13 (two), 0.15, 0.16, 0.17 (three), 019 and 0.21 mg/kg.

The Meeting combined the data from Brazil and the USA, giving levels, in ranked order, of: 0.02, 0.03, 0.06, 0.07 (two), 0.08, 0.10, 0.11 (four), <u>0.12</u> (four), 0.13 (two), 0.15, 0.16, 0.17 (three), 019 and 0.21 mg/kg.

The Meeting estimated a maximum residue level of 0.3 mg/kg, an STMR of 0.12 mg/kg and a highest reside level of 0.21 mg/kg for tomato.

Chili pepper

GAP was reported for Brazil (maximum rate of 0.1 kg ai/ha and 3-day PHI), Canada (fruiting vegetable), the Republic of Korea and the USA (maximum rate of 0.224 kg ai/ha, six applications and 0-day GAP).

Four trials were reported from Brazil at 0.15 kg ai/ha and three at 0.3 kg ai/ha, which did not correspond to GAP.

The levels of pyraclostrobin residues in chili peppers in trials conforming to GAP in the USA were 0.14, 0.22 and 0.82 mg/kg.

The Meeting agreed that no maximum residue level for chili pepper could be established.

Fruiting vegetables

GAP was reported for Canada (fruiting vegetable, cucurbits) and the USA (squash summer), at a rate of application of 0.22 kg ai/ha, four applications and a 0-day PHI.

Summer squash

In six trials conducted according to GAP on summer squash in the USA, the levels of residues, in ranked order, were: 0.03, 0.07, <u>0.14</u>, <u>0.17</u> and 0.18 mg/kg.

The Meeting estimated a maximum residue level of 0.3 mg/kg, an STMR of 0.15 mg/kg and a highest reside level of 0.18 mg/kg for summer squash.

Cucumber

Four trials were carried out on cucumber in Brazil at 0.1 kg ai/ha and three trials at 0.2 kg ai/ha, in accordance with GAP in Brazil (0.1 kg ai/ha, 3-day PHI). The pyraclostrobin residue levels were < 0.02 mg/kg.

The Meeting agreed that the data were insufficient, and no maximum residue level could be recommended.

Lettuce

No GAP was provided. Five trials in the USA were conducted at 0.22 kg ai/ha. The Meeting agreed that no maximum residue level for lettuce could be recommended.

Legume vegetables and pulses

Beans

GAP for beans in Brazil is a maximum rate of 0.075 kg ai/ha, three applications and a 14-day PHI; that in Canada is a maximum rate of 0.1kg ai/ha, two applications and a 30-day PHI; and that in the USA is a maximum rate of 0.2 kg ai/ha, two applications and a 30-day PHI.

In 10 trials conducted at 0.224 kg ai/ha, the residue levels 21 days after application were < 0.02 (eight), 0.04 and 0.10 mg/kg.

The Meeting estimated a maximum residue level of 0.2 mg/kg, an STMR of 0.02 mg/kg and a highest reside level of 0.10 mg/kg for dry beans.

The results of nine trials on snap beans were presented, but no GAP was available. The Meeting agreed that no maximum residue level for snap beans could be established.

Pyraclostrobin

Lentils

GAP was reported for Canada at a maximum rate of 0.1 kg ai/ha, two applications and a 30-day PHI. GAP in the USA is a maximum rate of 0.22 kg ai/ha with two applications.

Three trials were carried out in Canada, and three were conducted in the USA at a rate of 0.224 kg ai/ha. Pyraclostrobin residue levels in lentils in trials conforming to GAP in the USA were, in ranked order: 0.03, 0.08, <u>0.11</u>, <u>0.15</u>, 0.17 and 0.39 mg/kg.

The Meeting estimated a maximum residue level of 0.5 mg/kg, an STMR of 0.13 mg/kg and a highest reside level of 0.39 mg/kg for lentils.

Peas, dry

GAP in Canada for dry field peas is a maximum rate of 0.1 kg ai/ha, two applications and a 30-day PHI. That in the USA is a maximum rate of application of 0.224 kg ai/ha and a 30-day PHI. Six trials were conducted in Canada and two in the USA at the rate of 0.224 kg ai/ha.

The levels of pyraclostrobin residues in peas (dry) in trials conducted according to GAP in the USA, in ranked order, were: < 0.02 (two), 0.04, <u>0.05</u>, <u>0.09</u>, 0.13, 0.14, and 0.20 mg/kg.

The Meeting estimated a maximum residue level for peas, dry, of 0.3 mg/kg, an STMR of 0.07 mg/kg and a highest reside level of 0.20 mg/kg.

Peanut

GAP was reported for Argentina, Brazil and the USA. The critical GAP was that of the USA, which requires a maximum rate of 0.274 kg ai/ha, five applications and a 14-day PHI.

In four trials conducted in Brazil and 12 in the USA, no residues were detected in nutmeat (< 0.02 mg/kg).

The pyraclostrobin residue levels in peanut in trials conforming to GAP in Brazil and the USA were < 0.02 mg/kg or < 0.025 mg/kg (one). The Meeting estimated a maximum residue level of 0.05* mg/kg, an STMR of 0.02 mg/kg and a highest reside level of 0.025 mg/kg for peanut.

Soya bean

GAP was reported for Argentina, Brazil and Paraguay. GAP in Brazil is a maximum rate of 0.08 kg ai/ha with two applications and a 14-day PHI. One trial was conducted in Argentina and eight in Brazil (only four at GAP). In Argentina, the residue level was 0.03 mg/kg. In Brazil, results were presented only for grain. No residues were detected (< 0.02 mg/kg), even after application at 0.1 kg ai/ha.

The Meeting agreed that no maximum residue level for soya bean could be established.

Root and tuber vegetables

Carrot

GAP was reported for Brazil, Canada and the USA. The rate and number of applications are the same in Canada and the USA (0.22 kg ai/ha, three applications), but the PHI is 3 days in Canada and 0 days in the USA. In Brazil, the rate of application is lower (0.1 kg ai/ha) and the PHI is 7 days. One trial was conducted in Brazil and eight in the USA, only six of which were at GAP.

The levels of pyraclostrobin residues in carrots in trials conducted according to GAP in Brazil and the USA, in ranked order, were: 0.03 (two), 0.04, <u>0.12</u> (two), 0.15 and 0.24.mg/kg. The Meeting estimated a maximum residue level of 0.50 mg/kg, an STMR of 0.12 mg/kg and a highest reside level of 0.24 mg/kg for carrots.

Radish

GAP in the USA is a maximum application rate of 0.224 kg ai/ha, three applications and a 0-day PHI). The same GAP is applicable to horseradish. Five trials were carried out in the USA, and the values for radish tops and root were reported.

The levels of pyraclostrobin residues in radishes were 0.05, 0.07, <u>0.08</u>, 0.23 and 0.30 mg/kg. The Meeting estimated a maximum residue level of 0.50 mg/kg, an STMR of 0.08 mg/kg and a highest reside level of 0.30 mg/kg for radish.

The residue levels in radish tops were 7.5, 9.6, 9.9, 12 and 15 mg/kg The Meeting estimated a maximum residue level of 20 mg/kg, an STMR of 9.9 mg/kg and a highest reside level of 15 mg/kg for radish tops.

Sugar beet

GAP in Canada and the USA is the same, with a maximum rate of 0.22 kg ai/ha, four applications and a 7-day PHI. In 12 trials conducted in USA according to GAP, the pyraclostrobin residue levels in sugar beet were: < 0.02 (two), 0.02, 0.03 (two), <u>0.04</u> (two), 0.06, 0.08 (two), 0.11 and 0.13 mg/kg. The Meeting estimated a maximum residue level of 0.2 mg/kg, an STMR of 0.04 mg/kg and a highest reside level of 0.13 mg/kg for sugar beet.

Potato

GAP was reported for Brazil, Canada and the USA. GAP in Brazil is a maximum rate of 0.1 kg ai/ha with five applications and a 3-day PHI, and GAP in the USA is a maximum rate of application of 0.219 kg ai/ha with six applications and a 3-day PHI.

In trials conducted according to GAP in Brazil, Canada and the USA, no residues were detected (< 0.02 mg/kg).

The Meeting estimated a maximum residue level for potatoes of 0.02* mg/kg and STMR and highest reside values of 0.02 mg/kg.

Cereal grains

Oats

GAP was reported for Brazil, Denmark, Estonia, France, Ireland, Latvia, Lithuania and the United Kingdom. The maximum rate of application was around 0.2 kg ai/ha with one to two applications and a PHI of 30 days.

Eight trials were conducted according to GAP in Brazil at 0.166, 0.2, 0.333 or 0.4 kg ai/ha. The levels of pyraclostrobin residues in oat grain were, in ranked order: 0.04, 0.05, 0.06, <u>0.14</u>, <u>0.20</u>, 0.23, 0.25 and 0.42 mg/kg. The Meeting estimated a maximum residue level of 0.5 mg/kg, an STMR of 0.17 mg/kg and a highest reside level of 0.42 mg/kg for oat grain.

Wheat

GAP was reported from Argentina, Belgium, Brazil, Canada, Denmark, Estonia, France, Germany, Ireland, Latvia, Lithuania, The Netherlands, Switzerland and the United Kingdom. The rate of application was 0.20–0.25 kg ai/ha with two applications and a 35-day PHI. GAP in the USA is a maximum of 0.22 kg ai/ha and a 40-day PHI.

In Brazil, eight trials were conducted according to GAP, four at 0.167 kg ai/ha and four at 0.2 kg ai/ha; 13 trials exceeded GAP: four at 0.3 kg ai/ha, three at 0.335 kg ai/ha, three at 0.4 kg ai/ha and three at 0.6 kg ai/ha. Thirty trials were conducted in Europe: one in Denmark, eight in France, five in Germany, two in The Netherlands, nine in Spain, one in Sweden and four in the United Kingdom. In North America, 11 trials were conducted in Canada and 23 in the USA according to US GAP.

The levels of pyraclostrobin residues in wheat grain in the trials conducted according to GAP in Brazil were: 0.02, 0.03 and 0.04 (two) mg/kg.

In all the European trials, samples of whole plant without roots were taken directly after the last application. On the third sampling day, at the proposed PHI of 35 days, various samples were taken, depending on ripening, with ears taken in 30 trials and grain in 27 trials. The pyraclostrobin residue levels in wheat grain in trials that conformed to GAP were: < 0.02 (22), 0.03, 0.04 (two), 0.05 and 0.09 mg/kg.

The residue levels in wheat grain in 11 trials in Canada and 23 trials in the USA that conformed to GAP were < 0.02 mg/kg.

The levels of pyraclostrobin residues in wheat grain in GAP trials in Brazil, Europe, Canada and the USA were of the same order of magnitude, and the Meeting decided that the data could be pooled. The residue levels, in ranked order, were: < <u>0.02</u> (56), 0.02, 0.03 (two), 0.04 (four), 0.05 and 0.09 mg/kg. The Meeting estimated a maximum residue level for wheat grain of 0.2 mg/kg, an STMR of 0.02 mg/kg and a highest reside level of 0.09 mg/kg.

Barley

GAP was reported for Belgium, Brazil, Canada, Denmark, Estonia, France, Germany, Ireland, Latvia, Luxembourg, Macedonia, Switzerland, the United Kingdom and the USA. GAP in Europe is a rate of application of 0.20–0.25 kg ai/ha, two applications and a 30–35-day PHI. Two trials were conducted according to GAP in Belgium, seven in France, four in Germany, three in Spain, five in Sweden and four in the United Kingdom, for a total of 25 trials.

The pyraclostrobin residue levels in barley grain in trials corresponding to GAP in Europe were, in ranked order: < 0.02 (six), 0.02 (two), 0.03 (six), 0.04 (four), 0.05 (two), 0.06, 0.07, 0.08, 0.09, 0.10, 0.29 and 0.32 mg/kg.

A total of 14 trials were carried out in Brazil, but only eight conformed to GAP. The residue levels in barley grain in the latter trials were, in ranked order: 0.04, 0.05, 0.06, 0.07, 0.08 (three) and 0.09 mg/kg.

In the 26 trials conducted in the USA on barley grain according to GAP (0.22 kg ai/ha), the pyraclostrobin residue levels, in ranked order, were: < 0.02 (19), 0.03 (three), 0.05 (two), 0.07 and 0.14 mg/kg.

As GAP in Brazil, Europe and the USA is similar and the residue levels were in the same range, the results were combined. The levels, in ranked order, were: < 0.02 (25), 0.02 (two), <u>0.03</u> (nine), 0.04 (five), 0.05 (five), 0.06 (two), 0.07 (three), 0.08 (four), 0.09 (two), 0.10, 0.14, 0.29 and 0.32 mg/kg.

The Meeting estimated a maximum residue level for barley grain of 0.5 mg/kg, an STMR of 0.03 mg/kg and a highest reside level of 0.32 mg/kg.

Maize

GAP in Brazil allows two applications of 0.15 kg ai/ha or 0.1 kg ai/ha with a 45-day PHI on maize.

Four trials were conducted at 0.2 kg ai/ha and four at 0.133 kg ai/ha. The residue levels in trials conforming to GAP were < 0.02 mg/kg.

The Meeting estimated a maximum residue level for maize of 0.02* mg/kg and STMR and highest reside values of 0.02 mg/kg.

Rye

GAP in the USA allows two applications of 0.22 kg ai/ha with a 40-day on rye. In five trials conducted at 0.22 kg ai/ha but with a PHI of about 60 days, the pyraclostrobin residue levels in rye were < 0.02 mg/kg at 60 days.

The Meeting agreed that no maximum residue level for rye could be estimated.

Tree nuts

GAP was reported from the USA for beechnut, Brazil nut, butter nut, cashew, macadamia nut, pecan, walnut and pistachio. The rate of application was 0.13 kg ai/ha, with four applications and a waiting period of 14 days (pecan and pistachio). For almond, the new GAP was 0.13 kg ai/ha, with four applications and a waiting period of > 100 days.

Almond: Ten trials were carried out in the USA with a 120-day PHI. The results for nutmeat were < 0.02 mg/kg. The Meeting estimated a maximum residue level of 0.02*mg/kg, an STMR of 0.02 mg/kg and a highest reside level of 0.02 mg/kg.

Pecan: Ten trials were conducted in the USA according to GAP. Pyraclostrobin residue levels in pecan were < 0.02 mg/kg. The Meeting estimated a maximum residue level of 0.02*mg/kg, an STMR of 0.02 mg/kg and a highest reside level of 0.02 mg/kg for pecan.

Pistachio: Six trials were carried out in the USA according to GAP. The pyraclostrobin residue levels were: 0.02 (two), 0.16, 0.27, 0.44 and 0.45 mg/kg. The Meeting estimated a maximum residue level of 1 mg/kg, an STMR of 0.22 mg/kg and a highest reside level of 0.45 mg/kg for pistachio.

Coffee

GAP in Brazil is a maximum rate of 0.2 kg ai/ha with two applications and a 45-day PHI. Four trials were conducted at 0.175 kg ai/ha and three at 0.35 kg ai/ha. The pyraclostrobin residue levels were < 0.02 (two), 0.03 and 0.15 mg/kg.

The Meeting agreed that no maximum residue level for coffee could be estimated

Animal feedstuffs

Fodder beet leaves and tops

GAP is the same for Canada and the USA, with a maximum rate of 0.22 kg ai/ha, four applications and a 7-day PHI. In 12 trials conducted in the USA according to GAP, the levels of pyraclostrobin residues, in ranked order, were: 0.28, 1.3, 1.4, 1.5 (two), 1.6, 1.7, 2.0, 2.6, 2.8, 3.9 and 5.3 mg/kg.

Eight trials were conducted in Europe and reported, but the registration is pending.

The Meeting estimated a maximum residue level of 10 mg/kg, an STMR of 1.64 mg/kg and a highest reside level of 5.3 mg/kg for sugar beet tops. On a dry basis, the maximum residue level was 50 mg/kg, the STMR was 7.1 mg/kg and the highest reside level was 23 mg/kg.

Peanut hay

GAP was reported for Argentina, Brazil and the USA. The critical GAP is 0.274 kg ai/ha with five applications and a 14-day PHI. In 12 trials conducted according to GAP in the USA, the pyraclostrobin residue levels in peanut hay, in ranked order, were: 1.5, 3.3, 4.0, 4.8, 4.9, 9.0, 15, 18, 19 (two) and 24 mg/kg.

The Meeting estimated a maximum residue level of 50 mg/kg, an STMR of 9.0 mg/kg and a highest reside level of 24 mg/kg for peanut hay.

On the basis of the dry matter, which is listed as 85% in the *FAO Manual*, the STMR is equivalent to 11 mg/kg and the highest reside level to 29 mg/kg. These values were used to calculate the animal burden.

Pea hays and vines

GAP was reported for Canada (dried field peas, 0.1 kg ai/ha with two applications and a 30-day PHI) and the USA (0.22 kg ai/ha and a 30-day PHI). Six trials were conducted in Canada and two in the USA at an application rate of 0.224 kg ai/ha.

The levels of pyraclostrobin residues in pea vines in trials conforming to GAP in the USA were: 3.3, 3.8, 4.2, 5.0, 5.1, 5.5 (two) and 7.0 mg/kg.

The Meeting estimated a maximum residue level for pea vines of 10 mg/kg, an STMR of 5.1 mg/kg and a highest reside level of 7.0 mg/kg.

On the basis of the dry matter, which is listed as 25% in the *FAO Manual*, the maximum residue level was estimated at 40 mg/kg, the STMR at 20 mg/kg and the highest reside level at 28 mg/kg in pea vine. These values were used to calculate the animal burden.

The levels of pyraclostrobin residues in pea hay in trials conforming to GAP in the USA were: 4.9, 5.3, 6.4, 7.2, 7.5, 9.2, 12 and 18 mg/kg.

The Meeting estimated a maximum residue level for pea hay of 20 mg/kg, an STMR of 6.8 mg/kg and a highest reside level of 18 mg/kg.

On the basis of the dry matter, which is listed as 88% in the *FAO Manual*, the maximum residue level was estimated at 30 mg/kg, the STMR at 7.8 mg/kg and the highest reside level at 20 mg/kg in pea hay.

Barley straw, hay (fodder) and haulms

GAP was reported for Belgium, Brazil, Canada, Denmark, Estonia, France, Germany, Ireland, Latvia, Luxembourg, Macedonia, Switzerland, the United Kingdom and the USA. The maximum rate of application is 0.2–0.25 kg ai/ha with two applications and a 30–35-day PHI. In Europe, 25 trials were conducted, with two in Belgium, seven in France, four in Germany, three in Spain, five in Sweden and four in the United Kingdom.

The levels of pyraclostrobin residues in barley straw in trials that complied with GAP in Europe were: 0.48, 0.66, 0.78 (two), 0.72, 0.84, 0.99, 1.0, 1.7 (two), 1.8, 2.0, 2.6 (two), 2.8 (three), 3.9, 4.4 (two), 4.8, 4.9, 5.7, 5.8 and 6.9 mg/kg.

The levels of pyraclostrobin residues in barley straw in trials that complied with GAP in Canada and the USA were: 0.09, 0.12 (two), 0.26, 0.30 (two), 0.31, 0.32 (two), 0.39, 0.45, 0.52, 0.57, 0.82, 1.1, 1.3, 1.4, 1.5 (two), 1.9, 2.4, 2.8 and 4.0 mg/kg.

The Meeting agreed to combine the above results for estimating a maximum residue level for barley straw. The residue levels, in ranked order, were: 0.09, 0.12 (two), 0.26, 0.30 (two), 0.31, 0.32 (two), 0.39, 0.45, 0.48, 0.52, 0.57, 0.66, 0.72, 0.78 (two), 0.82, 0.84, 0.99, 1.0, 1.1, <u>1.3</u>, <u>1.4</u>, 1.5 (two), 1.7 (two), 1.8, 1.9, 2.0, 2.4, 2.6 (two), 2.8 (four), 3.9, 4.0, 4.4 (two), 4.8, 4.9, 5.7, 5.8 and 6.9 mg/kg.

The levels of pyraclostrobin residues in barley haulms in trials that complied with GAP in Europe were: 0.41, 0.53, 0.54, 0.58, 0.72, 0.73, 0.74, 0.87, 0.88, 0.98, 1.2, 1.3, 1.4 (two), 1.5, 1.6, 1.7, 1.8, 2.5, 3.2, 3.4, 4.1, 4.3, 6.6 and 7.6 mg/kg.

The levels of pyraclostrobin residues in barley hay in trials that complied with GAP in Canada and the USA were: 0.93, 0.96, 1.0 (two), 1.1, 1.2, 1.3, 1.5, 1.6 (three), <u>1.9</u>, <u>2.1</u>, 2.2 (two), 2.5, 2.8, 3.2 (two), 3.6, 3.7, 12 (two), 17 and 19 mg/kg.

Wheat straw, hay (fodder) and haulms

GAP was reported for Argentina, Belgium, Brazil, Canada, Denmark, Estonia, France, Germany, Ireland, Latvia, Lithuania, The Netherlands, Switzerland, the United Kingdom and the USA. The rate of application is 0.20–0.25 kg ai/ha with two applications and a 35-day PHI.

In 27 trials in Europe, samples of whole plant without roots were taken directly after the last application, and samples of haulms and straw were taken about 3 weeks after the last application. On the third sampling day, which was at the proposed 35-day PHI, various samples were taken, depending on ripening; in 27 trials, haulms and straw were taken.

The levels of pyraclostrobin residues in wheat straw in trials that conformed to GAP in Europe were: 0.67, 0.75, 0.87, 1.2, 1.4, 1.5, 1.6, 1.7, 1.7 (two), 1.8, 1.9 (two), 2.0 (two), 2.1, 2.2, (five), 2.3, 2.5, 3.2, 5.0, 5.5 and 5.7 mg/kg.

The levels of pyraclostrobin residues in wheat straw in trials that conformed to GAP in Canada and the USA were: 0.03, 0.06, 0.07, 0.09, 0.10 (two), 0.11, 0.12, 0.13 (two), 0.15, 0.20, 0.21, 0.23, 0.24, 0.32, 0.34, 0.37, 0.52, 0.56, 0.74, 0.85, 0.90, 0.95, 1.1, 1.6, 1.7, 2.2, 3.5, 3.8 and 4.1 mg/kg.

The Meeting agreed to combine the above results for estimating a maximum residue level for wheat straw. The residue levels, in ranked order, were: 0.03, 0.06, 0.07, 0.09, 0.10 (two), 0.11, 0.12, 0.13 (two), 0.15, 0.20, 0.21, 0.23, 0.24, 0.32, 0.34, 0.37, 0.52, 0.56, 0.67, 0.74, 0.75, 0.85, 0.87, 0.90, 0.95, 1.1, 1.2, 1.4, 1.5, 1.6 (two), 1.7 (four), 1.9 (three), 2.0 (two), 2.1, 2.2 (six), 2.3, 2.5, 3.1, 3.5, 3.8, 4.1, 5.0, 5.5 and 5.7 mg/kg.

The levels of pyraclostrobin residues in wheat haulms in trials that conformed to GAP in Europe were: 0.50, 0.52, 0.56, 0.62, 0.74, 0.75, 0.79, 0.81, 0.84, 0.85, 0.89, 0.92, 0.94, 0.96, 0.98, 0.99, 1.0 (two), 1.1, 1.2, 1.3 (two), 1.4 (two), 1.5, 1.6, 1.9, 2.7 (two) and 3.2 mg/kg.

The levels of pyraclostrobin residues in wheat hay in trials that conformed to GAP in Canada and the USA were: 0.21, 0.24, 0.27, 0.43, 0.46, 0.49, 0.54, 0.72, 0.75, 0.83, 0.89, 0.91, 0.93, 0.95, 1.0 (two), 1.1, 1.2, 1.4 (two), 1.5 (two), 1.6, 1.8 (two), 1.9, 2.0, 2.2 (two), 2.3, 3.0, 3.1 and 4.6 mg/kg.

Rye straw

GAP was reported from the USA at a rate of application of 0.2–0.25 kg ai/ha with two applications and a 40-day PHI. Five trials were conducted but with a longer PHI. The levels of pyraclostrobin residues were 0.11, 0.14, 0.17, 0.27 and 0.30 mg/kg.

The Meeting agreed that no maximum residue level for rye straw could be estimated.

The Meeting agreed to combine the results for barley and wheat straw (106 trials) in estimating a maximum residue level for cereal straw. The residue levels, in ranked order, were: 0.03, 0.06, 0.07, 0.09 (two), 0.10 (two), 0.11, 0.12 (three), 0.13 (two), 0.15, 0.20, 0.21, 0.23, 0.24, 0.26, 0.30 (two), 0.31, 0.32 (three), 0.34, 0.37, 0.39, 0.45, 0.48, 0.52 (two), 0.56, 0.57, 0.66, 0.67, 0.72, 0.74, 0.75, 0.78 (two), 0.82, 0.84, 0.85, 0.87, 0.90, 0.95, 0.99, 1.03, 1.1 (two), 1.2, 1.3, 1.4 (two), 1.5 (three), 1.6 (two), 1.7 (six), 1.8 (two), 1.9 (three), 2.0 (two), 2.1 (two), 2.2 (six), 2.3, 2.4, 2.5, 2.6 (two), 2.8 (four), 3.1, 3.5, 3.8, 3.9, 4.0, 4.10, 4.4 (two), 4.8, 4.9, 5.0, 5.5, 5.7 (two), 5.8 and 6.9 mg/kg.

The Meeting agreed to combine the results for barley and wheat fodder (59 trials) in estimating a maximum residue level for cereal fodder. The residue levels, in ranked order, were: 0.21, 0.24, 0.27, 0.43, 0.46, 0.49, 0.54, 0.72, 0.75, 0.83, 0.89, 0.91, 0.93 (two), 0.95, 0.96, 1.0 (four), 1.1 (two), 1.2 (two), 1.3, 1.4 (two), 1.5 (three), 1.6 (four), 1.8 (two), 1.9 (two), 2.0, 2.1, 2.2 (four), 2.3, 2.5, 2.8, 3.0, 3.1, 3.2 (two), 3.6, 3.7, 4.6, 11, 12, 17 and 19 mg/kg.

Allowing for the standard 88% dry matter for cereal straw and fodder (*FAO Manual* p. 49), the Meeting estimated a maximum residue level of 30 mg/kg, an STMR of 1.69 mg/kg and a highest reside level of 21.7 mg/kg. The highest reside level was taken into account in calculating the animal dietary burden.

Almond hulls

GAP was reported from the USA with 10 trials conducted according to GAP. The levels of pyraclostrobin residues in almond hulls were: < 0.02 (two), 0.11, 0.16, 0.19, 0.21, 0.47, 0.55, 0.87 and 1.3 mg/kg.

The Meeting estimated a maximum residue level of 2 mg/kg, an STMR of 0.20 mg/kg and a highest reside level of 1.34 mg/kg for almond hulls. The highest reside level was taken into account in calculating the animal dietary burden.

Fate of residues during processing

Studies were conducted on grapes, barley and wheat, and the respective intermediate end- and waste products were analysed. For grapes, the data covered whole grapes, cold must, heated must, wet pomace, wine from cold must, wine from heated must, juice and raisins. For barley, the data covered pearling dust, pot barley, malt, malt germs, spent grain, trub (flocks), beer yeast and beer. For wheat, flour, bran, middlings, shorts and germ were analysed.

The processing factors for total residues in the transformation from *grape* to must, wine and juice were < 1 (0.08–0.013), indicating that the residues did not concentrate. Concentration factors of 2.4–5.6 were calculated for residues in processing from whole grape to pomace; the concentration factor for total residues in processing from whole grapes to raisins was 3.1, which may be due to loss of water during processing.

In the processed fractions obtained for pot *barley* and beer production, such as pearling dust, malt, malt germs, spent grain, trub (flocks) and beer yeast, none of which are meant for consumption, the residues of pyraclostrobin showed some concentration, with factors ranging from 1.29 to 7.86. In the final products to be consumed, such as pot barley and beer, however, no concentration of pyraclostrobin residues was observed, as expressed by processing factors of < 1 (< 0.6). The processing factor from barley to beer cannot be calculated owing to the low contamination of barley, but the transfer factor can be assumed to be low.

The processing factors for total residues in the transformation from *wheat* grain to all processed fractions were < 1 (< 0.6), indicating that the residues did not concentrate. The transfer factor for wheat germ was 0.8. The processing factors and STMR-P values for all the commodities investigated were:

Commodity	Processing factor	STMR-P (mg/kg)
Grape juice	0.013	0.005
Wine	< 0.1	< 0.044
Must	0.15	0.07
Wet pomace	2.4–5.6	2.46
Raisin	3.1	1.36
Malt	1	0.03
Beer	< 0.6	< 0.025
Wheat flour	< 0.6	< 0.01
Wheat bran	< 0.6	< 0.01
Wheat germ	0.8	0.016

Residues in animals commodities

Dietary burden of farm animals

The Meeting estimated the dietary burden of pyraclostrobin residues in farm animals on the basis of the diets listed in Appendix IX of the *FAO Manual*. The percentage of dry matter is taken as 100% when MRLs and STMR values are expressed as dry weight.

Estimated maximum dietary burden of farm animals

Commodity	Group	Residue (mg/kg highest residue)	Residue (mg/kg on dry matter basis)[a]	Dietary content (%) Beef cattle	Dairy cattle	Poultry	Residue contribution (mg/kg) Beef cattle	Dairy cattle	Poultry
Almond hulls	AM	1.34	1.34						
Barley grain	GC	0.36	0.36	50	40	75	0.18		0.27
Cereal fodder	AS	21.7	21.7	25	60/50		5.4	10.8	
Sugar beet	AB	0.15	0.15						
Peanut hay	AL	28.8	28.8	25	50		7.28	14.4	
Pea hay	AL	20.5	20.5						
Pea vines	AL	28	28	25	50				
Fodder beet leaves	AV	23	23						
Total				100	100	75	12.9	25.2	0.27

[a] 100% dry matter for all commodities

Estimated median dietary burden of farm animals

Commodity	Group	Residue (mg/kg on STMR basis)	Residue (mg/kg on dry matter basis)[a]	Dietary content (%) Beef cattle	Dairy cattle	Poultry	Residue contribution (mg/kg) Beef cattle	Dairy cattle	Poultry
Almonds hulls	AM	0.2	0.22	10	10				
Barley grain	GC	0.034	0.034	50	40	75	0.017	0.014	0.025
Cereal fodder	AS	1.69	1.69	10	60				
Fodder beet tops	AV	7.1	7.1	20	10		1.42	0.71	
Sugar beet	AB	0.045	0.045	20	20				
Peanut hay	AL	10.6	10.6	25	50				
Pea hay	AL	7.75	7.75	25	50				
Pea vines	AL	20.2	20.2	25	50		5.06	10.1	
Total				95	100	75	6.5	10.8	0.025

[a] 100% dry matter for all commodities

The dietary burdens of pyraclostrobin for estimates of STMR and highest reside level values in animal commodities (residue levels in animal feeds expressed as dry weight) are, respectively, 6.5 mg/kg and 12.9 mg/kg for beef cattle, 10.8 mg/kg and 25.2 mg/kg for dairy cattle and 0.025 mg/kg and 0.27 mg/kg for poultry.

Feeding studies

In one feeding study, dairy cows were given feed containing pyraclostrobin at 0, 8.8, 27.2 or 89.6 mg/kg for 28 days. No residues of pyraclostrobin were detected in milk, meat, fat, kidney or tissues from the group given the concentration relevant to normal agricultural conditions (27.2 mg/kg) or at the other two concentrations. Low levels of pyraclostrobin metabolites might occur in liver.

Pyraclostrobin

The Meeting decided not consider these studies, as pyraclostrobin is fat-soluble and no residues were detected. The study of metabolism in goats, summarized below, was used to estimate residues in animal products.

Tissue	Residue (mg/kg)				Value taken into account	Dietary burden estimate (ppm)	
	At 12 ppm		At 50 ppm			25.2	10.8
	Chlorophenyl label	Tolyl label	Chlorophenyl label	Tolyl label			
Milk[a]	0.012	0.01	0.067	0.027	0.047	0.0236	0.01
Muscle	0.01	0.01	0.089	0.048	0.089	0.044	0.009
Fat	0.069	0.061	0.82	0.32	0.82	0.41	0.063
Liver	0.008[b]	0.006[b]	0.021	0.07	0.07	0.035	0.007
Kidney	0.01[b]	0.007[b]	0.074[b]	0.073[b]	0.074	0.037	0.009

[a] Mean values; comprises pyraclostrobin and metabolite 500M07
[b] Comprises pyraclostrobin and metabolite 500M07 at the lower dose for liver and at both doses for kidney

Maximum residue levels

On the basis of the estimated residue levels at the calculated dietary burdens, the Meeting recommended a maximum residue level of 0.03 mg/kg and an STMR of 0.01 mg/kg for milk.

The Meeting recommended a maximum residue level of 0.5 mg/kg for meat (fat) of mammals other than marine mammals and for edible offal, an STMR of 0.008 mg/kg and a highest reside level of 0.037 mg/kg for edible offal, an STMR of 0.009 mg/kg and a highest reside level of 0.044 mg/kg for muscle, and an STMR of 0.063 mg/kg and a highest reside level of 0.41 mg/kg for fat.

No feeding study was performed in chickens. The Meeting noted that in the study of metabolism in laying hens, pyraclostrobin was not detected in tissues (< 0.002 mg/kg) or eggs (< 0.002 mg/kg) at a feeding level of 12 mg/kg, which was 30 times higher than the calculated dietary burden (0.27 ppm).

The Meeting agreed that it is unlikely that pyraclostrobin residues will be detected in the products of poultry fed commodities treated with this compound. The Meeting estimated a maximum residue level of 0.05* mg/kg and STMR and highest reside level values of 0 for pyraclostrobin in eggs, meat (fat) and edible offal of poultry.

DIETARY RISK ASSESSMENT

Long-term intake

The IEDI of pyraclostrobin calculated on the basis of the recommendations made at th present Meeting for the five GEMS/Food regional diets represented 0–3% of the ADI (0–0.03 mg/kg per day in a 2-year study in rats).

The Meeting concluded that the long-term intake of residues of pyraclostrobin resulting from uses that have been considered by the JMPR is unlikely to present a public health concern.

Short-term intake

The IESTI of pyraclostrobin calculated on the basis of the recommendations made at the present Meeting represented 0–90% of the ARfD (0–0.05 mg/kg per day in a study on developmental toxicity in rabbits) for children and 0–40% of the ARfD for the general population.

The Meeting concluded that the short-term intake of residues of pyraclostrobin resulting from uses that have been considered by the JMPR is unlikely to present a public health concern.

4.27 SPICES

RESIDUE AND ANALYTICAL ASPECTS

The issue of setting maximum residue levels for spices on the basis of the results of monitoring was discussed by the CCPR on several occasions. The 2002 JMPR prepared guidelines for the format of submission of monitoring data for evaluation (ALINORM 02/24). At its Thirty-sixth Session, the CCPR proposed to divide the commodity group 028 into subgroups on the basis of the part of the plant from which they are obtained (seeds, fruits or berries, roots or rhizomes, bark, buds, arils and flower stigmas) and proposed that maximum residue levels for pesticides that had been evaluated within the Codex system should be set for the subgroups instead of for each pesticide–spice combination (ALINORM 04/24, para 236). Furthermore, the maximum residue levels for dry chili peppers should be set on the basis of the existing maximum residue levels for peppers, taking into account the processing and dehydration factors as appropriate (ALINORM 04/24, para 242).

The Government of India, through the Indian Spice Board, the American Spice Trade Association, the European Spice Association and the Government of Egypt provided data resulting from pesticide monitoring programmes on spices during the period 1996–2003. The Delegation of South Africa coordinated the compilation of data for submission to the JMPR.

The JMPR reviewed the residue monitoring data provided and estimated maximum residue levels for the spice subgroups. The residue data on cumin from Egypt indicated that several pesticides might be applied post-harvest. Similarly, the residue levels of malathion and profenofos on anise seed were much higher than those of other pesticides, indicating possible post-harvest use of malathion and profenofos. The Meeting concluded that post-harvest use of pesticides on spices should be regulated by national governments, and monitoring data should not be used for estimating maximum residue levels reflecting post-harvest use. Consequently, the results of residue trials that suggested post-harvest use were not included in the 2004 evaluation.

The Meeting emphasized that the fact that it has estimated maximum, median and high residue values does not mean that it has approved use of the compounds on spices and chili peppers.

The Meeting noted that poppy seed (SO 698), mustard seed (SO 90) and sesame seed (SO700) are used as major food ingredients in several countries. It therefore considered it more appropriate to keep them in the oil seeds group (A023) and to remove them from group A028. The recommended maximum residue levels for the seed subgroup of spices does not include these seeds.

The large number of residue values for some pesticide–commodity combination allowed proper statistical treatment of the monitoring data. The principles applied are discussed in detail under section 2.6 of the General Considerations.

In addition, the following major principles were followed in evaluating the residue data for estimation of maximum, median and high residue values:

- All the residue data were considered; no data point was excluded as an outlier.
- Residue values reported as '0' were replaced by 'below the limit of quantification' (LOQ).
- Maximum, median and high residue levels were recommended when the database allowed estimation of > 95th percentile of the residue population at a 95% confidence (probability) level. That required a minimum of 58–59 samples. Use of the 95% confidence interval is recommended, as it is used for estimating the variability factor in short-term exposure assessment and is generally applied in biometry.
- As very few data were available for commodities derived from bark, buds, arils and flower stigmas, these spices are summarized together in the tables. Maximum residue levels could be estimated in only a few cases
- When residues were detectable, estimation of the maximum residue level for the subgroup also took into account the number of residue data and residue levels for the particular pesticide–commodity combination.
- A maximum residue level was proposed at the limit of determination when no residues were detectable, even if the minimum sample requirements (59) was not met for satisfying the specified probability (> 95th percentile) and confidence (95%) interval for any subgroup.
- When residues were undetectable and different LOQ values were reported for a particular pesticide from the different data sources, the maximum residue level was proposed at the highest LOQ provided for the pesticide. As there was no evidence for nil residues, the median was calculated from the values corresponding to the reported LOQs. The high residue level was considered to be equal to the highest reported LOQ. Addition of * after a residue value does not necessarily indicate that residues will not occur in detectable amounts if a more sensitive method is used.
- The estimated median and high residue values can be used in the same way as the STMR and highest residue values obtained from supervised trials for estimating long-term and short-term intake of residues.
- A substantial proportion of random samples did not contain detectable residues, indicating that the sampled lots had probably not been treated with or exposed to the given pesticide. Therefore, the median residue values were derived from the detected residue levels. Long-term intake was calculated from the residue data for that commodity that made the largest contribution to intake and the percentage of the treated proportion of that crop.
- When no residues were detected, the median level of residues was taken as equivalent to the median LOQ, as a conservative estimate.
- The short-term intake calculations were performed with the highest residue value for the pesticide measured in any spice sample, after the necessary adjustments for consumption figures described in the evaluation.

Sampling and analytical methods

The samples were taken from randomly selected lots either before export or upon arrival in the importing country. No information was provided on the sampling procedures, the size or the mass of samples.

The samples were analysed by multi-residue procedures, resulting in average recoveries of 70–120%, with the exception of ethion, with a recovery of 150%. The LOQs ranged from 0.01 mg/kg to 0.5 mg/kg. The difference in the LOQ values reported by different laboratories was sometimes 10-fold. Data on the reproducibility of methods and other performance parameters were provided in only a few cases.

Results of monitoring studies

Acephate

The 225 samples analysed comprised seeds (79), fruits or berries (77), roots or rhizomes (42) and bark, buds and arils (27). All the results were below the LOQ: 0.2 mg/kg for data from the American Spice Trade Association and India and 0.02 mg/kg for data from the European Spice Association, regardless of the subgroup in which the spices were classified.

The Meeting estimated a maximum residue limit of 0.2 (*) mg/kg and median and high residue values of 0.2 mg/kg for spices.

As none of the samples contained detectable residues, no factor can be introduced into calculations of long-term intake to take into account the proportion of samples containing detectable residues.

Azinphos-methyl

The 260 samples analysed comprised seeds (86), fruits or berries (92), roots or rhizomes (46) and bark, buds and arils (36).

All the results were below the limit of determination: 0.5 mg/kg for data from the American Spice Trade Association and India and 0.1 mg/kg for data from the European Spice Association.

The Meeting estimated a maximum residue level of 0.5 (*) mg/kg, a median residue level of 0.1 mg/kg and a high residue level of 0.5 mg/kg for spices.

As none of the samples contained detectable residues, no factor can be introduced into calculations of long-term intake to take into account the proportion of samples containing detectable residues.

Chlorpyrifos

The 2632 samples analysed comprised seeds (2165), fruits or berries (155), roots or rhizomes (270) and bark, buds and arils (42). The LOQ was 0.05 mg/kg in all data sources.

Detectable residues were found in celery seed, coriander and cumin. The levels, in ranked order, were: 0.12, 0.15, 0.16, 0.18, 0.25, 0.26, 0.41, 0.44, 0.54 and 0.54 mg/kg. Detectable residues were found in one of 18 samples of anise seeds in Canada and 78 of 744 samples in Egypt. The levels found were, in ranked order: 0.01 (two), 0.05 (12), 0.06 (seven), 0.07 (eight), 0.08 (nine), 0.09 (four), 0.1 (two), 0.11 (three), 0.13 (six), 0.14 (six), 0.15, 0.17 (two), 0.18, 0.21 (three), 0.22, 0.23, 0.32, 0.33, 0.36, 0.5 (two), 0.54, 0.9, 1.8, 2, 3 and 3.6 mg/kg. For fennel, 1228 samples were analysed, and 18 contained detectable residues. The levels, in ranked order, were: 0.05 (two), 0.06, 0.07, 0.08, 0.1 (two), 0.11 (five), 0.13, 0.14, 0.15, 0.22 (two) and 1.4 mg/kg.

The Meeting concluded that the levels of residues in seeds are comparable and estimated a maximum residue level of 5 mg/kg, a median residue level of 0.09 mg/kg and a high residue level of 3.6 mg/kg (based on residue data for anise seed) for the seed subgroup of spices.

Detectable residues were measured in cardamom and pepper in the fruit subgroup, at levels, in ranked order, of: 0.05 (two), 0.06, 0.07, 0.08, 0.11, 0.20, 0.54 and 0.71 mg/kg.

The Meeting estimated a maximum residue level of 1 mg/kg, a median residue level of 0.05 mg/kg and a high residue level of 0.71 mg/kg for the fruit subgroup.

Detectable residues were found in ginger and turmeric in the roots subgroup, at levels, in ranked order, of: 0.05, 0.13, 0.24, 0.28, 0.31, 0.37 and 0.72 mg/kg.

The Meeting estimated a maximum residue level of 1 mg/kg, a median residue level of 0.05 mg/kg and a high residue level of 0.72 mg/kg for the roots subgroup.

Detectable residues were found (> 0.05 mg/kg) in the subgroups of bark, buds and arils in three studies. As only 42 measurements were available, however, maximum residue levels could not be estimated for these subgroups.

The Meeting recommended use of a high residue level of 3.6 mg/kg for calculation of short-term intake, and a median residue level of 0.09 mg/kg and a factor of 0.1 to take into account the proportion of samples containing detectable residues for calculating long-term intake.

Chlorpyrifos-methyl

The 1822 samples analysed comprised seeds (1432), fruits or berries (80), roots or rhizomes (142) and bark, buds and arils (25). The LOQ was 0.05 mg/kg in all data sources.

Detectable levels of residues were found in 68 of 983 samples of anise seed, one of 49 samples of celery seed (0.01 mg/kg), one of 19 samples of coriander (0.01 mg/kg) and four of 345 samples of fennel (0.07, 0.08, 0.12 and 0.16 mg/kg). The combined residue levels were, in ranked order: 0.02 (two), 0.05 (two), 0.06 (four), 0.07 (three), 0.08 (seven), 0.09 (eight), 0.1 (five), 0.11, 0.12 (three), 0.13 (three), 0.14, 0.15 (two), 0.16 (five), 0.17, 0.18 (two), 0.19, 0.2 (two), 0.22 (three), 0.24 (two), 0.25, 0.28, 0.3, 0.31, 0.32, 0.38 and 0.39 mg/kg.

The Meeting estimated a maximum residue level of 1 mg/kg, a high residue level of 0.39 mg/kg and a median residue level of 0.05 mg/kg for the seed subgroup of spices.

Three of 156 samples of caraway contained detectable residues, at: 0.06, 0.1 and 0.12 mg/kg.

The Meeting estimated a maximum residue level of 0.3 mg/kg, a high residue level of 0.12 mg/kg and a median residue level of 0.1 mg/kg for the fruits subgroup of spices.

Detectable residues were found in ginger and turmeric in the roots subgroup, at levels, in ranked order, of: 0.013 (12), 0.017, 0.03 (two), 0.031, 0.034 (two), 0.037, 0.038, 0.05, 0.06, 0.07, 0.071, 0.073, 0.082, 0.089, 0.091, 0.098, 0.14 (three), 0.16, 0.29, 0.39 and 2.9 mg/kg.

The Meeting estimated a maximum residue level of 5 mg/kg, a high residue level of 2.9 mg/kg and a median residue level (on the basis of data for ginger) of 0.77 mg/kg for the roots subgroup.

Lack of sufficient data prevented estimation of maximum residue levels for the bark, buds and arils subgroups.

The Meeting recommended use of a high residue level of 2.9 mg/kg for calculating short-term intake, and a median residue level of 0.77 mg/kg and a factor of 1 to take into account the proportion of samples containing detectable residues for calculating long-term intake.

Cypermethrin

The 174 samples analysed comprised seeds (38), fruits or berries (65), roots or rhizomes (58) and bark, buds and arils (13). The LOQ was 0.05 mg/kg in all data sources.

Commodities in the seed subgroup contained detectable residues at levels of 0.076–0.93 mg/kg.

One of 57 samples of pepper contained detectable residues (0.065 mg/kg), but none were found in eight samples of other commodities in the fruits subgroup. The Meeting estimated a maximum residue level of 0.1 mg/kg, a high residue level of 0.05 mg/kg and a median residue level of 0.05 mg/kg for the fruits or berries subgroup of spices.

Three of 58 samples of commodities in the subgroup of roots or rhizomes contained detectable residues, at levels of: 0.05, 0.11 and 0.12 mg/kg.

The Meeting estimated a maximum residue level of 0.2 mg/kg, a high residue level of 0.12 mg/kg and a median residue level of 0.11 mg/kg for roots or rhizomes.

The limited database was considered insufficient for estimating maximum residue levels for the other subgroups.

The Meeting recommended use of a high residue level of 0.12 mg/kg for calculating short-term intake, and a median residue level of 0.11 mg/kg and a factor of 0.04 to take into account the proportion of samples containing detectable residues in calculating long-term intake.

Diazinon

The 1948 samples analysed comprised seeds (1559), fruits or berries (115), roots or rhizomes (234) and bark, buds and arils (40). The LOQ was 0.1 mg/kg for the data from the American Spice Trade Association and India and 0.05 mg/kg for those from Egypt and Europe.

The residue levels in spices from fruits and from bark, buds and arils were below the LOQ (0.1 mg/kg).

Detectable residues were found in 69 of 667 samples of anise, at levels, in ranked order, of: 0.05, 0.06 (four), 0.07, 0.08 (two), 0.09 (three), 0.1 (two), 0.11, 0.12 (two), 0.13, 0.14 (four), 0.15 (three), 0.16 (two), 0.17 (four), 0.18 (two), 0.19 (three), 0.21, 0.22, 0.24, 0.28, 0.3, 0.32 (two), 0.33 (two), 0.35, 0.37, 0.39, 0.41, 0.42 (two), 0.47, 0.473, 0.48, 0.51, 0.59, 0.6, 0.82, 0.88, 0.9, 1.1 (two), 1.2, 1.3, 1.8 (two), 2.1, 2.7, 3.5 and 3.6 mg/kg. Detectable residues were found in 31 of 734 samples of fennel seed, at levels, in ranked order, of: 0.05 (two), 0.06 (three), 0.07 (three), 0.08 (two), 0.1 (three) 0.12, 0.17 (two), 0.19, 0.2, 0.21, 0.23, 0.24 (two), 0.26, 0.45, 0.59, 0.65, 0.72, 0.76, 0.77, 1.2 and 1.7 mg/kg. Detectable residues were measured in celery and cumin seeds at levels of 0.1 (two), 0.14 and 0.29 mg/kg.

The Meeting estimated a maximum residue level of 5 mg/kg, a high residue level of 3.6 mg/kg and a median residue level (on the basis of data on anise seed) of 0.19 mg/kg for the seed subgroup.

Detectable residues were measured in two samples of turmeric, at levels of 0.23 and 0.26 mg/kg. No residues were measured in fruit and rhizome spices.

The Meeting estimated a maximum residue level of 0.5 mg/kg, a high residue level of 0.26 mg/kg and a median residue level of 0.05 mg/kg for the roots or rhizomes subgroup of spices, and a maximum residue level of 0.1 (*) mg/kg, a high residue level of 0.1 mg/kg and a median residue level of 0.05 mg/kg for fruits. No recommendation could be made for the bark, buds and arils subgroups.

The Meeting recommended use of a high residue level of 3.6 mg/kg for calculating short-term intake, and a median residue level of 0.19 mg/kg and a factor of 0.1 to take into account the proportion of samples containing detectable residues in calculating long-term intake.

Dichlorvos

The 277 samples analysed comprised seeds (100), fruits or berries (93), roots or rhizomes (48) and bark, buds and arils (36).

The residue levels in all samples were below the LOQ (0.1 mg/kg). The Meeting estimated a maximum residue level of 0.1 (*) mg/kg and high and median residue levels of 0.1 mg/kg for residues of dichlorvos on all spices.

As none of the samples contained detectable residues, no factor can be used in long-term intake calculations to take into account the proportion of samples containing detectable residues.

Spices

Dicofol

The 416 samples analysed comprised seeds (67), fruits or berries (85), roots or rhizomes (230) and bark, buds and arils (34). The LOQ was 0.05 mg/kg for all data sources.

No residues were detected in the seeds subgroup. One of 42 pepper samples contained residue at the LOQ.

Detectable residues were found in ginger and turmeric in the roots subgroup, at levels, in ranked order, of: 0.02. 0.035, 0.036 and 0.05 mg/kg.

The Meeting estimated a maximum residue level of 0.05 (*) mg/kg and high and median residue levels of 0.05 mg/kg for the seed subgroup, and a maximum residue level of 0.1 mg/kg and high and median residue levels of 0.05 mg/kg for the fruit, rhizomes and roots subgroups.

Four samples of spices in the bark, buds and arils subgroups contained residues at the LOQ (0.05 mg/kg). The data did not allow estimation of maximum residue levels for these subgroups.

The Meeting recommended use of a high residue level of 0.05 mg/kg for calculating short-term intake, and a median residue level of 0.05 mg/kg and a factor of 0.03 to take into account the proportion of samples containing detectable residues in calculating long-term intake.

Dimethoate

The 2613 samples analysed comprised seeds (2121), fruits or berries (381), roots or rhizomes (75) and bark, buds and arils (36).

Detectable residues were found in 61 of 744 samples of anise seeds at levels, in ranked order, of: 0.05 (three), 0.06 (five), 0.07 (three), 0.08 (two), 0.09 (two), 0.1 (two), 0.12 (three), 0.13 (three), 0.14, 0.15 (five), 0.17 (two), 0.18, 0.24 (two), 0.25 (four), 0.27 (two), 0.28 (two), 0.29, 0.29, 0.32, 0.35, 0.36, 0.39, 0.41, 0.42, 0.43, 0.44, 0.46, 0.53 (two), 0.57, 0.62, 0.9, 1.4, 2.5 and 3 mg/kg.

Detectable residues were found in 69 of 1284 samples of fennel at levels, in ranked order, of: 0.03, 0.05 (four), 0.06 (four), 0.07 (six), 0.08 (four), 0.09 (four), 0.1 (three), 0.11, 0.12, 0.13, 0.14 (three), 0.15 (four), 0.16, 0.18 (two), 0.2 (two), 0.21, 0.25 (two), 0.3, 0.32 (three), 0.33 (two), 0.34, 0.35, 0.37, 0.38 (two), 0.43 (two), 0.51 (four), 0.53 (two), 0.54, 0.94, 1.1 and 1.4 (two) mg/kg. Residues were not detected in the other seed samples (0.05–0.1 mg/kg).

The Meeting estimated a maximum residue level of 5 mg/kg, a high residue level of 3 mg/kg and a median residue level of 0.17 mg/kg (on the basis of data on anise seed) for the seeds subgroup.

Five of 277 samples of caraway samples from Egypt contained residues, at levels of: 0.08, 0.17 (two) and 0.22 (two) mg/kg.

The Meeting estimated a maximum residue level of 0.5 mg/kg, a high residue level of 0.22 mg/kg and a median residue level of 0.05 mg/kg for the fruits subgroup.

No residue was detected in 75 samples of root and rhizome spices. The Meeting estimated a maximum residue level of 0.1 (*) mg/kg and high and median residue levels of 0.1 mg/kg for the roots and rhizomes subgroup.

No limits could be estimated for the bark, buds and arils subgroups.

The Meeting recommended use of a high residue level of 3 mg/kg for calculating short-term intake, and a median residue level of 0.17 mg/kg and a factor of 0.08 to take into account the proportion of samples containing detectable residues in calculating long-term intake.

Disulfoton

The 223 samples analysed comprised seeds (67), fruits or berries (66), roots or rhizomes (69) and bark, buds and arils (21).

As all the residue levels were below the LOQ (0.05 mg/kg), the Meeting estimated a maximum residue level of 0.05 (*) mg/kg and high and median residue levels of 0.05 mg/kg for disulfoton on spices. As none of the samples contained detectable residues, no factor could be used in calculating long-term intake to take into account the proportion of samples containing detectable residues.

Endosulfan

The 981 samples analysed comprised seeds (331), fruits or berries (208), roots or rhizomes (401) and bark, buds and arils (41). The LOQ was 0.03 mg/kg for the trials reported by the American Spice Trade Association and India and 0.1 mg/kg for data from Europe.

Data were provided for α-endosulfan, β-endosulfan, endosulfan sulfate, as well as for total endosulfan residues. The maximum residue levels were estimated on the basis of the results of monitoring for total endosulfan residues.

In the seeds subgroup, detectable residues were measured in celery seed, coriander and dill seed, at levels in ranked order of: 0.035 (two), 0.04, 0.1 (11), 0.12, 0.14, 0.20, 0.34, 0.45 and 0.63 mg/kg.

The Meeting estimated a maximum residue level of 1 mg/kg, a high residue level of 0.63 mg/kg and a median residue level of 0.03 mg/kg for the seeds subgroup of spices.

In the fruits subgroup, detectable residues were measured in cardamom and pepper, at levels in ranked order of: 0.03, 0.04, 0.075, 0.08, 0.09 (four), 0.10, 0.11, 0.12 (three), 3.1 and 3.2 mg/kg.

The Meeting estimated a maximum residue level of 5 mg/kg, a high residue level of 3.2 mg/kg and a median residue level of 0.12 mg/kg (on the basis of data for pepper) for the fruits subgroup of spices.

In the subgroup of roots or rhizomes, detectable residues were measured in ginger and turmeric at levels, in ranked order, of: 0.04 (two), 0.06, 0.08, 0.1 and 0.24 mg/kg.

The Meeting estimated a maximum residue level of 0.5 mg/kg, a high residue level of 0.24 mg/kg and a median residue level of 0.1 mg/kg for the subgroup roots or rhizomes.

The 41 samples in the bark, buds and arils subgroups did not contain detectable residues (< 0.1 mg/kg).

The Meeting recommended use of a high residue level of 3.2 mg/kg for calculating short-term intake, and a median residue level of 0.12 mg/kg and a factor of 0.05 to take into account the proportion of samples containing detectable residues in calculating long-term intake.

Ethion

The 754 analysed comprised seeds (190), fruits or berries (155), roots or rhizomes (367) and bark, buds and arils (42). The LOQ was 0.1 mg/kg for data from the American Spice Trade Association and India and 0.05 mg/kg for data from the European Spice Association.

In the seeds subgroup, detectable residues were measured in anise, coriander and cumin at levels, in ranked order, of 0.11, 0.13, 0.21 and 1.8 mg/kg.

The Meeting estimated a maximum residue level of 3 mg/kg, a high residue level of 1.8 mg/kg and a median residue level of 0.1 mg/kg for the seeds subgroup of spices.

In the fruits subgroup, detectable residues were measured in cardamom and pepper (black, white and pink) at levels, in ranked order, of 0.12, 0.33 and 3.1 mg/kg.

The Meeting estimated a maximum residue level of 5 mg/kg, a high residue level of 3.1 mg/kg and a median residue level of 1.7 mg/kg (on the basis of data on pepper) for the fruits subgroup of spices.

In the subgroup of roots or rhizomes, detectable residues were measured in ginger, at levels of 0.11 and 0.15 mg/kg.

The Meeting estimated a maximum residue level of 0.3 mg/kg, a high residue level of 0.15 mg/kg and a median residue level of 0.05 mg/kg for the roots or rhizomes subgroup of spices.

The 42 samples in the bark, buds and arils subgroups did not contain detectable residues (< 0.05– < 0.1 mg/kg)

The Meeting recommended use of a high residue level of 3.1 mg/kg for calculating short-term intake, and a median residue level of 1.7 mg/kg and a factor of 0.02 to take into account the proportion of samples containing detectable residues in calculating long-term intake.

Fenitrothion

The 2424 samples analysed comprised seeds (1920), fruits or berries (230), roots or rhizomes (234) and bark, buds and arils (40). The limit of determination was 0.1 mg/kg for data from the American Spice Trade Association and India, and 0.05 mg/kg for data from Egypt and the European Spice Association.

Detectable residues were found in 22 of 756 samples of anise seeds in Egypt, at levels of: 0.05, 0.08 (two), 0.1, 0.12 (two), 0.13, 0.15, 0.25, 0.4 (two), 0.41 (two), 0.87, 0.88, 1 (three), 1.4 (two), 2 and 5.4 mg/kg. No residues was detected in 18 samples from other sources (< 0.1 mg/kg). Detectable residues were measured in celery seed, cumin and coriander at levels of 0.17, 0.19 and 1.5 mg/kg, respectively.

The Meeting estimated a maximum residue level of 7 mg/kg, a high residue level of 5.4 mg/kg and and a median residue level of 0.4 mg/kg (on the basis of data for anise seed) for the seed subgroup of spices.

Detectable residues were measured in caraway in the fruit subgroup, at levels of: 0.05, 0.1, 0.12, 0.22 and 0.4 mg/kg.

The Meeting estimated a maximum residue level of 1 mg/kg, a high residue level of 0.4 mg/kg and a median residue level of 0.05 mg/kg for the fruit subgroup of spices.

All the samples in the root subgroup of spices contained residues at levels below the LOQ. The Meeting estimated a maximum residue level of 0.1 (*) mg/kg, a high residue level of 0.1 mg/kg and a median residue level of 0.05 mg/kg for the root subgroup of spices.

All samples in the bark, buds and arils subgroups contained residues at levels below the LOQ. The database did not allow estimation of a maximum residue level.

The Meeting recommended use of a high residue level of 5.4 mg/kg for calculating short-term intake, and a median residue level of 0.4 mg/kg and a factor of 0.03 to take into account the proportion of samples containing detectable residues in calculating long-term intake.

Iprodion

The 339 samples analysed comprised seeds (93), fruits or berries (38), roots or rhizomes (192) and bark, buds and arils (16). The LOQ was 0.05 mg/kg in all the data sources.

No residues were detectable in seeds. The Meeting estimated a maximum residue level of 0.05 (*) mg/kg and high and median residue levels of 0.05 mg/kg for the seed subgroup of spices.

One of 92 samples of ginger contained residue at the LOQ level (0.05 mg/kg). The Meeting estimated a maximum residue level of 0.1 mg/kg and high and median residue levels of 0.05 mg/kg for the root subgroup of spices.

Detectable residues were also found in the other subgroups, but the database was insufficient for estimating maximum residue levels.

The Meeting recommended use of a high residue level of 0.05 mg/kg for calculating short-term intake, and a median residue level of 0.05 mg/kg and a factor of 0.01 to take into account the proportion of samples containing detectable residues in calculating long-term intake.

Malathion

The total number of samples analysed (581) included seeds (185), fruits or berries (115), roots or rhizomes (234) and bark, buds and arils (47). The limit of determination was 0.1 mg/kg for data from American Spice Trade Association and India, and 0.05 mg/kg for data from Egypt and the European Spice Association.

In the seed subgroup, detectable residues were measured in anise, celery seed, cumin, fennel seed and nutmeg. The levels, in ranked order, were: 0.16 (four), 0.18, 0.22, 0.32, 0.38, 0.48, 0.58 and 0.86 mg/kg. As the data from Egypt indicated post-harvest use, they were not taken into consideration.

The Meeting estimated a maximum residue level of 2 mg/kg, a high residue level of 0.86 mg/kg and a median residue level of 0.48 mg/kg (on the basis of data on celery seed) for the seed subgroup of spices.

In the fruit subgroup, detectable residues were measured in three of 66 samples of pepper, at levels of 0.1, 0.42 and 0.48 mg/kg, and in 17 of 307 samples of caraway at levels of: 0.1, 0.3, 0.05 (two), 0.06 (two), 0.07 (four), 0.09 (two), 0.19, 0.26, 0.31, 0.33 and 0.46 mg/kg. The combined residue levels, in ranked order, were: 0.1, 0.3, 0.05 (two), 0.06 (two), 0.07 (four), 0.09 (two), 0.1, 0.19, 0.26, 0.31, 0.33, 0.42, 0.46 and 0.48 mg/kg.

The Meeting estimated a maximum residue level of 1 mg/kg, a high residue level of 0.48 mg/kg and a median residue level of 0.05 mg/kg for the fruit subgroup of spices.

In the subgroup of roots or rhizomes, detectable residues were measured in ginger and turmeric, at levels of 0.1, 0.12 and 0.16 mg/kg.

The Meeting estimated a maximum residue level of 0.5 mg/kg, a high residue level of 0.16 mg/kg and a median residue level of 0.05 mg/kg for the root or rhizome subgroup of spices.

The ranked order of the detectable residue levels in 47 samples from the bark, buds and arils subgroups was: 0.1 (four), 0.12, 0.14, 0.3, 0.48, 0.96, 1.02, 1.04, 1.88, 1.96 and 2 mg/kg.

The Meeting considered that the database was insufficient for estimating maximum residue levels.

The Meeting recommended use of a high residue level of 0.86 mg/kg for calculating short-term intake, and a median residue level of 0.48 mg/kg and a factor of 0.06 to take into account the proportion of samples containing detectable residues in calculating long-term intake.

Metalaxyl

The 1306 samples analysed in Egypt comprised anise seeds (411) and fennel seeds (895). The LOQ was 0.05 mg/kg.

Six of 411 samples of anise seeds contained detectable residues, at levels of: 0.2, 0.22, 0.4, 0.47, 0.64 and 0.65 mg/kg. Four of 895 samples of fennel contained detectable residues, at levels of: 0.17, 0.29, 0.4 and

3.2 mg/kg. The combined residue levels, in ranked order, were: 0.17, 0.2, 0.22, 0.29, 0.4 (two), 0.47, 0.64, 0.65 and 3.2 mg/kg.

The Meeting estimated a maximum residue level of 5 mg/kg, a high residue level of 3.2 mg/kg and a median residue level of 0.43 mg/kg for the seed subgroup of spices.

For calculation of long-term intake, the Meeting recommended use of a correction factor of 0.015 to take into account the proportion of samples containing detectable residues.

Methamidophos

The 260 samples analysed comprised seeds (96), fruits or berries (92), roots or rhizomes (46) and bark, buds and arils (36). All the residue levels were below the limit of determination (0.1mg/kg for data from the American Spice Trade Association and India and 0.01 mg/kg for data from the European Spice Association). The data on cumin from Egypt, which indicated post-harvest use, were not taken into consideration.

The Meeting estimated a maximum residue level of 0.1 (*) mg/kg, a high residue level of 0.1 mg/kg and a median residue level of 0.01 mg/kg for spices.

Mevinphos

The 554 samples analysed comprised seeds (158), fruits or berries (114), roots or rhizomes (232) and bark, buds and arils (40). The LOQ was 0.2 mg/kg for data from the American Spice Trade Association and India, and 0.05 mg/kg for data from the European Spice Association.

In the seed subgroup, detectable residues were measured in celery seed at 2.9 mg/kg. The Meeting estimated a maximum residue level of 5 mg/kg, a high residue level of 2.9 mg/kg and a median residue level of 0.05 mg/kg (on the basis of data on celery seed) for the seed subgroup of spices.

In the fruit subgroup, no detectable residues were measured. The Meeting estimated a maximum residue level of 0.2 (*) mg/kg, a high residue level of 0.2 mg/kg and a median residue level of 0.05 mg/kg for the fruit subgroup of spices.

In the subgroup of roots or rhizomes, detectable residues were measured in ginger and turmeric. The levels, in ranked order, were: 0.2, 0.21, 0.22, 0.24, 0.27, 0.3, 0.31, 0.34, 0.37, 0.39, 0.4, 0.41 and 0.47 mg/kg. The Meeting estimated a maximum residue level of 1 mg/kg, a high residue level of 0.47 mg/kg and a median residue level of 0.31 mg/kg for the root or rhizome subgroup of spices.

The 40 samples from bark, buds and arils did not contain detectable residues (< 0.2 mg/kg). The database did not allow estimation of maximum residue levels.

The Meeting recommended use of a high residue level of 2.9 mg/kg for calculating short-term intake, and a median residue level of 0.05 mg/kg and a factor of 0.02 to take into account the proportion of samples containing detectable residues in calculating long-term intake.

Parathion

The 329 samples analysed comprised seeds (114), fruits or berries (105), roots or rhizomes (74) and bark, buds and arils (36). The LOQ was 0.1 mg/kg for all data sources.

Seed spices did not contain detectable residues. The Meeting estimated a maximum residue level of 0.1 (*) mg/kg and high and median residue levels of 0.1 mg/kg for the seed subgroup of spices.

One sample of pepper and one sample in the root and rhizome subgroup contained residues at the LOQ. The Meeting estimated a maximum residue level of 0.2 mg/kg and high and median residue levels of 0.1 mg/kg for the fruit and the root and rhizome subgroups of spices.

Three samples in the bark subgroup contained residues at the LOQ, indicating that detectable residues might occur in these commodities. The database was insufficient for estimating maximum residue levels.

The Meeting recommended use of a high residue level of 0.1 mg/kg for calculating short-term intake, and a median residue level of 0.1 mg/kg and a factor of 1 to take into account the proportion of samples containing detectable residues in calculating long-term intake.

Parathion-methyl

The 821 samples analysed comprised seeds (359), fruits or berries (155), roots or rhizomes (265) and bark, buds and arils (42). The LOQ was 0.1 mg/kg for data from the American Spice Trade Association and India and 0.05 mg/kg for data from Egypt and the European Spice Association.

In the seed subgroup, detectable residues were measured in anise, celery seed, coriander, cumin and fennel seed. The levels, in ranked order, were: 0.1, 0.13, 0.14, 0.16, 0.2, 0.21, 0.23, 0.24 (two), 0.31, 0.33, 0.36, 0.37, 0.38, 0.43, 0.51, 0.86, 0.97, 1.1, 1.2 and 2.4 (two) mg/kg.

The Meeting estimated a maximum residue level of 5 mg/kg, a high residue level of 2.4 mg/kg and a median residue levels of 0.43 mg/kg (on the basis of data on cumin seed) for the seed subgroup of spices.

In the fruit subgroup, detectable residues were measured in pepper at levels of 0.1 and 3.4 mg/kg.

The Meeting estimated a maximum residue level of 5 mg/kg, a high residue level of 3.4 mg/kg and a median residue level of 0.1 mg/kg for the fruit subgroup of spices.

In the subgroup of roots or rhizomes, detectable residues were measured in ginger and turmeric. The levels, in ranked order, were: 0.1, 0.13, 0.14 (two), 0.17, 0.24, 0.3, 1.2 (two), 1.5 and 1.7 mg/kg.

The Meeting estimated a maximum residue level of 3 mg/kg, a high residue level of 1.7 mg/kg and a median residue level of 0.24 mg/kg for the root or rhizome subgroup of spices.

Three of the 42 samples from the bark, buds and arils subgroups contain detectable residues at the LOQ (< 0.1 mg/kg) The database did not allow estimation of maximum residue levels.

The Meeting recommended use of a high residue level of 3.4 mg/kg for calculating short-term intake, and a median residue level of 0.43 mg/kg and a factor of 0.25 to take into account the proportion of samples containing detectable residues in calculating long-term intake.

Permethrin

A total of 160 samples of various spices, including 68 seeds, were analysed for residues of permethrin. No residues were detected. The Meeting estimated a maximum residue level of 0.05 (*) mg/kg and high and median residue levels of 0.05 mg/kg.

As none of the samples contained detectable residues, no factor can be used in calculating long-term intake to take into account the proportion of samples containing detectable residues.

Phenthoate

Ten of 415 samples of anise seed contained detectable residues, at levels of: 0.05, 0.19, 0.49, 1.1 (two), 1.3, 1.4 (two), 5 and 5.2. The LOQ was 0.05 mg/kg.

The Meeting estimated a maximum residue level of 7 mg/kg, a high residue level of 5.2 mg/kg and a median residue level of 1.2 mg/kg for the seed subgroup of spices.

For the calculation of long-term intake, the Meeting recommended use of a correction factor of 0.024 to take into account the proportion of samples containing detectable residues.

Phorate

The 336 samples analysed comprised seeds (115), fruits or berries (117), roots or rhizomes (75) and bark, buds and arils (29). The LOQ was 0.1 mg/kg for data from the American Spice Trade Association and India, and 0.05 mg/kg for data from the European Spice Association.

Of the samples analysed, only two of cumin showed detectable residues of phorate, at 0.12 and 0.3 mg/kg.

The Meeting estimated a maximum residue level of 0.5 mg/kg, a high residue level of 0.3 mg/kg and a median residue level of 0.21 mg/kg for the seed subgroup of spices, and a maximum residue level of 0.1 (*) mg/kg and high and median residue levels of 0.1 mg/kg for the fruit and the root and rhizome subgroups. The data were insufficient for estimating maximum residue levels for the other subgroups.

The Meeting recommended use of a high residue level of 0.3 mg/kg for calculating short-term intake, and a median residue level of 0.21 mg/kg and a factor of 0.06 to take into account the proportion of samples containing detectable residues in calculating long-term intake.

Phosalone

The 607 samples analysed comprised seeds (176), fruits or berries (80), roots or rhizomes (226) and bark, buds and arils (25). The origin of the samples was not given in several cases. The LOQ was 0.05 mg/kg for all data sources.

Samples of anise, celery and cumin contained detectable residues at levels of 0.1, 0.95 and 0.25 mg/kg, respectively.

The Meeting estimated a maximum residue level of 2 mg/kg, a high residue level of 0.95 mg/kg and a median residue level of 0.25 mg/kg for the seed subgroup of spices.

Three of 44 samples of pepper contained detectable residues, at levels of 0.05, 0.85 and 0.89 mg/kg.

The Meeting estimated a maximum residue level of 2 mg/kg, a high residue level of 0.89 mg/kg and a median residue level of 0.85 mg/kg for the fruit subgroup of spices.

Detectable residues were found in ginger and turmeric in the root subgroup. The levels, in ranked order, were: 0.05, 0.14, 0.22, 0.27, 0.31, 0.4, 0.49, 0.5 and 1.49 mg/kg.

The Meeting estimated a maximum residue level of 3 mg/kg, a high residue level of 1.49 mg/kg and a median residue level of 0.31 mg/kg for the root and rhizome subgroup.

Residues were also detected in spices in the bark, buds and arils subgroups. The data available were, however, insufficient to allow estimation of maximum residue levels.

The Meeting recommended use of a high residue level of 1.5 mg/kg for calculating short-term intake, and a median residue level of 0.85 mg/kg and a factor of 0.07 to take into account the proportion of samples containing detectable residues in calculating long-term intake.

Pirimicarb

In Egypt, 129 of 484 samples of anise seed and 54 of 824 samples of fennel seed contained detectable residues, at levels of: 0.05 (11), 0.06 (22), 0.07 (nine), 0.08 (nine), 0.09 (14), 0.1 (11), 0.12 (six), 0.13 (seven), 0.14 (eight), 0.15 (seven), 0.16 (six), 0.17 (four), 0.18 (three), 0.19 (two), 0.2 (three), 0.22 (five), 0.23, 0.24, 0.26, 0.27 (three), 0.28 (four), 0.29 (two), 0.31, 0.33 (three), 0.34 (two), 0.35, 0.37 (two), 0.38, 0.39, 0.41, 0.42 (two), 0.43, 0.44, 0.45, 0.47 (two), 0.53 (two), 0.54 (two), 0.58 (two), 0.59 (two), 0.6 (two), 0.64, 0.67, 0.69, 0.7 (two), 0.8, 0.84, 0.93, 0.94, 1.2, 1.4 (two), 1.5 and 3 mg/kg. The LOQ was 0.05 mg/kg.

The Meeting estimated a maximum residue level of 5 mg/kg, a high residue level of 3 mg/kg and a median residue level of 0.14 mg/kg for the seed subgroup of spices.

For the calculation of long-term intake, the Meeting recommended use of a correction factor of 0.27 to take into account the proportion of samples containing detectable residues.

Pirimiphos-methyl

The 1314 samples analysed comprised seeds (1137), fruits or berries (94), roots or rhizomes (47) and bark, buds and arils (36). The LOQ was 0.1 mg/kg for data from the American Spice Trade Association and India and 0.05 mg/kg for those from Egypt and Europe.

In the seed subgroup, detectable residues were measured in 16 of 492 samples of anise, at levels of: 0.05, 0.06, 0.07, 0.08, 0.12, 0.17, 0.18, 0.19, 0.27, 0.32, 0.47, 0.58, 0.6, 0.61, 0.63 and 1.8 mg/kg); in five of 556 samples of fennel, at levels of: 0.05, 0.07, 0.08, 0.1 and 0.11 mg/kg) and in one sample of nutmeg at 0.1 mg/kg.

The Meeting estimated a maximum residue level of 3 mg/kg, a high residue level of 1.8 mg/kg and a median residue level of 0.23 mg/kg (on the basis of data for anise) for the seed subgroup of spices.

In the fruit subgroup, detectable residues were measured at the LOQ in cardamon (0.18 mg/kg) and pepper (0.1 mg/kg, two samples).

The Meeting estimated a maximum residue level of 0.5 mg/kg and high and median residue levels of 0.1 mg/kg for the fruit subgroup of spices.

Residues were also detected at the LOQ in one sample in the root subgroup and in the bark, buds and arils subgroups. The data were insufficient for estimation of maximum residue levels.

The Meeting recommended use of a high residue level of 1.8 mg/kg for calculating short-term intake, and a median residue level of 0.23 mg/kg and a factor of 0.03 to take into account the proportion of samples containing detectable residues in calculating long-term intake.

Quintozene

The 550 samples analysed comprised seeds (163), fruits or berries (111), roots or rhizomes (236) and bark, buds and arils (40). The LOQ was 0.01 mg/kg for all data sources.

In the seed subgroup, detectable residues were measured in coriander, cumin, and fennel seed. The levels, in ranked order, were: 0.01 (two), 0.02 (two) and 0.05 mg/kg.

The Meeting estimated a maximum residue level of 0.1 mg/kg, a high residue level of 0.05 mg/kg and a median residue level of 0.01 mg/kg for the seed subgroup of spices.

In the fruit subgroup, detectable residues were measured at the LOQ in one sample of pepper and one of vanilla.

The Meeting estimated a maximum residue level of 0.02 mg/kg and high and median residue levels of 0.01 mg/kg for the fruit subgroup of spices.

In the subgroup of roots or rhizomes, detectable residues were measured in ginger and turmeric. The levels, in ranked order, were: 0.01, 0.04, 0.05 (two), 0.08 and 1.2 mg/kg.

The Meeting estimated a maximum residue level of 2 mg/kg, a high residue level of 1.2 mg/kg and a median residue level of 0.05 mg/kg (on the basis of data on turmeric) for the root and rhizome subgroup of spices.

In the subgroup of bark, buds and arils, detectable residues were measured at the LOQ in cloves, cassia and cinnamon.

The database was insufficient for estimation of maximum residue levels.

The Meeting recommended use of a high residue level of 1.2 mg/kg for calculating short-term intake, and a median residue level of 0.05 mg/kg and a factor of 0.035 to take into account the proportion of samples containing detectable residues in calculating long-term intake.

Vinclozolin

The 442 samples analysed comprised seeds (110), fruits or berries (80), roots or rhizomes (227) and bark, buds and arils (25). The LOQ was 0.05 mg/kg for all data sources.

None of the sample contained detectable residues.

The Meeting estimated a maximum residue level of 0.05 (*) mg/kg and high and median residue levels of 0.05 mg/kg for spices.

As none of the samples contained detectable residues, no factor can be used in calculating long-term intakes to take into account the proportion of samples containing detectable residues.

Other pesticides

The results of monitoring were submitted for aldicarb, amitraz, bendiocarb, bifenthrin, captan, carbaryl, carbofuran, chlorothalonil, chlorpropham, cyfluthrin, demeton, dicloran, dicrotophos, diphenylamine, esfenvalerate, ethoprop, ethoxyquin, fenamiphos, fenarimol, fenpropathrin, fenthion, fenvalerate, folpet, imazalil, isophenphos, methoicarb, methomyl, myclobutanil, monocrotophos, *ortho*-phenylphenol, oxamyl, phosmet, phosphamidon, propiconazole, propoxur, pyrethrins, terbufos, thiabendazole, triadimefon, trichlorfon and triforin. The amount of data on residues did not, however, meet the minimum requirements for estimating maximum residue levels.

Estimation of maximum residue levels for pesticide residues in and on dry chili peppers

The database of the US Department of Agriculture indicates that the water content of various fresh peppers ranges from 88% to 94%. The average dehydration factor, derived on the basis of the assumption of complete loss of water, is 11.6. A similar approximate value (11.3) can be obtained by taking into account the water content of dried peppers. As dried peppers always contain 5–10% water, the rounded value of 10 used by the spice trade industry can be considered realistic for estimating the concentration of pesticide residues when it is assumed that all the residues present in fresh peppers remain in the dried peppers. As no processing studies were available, in accord with the request of the CCPR, the Meeting used the default value of 10 for estimating maximum residue levels for pesticide residues in dried chili peppers.

Further work or information

Before maximum residue levels can be estimated, additional data are required on residues in subgroups of spices for which insufficient data were available for evaluation by the present Meeting.

DIETARY RISK ASSESSMENT

The Meeting evaluated data on residues of 28 pesticides based on monitoring and estimated maximum reside levels for 47 pesticides in or on dried chili peppers on the basis of MRLs established for fresh sweet and chili peppers. The intakes of the pesticides from spices and chili were compared to existing ADI and ARfD values only; intakes arising from other uses of the compounds were not considered.

Details of the intake estimations are given in Annex 5 to this report.

4.28 SPINOSAD (203)

RESIDUE AND ANALYTICAL ASPECTS

Spinosad was first evaluated by the 2001 JMPR, which established an ADI of 0–0.02 mg/kg bw. An ARfD was judged to be unnecessary. MRLs were recommended for fruits, vegetables, nuts, oil seeds, cereal grains, animal feeds and animal commodities. Questions about the MRL for milk were raised by the CCPR at its Thirty-fifth and Thirty-sixth Sessions, and the JMPR was requested to consider further how MRLs for milk and milk fat should be expressed.

The Meeting received information on registered uses and data from supervised residue trials on grapes and stored grain. Information on direct uses of spinosad on sheep for control of blowfly and lice and supporting residue data were also received.

Methods of analysis

An immunoassay method previously evaluated by the 2001 JMPR was used in the supervised trials on grapes.

The analytical method used for analysis of spinosad residues in cereal grains and processed products was based on previously evaluated methods. Samples were extracted with acetonitrile and water and the extracts cleaned up on a strong cation-exchange column. Spinosyns A, D, K, B and N-demethyl D were eluted with dilute ammonium acetate in acetonitrile and methanol, ready for analysis by HPLC with mass selective detection. The LOQ was 0.01 mg/kg.

The analytical methods used for fat, muscle, liver, kidney and bovine milk were similar in principle to the above method, but with variations in clean-up depending on the substrate. The LOQ for milk was 0.005 mg/kg, and that for the other substrates was 0.01 mg/kg.

Results of supervised trials on crops

Grape

The Meeting received the results of supervised trials for use of spinosad on grapes in the USA. The samples were analysed by immunoassay method GRM 96.11, which was evaluated by the JMPR previously.

In the USA, spinosad may be applied to grapes at 0.14 kg ai/ha with a maximum seasonal application of 0.49 kg ai/ha and harvesting 7 days after the final application. In 12 trials in the USA that conformed substantially to the registered use, the residue levels were: < 0.01, 0.02, 0.03, 0.05, 0.077, 0.082, 0.086, 0.13, 0.17, 0.22, 0.23 and 0.39 mg/kg.

The residue levels of spinosad in grapes in supervised trials in France, Italy and Spain are recorded in Table 39 (p. 761) of the JMPR Residue Evaluations for 2001. Spinosad may be used on grapes in Cyprus at 0.072 kg ai/ha with a PHI of 7 days. The conditions used in trials in France, Italy and Spain, where the application rate was 0.060 kg with a PHI of 5 days, were considered sufficiently similar to those of Cyprus GAP. The residue levels, determined by an HPLC method, in grapes in two trials in France (0.01 and 0.03 mg/kg), one trial in Italy (0.09 mg/kg) and one trial in Spain (0.19 mg/kg) were, in ranked order: 0.01, 0.03, 0.09 and 0.19 mg/kg. The levels in the same samples determined by the immunoassay method were: 0.02, 0.04, 0.15 and 0.24 mg/kg.

The Meeting combined the data from Europe and the USA obtained by the immunoassay method. The residue levels in the 16 trials, in ranked order, median underlined, were: < 0.01, 0.02 (two), 0.03, 0.04, 0.05, 0.077, <u>0.082</u>, <u>0.086</u>, 0.13, 0.15, 0.17, 0.22, 0.23, 0.24 and 0.39 mg/kg.

The Meeting estimated a maximum residue level of 0.5 mg/kg and an STMR value for spinosad in grapes of 0.084 mg/kg.

Fate of residues during storage

In a series of trials with cereal grain (barley, maize, oats, rice and wheat) in the USA, spinosad was applied to grain at a target rate of 1 g ai/t. Most of the trials were small-scale, only 18–23 kg grain being treated; in two larger trials, 9.9 t of maize and 30.9 t of wheat were treated. The duration of storage was 3–11 months at ambient temperatures. Samples were analysed by HPLC with mass spectromtery detection.

The residue levels in grain immediately after treatment represented 43–91% of the target application rate, reflecting the efficiency of application in the experiments. In the two larger trials, the initial residue levels were 77% and 87% of the target rates. The residue levels declined very slowly, if at all. The highest residue level in each trial was taken, whether at day 0 or after 11 months' storage. The trial in which a dose rate of 1.6 g ai/t was used was excluded as being outside GAP. The residue levels in the 20 trials were: 0.43, 0.45, 0.47, 0.58, 0.59, 0.63, 0.67, 0.69 (three), 0.70, 0.75, 0.79, 0.81, 0.86, 0.90, 0.91 (two), 0.93 and 0.95 mg/kg.

In three further trials in the USA in which wheat in storage batches of 135–225 t was treated at 1 g ai/t for storage and processing, the spinosad residue levels after 6 months' storage were 0.52, 0.73 and 0.79 mg/kg.

The residue levels in the 23 trials, in ranked order, were: 0.43, 0.45, 0.47, 0.52, 0.58, 0.59, 0.63, 0.67, 0.69 (three), <u>0.70</u>, 0.73, 0.75, 0.79 (two), 0.81, 0.86, 0.90, 0.91 (two), 0.93 and 0.95 mg/kg.

The Meeting estimated a maximum residue level of 1 mg/kg and an STMR for spinosad on cereal grains of 0.70 mg/kg on the basis of post-harvest use. The Meeting withdrew its previous recommendations for maize (0.01* mg/kg) and sorghum (1 mg/kg), to be replaced by the recommendation for cereal grains.

Fate of residues during processing

Three trials in France and one in Italy on processing of grapes to pomace and wine are summarized on pp. 823–824 of the JMPR Residue Evaluations of 2001.

In these trials, the levels of spinosad residues were below the LOQ (0.01 mg/kg) in all wine samples. As the residue levels in grapes were low (< 0.01–0.03 mg/kg), the best estimate of the processing factor for wine is < 0.33. Processing factors of 3.3 and 1.4 for juice and 1.6 for raisins produced from grapes were calculated in two trials in the USA. Juice was produced on a very small scale, with manual crushing, pressing and straining of about 1 kg of grapes, and this was not considered representative of a commercial process.

Processing studies on cereals were provided from the USA, comprising milling of maize (two trials), rice (two trials) and wheat (one trial) and three trials of wheat milling and baking. Spinosad residues were found essentially on the outside of the grain and were strongly concentrated in the aspirated grain fraction from the milling of maize and wheat. The residue levels in grits and flour were much lower than those in the grain. Most of the residues on rice remained with the husk and bran, with little occurring on white rice.

The following processing factors were calculated from the results of the trials. The factors are mean values, excluding those calculated from undetectable results, except for wine in which no residues were detected.

Commodity	Product	Processing factor	No. of trials
Grapes	Wine	< 0.33	4
	Raisins	1.6	1
Maize	Grits	0.082	2
	Flour	0.19	2
	Oil, dry milling	0.28	2
	Oil, wet milling	1.1	2
Rice	Hulls	2.8	2
	Bran	0.79	2
	Brown rice	0.11	2
	White rice	0.022	2
Wheat	Bran	2.0	4
	Shorts	1.2	4
	Flour	0.26	4
	Baked bread	0.14	3

The Meeting used the processing factors for wine, raisins and cereals to estimate STMR-Ps for processed commodities.

The processing factor for wine (< 0.33) was applied to the STMR for grape (0.084 mg/kg) to calculate an STMR-P of 0.028 mg/kg for wine.

The processing factor for raisins (1.6) was applied to the highest residue level in grapes (0.39 mg/kg) and the STMR for grape (0.084 mg/kg) to calculate a highest residue level of 0.62 mg/kg and an STMR-P of 0.13 mg/kg for raisins.

The Meeting estimated a maximum residue level of 1 mg/kg and an STMR-P value of 0.13 mg/kg for spinosad on dried grapes (currants, raisins and sultanas).

The processing factors for processed cereal fractions were applied to the STMR for cereal grains (0.70 mg/kg) to calculate the following STMR-P values: grits, 0.057 mg/kg; maize flour, 0.13 mg/kg; maize oil, 0.77 mg/kg; rice hulls, 2.0 mg/kg; rice bran, 0.55 mg/kg; brown rice, 0.077 mg/kg; white rice, 0.015 mg/kg; wheat bran, 1.4 mg/kg; wheat flour, 0.18 mg/kg; and white bread, 0.098 mg/kg.

The processing factor for wheat bran (2.0) was applied to the highest residue level in cereals grain (0.95 mg/kg) to calculate a highest residue level of 1.9 mg/kg for wheat bran.

The Meeting estimated a maximum residue level for spinosad on wheat bran of 2 mg/kg.

Spinosad

Residues in animal commodities

Direct treatment of animals

The Meeting received information on residue levels occurring in the tissues of sheep treated with spinosad in a plunge dip, by application of a pour-on formulation and by application of an aerosol to fly-strike wounds.

Sheep (25) were treated in a plunge dip containing 20 mg ai/l spinosad prepared from a 25 g/l suspension concentrate in a supervised trial in line with Australian guidelines and registered uses in Australia in 2000. Groups of animals were slaughtered for tissue collection on days 5, 15, 35, 49 and 63 after treatment. Samples were analysed by HPLC-MS after a conventional extraction and clean-up procedure evaluated by the 2001 JMPR. The highest levels of residues of spinosad in tissues 5 or 15 days after treatment were: 0.014 mg/kg in liver, 0.011 mg/kg in kidney, 0.011 mg/kg in muscle, 0.032 mg/kg in back fat and 0.094 mg/kg in peri-renal fat.

Two plunge dip trials on sheep in Australia were reported by the 2001 JMPR (Residue Evaluations, Table 73, pp. 813–814). The dip concentration was 10 mg ai/l. The highest tissue concentrations of spinosad residues were: < 0.01 mg/kg in liver, 0.014 mg/kg in kidney, < 0.01 mg/kg in muscle, 0.033 mg/kg in back fat and 0.042 mg/kg in peri-renal fat.

The data on residues from pour-on trials on sheep could not be evaluated because spinosad pour-on uses on sheep are not registered.

A spinosad aerosol spray is registered in Australia for treating fly-strike wounds on sheep. A typical wound of 200 cm^2 should take 6 s to treat. In a trial of the aerosol formulation in Australia in 2002, 14 sheep with fly-strike lesions measuring 108–1600 cm^2 were treated with spinosad according to the proposed label instructions and were slaughtered 2 and 7 days after treatment for tissue collection. The aerosol product contained 4 mg/g spinosad and 0.8 mg/g chlorhexidine digluconate (The registered product has 2.8 mg/g spinosad and 0.39 mg/g chlorhexidine digluconate.) and was delivered at a rate of 1.54 g of formulation per second. The animals were clipped around the fly-strike area, the area was measured and the dose was calculated at a rate of 1 s of aerosol spray per 40 cm^2 of affected area. The highest residue levels were: 0.04 mg/kg in liver, 0.03 mg/kg in kidney, 0.03 mg/kg in muscle, 0.14 mg/kg in back fat and 0.20 mg/kg in peri-renal fat.

The Meeting noted that the aerosol treatment resulted in higher residue levels in tissues than the plunge dip or the previously evaluated jetting treatment.

The Meeting estimated maximum residue levels for spinosad of 0.3 (fat) mg/kg in sheep meat and 0.1 mg/kg in edible offal of sheep.

Maximum residue levels

Spinosad residues can occur in meat and milk after direct use on animals or from residues in animal feeds.

The 2001 JMPR evaluated a feeding study with dairy cows, compiled a dietary burden for farm animals and estimated maximum residue levels of 2 mg/kg in cattle meat (fat), 0.5 mg/kg in cattle kidney and 0.5 mg/kg in cattle liver It estimated STMRs of 0.32 mg/kg in cattle fat, 0.010 mg/kg in cattle meat, 0.032 mg/kg in cattle kidney and 0.064 mg/kg in cattle liver. These estimates for cattle commodities were superseded by estimates derived from direct treatment of cattle, which resulted in higher residue levels. The MRL recommendations associated with direct treatment of cattle were: 3 mg/kg (fat) in cattle meat, 1 mg/kg in cattle kidney and 2 mg/kg in cattle liver.

The 2002 JMPR[1] introduced a policy of recommending maximum residue levels for mammalian meat and offal rather than MRLs for cattle meat and offal when residues occurred in feed and a suitable study of cattle feeding was available. In the light of this policy, the Meeting recommended that the 2001 recommendations be reviewed.

The current Meeting proposed maximum residue levels for mammalian meat and offal based on the results of the feeding study in dairy cows and the corresponding dietary burden. None of the recommendations for MRLs by the current Meeting change the previously estimated dietary burden of spinosad residue in cattle. The MRLs for cattle meat, liver and kidney were retained because they are related to direct treatment, which produces higher residue levels than occur from feed. Therefore, the MRLs for mammalian meat and offal should have the qualification 'except cattle'.

The Meeting estimated a maximum residue level of 2 (fat) mg/kg for 'Meat (from mammals other than marine mammals) [except cattle]' and associated STMRs of 0.01 mg/kg for meat and 0.32 mg/kg for fat.

The Meeting estimated a maximum residue level of 0.5 mg/kg for 'Edible offal (mammalian) [except cattle]' and associated STMRs of 0.064 mg/kg for liver and 0.032 mg/kg for kidney.

The Meeting withdrew the current recommendations for sheep meat (0.01* (fat) mg/kg) and edible offal of sheep (0.01* mg/kg), which are superseded by the recommendations for mammalian meat and offal. The Meeting also noted that residue levels resulting from direct treatment of sheep by jetting, plunge dipping and aerosol treatment of wounds did not exceed the maximum residue levels resulting from feed residues. There is no separate MRL recommendation for sheep related to these direct uses.

The CCPR expressed concern about the MRL for spinosad in milk, the levels of spinosad in milk fat and how MRLs might best be expressed for partially fat-soluble compounds in milk. (See also general report item 2.7 on fat-soluble pesticide residues in milk.)

The 2001 JMPR reported that, after direct treatment of dairy cows with spinosad, residues were measured in 119 samples of milk and cream and that the mean quotient of the concentration in cream divided by the concentration in milk was 4.2. A plot of the same residue levels in whole milk against those in cream showed that the residue level in milk was approximately 24% of that in cream (line of best fit through the origin). (See figure in section 2.7.)

The levels of spinosad residues in milk and cream from a feeding study in dairy cows are summarized in Table 79 of the JMPR Residue Evaluations of 2001. The mean quotient of the concentration in cream divided by the concentration in milk from cows at feeding levels of 1, 3 and 10 ppm was 4.0, in good agreement with the results for direct treatment.

The MRL for milk (1 mg/kg) was estimated on the basis of the highest residue level in milk, 0.65 mg/kg, after direct treatment. The calculated concentration in cream would then be 0.65 × 4.2 = 2.7 mg/kg. On the assumption that cream is approximately 50% fat, the concentration in fat would be about 5 mg/kg.

The Meeting estimated a maximum residue level for spinosad residues in cattle milk fat of 5 mg/kg.

DIETARY RISK ASSESSMENT

Long-term intake

The evaluation of spinosad resulted in recommendations for new MRLs and STMR values for raw and processed commodities. Data on consumption were available for 42 food commodities from this and previous evaluations and were used to calculate dietary intake. The results are shown in Annex 3.

---

[1] JMPR Report. 2002. 2.11. Maximum residue levels for animal commodities—group MRLs.

The IEDIs in the five GEMS/Food regional diets, based on estimated STMRs were 9-30% of the ADI (0-0.02 mg/kg bw). The Meeting concluded that long-term intake of residues of spinosad from uses that have been considered by the JMPR is unlikely to present a public health concern.

Short-term intake

The 2001 JMPR concluded that it was unnecessary to establish an ARfD for spinosad. The Meeting therefore concluded that short-term dietary intake of spinosad residues is unlikely to present a risk to consumers.

4.29 TRIADIMENOL (168) AND TRIADIMEFON (133)

TOXICOLOGY

The toxicity of triadimenol (((1RS,2RS;1RS,2SR)-1-(4-chlorophenoxy)-3,3-dimethyl-1-(1*H*-1,2,4-triazol-1-yl)butan-2-ol), a triazole fungicide, was evaluated by the 1989 JMPR, when an ADI of 0–0.05 mg/kg bw was established based on a NOAEL of 5 mg/kg bw per day in a two-generation study in rats. As currently manufactured, triadimenol is an 80:20 mixture of the diastereoisomers A (1RS,2SR) and B (1RS,2RS). Older studies of toxicity in the database were performed with 60:40 mixtures.

Triadimefon is closely chemically related to triadimenol, with which it shares some similar metabolic pathways in animals. The toxicity of triadimefon ((RS)-1-(4-chlorophenoxy)-3,3-dimethyl-1-(1*H*-1,2,4-triazol-1-yl)butan-2-one) was evaluated by the JMPR in 1981, 1983 and 1985. An ADI of 0–0.03 mg/kg bw was established based on a NOAEL of 50 ppm, equivalent to 2.5 mg/kg bw per day, in a 2-year study in rats.

Although triadimenol and triadimefon are independent active ingredients, on the basis of their close chemical and toxicological relationship they were re-evaluated together by the present Meeting within the periodic review programme of the CCPR. Triadimenol and triadimefon act as fungicides by blocking fungal ergosterol biosynthesis. The mechanism of action of these fungicides is inhibition of demethylation.

Triadimenol

In rats, radiolabelled triadimenol is rapidly absorbed from the gastrointestinal tract, with radioactivity reaching peak concentrations in most tissues between 1 and 4 h after dosing. Up to 90% of the administered dose was excreted, with an elimination half-life for the radiolabel of between 6 and 15 h. Excretion was essentially complete within 96 h. After 5–6 days, radioactivity in most organs was below the limits of quantification.

Renal excretion accounted for up to 21% of the orally administered dose in males and up to 48% in females. The remainder was found in the faeces. In bile-duct cannulated males 93% of the administered dose was recovered in the bile and only 6% in the urine. Thus a substantial amount of the administered dose undergoes enterohepatic recycling. Radioactivity in expired air was negligible.

Triadimenol was extensively metabolized, predominantly by oxidation of one of the *t*-butyl methyl groups to give hydroxy or carboxy derivatives. The putative intermediate triadimefon has not been isolated. Cleavage of the chloro-phenyl and the triazole group was of minor significance. In the urine and faeces most of the metabolites were not conjugated, but in bile the metabolites were found to be extensively glucuronidated.

Triadimenol has low to moderate acute toxicity. The acute oral LD_{50} both in mice and rats was in the range of 700 to 1500 mg/kg bw, with increasing toxicity for increasing isomer ratios A (1RS,2SR):B (1RS,2RS). This finding was supported by an oral LD_{50} of 579 mg/kg bw for isomer A and 5000 mg/kg bw for isomer B tested separately. In rats, the dermal LD_{50} was > 5000 mg/kg bw and the LC_{50} upon inhalation was > 0.954 mg/l of air (after an exposure of 4 h).

Triadimenol is not an eye or skin irritant in rabbits and is not a sensitizer in the maximization test in guinea-pigs.

In short-term studies in mice, rats and dogs, the main effect of triadimenol was on the liver.

In a study comparing the 80:20 and 60:40 isomer mixtures, rats were treated for 28 days by gavage. Both isomer compositions slightly increased motor activity at ≥ 45 mg/kg bw per day and induced mixed function oxidase activity and reversibly increased liver weight at 100 mg/kg bw per day. In mice fed diets containing triadimenol at a concentration of 160 to 4500 ppm for 13 weeks, one out of ten males at 4500 ppm died. In both sexes at ≥ 1500 ppm, there were increased liver weights accompanied by increased alanine aminotransferase and aspartate aminotransferase activities. Reduced erythrocyte volume fraction and increased mean corpuscular haemoglobin concentration were observed in females at the highest dose. The NOAEL was 500 ppm, equal to 76.8 mg/kg bw per day.

In two 3-month feeding studies in rats, liver weights were increased at ≥ 600 ppm (< 10% at 600 ppm), with cellular hypertrophy at 3000 ppm. Liver enzyme activities in serum were not increased. In one study, at the highest dose of 2400 ppm, kidney and ovary weights were also increased. At the highest doses in both studies, there were slight changes in some haematology parameters. The lowest NOAEL after oral administration in the short-term studies in rats was 600 ppm, equal to 39.6 mg/kg bw per day. In a 3-week study in rats treated by inhalation, no effects were observed at up to the highest dose of 2.2 mg/l of air.

In a 3-week study in rabbits, dermal application of triadimenol did not cause any dermal or systemic reactions at the highest dose tested, 250 mg/kg bw per day.

In a 3-month, a 6-month and a 2-year study, dogs were given diets containing triadimenol at concentrations of up to 2400 ppm. The only significant findings were decreased body-weight gain at 2400 ppm, liver and kidney weight increases at the highest doses and increased P450 levels. The overall NOAEL was 600 ppm, equal to 21.1 mg/kg bw per day.

In two long-term studies, mice were given diets containing triadimenol at a concentration of up to 2000 ppm. In one study, Crl:CD-1(ICR)BR mice were kept for 80 weeks and in the other study CF_1/WF 74 mice were kept for 2 years. At 2000 ppm, reduced body-weight gains were recorded and liver weights were increased, as were testes weights in one study. Additionally, liver enzyme activity was higher. In one study, histopathological examination of the liver showed more basophilic foci at ≥ 80 ppm, predominantly in males, but there was a poor dose–response relationship and similar values have been reported in control groups in other studies. Hepatocellular hypertrophy and single cell necrosis were found at ≥ 400 ppm. At 2000 ppm, additional histopathological changes to the liver were reported. At the intermediate dose, 400 ppm, but not at the highest dose, males had slightly more liver adenomas and carcinomas. There was no clear dose–response relationship, and values were within the historical control range of 6–17%. In females at the highest dose, two out of 50 animals had luteomas; this was within the range for historical controls of 0.9–10%. In the other study, females at the intermediate and highest dose had more liver adenomas and in both sexes at the highest dose, the incidences of liver hyperplastic nodules and thyroid cystic alterations were increased. The increase in liver adenomas is a common finding in mice, which is considered to be of questionable relevance for humans. The overall NOAEL was 500 ppm, equal to 140 mg/kg bw per day.

In a long-term feeding study in rats, at the highest concentration of 2000 ppm reduced body-weight gain was found in both sexes, as were changes in the weights of a number of organs, including spleen, lung and testes. However, there was a poor relationship with dose. In females, kidney, liver and ovarian weights were higher at the highest dose. In both sexes at 2000 ppm, the activities of liver enzymes (alanine aminotransferase and aspartate aminotransferase in both sexes and glutamate dehydrogenase in males) were slightly increased. At the highest dose, minor changes in haematology parameters were at the borderline of the physiological range at some time-points. There was no histopathological evidence for any non-neoplastic or neoplastic changes. The NOAEL was 500 ppm, equal to 25 mg/kg bw per day.

In a series of studies of genotoxicity in vitro and in vivo, triadimenol consistently gave negative results. The Meeting concluded that triadimenol is unlikely to be genotoxic.

In view of the lack of genotoxicity observed and the finding of liver tumours only in female mice and only at concentrations at which liver toxicity was observed, the Meeting concluded that triadimenol is not likely to pose a carcinogenic risk to humans.

To study reproductive performance during exposure to triadimenol, two- and three-generation feeding studies were performed in rats given diets containing triadimenol at concentrations of up to 500 ppm and up to 2000 ppm, respectively. In the study in which the higher doses were administered, matings in all three generations consistently showed reduced fertility at ≥ 500 ppm; in F_0 matings, this finding was observed at 125 ppm. Reduced viability was observed in F_1 pups of both matings at 2000 ppm, F_2 pups from the first mating at ≥ 500 ppm and F_2 pups of the second mating at 2000 ppm. All F_3 pups from the first mating died at ≥ 500 ppm, but not those from the second mating. At 500 ppm, increased testicular and ovarian weights were observed in F_{1b} parents in the study in which lower doses were administered and increased testicular weights in the F_{2b} parents at 2000 ppm. The lowest NOAEL in these studies was 100 ppm, equal to 8.6 mg/kg bw per day

Several studies of developmental toxicity were performed in rats, over a dose range of 5 to 120 mg/kg bw per day. In one study, an increase in supernumerary lumbar ribs was found at ≥ 25 mg/kg bw per day, and in another study there was an increase in postimplantation losses at 120 mg/kg bw per day. In three out of the four studies, increased placental weights were noted at doses of 30 to 100 mg/kg bw per day. Such effects have been reported with other azoles. Triadimenol did not induce malformations in studies of developmental toxicity and clear NOAELs for developmental toxicity could be established; the lowest NOAEL was 15 mg/kg bw per day.

The NOAEL for offspring toxicity in rabbits was 4 mg/kg bw per day, on the basis of slightly increased postimplantation losses at the maternally toxic dose of 200 mg/kg bw per day.

Clinical signs (general restlessness, alternating phases of increased and reduced motility, aggressivity) observed during tests for acute toxicity suggested possible effects on the central nervous system.

The Meeting concluded that the existing database on triadimenol was adequate to characterize the potential hazards to fetuses, infants and children.

A medical survey of personnel working in the production of triadimenol gave no indication of any substance-related effects.

Toxicological evaluation

Although a series of tests for acute neurotoxicity in mice were available, a NOAEL for triadimenol for neurotoxicity could not be identified because of technical shortcomings in these studies. As triadimenol is closely related to triadimefon in terms of chemical structure and toxicological effects, and in the view of the lack of sound studies of neurotoxicity with triadimenol, the Meeting concluded that studies of neurotoxicity performed with triadimefon could serve as a basis for derivating an ADI and an ARfD for triadimenol. This was supported by evidence for similar neurotoxic potential in a published study of acute toxicity with triadimenol and triadimefon.

The Meeting established an ADI of 0–0.03 mg/kg based on the NOAEL of 3.4 mg/kg bw per day for hyperactivity in a study of neurotoxicity with triadimefon in a 13-week feeding study in rats, and with a safety factor of 100.

The Meeting established an ARfD of 0.08 mg/kg bw on the basis of the NOAEL of 2 mg/kg bw for hyperactivity in a study of acute neurotoxicity in rats treated with triadimefon by gavage. A safety factor of 25 was applied because the effects were C_{max}-dependent and reversible (see comments on triadimefon).

A toxicological monograph was prepared for triadimenol and triadimefon.

Triadimefon

In a study on the absorption, distribution, metabolism and excretion of triadimefon in rats, the dose given and pretreatment with non-labelled triadimefon did not significantly affect excretion and metabolism patterns. In males about one-third and in females about two-thirds of the administered dose was excreted in the urine and vice versa in the faeces. After 96 h, 2% of the radioactivity remained in females and 9% in males, with the highest residue levels found in liver and kidneys.

The metabolism of triadimefon starts either by direct oxidation of a *t*-butyl methyl group to the hydroxy or the carboxy compound with subsequent glucuronidation, or these steps are preceded by reduction of the keto group of triadimefon to the putative intermediate, triadimenol. Therefore, many of the metabolites found in triadimenol metabolism studies are also found with triadimefon. Nevertheless, the metabolism of triadimefon in rats provides a pathway for demethylation of the *t*-butyl group, which is not seen with triadimenol. This might be owing to very low biotransformation of triadimenol via triadimefon as intermediate.

The acute oral LD_{50} in mice and rats was in the range of 363 to 1855 mg/kg bw. The dermal LD_{50} was > 5000 mg/kg bw and the LC_{50} on inhalation was > 3.27 mg/l of air.

In rabbits, a few treatment-related effects including skin and eye irritation were recorded, but the irritation potential of triadimefon was very low. In guinea-pigs, technical-grade triadimefon of low purity was a sensitizer in the Büehler test for skin sensitization. However, purified triadimefon did not have any sensitizing potential in guinea-pigs in the Magnusson & Kligman maximization test, even after induction with technical-grade triadimefon of low purity.

In short-term studies in rats and dogs, the main effects of triadimefon were on the liver.

In three short-term studies in rats (treated by gavage at doses of up to 30 mg/kg bw per day for 30 days, by gavage at doses of up to 25 mg/kg bw per day for 4 weeks and given diets containing triadimefon at concentrations of up to 2000 ppm for 12 weeks) the overall NOAEL was 150 mg/kg bw per day, the highest dose tested.

In two studies in dogs fed diets containing triadimefon for 13 weeks and 2 years, the highest concentrations administered were 2400 ppm and 2000 ppm, respectively. Body-weight decreases, relative liver weight increases and liver enzyme induction were observed predominantly in the group receiving the highest dose, and, in the short-term study only, there were also effects on haematology parameters. The overall NOAEL in these studies was 600 ppm, equal to 17.3 mg/kg bw per day, in the 2-year study.

The dermal application of triadimefon at 1000 mg/kg bw per day to rats for 3 weeks (6 h per day for 5 days per week) caused diffuse acanthosis at the application site and increased activity and reactivity. The NOAEL was 300 mg/kg bw per day. The dermal application of triadimefon at 50 and 250 mg/kg bw per day to rabbits for 4 weeks (5 days per week) caused mild erythema at the application sites. Rats exposed by inhalation to triadimefon at 0.3 mg/l of air had reduced body-weight gain and increased liver weights.

In two 2-year feeding studies in mice, severely decreased body-weight gains, changes in several haematology parameters and increased liver weights and increased enzyme activity were observed at the highest dietary concentration of 1800 ppm. Starting at 300 ppm, histopathological changes, including nodular changes, hypertrophy and single cell necrosis, were found in the liver. These effects were more pronounced at the highest dose, and in one study an increase in hepatocellular adenomas was also reported. In the other study, a re-examination of histopathology slides led to re-classification of findings for adenomas and carcinomas. Owing to incomplete re-examination, a final conclusion on whether the incidences were increased or not was not possible. However, liver adenomas in the presence of liver toxicity in mice are generally not believed to be of toxicological concern for humans.

The lowest NOAEL was 50 ppm in the feed, equal to 13.5 mg/kg bw per day, on the basis of nodular changes and single cell necrosis in the liver at 300 ppm.

With the exception of behavioural changes and severe histopathological lesions in several organs observed in one study at the highest dose of 5000 ppm, the toxicological profile in two 2-year feeding studies in rats was very similar to that of the studies in mice. After 23 weeks of exposure to the highest dose at 5000 ppm, animals showed violent activity and refused the feed and became moribund. The surviving animals in this group were terminated at week 39. They showed haemorrhagic lesions in the stomach mucosa, blood-filled and dilated alveolar vessels, degenerative processes in proximal kidney tubules of females, atrophied spleens with signs of decreased haematopoesis, some giant spermatids in testes and decreased haematopoesis in the bone marrow of males. At the lower dietary concentrations of 1800 and 500 ppm, reduced body-weight gains, increased liver weights and mildly increased liver enzyme activities were recorded. In one study, ovary weights were higher and adrenal weights lower. Mild effects on haematology were found in both studies. In one study at the highest dietary concentration of 1800 ppm, a marginal increase in thyroid cystic hyperplasias and more thyroid follicular adenomas (five versus zero for both sexes taken together) were found. When compared with historical controls, this effect was not significant. The overall NOAEL was 300 ppm, equal to 16.4 mg/kg bw per day.

In a series of studies of genotoxicity in vitro and in vivo, all results were consistently negative. The Meeting concluded that triadimefon is unlikely to be genotoxic

In view of the lack of genotoxicity and the finding only of liver adenomas in mice and equivocal changes in thyroid follicular adenomas in rats at concentrations at which organ toxicity was observed, the Meeting concluded that triadimefon is not likely to pose a carcinogenic risk to humans.

In two related multigeneration studies, rats received diets containing triadimefon at concentrations of up to 1800 ppm. Maternal and pup weight development was reduced at doses of ≥ 300 ppm and, in the first generation at the highest dose, the viability of the pups was reduced. At the highest dose, two matings of the F_1 animals to give F_2 generation pups resulted in one female becoming pregnant in the first mating and none in the second. In the second study, again at 1800 ppm, the fertility of the F_0 generation was not affected, but that of the F_1 generation was, albeit not to the same extent as in the first study. Viability and pup weights were reduced. In a cross mating in which only one sex was exposed to triadimefon, only the matings with exposed males gave significantly reduced fertility, correlating with reduced insemination indices. Therefore, reduced fertility seemed to have resulted mainly from impaired mounting willingness of exposed males. In males at the highest dose, the concentration of testosterone was double that in control males, and testes weights were increased. However, no correlation between individual testosterone levels and spermiograms and mating willingness was observed, although reduced mating willingness did appear to correlate with reduced body weight. It appears that prenatal, but not postnatal, exposure of males affected mating willingness. The lowest NOAEL was 50 ppm, equivalent to 3.75 mg/kg bw per day, based on a LOAEL of 1800 ppm for reproductive effects.

In studies of developmental toxicity in rats treated by inhalation (one study) and by gavage (two studies), inhalation exposure at air concentrations of up to 0.114 mg/l of air on days 6–15 of gestation did not result in any findings indicative of developmental toxicity. In the studies of rats treated by gavage, however, supernumerary ribs in one study at 90 mg/kg bw per day, increased placental weights at 100 mg/kg bw per day, and cleft palates at doses of ≥ 75 mg/kg bw per day were found. These doses also reduced the body-weight gains of dams by up to 50% over the exposure period, but not when averaged over the whole gestation period. In four studies in rabbits, body-weight loss in dams was observed at a dose of ≥ 30 mg/kg bw per day. Over the dose range of 60 to 120 mg/kg bw per day, increased litter losses, and caudal vertebrae malformations and cleft palates were found either in one or the other study and delayed ossification and scapula malformations were observed in both studies. Additionally, in one study, the uncommon finding of umbilical hernia was recorded in pups at 60 and 80 mg/kg bw per day. Scapula deformations were also found at 40 mg/kg bw per day, the lowest dose tested in the study. Overall, the lowest NOAEL for offspring toxicity was 20 mg/kg bw on the basis of scapula deformations at 40 mg/kg bw in rabbits.

Several studies provide evidence that triadimefon has neurotoxic potential. In a study in which single doses of triadimefon were administered by gavage and in a 13-week feeding study, several signs of hyperactivity, increased motility and stereotypic behaviour were found. The NOAEL in the former study was 2 mg/kg bw on the basis of reversible neurotoxic effects at 35 mg/kg bw. These were considered to be C_{max}-dependent effects in view of the fact that a dose of 54.6 mg/kg bw per day in the short-term feeding study caused similar effects only after several days. The NOAEL for this study was 50 ppm, equivalent to 3.4 mg/kg bw. In a comparative study of acute neurotoxicity in Long Evans rats treated by gavage with a group of 14 triazoles or structurally related compounds, hyperactivity at 100 mg/kg bw, but not at 50 mg/kg bw, was recorded for both triadimenol and triadimefon. In this study, the dose–response curves for triadimenol and triadimefon were very similar, suggesting a common mechanism of neurotoxicity.

The Meeting concluded that the existing database on triadimefon was adequate to characterize the potential hazards to fetuses, infants and children.

A medical survey of the personnel working in the production of triadimefon gave no indication of any substance-related effects.

Toxicological evaluation

The Meeting established an ADI of 0–0.03 mg/kg bw on the basis of the NOAEL of 3.4 mg/kg bw per day for hyperactivity in a study of neurotoxicity in rats fed with triadimefon and a safety factor of 100.

The Meeting established an ARfD of 0.08 mg/kg bw based on the NOAEL of 2 mg/kg bw for hyperactivity in a study of acute neurotoxicity in rats given triadimefon by gavage. A safety factor of 25 was used since the effects were C_{max}-dependent and reversible.

A toxicological monograph was prepared for triadimenol and triadimefon.

Plant metabolites of triadimefon, triadimenol and other triazole fungicides

Triazole, triazolylalanine and triazole acetic acid are plant metabolites of several triazole fungicides, including triadimenol and triadimefon.

After oral administration of triazole, triazolylalanine and triazole acetic acid to rats, these compounds are rapidly and completely absorbed. Urinary excretion is the main excretion pathway for 90% or more of the administered dose, and only a few percent are found in the faeces. Except for triazolylalanine, which is metabolized to a minor extent to *N*-acetyltriazolylalanine, these compounds are virtually not metabolized and are excreted unchanged. Owing to rapid and complete excretion, there is no potential for accumulation in the body for any of these plant metabolites.

The acute oral toxicity of all three compounds is low, with LD_{50}s of > 5000 mg/kg bw, except for triazole, with an LD_{50} of 1649 mg/kg bw.

Only a few tests for genotoxicity have been performed on triazole and triazole acetic acid and all gave negative results. Triazolylalanine was more extensively tested; only one test for cell transformation in vitro gave a positive result, while the results of another similar test and all other tests were negative.

In a 3-month feeding study in rats, triazole induced fat deposition in the liver and changes in haematological parameters at the highest dose of 2500 ppm. In 3-month feeding studies in rats and, the only effect of triazolylalanine was to reduce body-weight gain at the highest dose of 20 000 ppm. No effects were recorded in a 2-week study in rats fed with triazole acetic acid at the highest dose of 8000 ppm.

In a study of developmental toxicity with triazole in rats, at ≥ 100 mg/kg bw per day fetuses showed increased incidence of undescended testicles and at 200 mg/kg bw per day malformations of the hind legs were found. In studies of reproductive and developmental toxicity with triazolylalanine in rats, only very minor effects on pups, indicative of general toxicity, such as reduced birth weights and retarded ossification processes were found at high doses. There were no studies of reproductive and developmental toxicity with triazole acetic acid.

Since triazolylalanine and triazole acetic acid were of low systemic toxicity and developmental effects with triazole occur at doses of ≥ 100 mg/kg bw per day, these metabolites were judged not to pose an additional risk to humans.

Levels relevant to risk assessment of triadimenol*

Species	Study	Effect	NOAEL	LOAEL
Mouse	80-week study of toxicity and carcinogenicity[a]	Toxicity	500 ppm, equal to 140 mg/kg bw per day	2000 ppm, equal to 620 mg/kg bw per day
		Carcinogenicity	500 ppm, equal to 140 mg/kg bw per day	2000 ppm, equal to 620 mg/kg bw per day
Rat	2-year study of toxicity and carcinogenicity[a]	Toxicity	500 ppm, equal to 25 mg/kg bw per day	2000 ppm, equal to 105 mg/kg bw per day
		Carcinogenicity	2000 ppm, equal to 105 mg/kg bw per day[c]	—
	Two-generation study of reproductive toxicity[a]	Parental toxicity	100 ppm, equal to 8.6 mg/kg bw per day	500 ppm, equal to 43.0 mg/kg bw per day
		Pup toxicity	100 ppm, equal to 8.6 mg/kg bw per day	500 ppm, equal to 43.0 mg/kg bw per day
	Developmental toxicity[b]	Maternal toxicity	25 mg/kg bw per day	60 mg/kg bw per day
		Embryo- and fetotoxicity	15 mg/kg bw per day	25 mg/kg bw per day
Rabbit	Developmental toxicity[b]	Maternal toxicity	40 mg/kg bw per day	200 mg/kg bw per day
		Embryo- and fetotoxicity	40 mg/kg bw per day	200 mg/kg bw per day
Dog	13-week study of toxicity[a]	Toxicity	600 ppm equal to 21.1 mg/kg bw per day	2400 ppm equal to 85.9 mg/kg bw per day

* See comments on triadimefon
[a] Diet
[b] Gavage
[c] Highest dose tested

Estimate of acceptable daily intake for humans

0–0.03 mg/kg bw

Estimate of acute reference dose

0.08 mg/kg bw

Studies that would provide information useful for the continued evaluation of the compound

Further observations in humans

Critical end-points for establishing guidance values for exposure to triadimenol

Absorption, distribution, metabolism and excretion in animals

Rate and extent of oral absorption	Rapid (peak within 1.5 h); > 90%
Distribution	Widely distributed
Potential for accumulation	Low, half lives of 6–15 h
Rate and extent of excretion	79–90% within 24 h
Metabolism	Very extensive; predominantly oxidation of *t*-butyl methyl group
Toxicologically significant compounds (animals, plants and the environment)	Triadimenol, triadimefon, triazole

Acute toxicity

Rat, LD_{50}, oral	579–5000 mg/kg bw (varies with isomer composition)
Rat, LD_{50}, dermal	> 5000 mg/kg bw
Rat, LC_{50}, inhalation	> 0.95 mg/l
Rabbit, dermal irritation	Not irritating
Rabbit, eye irritation	Not irritating
Skin sensitization	Not sensitizing (Magnusson & Kligman maximization test)

Short-term studies of toxicity

Critical effects	Liver toxicity (2-year study in dogs)
Lowest NOAEL	21.1 mg/kg bw

Genotoxicity   Negative results in vitro and in vivo

Long-term studies of toxicity and carcinogenicity

Critical effects	Body and organ weight changes (2-year study in rats)
Lowest NOAEL	25 mg/kg bw
Carcinogenicity	Liver adenomas in female mice; unlikely to pose a carcinogenic risk to humans

Reproductive toxicity

Critical effects	Increased ovary and testes weights (rat)
Lowest reproductive NOAEL	8.6 mg/kg bw
Critical effects	Increased supernumerary lumbar ribs; not teratogenic (rat)
Lowest developmental NOAEL	15 mg/kg bw

Neurotoxicity/delayed neurotoxicity

Critical effects at LOAEL	See triadimefon
Lowest NOAEL	See triadimefon

Other toxicological studies   Metabolites are of no greater toxicological concern than the parent

Medical data   No effects on health in manufacturing personnel

Summary	Value	Study	Safety factor
ADI	0–0.03 mg/kg bw	Rat, short-term study of neurotoxicity with triadimefon (see triadimefon)	100
ARfD	0.08 mg/kg bw	Rat, study of acute neurotoxicity with triadimefon (see triadimefon)	25

Levels relevant to risk assessment of triadimefon

Species	Study	Effect	NOAEL	LOAEL
Mouse	21-month study of toxicity and carcinogenicity[a]	Toxicity	50 ppm, equal to 13.5 mg/kg bw per day	300 ppm, equal to 76 mg/kg bw per day
		Carcinogenicity	300 ppm, equal to 76 mg/kg bw per day	1800 ppm, equal to 550 mg/kg bw per day
Rat	105-week study of toxicity and carcinogenicity[a]	Toxicity	300 ppm, equal to 16.4 mg/kg bw per day	1800 ppm, equal to 114 mg/kg bw per day
		Carcinogenicity	1800 ppm, equal to 114 mg/kg bw per day[c]	—
	Two-generation study of reproductive toxicity[a]	Parental toxicity	300 ppm, equal to 22.8 mg/kg bw per day	1800 ppm, equal to 136.8 mg/kg bw per day
		Pup toxicity	300 ppm, equal to 22.8 mg/kg bw per day	1800 ppm, equal to 136.8 mg/kg bw per day
	Developmental toxicity[b]	Maternal toxicity	10 mg/kg bw per day	30 mg/kg bw per day
		Embryo- and fetotoxicity	30 mg/kg bw per day	90 mg/kg bw per day
	Acute neurotoxicity[b]	Neurotoxicity	2 mg/kg bw	35 mg/kg bw
	13-week study of neurotoxicity[a]	Neurotoxicity	50 ppm, equivalent to 3.4 mg/kg bw per day	800 ppm, equivalent to 54.6 mg/kg bw per day
Rabbit	Developmental toxicity[b]	Maternal toxicity	10 mg/kg bw per day	30 mg/kg bw per day
		Embryo- and fetotoxicity	20 mg/kg bw per day	50 mg/kg bw per day
Dog	2-year study of toxicity[a]	Toxicity	300 ppm equal to 11.7 mg/kg bw per day	200 ppm equal to 48.8 mg/kg bw per day

[a] Diet
[b] Gavage
[c] Highest dose tested

Estimate of acceptable daily intake for humans

    0–0.03 mg/kg bw

Estimate of acute reference dose

    0.08 mg/kg bw

Studies that would provide information useful for the continued evaluation of the compound

    Further observations in humans

Critical end-points for setting guidance values for exposure to triadimenol

Absorption, distribution, metabolism and excretion in animals	
Rate and extent of oral absorption	≥ 28% in females, ≥ 67% in males as urinary excretion
Distribution	Widely distributed in kidneys and liver
Potential for accumulation	Low
Rate and extent of excretion	90–98% excretion within 96 h
Metabolism	Very extensive; predominantly oxidation of tert-butyl methyl group
Toxicologically significant compounds (plants, animals and the environment)	Triadimenol, triadimefon, triazole
Acute toxicity	
Rat, LD_{50}, oral	363–1855 mg/kg bw
Rat, LD_{50}, dermal	> 5000 mg/kg bw
Rat, LC_{50}, inhalation	> 3.27 mg/l
Rabbit, dermal irritation	Not irritating
Rabbit, eye irritation	Not irritating
Skin sensitization	Technical-grade triadimefon is sensitizing, purified triadimefon is not sensitizing (Büehler, and Magnusson & Kligman maximization tests)
Short-term studies of toxicity	
Critical effects	Liver effects (dog)
Lowest NOAEL	17.3 mg/kg bw
Genotoxicity	Negative in vitro and in vivo
Long-term studies of toxicity and carcinogenicity	
Critical effects	Liver nodular changes, hypertrophy and single cell necrosis
Lowest NOAEL	13.5 mg/kg bw per day
Carcinogenicity	Liver adenomas in mice; unlikely to pose a carcinogenic risk to humans
Reproductive toxicity	
Critical effects	Impaired reproductive performance (rat)
Lowest reproductive NOAEL	22.8 mg/kg bw per day
Critical effects	Scapula malformations at maternal toxic doses (rabbit)
Lowest developmental NOAEL	20 mg/kg bw per day
Neurotoxicity/delayed neurotoxicity	
Critical effects	Increased activity in study of acute neurotoxicity after gavage administration (rat)
Lowest NOAEL	2 mg/kg bw
Critical effects	Increased activity in short-term feeding study (rat)
Lowest NOAEL	3.4 mg/kg bw
Other toxicological studies	Metabolites are of no greater toxicological concern than the parent
Medical data	No effects on health in manufacturing personnel

Summary	Value	Study	Safety factor
ADI	0–0.03 mg/kg	Rat, short-term study of neurotoxicity	100
ARfD	0.08 mg/kg	Rat, study of acute neurotoxicity	25

DIETARY RISK ASSESSMENT

Triadimenol

Long-term intake

Theoretical maximum daily intakes were calculated for the commodities of human consumption for which Codex MRLs existed (Annex 3). The theoretical maximum daily intakes in the five GEMS/Food regional diets represented 1–20% of the maximum ADI (0.03 mg/kg bw per day). The Meeting concluded that the long-term intake of residues of triadimenol resulting from uses of triadimenol and triadimefon considered by the JMPR is unlikely to present a public heath concern.

Short-term intake

An ARfD of 0.08 mg/kg bw was established for triadimenol by the present Meeting. IESTIs could not, however, be calculated, as the residues of the compound were evaluated by the Meeting before the procedures for estimation of STMR and highest residue values were established. Triadimenol was scheduled for periodic evaluation of residues in 2006, when the risk assessment would be finalized.

Triadimefon

Long-term intake

Theoretical maximum daily intakes were calculated for the commodities of human consumption for which Codex MRLs existed (Annex 3). The theoretical maximum daily intakes in the five GEMS/Food regional diets represented 1–6% of the maximum ADI (0.03 mg/kg bw per day). The Meeting concluded that the long-term intake of residues of triadimefon considered by the JMPR is unlikely to present a public heath concern.

Short-term intake

An ARfD of 0.08 mg/kg bw was established for triadimefon by the present Meeting. IESTIs could not, however, be calculated, as the residues of the compound were evaluated by the Meeting before the procedures for estimation of STMR and highest residue values were established. Triadimefon was scheduled for periodic evaluation of residues in 2006, when the risk assessment would be finalized.

4.30 TRIFLOXYSTROBIN (213)

TOXICOLOGY

Trifloxystrobin (methyl(*E*)-methoxyimino-{(*E*)-α-[1-(α,α,α,-trifluoro-*meta*-tolyl)ethylideneaminooxy]-*ortho*-tolyl}acetate) is a new broad-spectrum foliar fungicide, which is a synthetic analogue of the naturally occurring strobilurins. Trifloxystrobin has not been evaluated previously by the JMPR.

After oral administration, radiolabelled trifloxystrobin was rapidly and appreciably absorbed (66% of the administered dose) in rats of both sexes. The main route of elimination (63–84%) was in the faeces; some of the fecal elimination was via bile (30–45%) while only one-third or less of the administered dose was excreted in the urine, and none through expired air. There was almost complete degradation of trifloxystrobin after single low dose, at 0.5 mg/kg bw, but up to 45% was eliminated unchanged in the faeces after a high dose, at 100 mg/kg bw. The metabolite pattern in rats is very complex; about 35 metabolites were identified in the urine, faeces and bile. The main steps in the metabolic pathway include hydrolysis of the methyl ester to the corresponding acid, *O*-demethylation of the methoxyimino group yielding a hydroxyimino compound and oxidation of the ethylideneamino methyl group to a primary alcohol and then to the corresponding carboxylic acid. These steps are followed by a complex pattern of further, minor reactions. Cleavage between the glyoxylphenyl and trifluormethylphenyl moieties accounted for about 10% of the administered dose.

The metabolism of trifloxystrobin in plants is similar to that in animals and occurs primarily via cleavage of the methyl ester group to form CGA 321113 (*E,E*)-methoxyimino-{2-[1-(3-trifluoro methyl-phenyl)-ethylideneaminooxymethyl]-phenyl}-acetic acid. In the rat, this metabolite undergoes further hydroxylation and conjugation (glucuronide and sulfate) at the trifluoromethyl phenyl ring. In goat liver, taurine and glycine conjugates of CGA 321113 were the principal residue components (up to 28% of the total radioactive residues). Conjugated metabolites are generally less toxic and more rapidly excreted than an unconjugated parent compound. Being biotransformation products in the rat, CGA 321113 and its metabolites are assumed to have been adequately tested and accounted for in rats given trifloxystrobin. Also, CGA 321113 is not likely to be more toxic than trifloxystrobin.

Dermal absorption of trifloxystrobin in rats was low and slightly decreased with increasing dose. Compared with human epidermis, rat epidermis was 9 and 19 times more permeable in a test in vitro at a dose of 0.24 and 10.27 mg/cm^2, respectively. In a study of absorption in vivo, in which a low or a high dose of radiolabelled trifloxystrobin was applied to the shaved backs of male rats, the amount of recovered radioactivity in the blood was low, but the overall absorption was moderate, ranging from 5 to 10% in 24 h and increasing to 16% at 48 h.

Trifloxystrobin has low acute oral toxicity in rats and mice (LD$_{50}$, > 5000 mg/kg), low acute dermal toxicity in rats and rabbits (LD$_{50}$, > 2000 mg/kg), low acute inhalation toxicity in rats (LC$_{50}$, > 4.65 mg/l), is not a skin irritant in rabbits, is a moderate eye irritant in unwashed rabbit eyes but is not irritating in washed rabbit eyes. It is a skin sensitizer in guinea-pigs, according to the Magnusson & Kligman maximization test, but is not a skin sensitizer in guinea-pigs according to the Buehler test.

In studies of toxicity with repeated doses, slight decreases (5–10%) in body weight and/or body-weight gain were regarded as non-adverse in the absence of other effects.

In studies of repeated doses in mice, the liver and spleen were the principal target organs at the same or higher doses than those affecting body weight and food efficiency. In the 90-day study in male and female mice, liver weight was increased and there were findings on microscopy, including hepatocyte hypertrophy and focal or single cell necrosis. There were also increased incidences of extramedullary haematopoiesis in the spleen at doses of ≥ 315 mg/kg bw per day. The NOAEL for these effects was 77 mg/kg bw per day.

In a 90-day dietary study in rats, the NOAEL was 31 mg/kg bw per day on the basis of statistically significantly decreased body-weight gain of 20% and 40% in males and females, respectively, increased relative liver weights, changes in clinical chemistry and liver histopathology findings (mainly hepatocellular hypertrophy), in addition to atrophy of the pancreas at the next higher dose of 127 mg/kg bw per day.

At or above a daily dose of trifloxystrobin at 150 mg/kg bw per day for 3 months or 50 mg/kg bw per day for 1 year, dogs had episodes of diarrhoea, vomiting, reduced food intake, increased relative liver weight and hepatocyte hypertrophy, in addition to changes in clinical chemistry parameters indicative of liver toxicity and/or perturbed metabolism, dehydration, poor nutrition and possible starvation. Body weights were also affected. In the 3-month study, animals of both sexes had body-weight loss of about 0.4 kg and 2.8 kg at 150 and 500 mg/kg bw per day, respectively. In the 1-year study, body-weight gain in females at 50 and 200 mg/kg bw per day were decreased throughout the study, and at week 52 body-weight gain was about 20% below control values. The NOAELs were 30 and 5 mg/kg bw per day in the 3-month and 1-year studies, respectively.

The long-term study of toxicity and carcinogenicity with trifloxystrobin was evaluated in bioassays in mice and rats. In the 18-month dietary feeding study in mice, the NOAEL was 36 mg/kg bw per day on the basis of liver effects, including increased liver weight (both sexes) and increased single cell necrosis (males), in addition to impaired body-weight gain (females). There was no evidence of carcinogenicity in mice tested at adequate doses.

In the 2-year study in rats, the NOAEL was 30 mg/kg bw per day on the basis of statistically significantly retarded body-weight gain in males (11–17%) and females (17–27%) and decreased food consumption (by 4% and 8%, respectively) and increased relative weights of heart, liver and kidneys (each by about 20%) in females at the highest dose of 62 mg/kg bw per day. The overall incidence of tumours was lower in the treated animals. Benign adrenal medullary tumours (10% versus 0% in controls) and haemangioma in the mesenteric lymph nodes (10.2% versus 0% in controls) were increased in male rats at the highest dose tested. Incidences of the adrenal medullary tumours were within the range of incidences for historical controls. The incidence of haemangioma in the mesenteric lymph nodes in males of the high dose group was outside the range of historical control incidences. There was markedly reduced mortality in the group receiving the highest dose tested, and this may have contributed to the higher incidence of tumours in this group compared to controls. In ageing male rats of this strain, degenerative lesions associated with the mesenteric lymph nodes are common and are hard to distinguish from neoplastic lesions (haemangiomas). Some age-associated non-neoplastic findings, such as angiomatous hyperplasia of the mesenteric lymph nodes, were increased in males at the highest dose and the increases were correlated with decreased food intake and a lower body-weight development.

The Meeting concluded that there was no treatment-related carcinogenicity of any toxicological concern.

A wide range of assays for genotoxic potential with trifloxystrobin were conducted in vitro and in vivo, including testing for gene mutation, chromosomal damage and DNA repair. At or near cytotoxic doses and in the presence of metabolic activation, trifloxystrobin was weakly mutagenic at cytotoxic doses in the test for forward gene mutation in Chinese hamster V79 cells. Results were equivocal in the absence of metabolic activation. Metabolites of trifloxystrobin [CGA 357261 (*Z*, *E*-isomer), CGA 373466 and NOA 414412] were not mutagenic in the Ames test. The Meeting concluded that trifloxystrobin and its metabolites are not genotoxic.

Because of the absence of findings indicative of genotoxicity or carcinogenicity, the Meeting concluded that trifloxystrobin is unlikely to pose a carcinogenic risk to humans.

In the two-generation study in rats given trifloxystrobin at a dose of 55 or 111 mg/kg bw per day, pups in the F_1 and F_2 litters had retarded body-weight development during lactation. The NOAEL for parental toxicity was 3.8 mg/kg bw per day on the basis of findings at 55 mg/kg bw per day, i.e. reduced body weight and food consumption, in addition to histopathology findings in the liver and kidneys. The NOAEL for offspring toxicity was 3.8 mg/kg bw per day on the basis of retarded body-weight development during lactation. The NOAEL for reproductive toxicity was 111 mg/kg bw per day.

Trifloxystrobin was not teratogenic in rats and rabbits when tested at doses of up to 1000 and 500 mg/kg bw per day, respectively. In rats, the NOAEL for developmental toxicity was 100 mg/kg bw per day on the basis of increased incidences of enlarged thymus. In rabbits, the NOAEL for developmental toxicity was 250 mg/kg bw per day on the basis of increased incidences of skeletal anomalies in the form of fused sternebrae 3 and 4. Maternal toxicity in rats and rabbits was limited to reduced food consumption and body-weight loss at 100 and 250 mg/kg bw per day with NOAELs of 10 and 50 mg/kg bw per day, respectively. The developmental effects were considered to be a consequence of overall maternal toxicity.

The Meeting concluded that the existing database on trifloxystrobin was adequate to characterize the potential hazards to fetuses, infants and children.

In a study of acute oral neurotoxicity in rats given a single dose of trifloxystrobin at 2000 mg/kg bw, the functional observational battery revealed no indications for potential neurological or behavioural effects.

Toxicological evaluation

The Meeting established an ADI of 0–0.04 mg/kg bw based on the parental NOAEL of 3.8 mg/kg bw per day in a multigeneration study of reproductive toxicity in rats and a 100-fold safety factor. The LOAEL was 55 mg/kg bw per day on the basis of effects on body weight and food consumption, in addition to liver and kidney histopathology findings. This value is supported by the NOAEL of 5 mg/kg bw per day in the 1-year study in dogs.

The Meeting concluded that it was unnecessary to establish an ARfD for trifloxystrobin on the basis of its low acute toxicity and the fact that developmental effects were considered to be a result of severe maternal toxicity, which is related to decreased food intake rather than systemic toxicity. Also, the vomiting and diarrhoea observed in dogs were clearly related to local irritation, rather than systemic acute toxicity.

A toxicological monograph was prepared.

Levels relevant to risk assessment

Species	Study	Effect	NOAEL	LOAEL
Mouse	18-month study of toxicity and carcinogenicity[a]	Toxicity	300 ppm, equal to 36 mg/kg bw per day	1000 ppm, equal to 124 mg/kg bw per day
		Carcinogenicity	2000 ppm, equal to 246 mg/kg bw per day[b]	—
Rat	2-year studies of toxicity and carcinogenicity[a]	Toxicity	750 ppm, equal to 30 mg/kg bw per day	1500 ppm, equal to 62 mg/kg bw per day[b]
		Carcinogenicity	1500 ppm, equal to 62 mg/kg bw per day[b]	—
	Two-generation reproductive toxicity[a]	Parental toxicity	50 ppm, equal to 3.8 mg/kg bw per day	750 ppm, equal to 55 mg/kg bw per day
		Offspring toxicity	50 ppm, equal to 3.8 mg/kg bw per day	750 ppm, equal to 55 mg/kg bw per day
	Developmental toxicity[c]	Maternal toxicity	10 mg/kg bw per day	100 mg/kg bw per day
		Embryo- and fetotoxicity	100 mg/kg bw per day	1000 mg/kg bw per day
Rabbit	Developmental toxicity[c]	Maternal toxicity	50 mg/kg bw per day	250 mg/kg bw per day
		Embryo- and fetotoxicity	250 mg/kg bw per day	500 mg/kg bw per day
Dog	3-month study of toxicity[d,e]	Toxicity	30 mg/kg bw per day	150 mg/kg bw per day
	12-month study of toxicity[d]	Toxicity	5 mg/kg bw per day	50 mg/kg bw per day

[a] Diet
[b] Highest dose tested
[c] Gavage
[d] Gelatine capsule
[e] Two or more studies combined

Estimate of acceptable daily intake for humans

    0–0.04 mg/kg bw

Estimate of acute reference dose

    Unnecessary

Studies that would provide information useful for continued evaluation of the compound

    Further observations in humans

Critical end-points for setting guidance values for exposure to trifloxystrobin

Absorption, distribution, excretion and metabolism in animals	
Rate and extent of absorption	66% in 48 h
Distribution	Widely distributed; highest concentrations in blood, liver and kidneys
Potential for accumulation	No potential for accumulation.
Rate and extent of excretion	Within 48 h, 72–96% of the administered dose is eliminated in the urine and faeces
Metabolism in animals	Extensive: hydrolysis, *O*-demethylation, oxidation, conjugation, chain shortening and cleavage between glyoxylphenyl and trifluoromethyl moieties
Toxicologically significant compounds (plants, animals and the environment)	Parent compound, main acid metabolite is CGA 321113

Trifloxystrobin

Acute toxicity

Rat, LD$_{50}$, oral	> 5000 mg/kg bw
Rat, LD$_{50}$, dermal	> 2000 mg/kg bw
Rat, LC$_{50}$, inhalation:	> 4.6 mg/l
Rabbit, skin irritation:	Not irritating
Rabbit, eye irritation:	Not irritating
Skin sensitization	Sensitizer (Magnusson & Kligman test)

Short-term studies of toxicity

Target/critical effect	Body weight, food consumption, clinical signs, liver (pathology), kidney (weight), pancreas (atrophy), spleen (weight and pathology)
Lowest relevant oral NOAEL	5 mg/kg bw per day (1-year study in dogs)
Lowest relevant dermal NOAEL	≥ 1000 mg/kg bw per day (28-day study in rats)
Lowest relevant inhalation NOAEC	No relevant study

Genotoxicity

No genotoxic potential, negative results in vivo, one positive result in study in vitro at cytotoxic doses.

Long-term studies of toxicity and carcinogenicity

Target/critical effect	Body weight (mouse, rat), food consumption (rat), liver (mouse, rat)
Lowest relevant NOAEL	30 mg/kg bw per day (2-year study in rats)
Carcinogenicity	Unlikely to pose a carcinogenic risk to humans

Reproductive toxicity

Target/critical effect	Decreased body-weight gain of pups accompanied by delayed eye opening at parental toxic doses
Lowest relevant reproductive NOAEL	50 ppm (3.8 mg/kg bw per day)s
Developmental target/critical effect	Enlarged thymus (rat) and skeletal effects (rabbit) at maternally toxic doses
Lowest relevant developmental NOAEL	100 mg/kg bw per day (rat)

Neurotoxicity

No evidence of acute neurotoxicity in rats

Other toxicological studies

No evidence of replicative DNA synthesis in rat or mouse heptocytes after 3-months administration in diet

A range of metabolites had low acute oral toxicity and there was no evidence of genotoxic activity

Medical data

New active substance; limited data; some evidence of skin and eye irritation in three people during field trials (but 120 people without effects)

Summary	Value	Study	Safety factor
ADI	0–0.04 mg/kg bw	Rat, reproduction study, reduced body weight, liver and kidney effects	100
ARfD	Unnecessary	—	—

RESIDUE AND ANALYTICAL ASPECTS

The residue and analytical aspects of trifloxystrobin were considered for the first time by the present Meeting.

Trifloxystrobin, a member of the strobilurin group, is a broad-spectrum contact fungicide for foliar use; it has mesosystemic properties. The mode of action of strobilurins involves inhibition of mitochondrial respiration by blockage of the electron transfer chain. The fungicidal properties of trifloxystrobin are derived from the parent ester, and the acid (the main metabolite) is essentially inactive. Trifloxystrobin has registered uses on horticultural crops, vegetables and cereals in many countries.

IUPAC name: methyl (*E*)-methoxyimino-{(*E*)-α-[1-(α,α,α-trifluoro-*meta*-tolyl)ethylidene-aminooxy]-*ortho*-tolyl}acetate

Chemical Abstracts name: methyl (α*E*)-α-(methoxyimino)-2-[[[(*E*)-[1-[3-(trifluoromethyl)phenyl]-ethylidene]amino]oxy]methyl]benzeneacetate

The Meeting received information on the metabolism and environmental fate of trifloxystrobin, methods of residue analysis, stability in freezer storage, national registered use patterns, the results of supervised residue trials, the results of farm animal feeding studies, fate of residues in processing and national MRLs.

Trifloxystrobin is a white powder which melts at 73 °C. It is not highly volatile (vapour pressure, 3×10^{-6} Pa). It does not dissociate and is only slightly water-soluble (0.6 mg/l). The log P_{OW} is 4.5, suggesting that bioaccumulation may occur. Trifloxystrobin is hydrolytically stable at environmental pHs, but photochemical degradation was shown to occur. The active technical substance is not considered to be explosive or inflammable.

Metabolism

Animals

The metabolism of trifloxystrobin was investigated in rats, goats and poultry, and the metabolic pathways were comparable in the three species. The studies were performed with ^{14}C-trifloxystrobin labelled uniformly in one of the two phenyl rings, [glyoxylphenyl-U^{14}C]trifloxystrobin and [trifluoromethylphenyl-U^{14}C]trifloxystrobin, each compound being administered separately. The name [glyoxylphenyl-U^{14}C]trifloxystrobin was introduced during the development of trifloxystrobin to reflect the route of synthesis of radiolabelled material.

After oral administration to *rats* of each sex, rabiolabelled trifloxystrobin was rapidly and appreciably absorbed (35–65% of dose). Faeces was the main route of elimination (63–84%), some of which was through bile (30–45%) while only one-third or less of the administered dose was excreted in urine and none in expired air. There was near-complete degradation of trifloxystrobin after a single low dose of 0.5 mg/kg bw;

however, after a dose of 100 mg/kg bw, up to 45% was eliminated unchanged in faeces. The pattern of metabolites in rats is complex: about 35 metabolites were isolated from urine, faeces and bile and identified. The main steps in the metabolic pathway include hydrolysis of the methyl ester to the corresponding acid, *O*-demethylation of the methoxyimino group, yielding a hydroxyimino compound, and oxidation of the ethylideneamino methyl group to a primary alcohol and then to the corresponding carboxylic acid. These stepa are followed by a complex pattern of further, minor reactions. Cleavage between the two phenyl rings accounted for about 10% of the dose.

Lactating *goats* were given diets containing [glyoxylphenyl-U^{14}C]trifloxystrobin or [trifluoromethylylphenyl-U^{14}C]trifloxystrobin at an equivalent of 100 ppm for 4 days and were slaughtered 6 h after the last dose. Up to 20% of the applied dose was excreted in urine and 45% in faeces, while 0.05–0.08% of the total dose was eliminated in milk, corresponding to about 0.1 mg/kg trifloxystrobin equivalents, and a plateau was reached after 48 h.

Most tissue residues were found in liver, bile and kidney, accounting for 0.28–0.57%, 0.07–0.24% and 0.026–0.052% of the applied dose, respectively. These values correspond to 2.6–5.2 mg/kg, 29–77 mg/kg and 1.7–2.9 mg/kg as trifloxystrobin equivalents. Lower levels were found in fat, muscle and blood. The main components of the residue were the parent compound, its carboxylic acid, CGA 321113 (chemical name: *(E,E)*-methoxyimino-{2-[1-(3-trifluoromethyl-phenyl)ethylideneaminooxymethyl]phenyl}acetic acid) and taurine and glycine conjugates of CGA 321113. The amino acid conjugates were the main residue components in the liver (up to 28% TRR). These metabolites were not considered to be of toxicological concern. CGA 321113 was the main radioactive residue in muscle (up to 57% TRR) and kidney (up to 74% TRR), and trifloxystrobin was the principal component in milk (up to 74% TRR) and fat (up to 82% TRR).

Hens were given diets containing trifloxystrobin at an equivalent of 100 ppm in the diet for 4 days and were killed 6 h after the last dose. Up to 0.16% and 87% of the applied dose were eliminated in eggs and excreta, respectively. A plateau was not reached in egg yolk. The residue levels appeared to be increasing rapidly at the end of the study.

Eggs contained 0.1–0.2% of the applied dose. The maximum concentration in egg white was 0.56 mg/kg, and that in egg yolk was 2.3 mg/kg as trifloxystrobin equivalents. Lean meat contained 0.11–0.22% of the dose (0.13–0.35 mg/kg trifloxystrobin equivalents); skin and attached fat 0.14–0.39% (0.8–1.8 mg/kg); peritoneal fat, 0.07–0.21% (1.9–2.7 mg/kg); kidney, 0.11–0.25% (6–13 mg/kg); and liver, 0.28–0.68% (3.8–8.6 mg/kg). The TRR (including that in intestine and gizzard) was 78–91%.

Characterization of the radioactive tissue residues revealed that parent trifloxystrobin was a major residue in muscle (up to 28% TRR), fat and skin (up to 55% TRR) and egg yolk (up to 9% TRR) of laying hens. The carboxylic acid derivative (CGA 321113) was the main residue in egg white (up to 26% TRR) and liver (up to 5.1% TRR).

Plants

The metabolism of trifloxystrobin in plants was investigated in wheat, apples, cucumbers, sugar beet and peanuts with ^{14}C-trifloxystrobin applied by spray. Although the number of metabolite fractions differed in the different plants, the metabolic pathways in these the crops were comparable.

In mature *wheat*, the highest TRRs were found in straw (3.85 mg/kg trifloxystrobin equivalents), followed by husks (0.14 mg/kg) and grain (0.02 mg/kg). The composition of the TRRs was complex; trifloxystrobin and its isomers constituted less than 5%.

Studies on wheat showed that the absorption of trifloxystrobin by plants was relatively rapid, with about 15% of the TRR appearing within the first 24 h, 29% within 3 days and 44% within 14 days. Characterization of the surface radioactivity in wheat revealed that trifloxystrobin is relatively stable to photodegradation, accounting for up to about 80% of the surface radioactivity after 14 days. In contrast, absorbed residue appeared to undergo rapid degradation: the trifloxystrobin concentration declined exponentially, with an apparent half-life of 12 h. Up to 35 metabolite fractions were found in wheat, most of which constituted less than 1% TRR.

In *apple*, 14 days after treatment, the main residue component was the parent compound trifloxystrobin (*E,E* isomer), which, together with its *Z,Z*, *Z,E* and *E,Z* isomers, constituted about 92% of the residue.

In the leaves and fruits of *cucumber*, the residue consisted of trifloxystrobin (80–93% TRR), isomers of trifloxystrobin (2.3–3.8% TRR) and CGA 321113 (0.9–4.2% TRR).

In *sugar-beet*, the main compounds found, with both labels, in the tops and roots were trifloxystrobin and its *E,Z* and *Z,Z* isomers. They accounted for up to 69% TRR in tops (1.1 mg/kg trifloxystrobin) and 52% in roots (0.02 mg/kg trifloxystrobin). CGA 321113 represented up to 5.2% (0.073 mg/kg) and up to 11% (0.012 mg/kg) of the TRR in tops and roots, respectively.

In *peanut*, many metabolite fractions containing only one moiety of the parent molecule were detected, generally similar to those found in wheat. Extensive formation of sugar and malonyl sugar conjugates was found in most metabolite fractions. In vines, the percentage of extractable radioactive residues (acetonitrile:water) amounted to 91% TRR. Extractable residues represented up to 74% in mature hay and up to 53% in nutmeat. The unextracted residues were solubilized by hot extraction and sequential hydrolyses with cellulase, protease, HCl and NaOH. The radioactive residues that remained unextracted under these exhaustive conditions represented < 10% TRR.

In general, the metabolism of trifloxystrobin in crops is complex, owing to isomerization of the parent compound and its metabolites. Overall, the metabolism of trifloxystrobin is similar in all crops and involves the following steps:

- *cis–trans* isomerization of trifloxystrobin (*E,E*- isomer) to its *E,Z*-, *Z,Z* and *Z,E*- isomers
- hydrolysis of the methyl esters of the parent and its isomers to carboxylic acids
- *cis–trans* isomerization of the *E,E*-carboxylic acid CGA 321113
- hydroxylation of the trifluoromethylphenyl ring, followed by sugar conjugation
- oxidation of the methyl of the 2-ethylideneamino group with subsequent sugar conjugation
- cleavage of the N–O bridge, followed by oxidation of the trifluoromethylphenyl moiety to form the acetophenone derivative, with subsequent sugar conjugation
- cleavage of the N–O bridge, followed by oxidation of the glyoxylphenyl moiety, with eventual formation of phthalic acid
- formation of unextracted residues.

Environmental fate

Water–sediment systems

Because trifloxystrobin is used for foliar spray treatment and on paddy rice, only studies of hydrolysis and degradation in water–sediment systems were considered.

Trifloxystrobin is relatively stable hydrolytically under sterile neutral and weakly acid conditions, whereas under alkaline conditions hydrolytic degradation increases with increasing pH. The acid CGA 321113 formed under alkaline conditions is not degraded hydrolytically. No ring cleavage is observed at pH ≥ 5.

In biologically active aquatic systems such as a paddy rice plot, trifloxystrobin was rapidly degraded in both flooding water and paddy soil, with a maximum half-life of 2–5 days. As in sterile hydrolysis, the main product in a paddy rice field was the acid CGA 321113. While this metabolite is stable to sterile hydrolysis, it was rapidly degraded in the rice plot, with degradation half-lives of 7–8 days in flooding water and paddy soil. Besides CGA 321113, formed by biotic hydrolysis, isomerization of the parent compound and CGA 321113 occurred, resulting in formation of the parent *Z,E-* isomer CGA 357261 in small amounts and the acid *Z,E-* isomer CGA 373466 in large amounts. CGA 373466 degraded rapidly in the water layer, with a half-life of 4.2 days. A half-life with reasonable significance could not be estimated for CGA 357261 owing to the very low concentrations in the range of the LOQ.

The photolytic half-lives of trifloxystrobin in sterile aqueous buffered solutions at 25 °C under a xenon arc light (12 h light–12 h dark cycle) were 20.4 h at pH 5 and 31.5 h at pH 7. The corresponding predicted environmental half-lives in summer sunlight at a geographical latitude of 40° N were 1.1 and 1.7 days at pH 5 and pH 7, respectively.

Rotational crops

The Meeting received the results of confined crop rotation studies with ^{14}C-trifloxystrobin with both labels and from crop rotation trials with unlabelled trifloxystrobin. In some trials, a first crop was treated with trifloxystrobin, while in others bare ground was treated directly with trifloxystrobin as an extreme case for residues in soil from the first crop. The normal rotation was a first crop followed in rotation by a root crop (radish, turnip), a vegetable (lettuce, spinach) and a cereal (wheat). The rotation crops were sown or planted from 30 days to 1 year after the final treatment of the first crop or bare ground.

In a study with an exaggerated application rate of 2.2 kg ai/ha to bare soil, turnips, spinach and wheat were planted and components of each were analysed 30 and 120 days after application. The residue levels of trifloxystrobin equivalent were higher with the trifluoromethylphenyl label than the glyoxylphenyl label. The levels of trifluoromethylphenyl label (as trifloxystrobin equivalents) after 30/120 days were: 0.06/0.04 mg/kg in turnip leaves, 0.02/0.02 mg/kg in turnip roots, 0.25/0.26 mg/kg in spinach, 0.28/0.19 mg/kg in 25% mature wheat fodder, 0.14/0.10 mg/kg in mature wheat fodder, 0.17/0.20 mg/kg in wheat straw and 0.07/0.06 mg/kg in wheat grain. With the other label, only a small proportion of the TRR were usually identified or characterized; trifloxystrobin represented < 2%. CGA 321113 represented up to 8.5% of the TRR in turnip leaves and 17.5% in turnip roots (0.003 mg/kg). With the trifluoromethyl-^{14}C label, 37–100% of the TRR was identified or characterized. Trifloxystrobin, its conformational isomers and the acid CGA 321113 and its isomers were reported, all at < 0.01 mg/kg. Trifluoroacetic acid was found as a major degradation product in all crops, especially in wheat (up to 0.23 mg/kg in immature fodder and 0.12 mg/kg in straw), indicating breakdown of the trifluoromethylphenyl ring. As trifluoroacetic acid was observed only rarely as a plant metabolite in target crops after foliar application, it is likely that its precursor is formed in the soil or rhizosphere of the plants.

In unconfined rotational studies with unlabelled trifloxystrobin, no residues of trifloxystrobin (< 0.02 mg/kg) or CGA 321113 (< 0.02 mg/kg) were detected in any of the rotational crops at 30-day plant-back intervals, except in wheat straw and grain in one trial,.

The rotational crop studies suggest that trifloxystrobin itself and the acid CGA 321113 do not occur in rotational crops at levels ≥ 0.01 mg/kg.

Methods of analysis

The Meeting received descriptions and validation data for analytical methods for residues of trifloxystrobin and the metabolite CGA 321113 in crops and animal commodities. The methods rely on gas–liquid chromatography, HPLC and liquid chromatography with tandem mass spectrometry detection and generally achieve LOQs of 0.01–0.02 mg/kg in crop and animal matrices, except dry matrices such as hay, straw (LOQ, 0.05 mg/kg) and hops (LOQ, 0.1 mg/kg). The recoveries were in the range of 70–120% for both analytes.

In most of the field studies, the determination of trifloxystrobin and CGA 321113 in plant and animal commodities was based on extraction of the samples with acetonitrile and water (80:20, v/v), filtration, liquid–liquid partitioning with a three-solvent system (sodium chloride-saturated water, toluene and hexane), clean-up on a C18 solid extraction column, partitioning into methyl *tert*-butyl ether:hexane, concentration to dryness and dissolution in 0.1% polyethylene glycol in acetone for gas chromatographic analysis with nitrogen–phosphorus detection. The LOQ was 0.02 mg/kg in all matrices except peanut hay and cereal straw (0.05 mg/kg) and milk (0.01 mg/kg). This method (or with electron capture detection) was used in the rotational crop, storage stability and field trial studies. The nitrogen–phosphorus detection method is proposed as a monitoring method.

The standard multi-method DFG S 19 can be used for enforcement purposes for the determination of trifloxystrobin in all plant materials except hops.

Data on the extraction efficiency of the methods with weathered radiolabelled samples from the studies of metabolism in apples, cucumbers, peanuts, wheat grain and straw, and matrices from the animal metabolism studies were submitted. The amount of residue extracted was similar to that in the metabolism studies.

Stability of residues in stored analytical samples

The Meeting received information on the stability of trifloxystrobin and CGA 321113 in crops, farm animal commodities and processed commodities at freezer temperatures for 1–2 years. Trifloxystrobin and CGA 321113 were generally stable for the duration of testing, i.e. a decline in residue levels was not evident or was < 30%.

Definition of the residue

The metabolism of trifloxystrobin in animals is similar to that in plants and occurs primarily via cleavage of the methyl ester group to form CGA 321113. In plants, the main component of the residue is trifloxystrobin. CGA 321113 is the principal residue component in animal tissues (except fat). Trifloxystrobin is the principal residue in milk and fat.

The metabolite CGA 321113 accounts for about 30% of the terminal residue in some raw plant commodities (strawberries, leeks, Brussels sprouts, flowerhead brassicas, carrots, barley, wheat, maize, rice, hops, peanut fodder, barley straw, maize fodder and rice straw). Furthermore, trifloxystrobin can be hydrolysed to CGA 321113 during processing. In these cases, the nature of the residue in the processed product may differ somewhat from that in the raw agricultural commodity. Therefore, CGA 321113 should be included in the residue definition for risk assessment for plant commodities.

The Meeting agreed that the residue definition for enforcement purposes for plant commodities should be trifloxystrobin *per se*, and that for animal commodities should be the parent compound plus CGA 321113 (expressed as trifloxystrobin equivalents).

The Meeting agreed that the residue definition for consideration of dietary intake should consist of the parent compound plus CGA 321113 (expressed as trifloxystrobin equivalents), to cover the occurrence of residues in both plant and animal commodities as well as in processed products.

The log P_{OW} for trifloxystrobin is 4.5, which suggests that the compound is fat-soluble. As the levels of trifloxystrobin were higher in fat than in muscle, residues in fat are appropriate for controlling residues in meat. The Meeting agreed that the residues of trifloxystrobin are fat-soluble.

Definition of the residue in plant commodities for compliance with MRLs: trifloxystrobin.

Definition of the residue in plant commodities for estimation of dietary intake: sum of trifloxystrobin and CGA 321113, expressed as trifloxystrobin.

Trifloxystrobin

Definition of the residue in animal commodities for compliance with MRLs and estimation of dietary intake: sum of trifloxystrobin and CGA 321113, expressed as trifloxystrobin.

The residue is fat-soluble.

Results of supervised trials on crops

The Meeting received the results of supervised trials on citrus fruit, pome fruit, stone fruit, grapes, strawberries, bananas, leeks, head cabbage, Brussels sprouts, cauliflower and broccoli, cucumbers, melons, summer squash, tomatoes, peppers, Chinese cabbage, beans, soya beans, carrots, celeriac, potatoes, sugar-beet, celery, chicory, cereals (wheat, barley, maize, rice), almonds, tree nuts, cotton-seed, peanuts, coffee beans and hops; and on animal feed items such as almond hulls, cereal straws, peanut hay and sugar-beet tops. In most cases, the acid metabolite CGA 321113 was determined as well as the parent compound.

The sum of trifloxystrobin and CGA 321113 was calculated and expressed as trifloxystrobin on the basis of the relative molecular masses. A conversion factor of 1.036 is required to express CGA 321113 as trifloxystrobin. As CGA 321113 does not generally constitute a significant proportion of the residue in crops, when the levels of trifloxystrobin or CGA 321113 were below the LOQ, their sum was calculated as in the examples below.

Trifloxystrobin (mg/kg)	CGA 321113 (mg/kg)	Total (expressed as trifloxystrobin) (mg/kg)
< 0.02	< 0.02	< 0.02
< 0.02	0.03	0.05
0.10	< 0.02	0.10
0.92	0.16	1.1

Two sets of data are reported: trifloxystrobin *per se* for estimation of maximum residue levels and the sum of trifloxystrobin and CGA 321113 expressed as trifloxystrobin for estimation of STMRs.

Treatment with trifloxystrobin is limited to up to four applications per season, but in some trials on fruits (pome fruit, grapes, banana) and vegetables (cucurbits, sweet pepper, tomato) up to 10 treatments were made. To investigate the influence of the number of applications on the residue levels, two trials were conducted in apples. Trifloxystrobin was applied four times to apple trees at a rate of 0.12 or 0.15 kg ai/ha with spray intervals of 12–17 days. Samples of fruit were taken before and after each application. The average carry-over of residue (ratio of residue concentration before and after pesticide application) was approximately 40% and at the same level, suggesting that two applications are likely to result in higher residue levels than one application, but three or more applications should not produce residue levels significantly different from those resulting from two. The Meeting agreed that the residue at harvest is influenced only by the final three or four applications, and trials with more than four treatments were used to estimate maximum residue levels and STMRs.

Citrus fruit

The results of supervised trials for residues in orange, grapefruit and lemon were received from South Africa and the USA.

Use of trifloxystrobin as a foliar spray is registered in South Africa with two applications of 0.005 kg ai/hl and a PHI of 76 days. The six trials on orange that were submitted did not reflect GAP.

In the USA, trifloxystrobin may be used as a foliar spray on citrus fruit at three to four applications of 0.07–0.14 kg ai/ha with a 30-day PHI. As the level of CGA 321113 was below the LOQ in all samples, the data populations for enforcement and risk assessment purposes are identical. The levels of trifloxystrobin residues in whole fruit in trials approximating these conditions at the highest rate were: 0.05, 0.07, 0.08, 0.09, 0.10, 0.11, 0.12, 0.15 (two), 0.16, 0.17, 0.19, 0.21 and 0.23 mg/kg in orange; < 0.02, 0.03 (two), 0.04, 0.06, 0.08 (two) and 0.10 mg/kg in grapefruit; and < 0.02, 0.02, 0.09, 0.11, 0.13 and 0.22 mg/kg in lemon.

The Meeting agreed to combine these results in estimating a maximum residue level for citrus fruit. As no data were available on residues in the edible portion, the STMR was also estimated from the data for whole fruit. The combined concentrations in the 28 trials in the USA, in ranked order, were: < 0.02 (two), 0.02, 0.03 (two), 0.04, 0.05, 0.06, 0.07, 0.08 (three), <u>0.09</u> (two), <u>0.10</u> (two), 0.11 (two), 0.12, 0.13, 0.15 (two), 0.16, 0.17, 0.19, 0.21, 0.22 and 0.23 mg/kg for trifloxystrobin as well as for the sum of trifloxystrobin and CGA 321113 expressed as trifloxystrobin.

The Meeting estimated a maximum residue level of 0.5 mg/kg and an STMR value of 0.095 mg/kg for residues of trifloxystrobin in whole citrus fruits.

Pome fruit

Trials were conducted on apple and pear in Australia, Canada, Europe, South Africa and the USA.

In Australia, trifloxystrobin may be applied to *apples* in three applications of 0.005 kg ai/hl with a 35-day PHI. Because the number of treatments had little influence on the residue concentration, two trials with six applications of 0.0038 kg ai/hl were considered close enough to GAP to allow evaluation. The trifloxystrobin residue level in apples was < 0.04 mg/kg, and the level for the sum of trifloxystrobin and CGA 321113 (< 0.04 mg/kg) expressed as trifloxystrobin was 0.08 mg/kg.

GAP for use of trifloxystrobin on apples and pears is similar in many countries in Europe. In France and Italy, trifloxystrobin is registered for use on apples and pears up to a total of three applications at 0.0075 kg ai/hl with a PHI of 14 days. The trifloxystrobin residue levels in apples in four trials in France, one in Germany, one in Greece, six in Italy, three in Spain and nine in Switzerland, conducted according to appropriate GAP, were: 0.04 (two), 0.05 (four), 0.06, 0.07 (two), 0.08, 0.09 (two), 0.10 (two), 0.12, 0.13, 0.15, 0.17, 0.19 (two), 0.20, 0.21, 0.30 and 0.44 mg/kg. The residue levels of CGA 321113 were all below the LOQ.

One trial in Greece, one in Italy and two in Spain on *pears* were reported. The residue concentrations of both trifloxystrobin and the sum of trifloxystrobin and CGA 321113 expressed as trifloxystrobin were 0.06, 0.07, 0.11 and 0.12 mg/kg.

In Spain, four treatments at 0.0075–0.015 kg ai/hl with a PHI of 14 days are allowed on apples and pears. The residue concentration of both trifloxystrobin and the sum of trifloxystrobin and CGA 321113 expressed as trifloxystrobin in one Spanish trial on apples was 0.19 mg/kg.

In Germany, four treatments at 0.025 kg ai/ha per metre height of crown (0.075 kg ai/ha for a tree with a 3-m crown) and 0.005 kg ai/hl with a PHI of 14 days on apples and pears are allowed. The residue levels of trifloxystrobin and of the sum of trifloxystrobin and CGA 321113 expressed as trifloxystrobin in one trial on apples were both 0.11 mg/kg. The residue level of trifloxystrobin in one trial on pears was 0.17 mg/kg, and the level of the sum of trifloxystrobin and CGA 321113 expressed as trifloxystrobin was 0.19 mg/kg.

In Belgium and Luxembourg, four treatments with 0.085 kg ai/ha and a PHI of 14 days are allowed on apples and pears. The trifloxystrobin residue levels in apples in five trials in France, two in Germany and three in The Netherlands were: 0.03, 0.04, 0.05, 0.07 (four), 0.13 (two), 0.14 and 0.37 mg/kg. The levels of the sum of trifloxystrobin and CGA 321113 expressed as trifloxystrobin were 0.03, 0.04, 0.05, 0.07 (four), 0.13, 0.14, 0.15 and 0.41 mg/kg.

In South Africa, trifloxystrobin is registered for use on apples at up to three applications of 0.0038–0.005 kg ai/hl with a PHI of 7 days, and it is registered for use on pears at three applications of 0.0038 kg ai/hl with a PHI of 14 days. In two trials each on apples and pears, the residue levels of trifloxystrobin were 0.03, 0.04, 0.05 and 0.06 mg/kg, and the levels of the sum of trifloxystrobin and CGA 321113 expressed as trifloxystrobin were 0.03, 0.04 and 0.06 (two) mg/kg.

The use pattern in the USA allows spraying of trifloxystrobin in four applications of 0.105 kg ai/ha on apples and pears with a PHI of 14 days. The concentrations of trifloxystrobin residues in trials in Canada and the USA in apples were: 0.04, 0.09, 0.10, 0.12, 0.13, 0.14, 0.16 (two), 0.18 (three), 0.21, 0.24, 0.26, 0.31 and 0.37 mg/kg, and those in pears were: 0.07, 0.08, 0.09, 0.10 (two), 0.14, 0.15, 0.17, 0.22 and 0.23 mg/kg. The residue levels of the sum of trifloxystrobin and CGA 321113 expressed as trifloxystrobin in apples were: 0.04, 0.09, 0.10, 0.12, 0.13, 0.14, 0.16 (two), 0.18 (two), 0.21 (two), 0.24, 0.26, 0.31 and 0.37 mg/kg, and those in pears were: 0.07, 0.08, 0.09, 0.10, 0.14, 0.15, 0.17, 0.20, 0.22 and 0.23 mg/kg.

The Meeting agreed to combine the data sets on apples and pears from two trials in Australia 42 trials in Europe, four trials in South Africa and 26 trials in Canada and the USA. The residue concentrations of trifloxystrobin *per se*, in ranked order, were: 0.03 (two), < 0.04, 0.04 (five), 0.05 (six), 0.06 (three), 0.07 (eight), 0.08 (three), 0.09 (four), <u>0.10</u> (five), <u>0.11</u> (two), 0.12 (three), 0.13 (four), 0.14 (three), 0.15 (two), 0.16 (two), 0.17 (three), 0.18 (three), 0.19 (three), 0.20, 0.21 (two), 0.22, 0.23, 0.24, 0.26, 0.30, 0.31, 0.37 (two) and 0.44 mg/kg. The residue concentrations of the sum of trifloxystrobin and CGA 321113 expressed as trifloxystrobin, in ranked order, were: 0.03 (two), < 0.04, 0.04 (five), 0.05 (five), 0.06 (four), 0.07 (eight), 0.08 (three), 0.09 (four), 0.10 (four), <u>0.11</u> (two), 0.12 (three), 0.13 (three), 0.14 (three), 0.15 (three), 0.16 (two), 0.17 (two), 0.18 (two), 0.19 (four), 0.20 (two), 0.21 (three), 0.22, 0.23, 0.24, 0.26, 0.30, 0.31, 0.37, 0.41 and 0.44 mg/kg.

The Meeting estimated a maximum residue level of 0.7 mg/kg and an STMR value of 0.11 mg/kg for residues of trifloxystrobin in pome fruit.

Stone fruit

The results of supervised trials on residues of trifloxystrobin in apricots, cherries, peaches and plums were received from Europe and the USA. Trifloxystrobin is registered for use on apricots, nectarines, peaches, cherries and plums in Switzerland (three applications of 0.2 kg ai/ha, 0.013 kg ai/hl, 21-day PHI) and the USA (four applications of 0.14 kg ai/ha, 1-day PHI). In Spain, trifloxystrobin may be used on peaches and nectarines four times at 0.015 kg ai/hl with a 7-day PHI.

In 14 trials on *cherry* in six states of the USA in 1998, with four applications of 0.14 kg ai/ha and harvesting after 1 day, the concentrations of trifloxystrobin residues were: 0.26, 0.33, 0.34, 0.37, 0.38, 0.39, 0.53, 0.54, 0.55, 0.56, 0.58, 0.63, 0.66 and 0.84 mg/kg. The residue concentrations of the sum of trifloxystrobin and CGA 321113 expressed as trifloxystrobin were: 0.28, 0.37, 0.38 (two), 0.41, 0.42, 0.58, 0.59 (two), 0.61, 0.62, 0.69, 0.73 and 0.90 mg/kg

In two trials on *apricot* in Switzerland, which matched GAP, the trifloxystrobin residue levels were 0.14 and 0.28 mg/kg, and the total residue levels were 0.14 and 0.30 mg/kg.

Three trials conducted in southern Europe (France, Italy and Spain) on *peach* were evaluated against Spanish GAP. The trifloxystrobin residue levels were 0.14, 0.18 and 0.48 mg/kg, and the total residue levels were 0.15, 0.18 and 0.52 mg/kg. In 14 trials on peach in six states of the USA in 1998, with four applications of 0.14 kg ai/ha and harvesting after 1 day, the concentrations of trifloxystrobin residues were: 0.06, 0.18, 0.21, 0.25, 0.32, 0.34, 0.39, 0.41, 0.65, 0.82 (two), 0.89, 1.8 and 1.9 mg/kg. The residue concentrations of the sum of trifloxystrobin and CGA 321113 expressed as trifloxystrobin were: 0.06, 0.21 (two), 0.25, 0.32, 0.34, 0.39, 0.41, 0.70, 0.86, 0.88, 0.94, 1.9 and 2.0 mg/kg.

In nine trials on *plum* in four states of the USA in 1998, with four applications of 0.14 kg ai/ha and harvesting after 1 day, the concentrations of trifloxystrobin residues and of total residues were: 0.02, 0.06 (two), 0.09, 0.15, 0.19, 0.21 (two) and 0.53 mg/kg.

The Meeting agreed to combine the data from all the trials on residues in stone fruit. The combined results for trifloxystrobin were: 0.02, 0.06 (three), 0.09, 0.14 (two), 0.15, 0.18 (two), 0.19, 0.21 (three), 0.25, 0.26, 0.28, 0.32, 0.33, 0.34 (two), 0.37, 0.38, 0.39 (two), 0.41, 0.48, 0.53 (two), 0.54, 0.55, 0.56, 0.58, 0.63, 0.65, 0.66, 0.82 (two), 0.84, 0.89, 1.8 and 1.9 mg/kg. The residue concentrations of the sum of trifloxystrobin and CGA 321113 expressed as trifloxystrobin were: 0.02, 0.06 (three), 0.09, 0.14, 0.15 (two), 0.18, 0.19, 0.21 (four), 0.25, 0.28, 0.30, 0.32, 0.34, 0.37, <u>0.38</u> (two), 0.39, 0.41 (two), 0.42, 0.52, 0.53, 0.58, 0.59 (two), 0.61, 0.62, 0.69, 0.70, 0.73, 0.86, 0.88, 0.90, 0.94, 1.9 and 2.0 mg/kg.

The Meeting estimated a maximum residue level of 3 mg/kg and an STMR of 0.38 mg/kg for residues in stone fruit.

Berries and small fruit

Trials on *grape* were conducted in Australia, Canada, France, Germany, Greece, Italy, South Africa, Spain, Switzerland and the USA.

In Australia, trifloxystrobin may be used on grapes at 0.0075 kg ai/hl with a 35-day PHI after three applications. In trials in Australia matching GAP conditions, the trifloxystrobin residue levels were: < 0.02, 0.04, 0.08 and 0.09 (two) mg/kg. The residue concentrations of the sum of trifloxystrobin and CGA 321113 expressed as trifloxystrobin were: < 0.02, 0.04, 0.09 (two) and 0.11 mg/kg.

In Canada, trifloxystrobin may be used up to four times at 0.07 kg ai/ha and in the USA up to 0.14 kg ai/ha with a 14-day PHI. In two Canadian and 12 US trials matching GAP conditions, the trifloxystrobin residue levels were: 0.04, 0.06, 0.09, 0.16, 0.17, 0.21, 0.26, 0.28, 0.29, 0.33, 0.61, 0.62, 1.1 and 2.2 mg/kg. The residue concentrations of the sum of trifloxystrobin and CGA 321113 expressed as trifloxystrobin were: 0.04, 0.06, 0.09, 0.16, 0.17, 0.21, 0.26, 0.28, 0.33, 0.36, 0.63, 0.64, 1.2 and 2.2 mg/kg.

In Germany, registered use is three applications of 0.12 kg ai/ha with harvesting 35 days after the last treatment. Two trials in France, four in Germany and four in Switzerland with three applications of 0.13 kg ai/ha and a 35–36-day PHI matched this GAP. The trifloxystrobin residue levels were 0.03, 0.04 (three), 0.06 (two), 0.13, 0.14, 0.27 and 0.29 mg/kg. The residue concentrations of the sum of trifloxystrobin and CGA 321113 expressed as trifloxystrobin were: 0.04 (two), 0.05 (two), 0.06, 0.07, 0.16 (two), 0.29 and 0.30 mg/kg.

Trifloxystrobin is registered for use in South Africa at up to 0.011 kg ai/hl with a 14-day PHI. Residues in grapes in three trials with this use pattern were 0.11, 0.18 and 0.24 mg/kg for trifloxystrobin and 0.15, 0.22 and 0.38 mg/kg for total residues.

Trifloxystrobin is registered for use in Spain at four applications of 0.075 kg ai/ha with a 30-day PHI. In one trial in Greece, two in Italy and four in Spain approximating these conditions, the trifloxystrobin residue levels were 0.05, 0.08, 0.11, 0.13, 0.14, 0.28 and 0.36 mg/kg. The total residue levels were 0.05, 0.08, 0.11, 0.13, 0.14, 0.28 and 0.38 mg/kg.

In summary, the residue levels of trifloxystrobin *per se* in 39 trials in Australia, Europe, South Africa, Canada and the USA, in ranked order, were: < 0.02, 0.03, 0.04 (five), 0.05, 0.06 (three), 0.08 (two), 0.09 (three), 0.11 (two), 0.13 (two), 0.14 (two), 0.16, 0.17, 0.18, 0.21, 0.24, 0.26, 0.27, 0.28 (two), 0.29 (two), 0.33, 0.36, 0.61, 0.62, 1.1 and 2.2 mg/kg. The residue concentrations of the sum of trifloxystrobin and CGA 321113 expressed as trifloxystrobin were: < 0.02, 0.04 (four), 0.05 (three), 0.06 (two), 0.07, 0.08, 0.09 (three), 0.11 (two), 0.13, 0.14, <u>0.15</u>, 0.16 (three), 0.17, 0.21, 0.22, 0.26, 0.28 (two), 0.29, 0.30, 0.33, 0.36, 0.38 (two), 0.63, 0.64, 1.2 and 2.2 mg/kg.

The Meeting estimated a maximum residue level of 3 mg/kg and an STMR of 0.15 mg/kg for residues in grapes.

The Swiss pattern of use of trifloxystrobin on *strawberry* allows three spray applications at 0.25 kg ai/ha with a PHI of 14 days. In five trials matching GAP conditions, the residue levels of trifloxystrobin *per se* were: 0.04, 0.05, 0.06, 0.10 and 0.13 mg/kg. The residue concentrations of the sum of trifloxystrobin and CGA 321113 expressed as trifloxystrobin were 0.08, 0.09, 0.10, 0.14 and 0.18 mg/kg.

The Meeting estimated a maximum residue level of 0.2 mg/kg and an STMR of 0.10 mg/kg for residues in strawberries.

Banana

Trifloxystrobin may be used in Latin America on bananas at four applications of 0.09 kg ai/ha with a 0-day PHI. The results of supervised trials on residues in bagged and unbagged banana after up to 11 treatments at 0.01–0.16 kg ai/ha by aerial application were received from Colombia (four), Costa Rica (six), Ecuador (six), Guatemala (four), Honduras (two) and Mexico (two). The treatment was typical of that performed by commercial aerial sprayers from a fixed-winged airplane or a helicopter. In four trials in Martinique, bananas were treated six times at 0.09 kg ai/ha by a foliar backpack sprayer with aerial boom. In two trials in Puerto Rico, plants were treated 10 times at 0.1–0.13 kg ai/ha by spraying over the top, simulating aerial application. Although the actual treatment rate in some trials exceeded GAP by more than 30%, these trials were included in the evaluation because of the very low residue levels ($\leq$ LOQ). In whole bagged and unbagged fruit, the trifloxystrobin residue concentrations for estimation of the maximum residue level were: < 0.01 (four), < 0.02 (23), 0.02 and 0.03 (two) mg/kg. In the edible portion, the residue concentrations of the sum of trifloxystrobin and CGA 321113 expressed as trifloxystrobin were: < 0.01 (four) and < 0.02 (26) mg/kg.

The Meeting estimated a maximum residue level of 0.05 mg/kg and an STMR of 0.02 mg/kg for residues in bananas.

Leek

The Swiss pattern of use of trifloxystrobin on leeks allows three spray applications at 0.19 kg ai/ha with a PHI of 7 days. The levels of trifloxystrobin residues in leek in one trial in France, one in Germany, one in The Netherlands and two in Switzerland that met these conditions were: 0.08, 0.14, 0.15 and 0.40 (two) mg/kg. The corresponding total residues were: 0.13, 0.26, 0.31, 0.47 and 0.49 mg/kg.

The Meeting estimated a maximum residue level of 0.7 mg/kg and an STMR of 0.31 mg/kg for residues in leeks.

Brassica vegetables

The Swiss pattern of use of trifloxystrobin on broccoli, cauliflower, Brussels sprouts and head cabbage allows three spray applications at 0.13–0.25 kg ai/ha with a PHI of 7 days. The trials on head cabbage, Brussels sprouts, broccoli and cauliflower were evaluated together for mutual support.

One trial on *head cabbage* in Germany, one in The Netherlands and two in Switzerland matching maximum GAP, with a rate of 0.25 kg ai/ha, were submitted. The residue levels of trifloxystrobin *per se* were: 0.02 (two), 0.03 and 0.07 mg/kg, and those of total residues were 0.03, 0.04 (two) and 0.11 mg/kg.

Two trials on *Brussels sprouts* in France, one in Germany, two in Switzerland and one in the United Kingdom matching maximum GAP, with a rate of 0.25 kg ai/ha, were submitted. The residue levels of trifloxystrobin *per se* were: 0.10, 0.16, 0.18, 0.19, 0.20 and 0.35 mg/kg, and those of total residues were: 0.18, 0.22, 0.26, 0.27, 0.28 and 0.39 mg/kg.

One trial on *cauliflower* in Germany and two in Switzerland and two trials on broccoli in Germany and one in the United Kingdom, which matched GAP with a rate of 0.2–0.25 kg ai/ha, were submitted. The residue levels of trifloxystrobin *per se* were: < 0.01, < 0.02, 0.09, 0.13 (two) and 0.26 mg/kg, and those of total residues were: < 0.02, 0.04, 0.13, 0.16, 0.23 and 0.26 mg/kg.

The combined levels of trifloxystrobin residues in broccoli, cauliflower, Brussels sprouts and head cabbage were, in ranked order: < 0.01, < 0.02, 0.02 (two), 0.03, 0.07, 0.09, 0.10, 0.13 (two), 0.16, 0.18, 0.19, 0.20, 0.26 and 0.35 mg/kg. Those of total residues were: < 0.02, 0.03, 0.04 (three), 0.11, 0.13, 0.16, 0.18, 0.22, 0.23, 0.26 (two), 0.27, 0.28 and 0.39 mg/kg.

The Meeting estimated a maximum residue level of 0.5 mg/kg and an STMR of 0.17 mg/kg for residues in flowerhead brassica, Brussels sprouts and head cabbage.

Chinese cabbage

Trifloxystrobin is registered in Switzerland for use on Chinese cabbage up to three times at 0.25 kg ai/ha with a 7-day PHI. The concentrations of trifloxystrobin residues in Chinese cabbage in two Swiss trials were 0.01 and 0.33 mg/kg. The corresponding total residue levels were 0.01 and 0.35 mg/kg.

The Meeting concluded that there were insufficient data to estimate a maximum residue level and an STMR for residues in Chinese cabbage.

Fruiting vegetables

Cucurbits

Trifloxystrobin is registered in Switzerland for use on *cucumber* up to three times at 0.25 kg ai/ha and a 3-day PHI for indoor use. Eight trials conducted in glasshouses in Italy (one), The Netherlands (two), Spain (four) and Switzerland (one) approximated Swiss GAP. The trifloxystrobin residue levels were: 0.02, 0.03 (two), 0.04, 0.06, 0.07 and 0.14 (two) mg/kg. The total residue levels were 0.02, 0.03 (two), 0.04, 0.10, 0.12 and 0.17 (two) mg/kg.

Trifloxystrobin is registered in the USA for use on cucurbit vegetables such as chayote, Chinese waxgourd, citron melon, cucumber, gherkin, edible gourds, muskmelon, pumpkin, summer squash, winter squash and watermelon, as up to four applications of 0.14 kg ai/ha and a 0-day PHI for outdoor use. Eight outdoor trials on *cucumber* in seven states in accord with GAP conditions were reported. The trifloxystrobin residue levels were: 0.03, 0.04 (three), 0.05, 0.06, 0.17 and 0.22 mg/kg. The total residue levels were 0.03, 0.04 (three), 0.05, 0.06, 0.17 and 0.24 mg/kg.

The data from the indoor and outdoor trials on cucumbers could be combined as they were apparently for similar data populations. The residue levels of trifloxystrobin *per se* in the European and North American trials were, in ranked order: 0.02, 0.03 (three), 0.04 (four), 0.05, 0.06 (two), 0.07, 0.14 (two), 0.17 and 0.22 mg/kg. The corresponding total residue levels were 0.02, 0.03 (three), 0.04 (four), 0.05, 0.06, 0.10, 0.12, 0.17 (three) and 0.24 mg/kg.

In Italy, trifloxystrobin may be used on *melon* up to three times at 0.13 kg ai/ha with a 3-day PHI. Three trials in Italy and four in Spain complied with this use pattern. Both the trifloxystrobin and the total residue levels were: < 0.02 (two), 0.04, 0.07, 0.10, 0.11 and 0.19 mg/kg. Six trials in the USA on melons matching GAP for cucurbits resulted in trifloxystrobin and total residue levels of: 0.07, 0.10 (two), 0.11, 0.18 and 0.24 mg/kg.

In five trials in the USA on *summer squash* matching GAP for cucurbits, the trifloxystrobin and the total residue levels were: < 0.02, 0.09, 0.11, 0.15 and 0.23 mg/kg.

The Meeting decided to pool the data on cucumbers (16 trials), melons (13 trials) and summer squash (five trials) to estimate a maximum residue level for cucurbits. The trifloxystrobin residue levels were < 0.02 (three), 0.02, 0.03 (three), 0.04 (five), 0.05, 0.06 (two), 0.07 (three), 0.09, 0.10 (three), 0.11 (three), 0.14 (two), 0.15, 0.17, 0.18, 0.19, 0.22, 0.23 and 0.24 mg/kg. The corresponding total residue levels, in ranked order, were: < 0.02 (three), 0.02, 0.03 (three), 0.04 (five), 0.05, 0.06, 0.07 (two), 0.09, 0.10 (four), 0.11 (three), 0.12, 0.15, 0.17 (three), 0.18, 0.19, 0.23 and 0.24 (two) mg/kg.

The Meeting estimated a maximum residue level of 0.3 mg/kg and an STMR of 0.095 mg/kg for residues in cucurbits.

Sweet peppers

In the USA, trifloxystrobin may be used on sweet peppers four times at 0.14 kg ai/ha with a 3-day PHI. The levels of both trifloxystrobin and total residues in 12 outdoor trials in five states conducted according to GAP were: 0.03, 0.04 (two), 0.05 (two), <u>0.08</u>, <u>0.12</u> (three), 0.14 and 0.16 (two) mg/kg.

The Meeting estimated a maximum residue level of 0.3 mg/kg, and an STMR of 0.1 mg/kg.

Tomato

In the USA, trifloxystrobin may be used on tomatoes four times at 0.14 kg ai/ha with a 3-day PHI. The trifloxystrobin and the total residue levels in 18 outdoor trials in five states conducted according to GAP were: < 0.02 (two), 0.03, 0.06, <u>0.07</u> (five), <u>0.09</u> (three), 0.10, 0.13, 0.20, 0.29, 0.43 and 0.49 mg/kg.

The Meeting estimated a maximum residue level of 0.7 mg/kg and an STMR of 0.08 mg/kg for residues in tomatoes.

Legume vegetables and pulses

Beans (dry)

Trifloxystrobin may be used on beans in Brazil three times at 0.1–0.13 kg ai/ha with a 15-day PHI. Five trials with four applications at 0.15 kg ai/ha and a further five trials with three to four applications of 0.094 kg ai/ha and a 15-day PHI in Brazil approximately matched GAP. The trifloxystrobin residue levels were < 0.02 (four) and <u>≤ 0.05</u> (six) mg/kg. As CGA 321113 was not determined, the trials were not considered for evaluation.

The Meeting concluded that there were insufficient data to estimate a maximum residue level and an STMR for residues in beans (dry).

Soya beans (dry)

Trifloxystrobin may be used on soya beans in Brazil twice at 0.056–0.075 kg ai/ha with a 30-day PHI and once at 0.056 kg ai/ha with a 20-day PHI in Argentina. Three trials with two applications at 0.063 kg ai/ha, three trials with two applications at 0.094 kg ai/ha and a further three trials with two applications at 0.13 kg ai/ha and a 21–30-day PHI in Brazil were reported. The trifloxystrobin residue levels were all < 0.05 mg/kg. As CGA 321113 was not determined, the trials were not considered for evaluation.

The Meeting concluded that there were insufficient data to estimate a maximum residue level and an STMR for residues in soya bean (dry).

Root and tuber vegetables

Carrot

Trifloxystrobin is registered in Switzerland for use on carrots up to three times at 0.25 kg ai/ha with a 7-day PHI. The concentrations of trifloxystrobin residues in carrots in one trial in Belgium, one in Germany, two in The Netherlands and two in Switzerland conducted according to the Swiss use pattern were: < 0.02, 0.02 (two), 0.03 (two) and 0.04 mg/kg. The corresponding total residue levels were < 0.02, 0.02, <u>0.03</u>, <u>0.04</u> (two) and 0.08 mg/kg.

The Meeting estimated a maximum residue level of 0.1 mg/kg and an STMR of 0.035 mg/kg for residues in carrot.

Celeriac

Trifloxystrobin is registered in Switzerland for use on celeriac up to three times at 0.25 kg ai/ha with a 14-day PHI. The concentrations of trifloxystrobin residues in celeriac in two Swiss trials were 0.02 and 0.03 mg/kg. The corresponding total residue levels were 0.03 and 0.04 mg/kg.

The Meeting concluded that there were insufficient data to estimate a maximum residue level and STMR for residues in celeriac.

Potato

In the USA, trifloxystrobin may be used on potatoes six times at 0.14 kg ai/ha with a 7-day PHI. In 15 trials in 13 states conducted according to GAP, all the levels of trifloxystrobin and CGA 321113 residues in tubers were below the LOQ (0.02 mg/kg).

The Meeting estimated a maximum residue level of 0.02* mg/kg and an STMR of 0.02 mg/kg for residues in potato.

Sugar-beet

Italian GAP allows three treatments with an emulsifiable concentrate at rates of 0.11–0.15 kg ai/ha, and Swiss GAP allows one application at 0.15 kg ai/ha, both with a PHI of 21 days, for sugar-beet. In nine trials in France, two in Italy, two in Spain and one in Switzerland that matched GAP, the residue levels of trifloxystrobin *per se* and of total residues in the roots were: < 0.02 (13) and 0.02 mg/kg.

Trifloxystrobin is registered in the USA for use on sugar-beet up to three times at 0.12 kg ai/ha with a 21-day PHI. The trifloxystrobin residue levels in sugar-beet roots in 19 trials in seven states conducted in line with these conditions were: < 0.02 (11), 0.02 (three), 0.03 (three) and 0.04 (two) mg/kg. The corresponding total residue levels were < 0.02 (11), 0.02 (three), 0.03 (two), 0.04 (two) and 0.06 mg/kg.

In summary, the trifloxystrobin residue levels in the 14 trials in Europe and 19 in the USA were: < 0.02 (24), 0.02 (four), 0.03 (three) and 0.04 (two) mg/kg. The corresponding total residue levels were ≤ 0.02 (24), 0.02 (four), 0.03 (two), 0.04 (two) and 0.06 mg/kg.

The Meeting estimated a maximum residue level of 0.05 mg/kg and an STMR of 0.02 mg/kg for residues in sugar-beet.

Celery

Trifloxystrobin is registered in Switzerland for use on celery up to three times at 0.25 kg ai/ha with a 7-day PHI. The concentrations of trifloxystrobin and total residues in whole celery plants without roots in three Swiss trials conducted according to GAP were: 0.12, 0.18 and 0.21 mg/kg.

The Meeting concluded that there were sufficient data and estimated a maximum residue level of 1 mg/kg and an STMR of 0.18 mg/kg for residues in celery as a minor crop.

Witloof chicory

Trifloxystrobin is registered in Switzerland for use on chicory up to three times at 0.25 kg ai/ha with a 21-day PHI. The concentrations of trifloxystrobin residues in chicory leaves in two Swiss trials were 0.34 and 0.86 mg/kg. The corresponding total residue levels were 0.37 and 0.94 mg/kg. The trifloxystrobin and the total residue levels in the roots were both 0.02 mg/kg.

The Meeting concluded that there were insufficient data to estimate a maximum residue level and STMR for residues in witloof chicory.

Cereal grains

Barley

In some countries of Europe, trifloxystrobin is used twice at 0.25 kg ai/ha (France, Germany, United Kingdom) or 0.19 kg ai/ha (Belgium) or once at 0.15 kg ai/ha (Austria). The PHI is 35 days in Austria, Germany and the United Kingdom and 42 days in Belgium and France. In one trial in Denmark, 30 in France, six in Germany and two in the United Kingdom matching the appropriate European GAP, the trifloxystrobin residue levels in barley grain were < 0.02 (10), 0.02 (four), 0.03 (five), 0.04 (two), 0.05 (four), 0.06, 0.07 (three), 0.11 (four), 0.12 (two), 0.13 (two), 0.18 and 0.40 mg/kg. The corresponding total residue levels were < 0.02 (10), 0.02 (four), 0.03 (four), 0.04 (three), 0.05 (two), 0.07 (four), 0.09 (two), 0.11 (three), 0.13, 0.14, 0.15, 0.16, 0.17, 0.18 and 0.46 mg/kg.

The Meeting estimated a maximum residue level of 0.5 mg/kg and an STMR of 0.04 mg/kg for residues in barley. A highest residue level of 0.46 mg/kg was estimated for calculating the dietary burden of farm animals.

Wheat

Brazilian GAP allows two treatments at 0.075 kg ai/ha with a 30-day PHI. Six Brazilian trials were conducted with three applications at 0.15 kg ai/ha. In all these trials, the trifloxystrobin residue levels were below the LOQ of 0.05 mg/kg. As CGA 321113 was not determined, the trials were not considered for evaluation.

Trifloxystrobin may be used twice at 0.25 kg ai/ha in France Germany, Ireland and the United Kingdom; at 0.19 kg ai/ha in Austria, Belgium, Hungary, Italy, Luxembourg, Poland and Switzerland; and at 0.13 kg ai/ha in Slovakia, with PHIs of 35–45 days. In 26 trials in France, 10 in Germany, one in Sweden and two in Switzerland matching the appropriate European GAP, the trifloxystrobin residue levels in wheat grain were: < 0.02 (32), 0.02 (three), 0.03 (two), 0.05 and 0.14 mg/kg. The corresponding total residue levels were: < 0.02 (32), 0.02 (three), 0.03 (two), 0.07 and 0.20 mg/kg.

Trifloxystrobin is registered in the USA for use on wheat up to two times at 0.09 kg ai/ha with a 35-day PHI. In 33 trials in 11 states where these conditions were approximated, the trifloxystrobin and the total residue levels in wheat grain were: < 0.02 (30), 0.02 and 0.03 (two) mg/kg.

In summary, the trifloxystrobin residue levels in the 39 European and the 33 US trials were: < 0.02 (62), 0.02 (four), 0.03 (four), 0.05 and 0.14 mg/kg. The corresponding total residue levels were: ≤ 0.02 (62), 0.02 (four), 0.03 (four), 0.07 and 0.20 mg/kg.

The Meeting estimated a maximum residue level of 0.2 mg/kg and an STMR of 0.02 mg/kg for residues in wheat. A highest residue level of 0.2 mg/kg was estimated for calculating the dietary burden of farm animals.

Maize

Brazilian GAP allows two treatments at 0.075–0.1 kg ai/ha with a 30-day PHI. Three Brazilian trials at three times 0.1 kg ai/ha were conducted. In all the trials, the trifloxystrobin residue levels were below the LOQ of 0.05 mg/kg. As CGA 321113 was not determined, the trials were not considered for evaluation.

Trifloxystrobin is registered in the USA for use on maize up to three times at 0.11 kg ai/ha. The PHI for maize grain is not specified, but the product should not be applied after silking. In 24 field trials in 14 states where these conditions were approximated, the levels of trifloxystrobin residues in maize grain were < 0.02 (24) mg/kg, and the total residue levels were < 0.02 (23) and 0.05 mg/kg.

The Meeting estimated a maximum residue level and an STMR of 0.02 mg/kg for residues in maize. A highest residue level of 0.05 mg/kg was estimated for calculating the dietary burden of farm animals.

Rice

Brazilian GAP allows two foliar treatments at 0.10–0.13 kg ai/ha with a 15-day PHI. Five Brazilian trials with three applications at 0.15 kg ai/ha and a 14–18-day PHI were conducted. The levels of trifloxystrobin residues in rice grain with husk were: < 0.05, 0.05, 0.10, 0.13 and 0.22 mg/kg. As CGA 321113 was not determined, these trials were not included in the evaluation.

Trifloxystrobin is registered in the USA for use on rice up to two times at 0.17 kg ai/ha with a 35-day PHI. In 19 trials in five states where these conditions were approximated, the trifloxystrobin residue levels in rice grain with husk before processing were: < 0.02 (five), 0.03, 0.04 (two), 0.10, 0.11 (two), 0.12, 0.25, 0.30, 0.34, 0.56, 0.68, 2.4 and 3.4 mg/kg. The corresponding total residue levels were < 0.02 (five), 0.04, 0.06, 0.07, 0.13, <u>0.16</u>, 0.20, 0.21, 0.33, 0.41, 0.46, 0.63, 0.75, 2.5 and 3.4 mg/kg.

The Meeting estimated a maximum residue level of 5 mg/kg and an STMR of 0.16 mg/kg for residues in rice. A highest residue level of 3.4 mg/kg was estimated for calculating the dietary burden of farm animals.

Tree nuts

Trifloxystrobin is registered in the USA for use on beechnuts, brazil nuts, butternuts, cashew nuts, chestnuts, chinquapins, filberts, macadamia nuts and walnuts up to four times, and almonds up to three times at 0.14 kg ai/ha with a 60-day PHI. Three treatments at 0.091 kg ai/ha with a PHI of 30 days may be used on pecans.

In six trials on *almonds* in California in which these conditions were approximated, the trifloxystrobin and the total residue levels in almond nuts without shells were < 0.02 mg/kg.

In 11 trials on *pecans* in five states, with eight treatments at 0.14 kg ai/ha and a 30-day PHI, the trifloxystrobin and the total residue levels in pecan nuts without shells were < 0.02 mg/kg.

The Meeting estimated a maximum residue level of 0.02* mg/kg and an STMR of 0 mg/kg for residues in tree nuts.

Cotton-seed

Brazilian GAP allows up to three foliar treatments at 0.063–0.075 kg ai/ha with a 21-day PHI. Three Brazilian trials with three applications at 0.1 kg ai/ha and a 21-day PHI were conducted. The samples were not analysed for CGA 321113. In all three trials, the trifloxystrobin residue levels were below the LOQ of 0.05 mg/kg.

The Meeting concluded that there were insufficient data to estimate a maximum residue level and an STMR for cotton-seed.

Peanuts

Brazilian GAP allows three foliar treatments at 0.075 kg ai/ha with a 15-day PHI for peanuts. Three Brazilian trials with three applications at 0.10 kg ai/ha and a 15-day PHI were conducted. The trifloxystrobin residue levels in peanuts without shells were < 0.05 mg/kg. As CGA 321113 was not determined, these trials were not included in the evaluation.

Trifloxystrobin is registered in the USA for use on peanuts twice at an application rate of 0.13 kg ai/ha or six times at 0.064 kg ai/ha with a PHI of 14 days. In 22 trials with eight applications at 0.07 kg ai/ha and 12 trials with eight applications at 0.14 kg ai/ha were in seven states in 1996–98, the trifloxystrobin and total residue levels in kernels were all < 0.02 mg/kg.

The Meeting estimated a maximum residue level of 0.02* mg/kg and an STMR of 0 mg/kg for residues in peanuts.

Trifloxystrobin

Coffee beans

Brazilian GAP allows three treatments at 0.075–0.11 kg ai/ha with a 30-day PHI. Four Brazilian trials with three applications at 0.11 kg ai/ha were conducted. In all four trials, the trifloxystrobin residue levels were below the LOQ of 0.05 mg/kg. As CGA 321113 was not determined, these trials were not included in the evaluation.

The Meeting concluded that there were insufficient data to estimate a maximum residue level or an STMR for coffee beans.

Hops

German and Austrian GAP allows two treatments of hops at a spray concentration of 0.013 kg ai/hl and a PHI of 14 days. Five German trials with four to six applications of trifloxystrobin at a spray concentration of 0.013 kg ai/ha were reported. In cones harvested 14 days after the last treatment and dried, the trifloxystrobin residue levels were: 4.7, 5.4, 8.8, 16 and 26 mg/kg. The corresponding total residue levels were: 6.2, 6.7, 10, 18 and 29 mg/kg.

In the USA, trifloxystrobin may be used on hops four times at 0.14 kg ai/ha with a 14-day PHI. In three trials in two states with six treatments at 0.14 kg ai/ha and a 13–14-day PHI, the trifloxystrobin residue levels in dried cones were 4.5, 9.3 and 10 mg/kg. The corresponding total residue levels were 4.9, 9.9 and 11 mg/kg.

In summary, the trifloxystrobin residue levels in the three trials in Germany and the three in the USA were: 4.5, 4.7, 5.4, 8.8, 9.3, 10, 16 and 26 mg/kg, and the corresponding total residue levels were: 4.9, 6.2, 6.7, <u>9.9</u>, <u>10</u>, 11, 18 and 29 mg/kg.

The Meeting estimated a maximum residue level of 40 mg/kg and an STMR of 9.95 mg/kg for residues in hops, dry.

Animal feedstuffs

Almond hulls

Trifloxystrobin is registered in the USA for use on almonds up to three times at 0.14 kg ai/ha and a 60-day PHI. Six trials on almonds in California approximating these conditions were reported. The trifloxystrobin residue levels in hulls were: 0.25, 0.42, 0.72, 1.2, 1.6 and 1.8 mg/kg, and the total residue levels were: 0.25, 0.42, <u>0.75</u>, <u>1.2</u>, 1.6 and 1.9 mg/kg (fresh weight).

Allowing for the standard 90% dry matter for almond hulls (*FAO Manual*, p. 147), the Meeting estimated a maximum residue level of 3 mg/kg and an STMR of 1.08 mg/kg for almond hulls (dry weight). A highest residue level of 2.1 mg/kg was estimated for calculating the dietary burden of farm animals.

Peanut fodder

Trifloxystrobin is registered in the USA for use on peanuts with a maximum GAP of two applications at 0.13 kg ai/ha and a PHI of 14 days. In 12 trials with eight applications at 0.14 kg ai/ha in six states in 1996, the trifloxystrobin residue levels in peanut hay were: 0.19 (two), 0.25, 0.27, 0.29, 0.34, 0.46, 0.71, 0.84, 1.4, 3.4 and 3.7 mg/kg. The corresponding total residue levels were: 0.37, 0.40, 0.42, 0.47, 0.50, <u>0.63</u>, <u>0.82</u>, 1.1, 1.4, 2.1, 4.1 and 4.2 mg/kg (fresh weight).

Allowing for the standard 85% dry matter for peanut hay (*FAO Manual*, p. 148), the Meeting estimated a maximum residue level of 5 mg/kg and an STMR (dry weight) of 0.85 mg/kg for residues in peanut fodder. A highest residue level of 4.94 mg/kg was estimated for calculating the dietary burden of farm animals.

Barley straw and fodder, dry

Trifloxystrobin may be used twice at 0.25 kg ai/ha in France, Germany and the United Kingdom; twice at 0.19 kg ai/ha in Belgium; and once at 0.15 kg ai/ha in Austria. The PHI is 35 days in Austria,

Germany and the United Kingdom and 42 days in Belgium and France. The trifloxystrobin residue levels in barley straw in one trial in Denmark, 23 in France, six in Germany and two in the United Kingdom, matching appropriate European GAP were: 0.09, 0.15, 0.23, 0.31, 0.32, 0.33, 0.38 (two), 0.43, 0.49, 0.50, 0.53, 0.61, 0.64, 0.66, 0.68, 0.69, 0.72, 0.78, 0.81 (two), 0.91 (two), 0.93, 1.0, 1.1, 1.3, 1.5, 1.6, 1.8, 2.4 and 4.2 mg/kg. The corresponding total residue levels were: 0.09, 0.15, 0.30, 0.33, 0.38 (three), 0.48, 0.58, 0.64, 0.67, 0.68, 0.75, 0.77, 0.79, 0.80 (two), 0.86, 0.93, 0.94 (two), 1.1 (four), 1.5 (two), 1.7, 1.8, 1.9, 2.6 and 4.4 mg/kg (fresh weight).

Allowing for the standard 89% dry matter for barley straw (*FAO Manual*, p. 147), the Meeting estimated a maximum residue level of 7 mg/kg and an STMR value (dry weight) of 0.9 mg/kg for residues in barley straw and fodder, dry. A highest residue level of 4.9 mg/kg was estimated for calculating the dietary burden of farm animals.

Wheat straw and fodder, dry

Trifloxystrobin may be used on wheat at 0.25 kg ai/ha in France, Germany, Ireland and the United Kingdom; at 0.19 kg ai/ha in Austria, Belgium, Hungary, Italy, Luxembourg, Poland and Switzerland; and at at 0.13 kg ai/ha in Slovakia. The PHIs are 35–45 days. In 26 trials in France, 10 in Germany, one in Sweden and two in Switzerland that matched appropriate European GAP, the trifloxystrobin residue levels in wheat straw were: < 0.05, 0.07, 0.09 (two), 0.13, 0.16 (two), 0.17, 0.19 (two), 0.30, 0.31, 0.33, 0.34, 0.35, 0.38, 0.40, 0.50, 0.57, 0.59, 0.62, 0.70, 0.73, 0.76, 0.77, 0.81, 0.83, 0.85, 0.94, 0.99, 1.1, 1.3 (two), 1.4, 1.6, 1.8, 1.9, 2.3 and 2.5 mg/kg. The corresponding total residue levels were < 0.05, 0.07, 0.09, 0.13, 0.15, 0.16 (two), 0.24, 0.25, 0.26, 0.38 (three), 0.41, 0.42, 0.43, 0.51, 0.57, 0.77, 0.78 (two), 0.87, 0.89, 0.94, 0.95 (two), 1.0, 1.1, 1.2 (two), 1.3, 1.5, 1.6, 1.7, 1.8, 2.1, 2.6 (two) and 2.7 mg/kg (fresh weight).

Trifloxystrobin is registered in the USA for use on wheat up to two times at 0.09 kg ai/ha with a 35-day PHI. In 23 trials in eight states where these conditions were approximated, the trifloxystrobin residue levels in wheat straw were: 0.08 (two), 0.11 (two), 0.12 (two), 0.13, 0.14, 0.15, 0.17, 0.19 (two), 0.26, 0.27, 0.29, 0.31, 0.34, 0.51, 0.61, 0.96, 0.97, 1.4 and 1.9 mg/kg. The corresponding total residue levels were 0.08, 0.11 (three), 0.12 (two), 0.14, 0.15 (two), 0.19, 0.21, 0.22, 0.26, 0.29, 0.31, 0.36, 0.44, 0.51, 0.64, 1.0, 1.1, 1.6 and 2.4 mg/kg (fresh weight).

In summary, the trifloxystrobin residue levels in the 39 European and the 23 US trials, in ranked order, were: < 0.05, 0.07, 0.08 (two), 0.09 (two), 0.11 (two), 0.12 (two), 0.13 (two), 0.14, 0.15, 0.16 (two), 0.17 (two), 0.19 (four), 0.26, 0.27, 0.29, 0.30, 0.31 (two), 0.33, 0.34 (two), 0.35, 0.38, 0.40, 0.50, 0.51, 0.57, 0.59, 0.61, 0.62, 0.70, 0.73, 0.76, 0.77, 0.81, 0.83, 0.85, 0.94, 0.96, 0.97, 0.99, 1.1, 1.3 (two), 1.4 (two), 1.6, 1.8, 1.9 (two), 2.3 and 2.5 mg/kg. The total residue levels were: < 0.05, 0.07, 0.08, 0.09, 0.11 (three), 0.12 (two), 0.13, 0.14, 0.15 (three), 0.16 (two), 0.19, 0.21, 0.22, 0.24, 0.25, 0.26 (two), 0.29, 0.31, 0.36, 0.38 (two), 0.38, 0.41, 0.42, 0.43, 0.44, 0.51 (two), 0.57, 0.64, 0.77, 0.78 (two), 0.87, 0.89, 0.94, 0.95 (two), 1.0 (two), 1.1 (two), 1.2 (two), 1.3, 1.5, 1.6 (two), 1.7, 1.8, 2.1, 2.4, 2.6 (two) and 2.7 mg/kg (fresh weight).

Allowing for the standard 88% dry matter for wheat straw (*FAO Manual*, p. 149), the Meeting estimated a maximum residue level of 5 mg/kg and an STMR (dry weight) of 0.48 mg/kg for residues in wheat straw and fodder, dry. A highest residue level of 3.07 mg/kg was estimated for calculating the dietary burden of farm animals.

Maize fodder

Trifloxystrobin is registered in the USA for use on maize up to three times at 0.11 kg ai/ha. In 24 field trials in 14 states where these conditions were approximated and with a PHI of 30 days, the trifloxystrobin residue levels in maize stover were: 0.04, 0.32, 0.37, 0.42, 0.43, 0.53, 0.56, 0.64, 0.88, 0.96 (two), 1.0, 1.2, 1.5, 2.0, 2.1, 2.2 (two), 2.7, 2.9, 3.2, 3.9, 4.0 and 5.4 mg/kg. The corresponding total residue levels were 0.09, 0.37, 0.41, 0.47, 0.56, 0.65, 0.74, 0.80, 1.3, 1.4 (three), 1.5, 1.9, 2.4, 2.5, 2.6, 2.8, 3.5, 3.9, 4.4 (two), 4.5 and 7.1 mg/kg.

Allowing for the standard 83% dry matter for maize stover (*FAO Manual*, p. 147), the Meeting estimated a maximum residue level of 10 mg/kg and an STMR (dry weight) of 1.75 mg/kg for residues in maize fodder. A highest residue level of 8.55 mg/kg was estimated for calculating the dietary burden of farm animals.

Rice straw and fodder, dry

Trifloxystrobin is registered in the USA for use on rice up to two times at 0.17 kg ai/ha with a 35-day PHI. In 19 trials in five states where these conditions were approximated, the trifloxystrobin residue levels in rice straw were: 0.07, 0.25, 0.37, 0.42, 0.44, 0.50, 0.54, 0.57, 0.78, 1.0, 1.1, 1.3, 2.0, 2.4, 2.5, 2.6 (two), 5.3 and 6.1 mg/kg. The corresponding total residue levels were 0.07, 0.32, 0.45, 0.54, 0.60, 0.74, 0.83, 0.84, 1.0, 1.3, 1.6, 2.1, 2.6, 3.2 (three), 3.7, 5.5 and 7.3 mg/kg (fresh weight).

Allowing for the standard 90% dry matter for rice straw (*FAO Manual*, p. 149), the Meeting estimated a maximum residue level of 10 mg/kg and an STMR (dry weight) of 1.4 mg/kg for residues in rice straw and fodder, dry. A highest residue level of 8.1 mg/kg was estimated for calculating the dietary burden of farm animals.

Sugar-beet leaves or tops

Italian GAP allows three treatments at rates of 0.11–0.15 kg ai/ha, and Swiss GAP allows once at 0.15 kg ai/ha, both with a PHI of 21 days, for sugar-beets. The residue levels of trifloxystrobin *per se* in sugar-beet tops in nine trials in France, two in Italy, two in Spain and one in Switzerland matching GAP were: < 0.02 (three), < 0.05 (five), 0.05, 0.07, 0.09, 0.14, 0.33 and 0.44 mg/kg. The corresponding total residue levels were: < 0.02 (three), < 0.05 (five), 0.05, 0.09 (two), 0.17, 0.41 and 0.44 mg/kg (fresh weight).

Trifloxystrobin is registered in the USA for use on sugar-beet up to three times at 0.12 kg ai/ha with a 21-day PHI. In 19 trials in seven states where these conditions were matched, the trifloxystrobin residues in sugar-beet tops were 0.08 (two), 0.14, 0.17, 0.21, 0.23, 0.24, 0.26, 0.35, 0.54, 0.56, 0.61, 0.64, 0.72, 0.98, 1.6, 2.3 and 2.4 (two) mg/kg. The corresponding total residue levels were 0.08 (two), 0.14, 0.17, 0.23 (two), 0.24, 0.26, 0.39, 0.54, 0.61 (two), 0.66, 0.72, 0.98, 1.6, 2.4 and 2.5 (two) mg/kg (fresh weight).

The data sets from Europe and the USA appeared to be from different populations and were not combined. The Meeting agreed to estimate a maximum residue level and an STMR on the basis of the results of the trials in the USA.

Allowing for the standard 23% dry matter for sugar beet tops (*FAO Manual*, p. 147), the Meeting estimated a maximum residue level of 15 mg/kg and an STMR (dry weight) of 2.3 mg/kg for residues in sugar-beet leaves or tops. A highest residue level of 10.9 mg/kg was estimated for calculating the dietary burden of farm animals.

Fate of residues during processing

The Meeting received information on the fate and nature of trifloxystrobin residues under various conditions of hydrolysis. Trifloxystrobin is partially hydrolysed to CGA 321113 under conditions representative of baking, brewing and boiling (2.6%) and sterilization (22.5%). It was stable under conditions representative of pasteurization. Any possible effects of hydrolysis on the nature of the residue during processing are covered by the fact that the only relevant metabolite (CGA 321113) was determined in all the residue and processing trials.

The effect of processing on the level of residues of trifloxystrobin has been studied for barley, cabbage, cotton, grapes, hops, maize, oranges, peanuts, pome fruit, potatoes, rice, stone fruit, strawberries, sugar-beet, tomatoes and wheat. The processing factors shown below were calculated from the total residue levels (sum of trifloxystrobin and CGA 321113, expressed as trifloxystrobin).

Raw agricultural commodity	Processed product	No. of samples	Mean processing factor
Orange	Juice	5	< 0.19
	Oil	5	130
	Pulp, dry	5	3.4
Apple, pear	Juice	7	0.16
	Sauce and preserve	4	0.48
	Fruit, dried	2	0.39
	Pomace, wet	6	9.4
	Pomace, dried	1	25.6
Plum	Dried prune	4	1.5
Peach	Preserve	1	< 0.05
Grapes	Juice	14	0.24
	Must	27	0.46
	Wine	35	0.15
	Fruit, dried	4	2.3
	Pomace, wet	1	2.25
Strawberry	Preserve	2	0.29
	Jam	2	0.62
Tomato	Paste	5	1.6
	Puree	5	0.56
Potato	Flakes	2	< 0.42
	Chips	2	< 0.42
	Wet peel	2	2.3
Sugar-beet	White sugar	2	< 0.18
	Dried pulp	2	3.4
	Molasses	2	1.5
Barley	Beer	1	0.04
Wheat	Bran	2	2.7
	Germ	1	< 0.67
	Meal and flour	2	0.4
	Wholemeal	1	0.5
	Wholemeal bread	1	0.25
Rice	Polished grain	4	0.18
	Hull	4	3.2
	Bran	4	1.4
Hops	Spent hops	1	0.04
	Yeast	1	0.007
	Beer	1	< 0.001

Oranges were processed into juice, oil and dried pulp with processing factors of < 0.19, 130 and 3.4, respectively. On the basis of the STMR value of 0.095 mg/kg for whole citrus fruits, the STMR-Ps were 0.018 mg/kg for citrus juice and 12 mg/kg for oil. Allowing for the standard 91% dry matter, the Meeting estimated a maximum residue level of 1 mg/kg and an STMR-P of 0.35 mg/kg (0.095 × 3.4 × 1.0989) for residues in dried citrus pulp (dry weight).

Apples and *pears* were processed into juice, sauce or preserve, wet pomace, dry pomace and dried fruit, with processing factors of 0.16, 0.48, 9.4, 25.6 and 0.39, respectively. On the basis of the STMR value of 0.11 mg/kg for pome fruit, the STMR-P was 0.018 mg/kg for juice, 0.053 mg/kg for sauce, 0.053 mg/kg for preserve and 0.043 mg/kg for dried fruit of apple and pear. In the *FAO Manual* (Appendix IX), wet apple pomace is listed as animal feed. Allowing for the standard 40% dry matter, the Meeting estimated an STMR-P of 2.6 mg/kg (0.11 × 9.4 × 2.5) for residues in wet apple pomace (dry weight).

Peaches were processed into preserve (canned fruits) with a processing factor of 0.05. On the basis of the STMR value of 0.38 mg/kg for stone fruit, the STMR-P was 0.019 mg/kg for residues in canned fruits of peaches, nectarines and apricots.

Plums were processed into dried prunes with a processing factor of 1.5. On the basis of the STMR value of 0.38 mg/kg for stone fruit, the STMR-P was 0.57 mg/kg for dried prunes.

Grapes were processed into juice, must, wine and dried fruit (raisins) with processing factors of 0.24, 0.46, 0.15 and 2.3, respectively. On the basis of the STMR value of 0.15 mg/kg for grapes, the STMR-P was 0.036 mg/kg for juice, 0.07 mg/kg for must, 0.023 mg/kg for wine and 0.345 mg/kg for raisins (dried grapes). On the basis of the highest trifloxystrobin residue level of 2.2 mg/kg, the Meeting estimated a maximum residue level of 5 mg/kg for residues in raisins (dried grapes).

Strawberries were processed into preserve (canned fruits) and jam with processing factors of 0.29 and 0.62, respectively. On the basis of the STMR value of 0.10 mg/kg for strawberries, the STMR-P values were 0.029 mg/kg for residues in canned strawberries and 0.062 mg/kg for those in jam.

Head cabbage was cooked. Because residues were not detected in the raw commodity, a processing factor could not be calculated and the levels of residues in the processed commodity could not be estimated.

Tomatoes were processed into paste and puree with processing factors of 1.6 and 0.56, respectively. On the basis of the STMR value of 0.08 mg/kg for tomato, the STMR-Ps were 0.13 mg/kg for residues in tomato paste and 0.045 mg/kg for residues in puree.

Potatoes were processed into flakes, chips and wet peel with processing factors of 0.42, 0.42 and 2.3, respectively. On the basis of the STMR value of 0.02 mg/kg for residues in potatoes, the STMR-Ps were 0.008 mg/kg for residues in potato flakes and chips. In the *FAO Manual* (Appendix IX), wet peel (processed potato waste) is listed as animal feed. Allowing for the standard 15% dry matter, an STMR-P of 0.307 mg/kg (0.02 × 2.3 × 6.67) was estimated for potato wet peel (dry weight).

Sugar-beet was processed into white sugar, dried pulp and molasses with processing factors of 0.18, 3.4 and 1.5, respectively. On the basis of the STMR value of 0.02 mg/kg, the STMR-P for white sugar was 0.0036 mg/kg. In the *FAO Manual* (Appendix IX), sugar-beet dried pulp (88% dry matter) and molasses (75% dry matter) are listed as animal feeds. On the basis of the highest trifloxystrobin residue level of 0.04 mg/kg, the Meeting estimated maximum residue levels of 0.2 mg/kg (0.04 × 3.4 × 1.14) for sugar-beet dried pulp and 0.1 mg/kg (0.04 × 1.5 × 1.33) for sugar-beet molasses (dry weight). The estimated STMR-P values were 0.077 mg/kg (0.02 × 3.4 × 1.14) for sugar-beet dried pulp and 0.04 mg/kg (0.02 × 1.5 × 1.33) for sugar beet molasses (dry weight).

Wheat was processed into the milled by-products bran, flour, wholemeal, wholemeal bread and germ, with processing factors of 2.7, 0.4, 0.67, 0.33 and 0.67, respectively. On the basis of the STMR value of 0.02 mg/kg for wheat grain, the STMR-Ps were 0.008 for wheat flour, 0.01 for wholemeal, 0.005 for wholemeal bread and 0.013 for germ. In the *FAO Manual* (Appendix IX), bran is listed as an animal feed.

Allowing for the standard 88% dry matter for wheat milled by-products, the Meeting estimated an STMR-P of 0.062 mg/kg (0.02 × 2.7 × 1.14) for wheat bran, unprocessed (dry weight). On the basis of the highest trifloxystrobin residue level of 0.14 mg/kg, the Meeting estimated a maximum residue level of 0.5 mg/kg (0.14 × 2.7 × 1.14 = 0.43) for wheat bran, unprocessed (dry weight).

Maize was processed to meal, grits, flour and oil. Because residues were not detected in the raw commodity, processing factors could not be calculated and the residue levels in the processed commodities could not be estimated.

Rice with husk was processed into polished rice, bran and hulls with processing factors of 0.18, 1.4 and 3.2, respectively. On the basis of the STMR of 0.16 mg/kg for rice with husks, an STMR-P of 0.029 mg/kg was calculated for polished rice. In the *FAO Manual* (Appendix IX), rice bran and hulls are listed as animal feed. Allowing for the standard 90% dry matter, the Meeting estimated STMR-P values of 0.57 mg/kg (0.16 × 3.2 × 1.1) for rice hulls and 0.25 mg/kg (0.16 × 1.4 × 1.1) for rice bran, unprocessed (dry weight). On the basis of the highest trifloxystrobin residue level of 3.4 mg/kg in rice with husks, the Meeting estimated a maximum residue level of 7 mg/kg for residues in rice bran, unprocessed (dry weight).

Cotton was processed to refined oil. Because residues were not detected in the raw commodity, a processing factor could not be calculated and the residue levels in the processed commodities could not be estimated.

Peanuts were processed to meal and refined oil. Because residues were not detected in the raw commodity, a processing factor could not be calculated and the residue levels in the processed commodities could not be estimated.

Hops were processed for use in beer, with a processing factor of 0.001. On the basis of the STMR value of 9.95 mg/kg for dry hops, an STMR-P of 0.01 mg/kg was calculated for beer. Barley was processed into beer with a processing factor of 0.04. On the basis of the STMR value of 0.04 mg/kg, the STMR-P for beer was 0.0016 mg/kg. Because the STMR-P arising from residues in barley was lower, the Meeting estimated an STMR-P of 0.01 mg/kg for residues in beer, on the basis of residues in hops.

Residues in animal commodities

Dietary burden of farm animals

The Meeting estimated the dietary burden of trifloxystrobin residues in farm animals on the basis of the diets listed in Appendix IX of the *FAO Manual*. Calculation from highest residue and STMR-P values provides the levels in feed suitable for estimating MRLs, while calculation from STMR and STMR-P values for feed is suitable for estimating STMR values for animal commodities. The percentage dry matter is taken as 100% when the highest residue levels and STMRs are already expressed as dry weight.

Estimated maximum dietary burden of farm animals

Commodity	CC	Residue (mg/kg)	Basis	Dry matter (%)	Residue, dry weight (mg/kg)	Diet content (%) Beef cattle	Diet content (%) Dairy cattle	Diet content (%) Poultry	Residue contribution (mg/kg) Beef cattle	Residue contribution (mg/kg) Dairy cattle	Residue contribution (mg/kg) Poultry
Almond hulls	AM	2.1	HR	100	2.1						
Apple pomace, wet	AB	2.6	STMR-P	100	2.6	15			0.39		
Barley grain	GC	0.46	HR	88	0.528			40			0.211
Barley straw	AS	4.9	HR	100	4.9		60			2.94	
Citrus pulp, dried	AB	0.35	STMR-P	100	0.35						
Maize grain	GC	0.05	HR	88	0.057						

Trifloxystrobin

Commodity	CC	Residue (mg/kg)	Basis	Dry matter (%)	Residue, dry weight (mg/kg)	Diet content (%) Beef cattle	Diet content (%) Dairy cattle	Diet content (%) Poultry	Residue contribution (mg/kg) Beef cattle	Residue contribution (mg/kg) Dairy cattle	Residue contribution (mg/kg) Poultry
Maize fodder	AS	8.55	HR	100	8.55	25			2.14		
Peanut fodder (hay)	AL	4.94	HR	100	4.94						
Potato wet peel	AB	0.307	STMR-P	100	0.307						
Rice	GC	3.4	HR	88	3.86	40	30	60	1.54	1.16	2.32
Rice bran	CM	0.25	STMR-P	100	0.25						
Rice hulls	CM	0.57	STMR-P	100	0.57						
Rice straw and fodder, dry	AS	8.1	HR	100	8.1						
Sugar-beet leaves and tops	AV	10.9	HR	100	10.9	20	10		2.18	1.09	
Sugar-beet, dried pulp	AB	0.077	STMR-P	100	0.077						
Sugar-beet molasses	DM	0.04	STMR-P	100	0.04						
Wheat grain	GC	0.2	HR	89	0.225						
Wheat straw	AS	3.07	HR	100	3.07						
Wheat milled by-products (bran)	CM	0.062	STMR-P	100	0.062						
Total						**100**	**100**	**100**	**6.3**	**5.2**	**2.5**

Estimated mean dietary burden of farm animals

Commodity	CC	Residue (mg/kg)	Basis	Dry matter (%)	Residue/ dry weight (mg/kg)	Diet content (%) Beef cattle	Diet content (%) Dairy cattle	Diet content (%) Poultry	Residue contribution (mg/kg) Beef cattle	Residue contribution (mg/kg) Dairy cattle	Residue contribution (mg/kg) Poultry
Almond hulls	AM	1.08	STMR	100	1.08		10			0.108	
Apple pomace, wet	AB	2.6	STMR-P	100	2.6	40	20		1.04	0.52	
Barley grain	GC	0.04	STMR	88	0.045						
Barley straw	AS	0.9	STMR	100	0.9		60			0.54	
Citrus pulp, dried	AB	0.35	STMR-P	100	0.35						
Maize grain	GC	0.02	STMR	88	0.023						
Maize fodder	AS	1.75	STMR	100	1.75	25			0.4375		
Peanut fodder (hay)	AL	0.85	STMR	100	0.85	15			0.1275		
Potato, wet peel	AB	0.307	STMR-P	100	0.307						
Rice	GC	0.16	STMR	88	0.182			60			0.1092
Rice bran	CM	0.25	STMR-P	100	0.25			25			0.0625
Rice hulls	CM	0.57	STMR-P	100	0.57			15			0.086
Rice straw and fodder, dry	AS	1.4	STMR	100	1.4						
Sugar-beet leaves and tops	AV	2.3	STMR	100	2.3	20	10		0.46	0.23	
Sugar-beet dried pulp	AB	0.077	STMR-P	100	0.077						
Sugar-beet molasses	DM	0.04	STMR-P	100	0.04						
Wheat grain	GC	0.02	STMR	89	0.0225						

Commodity	CC	Residue (mg/kg)	Basis	Dry matter (%)	Residue/ dry weight (mg/kg)	Diet content (%) Beef cattle	Diet content (%) Dairy cattle	Diet content (%) Poultry	Residue contribution (mg/kg) Beef cattle	Residue contribution (mg/kg) Dairy cattle	Residue contribution (mg/kg) Poultry
Wheat straw	AS	0.48	STMR	100	0.48						
Wheat milled by-products (bran)	CM	0.062	STMR-P	100	0.062						
Total						100	100	100	2.1	1.4	0.26

The dietary burdens of trifloxystrobin for estimating maximum residue levels and STMRs for animal commodities (residue concentrations in animal feeds expressed as dry weight) are 6.3 and 2.1 mg/kg for beef cattle, 5.2 and 1.4 mg/kg for dairy cattle, 2.5 and 0.26 mg/kg for poultry.

Feeding studies

The Meeting received information on residues arising in tissues and milk of *dairy cows* dosed with trifloxystrobin in capsules at the equivalent of 2, 5.9 or 21 ppm in the diet for 28–30 days.

The sum of trifloxystrobin and CGA 321113 was calculated and expressed as trifloxystrobin on the basis of relative molecular masses. A conversion factor of 1.036 is required to express the CGA 321113 residues as trifloxystrobin equivalents. As this metabolite constitutes a significant proportion of the residue in animal products, when the level of trifloxystrobin or CGA 321113 was below its LOQ, the sum of trifloxystrobin and CGA 321113 was calculated, as in the examples below, and expressed as trifloxystrobin.

Trifloxystrobin (mg/kg)	CGA 321113 (mg/kg)	Total expressed as trifloxystrobin (mg/kg)
< 0.02	< 0.02	< 0.04
< 0.02	0.03	0.05
0.09	< 0.02	0.11

No residues (sum of trifloxystrobin and CGA 321113) were detectable in milk (< 0.02 mg/kg), round muscle (< 0.04 mg/kg) or tenderloin (< 0.04 mg/kg) from cattle given the highly exaggerated feeding level of 21 ppm. No residues of either parent or metabolite were found in liver, kidney or fat samples from animals at 2 or 5.9 ppm (total residue, < 0.04 mg/kg). Maximum residue levels of 0.09 mg/kg (total residue, 0.11 mg/kg) and 0.02 mg/kg (total residue, 0.04 mg/kg), detected as the metabolite CGA 321113 and expressed as trifloxystrobin, were found in liver and kidney, respectively, from cattle at 21 ppm; and maximum residue levels of 0.06 mg/kg (total residue, 0.08 mg/kg) and 0.05 mg/kg (total residue, 0.07 mg/kg), detected as intact trifloxystrobin, were found in perirenal fat and omental fat, respectively, from these animals.

The Meeting received information on the concentrations of residues in tissues and eggs from *laying hens* dosed with trifloxystrobin at the equivalent of 1.5, 4.5 or 15 ppm in the diet for 28 days. The hens were killed on day 29, and composite tissue samples of breast plus thigh, skin plus attached fat, peritoneal fat, and liver were taken. Eggs and tissues were analysed for trifloxystrobin and CGA 321113. No residues (total residues, < 0.04 mg/kg) were detected in any of the eggs, tissues or organs taken from hens at the highest dietary level of 15 ppm.

Maximum residue levels

The Meeting noted that no trifloxystrobin or CGA 321113 residues were detected in milk (total, < 0.02 mg/kg), muscle (< 0.04 mg/kg), kidney (< 0.04 mg/kg), liver (< 0.04 mg/kg) or fat (< 0.04 mg/kg) from animals dosed for 28 days at 5.9 ppm, which was close to the maximum dietary burdens of beef and dairy cattle (8.2 and 7.4 ppm). The highest residue level of trifloxystrobin was found in perirenal fat at

Trifloxystrobin

0.06 mg/kg (total residue, 0.08 mg/kg), and the highest level of CGA 321113 was found in liver at 0.09 mg/kg (total residue, 0.11 mg/kg) from animals given 21 ppm.

Dietary burden (ppm) Feeding level [ppm]	Trifloxystrobin total residue (mg/kg)								
	Milk (mean)	Muscle		Liver		Kidney		Fat	
		Highest	Mean	Highest	Mean	Highest	Mean	Highest	Mean
MRL beef cattle (6.3) [21]		(< 0.012) < 0.04		(0.033) 0.11		(0.012) 0.04		(0.024) 0.08	
MRL dairy cattle (5.2) [21]	(< 0.005) < 0.02								
STMR beef cattle (2.1) [21]		(< 0.004) < 0.04		(0.008) 0.08		(0.004) 0.04		(0.006) 0.06	
STMR dairy cattle (1.4) [21]	(< 0.001) < 0.02								

The maximum concentrations of residues expected in tissues are < 0.012 mg/kg in muscle, 0.033 mg/kg in liver, 0.012 mg/kg in kidney, 0.024 mg/kg in fat and < 0.005 mg/kg in milk. The mean extrapolated concentrations are < 0.004 mg/kg in muscle, 0.008 mg/kg in liver, 0.004 mg/kg in kidney, 0.006 mg/kg in fat and < 0.001 mg/kg in milk.

Taking into account the fat solubility of trifloxystrobin (the acid metabolite CGA 321113 is poorly soluble in fat), the Meeting estimated a maximum residue level of 0.05 mg/kg for the sum of trifloxystrobin and CGA 321113 in meat (fat) from mammals other than marine mammals on the basis of residue levels in trimmable fat, and a maximum residue level of 0.02* mg/kg for residues in milks. The estimated maximum residue levels are 0.05 mg/kg for liver and 0.04* mg/kg for kidney of cattle, goats, pigs and sheep.

The estimated STMR values are 0.006 mg/kg in fat, 0 mg/kg in muscle, 0.008 mg/kg in liver, 0.004 mg/kg in kidney and 0 in milks.

The Meeting noted that in the feeding study in laying hens, no trifloxystrobin or CGA 321113 residues (total residue, < 0.04 mg/kg) were detected in eggs, tissues or organs from hens at the highest feeding level of 15 ppm. As the maximum dietary burden of 2.5 mg/kg was much lower, the Meeting agreed that the expected level of trifloxystrobin and CGA 321113 residues in poultry tissues and eggs would be essentially 0.

The Meeting estimated maximum residue levels of 0.04* mg/kg for residues in eggs, poultry meat (fat) and edible offal. The Meeting recommended that the STMR values should be 0 in eggs, poultry meat, edible offal and fat.

DIETARY RISK ASSESSMENT

Long-term intake

The IEDIs of trifloxystrobin, on the basis of the STMRs estimated for 37 commodities, for the five GEMS/Food regional diets represented 1–2% of the ADI (Annex 3). The Meeting concluded that the long-term intake of residues of trifloxystrobin resulting from uses that have been considered by the JMPR is unlikely to present a public health concern.

Short-term intake

The 2004 JMPR decided that it was unnecessary to establish an ArfD. The present Meeting therefore concluded that the short-term intake of trifloxystrobin residues is unlikely to present a public health concern.

5. RECOMMENDATIONS

5.1 The Meeting developed a guidance document for deriving acute reference doses (ARfDs) (section 2.1). This document includes general considerations, summarizing all previous considerations of the Meeting on this topic, including a stepwise process for setting ARfDs and for selecting appropriate end-points and safety factors, as well as specific guidance for assessing particular toxicologic end-points of relevance for setting ARfDs. This document should be updated regularly, as further experience in establishing ARfDs is gained. The introduction of a specific test protocol for single-dose, oral toxicity studies will aid in establishing ARfDs when relevant.

5.2 The WHO Core Assessment Group occasionally establishes ARfDs for compounds that were not scheduled for toxicological evaluation, on the basis of data from previous evaluations, to facilitate acute dietary risk assessment. The Meeting decided to call these values 'interim ARfDs'. These interim ARfDs can be used in dietary risk assessments until they are replaced by a full evaluation, if this is considered necessary.

5.3 The pilot work-sharing project concluded and recommended the following:

- The availability of several national or regional evaluations was found to be useful by the WHO and FAO evaluators. Although problems were encountered, consideration of national or regional evaluations in the development of JMPR monographs had several benefits. FAO, WHO and OECD should consider means to facilitate the provision of national and regional evaluations for JMPR evaluators.

- Consideration of multiple national and regional evaluations should promote international harmonization of dossiers and evaluations.

- The evaluation process, including standardization of formats and development of guidelines, should be further harmonized at the international level. Good progress has been reached in toxicological evaluations, and more basic work is necessary to improve work-sharing for residue evaluations.

- A further JMPR pilot work-sharing project should have a more flexible procedure, which should be reviewed when there is greater harmonization of formats and evaluation procedures (harmonization of guidance documents).

5.4 The JMPR considered that extensive use of the interim MRL process might create a serious problem. As interim MRLs are limited to a period of four years, pesticides nominated under the process must be scheduled for and reviewed by the JMPR within this period. If there are many interim MRL pesticide nominations, evaluations might not be completed for some within the four-year period or other priorities, such as periodic review of pesticides and evaluations, might have to be severely curtailed, owing to the current limited resources of the JMPR.

5.5 For spices, the Meeting recommended that CCPR accept the principle of setting MRLs on the basis of monitoring data covering the 95th percentile of the residue population with 95% confidence. That decision would facilitate use of statistically based procedures for estimating maximum residue levels and acceptance of recommended limits. It should be noted, however, that when MRLs are set at the 95th percentile with 95% confidence, the residue levels might exceed the MRL in 5% of cases.

Monitoring data should not be used for estimating maximum residue levels reflecting post-harvest use, which results in significantly higher residue levels than foliar application or exposure to spray drift.

Recommendations

5.6 The Meeting decided that two maximum residue levels will be estimated for fat-soluble pesticides from now on, if the data permit: one for whole milk and one for milk fat. For enforcement purposes, the residue level in milk fat can be compared with the MRL for milk (fat), or the residue level in whole milk can be compared with the MRL for milk. When needed, maximum residue levels in milk products can be calculated from these two numbers, by taking into account both the fat content of the milk product and the contribution from the non-fat fraction.

5.7 For animal feed commodities, the Meeting agreed that the assumption of the 1986 JMPR used in developing guidance in this area, that "it was unrealistic to assume that the theoretical maximum residue level would be achieved and maintained in the rations of food-producing animals receiving feeds produced on the farm", should no longer apply. Hence, the concept of time for residues to reach a plateau level, rapid or slow, is also no longer required, and the process of estimating dietary burdens is simplified. The Meeting recognized that it is unlikely that the individual ingredients in mixed feeds produced from commercially available ingredients would all contain residues at the theoretical maximum level; in these cases, the STMR should be applied for each component.

A revision of the relevant text in the *FAO Manual*, taking into account the above, will appear on the FAO website.

5.8 The Meeting expressed interest in receiving spreadsheets and documentation for evaluating the statistical method for MRL estimation. Such systems might aid the evaluators but could never replace professional judgement, when available. Further developments are awaited.

5.9 Some of the recommendations of the York workshop on minimum data requirements and the global zoning report have been used by the JMPR and will continue to be considered as auxiliary advice. The Meeting concluded, however, that substantial additional work is required to make the recommendations generally applicable as guidance. Areas that require additional effort include: (1) defining significance in trade, perhaps with a table of commodities and their classification; (2) determining the criteria for zones and the number of zones for each commodity; (3) extending the list of commodity translations, for the purpose of recommending maximum residue levels for one commodity on the basis of data from field trials for another commodity; and (4) completing the list of representative crops for the purpose of recommending maximum residue levels for crop groups.

6. FUTURE WORK

The items listed below should be considered by the Meeting in 2006 and 2007. The compounds listed include those recommended as priorities by the CCPR at its Thirty-sixth and earlier sesssions and compunds scheduled for re-evaluation within the CCPR periodic review programme.

2006 JMPR

Toxicological evaluations

New compounds
Bifenazate
Dimethomorph
Quinoxyfen

Periodic re-evaluations
Cyfluthrin and β-cyfluthrin (157)
Cyromazine (169)
α- and ζ-Cypermethrin
Flusilazole (165)
Procymidone (136)
Profenofos (171)

Evaluations
Haloxyfop (194)
Pirimiphos-methyl (086): acute toxicity
Thiabendazole (065): acute toxicity
Thiophanate-methyl (077): acute toxicity

Residue evaluations

New compounds
Bifenazate
Dimethomorph
Quinoxyfen

Periodic re-evaluations
α- and ζ-Cypermethrin
Cypermethrin (118)
Pirimicarb (101)
Propamocarb (148)
Propiconazole (160)
Triadimefon (133) To be evaluated
Triadimenol (168) together

Evaluations
Propargite (113)

2007 JMPR

Toxicological evaluations

New compounds
Pyrimethanil
Zoxamide

Periodic re-evaluations
Azinphos-methyl (002)
λ-Cyhalothrin
Fentin (040)
Vinclozolin (159)

Residue evaluations

New compounds
Pyrimethanil
Zoxamide

Periodic re-evaluations
Clofentezine (156)
Permethrin (120)
Triazophos (143)
Triforine (116)

ANNEX 1

ACCEPTABLE DAILY INTAKES, SHORT-TERM DIETARY INTAKES, ACUTE REFERENCE DOSES, RECOMMENDED MAXIMUM RESIDUE LIMITS AND SUPERVISED TRIALS MEDIAN RESIDUE VALUES RECORDED BY THE 2004 MEETING

The following extracts of the results of the annual Joint FAO/WHO Meeting on Pesticide Residues (JMPR) are provided to make them accessible to interested parties at an early date.

The Meeting evaluated 31 pesticides. Three were new compounds, and 11 were re-evaluated within the periodic review programme of the Codex Committee on Pesticide Residues (CCPR). The Meeting allocated acceptable daily intakes (ADIs), acute reference doses (ARfDs) and, for the first time, interim ARfDs. Interm ARfDs were allocated for use in short-term dietary risk assessment of compounds that were not scheduled for toxicological evaluation. These values are derived ad hoc from data used at previous meetings. They can be used in dietary risk assessments until they are replaced by a full evaluation, if this is considered necessary.

The Meeting estimated maximum residue levels, which it recommended for use as maximum residue limits (MRLs) by the CCPR. It also estimated supervised trials median residue (STMR) and highest residue (HR) levels as a basis for estimation of the dietary intake of residues of the pesticides reviewed. Application of HR levels is explained in the report of the 1999 Meeting (section 2.4). The allocations and estimates are shown in the Table 1.

The Meeting also estimated, for the first time, maximum residue levels of various pesticides in spice subgroups specified by the CCPR from monitoring data, and in dried chilli peppers taking into account the MRLs for fresh chilli peppers and peppers. The recommended values are listed separately in Tables 2 and 3.

Pesticides for which the estimated dietary intakes might, on the basis of the available information, exceed their ADIs are marked with footnotes, as explained in detail in the report of the 1999 Meeting (section 2.2). Footnotes are also applied to specific commodities when the available information indicated that the ARfD of a pesticide might be exceeded when the commodity was consumed. It should be noted that these distinctions apply only to new compounds and those re-evaluated within the CCPR periodic review programme.

Table 1 includes the Codex reference numbers of the compounds and the Codex classification numbers (CCNs) of the commodities, to facilitate reference to the Codex maximum limits for pesticide residues (*Codex Alimentarius,* Vol. 2B) and other documents and working documents of the Codex Alimentarius Commission. Both compounds and commodities are listed in alphabetical order.

Apart from the abbreviations indicated above, the following qualifications are used in the Table.

* (following name of pesticide)	New compound
** (following name of pesticide)	Compound reviewed within CCPR periodic review programme
* (following recommended MRL)	At or about the limit of quantification
HR-P	Highest residue in a processed commodity, in mg/kg, calculated by multiplying the HR in the raw commodity by the processing factor
Po	The recommendation accommodates post-harvest treatment of the commodity.
PoP (following recommendation for processed foods (classes D and E in the Codex classification)	The recommendation accommodates post-harvest treatment of the primary food commodity.
STMR-P	An STMR for a processed commodity calculated by applying the concentration or reduction factor for the process to the STMR calculated for the raw agricultural commodity.
V (following recommendations for commodities of animal origin)	The recommendation accommodates veterinary uses.
W (in place of a recommended MRL)	The previous recommendation is withdrawn, or withdrawal of the recommended MRL or existing Codex or draft MRL is recommended.

Annex 1

Table 1. Recommended maximum residue levels, STMR and HR values and allocated ADI and ARfD values

Pesticide (Codex reference number)	CCN	Commodity	Recommended MRL mg/kg New	Recommended MRL mg/kg Previous	STMR or STMR-P, mg/kg	HR or HR-P mg/kg
Bentazone (172) ADI: 0–0.1 mg/kg bw ARfD: unnecessary						
Captan (007) ADI: 0–0.1 mg/kg bw ARfD: 0.3 mg/kg bw (for women of child-bearing age)						
Carbofuran (096) ADI: 0–0.002 mg/kg bw ARfD: 0.009 mg/kg bw	FC 0004	Orange, sweet, sour			0.1	0.05

Residue for estimation of dietary intake in plant and animal commodities: sum of carbofuran, 3-hydroxycarbofuran and conjugated 3-hydroxycarbofuran, expressed as carbofuran

Pesticide	CCN	Commodity	New	Previous	STMR	HR
Chlorpyrifos (017) ADI: 0–0.01 mg/kg bw ARfD: 0.1 mg/kg bw	SO 0691	Cotton-seed	0.3		0.07	
		Cotton-seed meal		< 0.01		
		Cotton-seed hulls			0.05	
	OC 0691	Cotton-seed oil, crude			0.10	
	OR 0691	Cotton-seed oil, refined	0.05		0.01	
	VR 0589	Potato	2		0.51	0.87
	GC 0649	Rice	0.5		0.12	
	CM 1205	Rice, polished			0.008	
	CM 0649	Rice, husked			0.016	
		Rice hulls			0.29	
	CM 1206	Rice bran, unprocessed			0.22	
	VD 0541	Soya bean (dry)	0.1		0.01	
		Soya bean meal			< 0.002	
	OC 0541	Soya bean oil, crude			0.004	
	OR 0541	Soya bean oil, refined	0.03		0.004	
	DT 1114	Tea, green, black (black, fermented and dried)	2		0.34	

Residue for compliance with MRLs and estimation of dietary intake in plant and animal commodities: chlorpyrifos
The residue is fat-soluble

Dimethipin (151)
ADI: 0–0.02 mg/kg bw
ARfD: 0.2 mg/kg bw

Dithiocarbamates	VC 0424	Cucumber	2 c, N, p	2 c, N		
Ferbam	MO 0105	Edible offal (mammalian)	0.1 C, m, p	0.1 C, m		
ADI: 0–0.003 mg/kg bw	PE 0112	Eggs	0.05(*) C, p	0.05(*) C		
Propineb	FB 0269	Grapes	5 C, m, n	5 C, m, n, p		

Pesticide (Codex reference number)	CCN	Commodity	Recommended MRL mg/kg New	Recommended MRL mg/kg Previous	STMR or STMR-P, mg/kg	HR or HR-P mg/kg
ADI: 0–0.007 mg/kg bw	MM 0095	Meat (from mammals other than marine mammals)	0.05(*) c, m, p	0.05(*) c, m		
Thiram	VC 0046	Melon (except watermelon)	0.5 C	0.5 C, p		
ADI: 0–0.01 mg/kg bw	ML 0106	Milks	0.05(*) c, m, p	0.05(*) c, m		
Ziram	VA 0385	Onion, bulb	0.5 C	0.5 C, p		
ADI: 0–0.003 mg/kg bw	TN 0672	Pecan	0.1(*) Z	0.1(*) T Z		
	VO 0445	Peppers, sweet	7 c, m, P	1 c, m,		
	FP 0009	Pome fruits	5 C, M, H, Z	5 C, M, p, H, Z		
	VR 0587	Potato	0.2 c, m, n, p	0.2 c, m, n		
	PM 0110	Poultry meat	0.1 c, p	0.1 C		
	PO 0111	Poultry, edible offal of	0.1 c, p	0.1 C		
	FS 0012	Stone fruit	7 h, p, Z	7 T h, Z		

Residue for compliance with MRLs and estimation of dietary intake in plant and animal commodities: total dithiocarbamates, determined as CS_2, evolved during acid digestion and expressed as mg CS_2/kg.

Notes:

1. Recommended MRLs refer to the total residues from the use of any or each of the dithiocarbamates.
2. Recommendations are based on trials with; n, maneb; m, metiram; c, mancozeb; p, propineb; h, thiram; z, ziram. Compounds shown in upper case as those on which the estimates of maximum residue levels are mainly based.
3. The information provided to the JMPR precludes an estimate that the dietary intake would be below the interim ARfD for propineb.

T: the 1996 JMPR recommended that the MRL should be designated as temporary pending review of data on environmental fate.

Pesticide (Codex reference number)	CCN	Commodity	Recommended MRL mg/kg New	Recommended MRL mg/kg Previous	STMR or STMR-P, mg/kg	HR or HR-P mg/kg
Ethoprophos** (149)	FI 0327	Banana	0.02	0.02(*)	0.02	0.02
ADI: 0–0.0004 mg/kg bw	VR 0574	Beetroot	W	0.02(*)		
ARfD: 0.05 mg/kg bw	VB 0041	Cabbage, head	W	0.02(*)		
	VC 0424	Cucumber	0.01	0.02(*)	0.01	0.01
	MO 0105	Edible offal (mammalian)	0.01(*)		Liver 0 Kidney 0	Liver 0 Kidney 0
	VC 0425	Gherkin	W	0.02*		
	FB 0269	Grape	W	0.02*		
	VL 0482	Lettuce, head	W	0.02(*)		
	GC 0645	Maize	W	0.02(*)		
	AS 0645	Maize fodder	W	0.02(*)		
	AF 0645	Maize forage	W	0.02(*)		
	MM 0095	Meat (from mammals other than marine mammals)	0.01(*)		Muscle 0 Fat 0	Muscle 0 Fat 0
	VC 0046	Melons, except watermelon	0.02	0.02(*)	0.005	0.012
	ML 0106	Milks	0.01(*)		0	
	VA 0385	Onion, bulb	W	0.02(*)		
	SO 0697	Peanut	W	0.02(*)		
	AL 0697	Peanut fodder	W	0.02(*)		
	VP 0063	Peas (pods and succulent = immature seeds)	W	0.02(*)		
	VO 0051	Peppers	W	0.02(*)		
	VO 0445	Peppers, sweet	0.05		0.005	0.044

Pesticide (Codex reference number)	CCN	Commodity	Recommended MRL mg/kg New	Recommended MRL mg/kg Previous	STMR or STMR-P, mg/kg	HR or HR-P mg/kg
	FI 0353	Pineapple	W	0.02(*)		
	AM 0353	Pineapple fodder	W	0.02(*)		
	AV 0353	Pineapple forage	W	0.02(*)		
	VR 0589	Potato	0.05	0.02(*)	0.01	0.03
	VD 0541	Soya bean (dry)	W	0.02(*)		
	AL 0541	Soya bean fodder	W	0.02(*)		
	GS 0659	Sugar cane	0.02	–	0.02	0.02
	VR 0508	Sweet potato	0.05	–	0.01	0.03
	VO 0448	Tomato	0.01(*)	–	0.005	0.01

Residue for compliance with MRLs and estimation of dietary intake in plant and animal commodities: ethoprophos

Pesticide (Codex reference number)	CCN	Commodity	Recommended MRL New	Previous	STMR or STMR-P, mg/kg	HR or HR-P mg/kg
Fenitrothion (037)[1]	FP 0226	Apple	0.5	W	0.04	0.41
ADI: 0–0.005 mg/kg bw	GC 0080	Cereal grain	10 (Po)[2,3]	10 (Po)	5	7.6
ARfD: 0.04 mg/kg bw	MO 0105	Edible offal (mammalian)	0.05(*)	–	Liver 0 Kidney 0	Liver 0 Kidney 0
	PE 0112	Egg	0.05(*)	–	0	0
	MM 0095	Meat (from mammals other than marine mammals)	0.05(*)	W	Muscle 0 Fat 0	Muscle 0 Fat 0
	ML 0106	Milks	0.01	W	0	
	FP 0230	Pear	W	W		
	PM 0110	Poultry meat	0.05(*)	–	Muscle 0 Fat 0	Muscle 0 Fat 0
	CM 1206	Rice bran, unprocessed	60	W	36	55
	CM 0649	Rice, husked			3.2	4.9
	CM 1205	Rice, polished	W	W	0.44	1.1
		Cooked husked rice			0.55	0.84
		Cooked polished rice			0.2	0.30
		Washed polished rice			0.23	0.35
		Cooked washed polished rice			0.1	0.15
	CM 0654	Wheat bran, unprocessed	30 (PoP)[1]	20 (PoP)	20	30
	CF 1211	Wheat flour	W	W	1.2	1.8
	CP 1211	White bread	W	W	0.5	0.76
	CP 1212	Wholemeal bread	W		1.9	2.9

Residue for compliance with MRLs and estimation of dietary intake in plant and animal commodities: Fenitrothion

[1] The information provided to the JMPR precludes an estimate that the dietary intake would be below the ADI.
[2] Proposal by 2003 JMPR, the recommendation is now considered to also cover pre-harvest use of fenitrothion.
[3] The information provided to the JMPR precludes an estimate that the dietary intake would be below the ARfD.

Fenpropimorph (188)
ADI: 0–0.003 mg/kg bw
ARfD: 0.2 mg/kg bw

Fenpyroximate (193)
ADI: 0–0.01 mg/kg bw
ARfD: 0.01 mg/kg bw

Annex 1

Pesticide (Codex reference number)	CCN	Commodity	Recommended MRL mg/kg New	Recommended MRL mg/kg Previous	STMR or STMR-P, mg/kg	HR or HR-P mg/kg
Fludioxonil* (211)	HH 0722	Basil	10		2.4	
ADI: 0–0.4 mg/kg bw	DH 0722	Basil, dry	50		20	
ARfD: unnecessary	VD 0071	Bean (dry)	0.07		0.02	
	VP 0061	Bean, except broad bean and soya bean (green pods and immature seeds)	0.3		0.04	
	VP 0062	Beans, shelled (succulent = immature seeds)	0.03		0.02	
	FB 0264	Blackberry	5		1.0	
	FB 0020	Blueberry	2		0.60	
	VB 0400	Broccoli	0.7		0.23	
	VB 0041	Cabbage, head	2		0.24	
	VR 0577	Carrot	0.7		0.20	
	GC 0080	Cereal grain	0.05(*)		0.02	
	HH 0727	Chive	10		2.8	
		Chive (dried)	50		22	
	FC 0001	Citrus fruit	7 (Po)		1.1	
	SO 0691	Cotton-seed	0.05(*)		0.05	
	VC 0424	Cucumber	0.3		0.06	
	FB 0266	Dewberry (including boysenberry and loganberry)	5		1.0	
	DF 0269	Dried grape (= currant, raisin and sultana)			0.31	
	MO 0105	Edible offal (mammalian)	0.05(*)		0	
	VO 0440	Egg plant (aubergine)	0.3		0.06	
	PE 0112	Egg	0.05(*)		0	
	FB 0269	Grape	2		0.28	
	JF 0269	Grape juice			0.26	
	FI 0341	Kiwifruit	15 (Po)		7.2	
	VL 0482	Lettuce, head	10		2.7	
	AF 0645	Maize forage (dry)	0.03(*)		0	
	MM 0095	Meat (from mammals other than marine mammals)	0.01(*)		0	
	VC 0046	Melon, except watermelon	0.03		0.02	
	ML 0106	Milks	0.01		0	
	VL 0485	Mustard greens	10		1.2	
	VA 0385	Onion, bulb	0.5		0.04	
	VA 0389	Onion, spring	5		0.59	
	FP 0230	Pear	0.7		0.21	
	VD 0072	Pea (dry)	0.07		0.02	
	VP 0063	Pea (pods and succulent = immature seeds)	0.3		0.04	
	VP 0064	Pea, shelled (succulent seeds)	0.03		0.02	
	VO 0455	Peppers, sweet	1		0.18	
	TN 0675	Pistachio nut	0.2		0.05	
		Plum juice			0.08	

Pesticide (Codex reference number)	CCN	Commodity	Recommended MRL mg/kg New	Recommended MRL mg/kg Previous	STMR or STMR-P, mg/kg	HR or HR-P mg/kg
		Plum preserve			0.40	
		Plum puree			0.64	
	VR 0589	Potato	0.02		0.01	
	PM 0110	Poultry meat	0.01(*)		0	
	PO 0111	Poultry, edible offal of	0.05(*)		0	
	DF 0014	Prune (dried plum)			0.96	
	SO 0495	Rape seed	0.02(*)		0.02	
	FB 0272	Raspberry, red, black	5		1.0	
	FS 0012	Stone fruit	5 (Po)		0.80	
	AS 0081	Straw and fodder (dry) of cereal grain	0.06(*)		0	
	FB 0275	Strawberry	3		0.27	
		Strawberry juice			0.06	
		Strawberry preserve			0.24	
		Strawberry jam			0.13	
	VC 0431	Squash, summer	0.3		0.06	
	VO 0447	Sweet corn (corn-on-the-cob)	0.01(*)		0.01	
	VO 0448	Tomato	0.5		0.12	
	JF 0448	Tomato juice			0.026	
		Tomato paste			0.17	
	VL 0473	Watercress	10		1.2	
		Wine (grape)			0.01	

Residue for compliance with MRLs and estimation of dietary intake in plant commodities: fludioxonil

Residue for compliance with MRLs and estimation of dietary intake in animal commodities: sum of fludioxonil and metabolites determined as 2,2-difluorobenzo[1,1]dioxole-4-carboxylic acid, expressed as fludioxonil

Fludioxonil is fat-soluble

Folpet (041)

ADI: 0–0.1 mg/kg bw

ARfD: 0.2 mg/kg bw (for women of childbearing age)

Glyphosate** (158)

ADI: 0–1 mg/kg bw

ARfD: unnecessary

Malathion (049)	FP 0226	Apple	0.5	W	0.11	0.37
ADI: 0–0.3 mg/kg bw	FC 0001	Citrus fruit	7	W	0.02	0.22
ARfD: 2 mg/kg bw	FB 0269	Grape	5	W	0.16	2.6
	FS 0247	Peach	W	W		

Residue for compliance with MRLs and estimation of dietary intake in plant and animal commodities: malathion

Metalaxyl-M* (212)	FP 0009	Apple	0.02(*)	(Metalaxyl) Pome fruit 1	0	
ADI: 0–0.08 mg/kg bw for metalaxyl-M + metalaxyl	FB 0269	Grape	1	1	0.14	
ARfD: unnecessary	VL 0482	Lettuce, head	0.5	2	0.02	

… Annex 1

Pesticide (Codex reference number)	CCN	Commodity	Recommended MRL mg/kg New	Recommended MRL mg/kg Previous	STMR or STMR-P, mg/kg	HR or HR-P mg/kg
	VA 0385	Onion, bulb	0.03	2	0.02	
	VO 0445	Peppers, sweet	0.5	Peppers 1	0.03	
	VO 0448	Tomato	0.2	0.5	0.045	
	VR 0589	Potato	0.02(*)	0.05(*)	0.02	
	VL 0502	Spinach	0.1	2	0.02	
	SO 0702	Sunflower seed	0.02(*)	0.05(*)	0	
	SB 0715	Cacao bean	0.02	0.2	0.02	
		Grape juice			0.050	
		Wine			0.092	

Residue (of metalaxyl including metalaxyl-M) for compliance with MRLs and estimation of dietary intake in plant commodities: metalaxyl

Residue (of metalaxyl including metalaxyl-M) for compliance with MRLs and estimation of dietary intake in animal commodities: sum of metalaxyl and metabolites containing the 2,6-dimethylaniline moiety, expressed as metalaxyl.

Notes

No new MRLs are recommended because all MRLs required for metalaxyl-M are covered by existing MRLs for metalaxyl. The values listed in the 'New' column are the estimated maximum residue levels for metalaxyl-M and, because they do not exceed the existing metalaxyl MRLs, are not intended to replace them.

Metalaxyl is a racemic mixture of (R) and (S) enantiomers. Metalaxyl-M is the (R) enantiomer.

No MRLs are currently recommended for animal commodities.

Pesticide (Codex reference number)	CCN	Commodity	New	Previous	STMR	HR
Methamidophos (100) ADI: 0–0.004 mg/kg bw ARfD: 0.01 mg/kg bw	VC 0424	Cucumber	W	1		
Methomyl (094) ADI: 0–0.02 mg/kg bw ARfD: 0.02 mg/kg bw	VO 0051	Peppers	0.7	1	0.105	0.44
	AM 0738	Mint hay	0.5	2	0.08	0.32

Residue for compliance with MRLs and estimation of dietary intake in plant and animal commodities: sum of methomyl and thiodicarb, expressed as methomyl

Oxydemeton methyl (166) ADI: 0–0.0003 [1] mg/kg bw ARfD: 0.002 mg/kg bw	FP 0226	Apple[2]		0.05	0.01	0.04
	JF 0226	Apple juice			0.01	
		Apple sauce			0.005	
	GC 0640	Barley[4]	0.02(*)	0.05(*)	0.01	
	AS 0640	Barley straw and fodder, dry[4]	0.1	2		
	VB 0041	Cabbage, head [2,3]		0.05(*)	0.03	0.05
	MF 0812	Cattle fat[3]		0.05(*)	0	0
	VB 0404	Cauliflower[4]	0.01(*)	W	0.01	0.01
	VD 0526	Common bean (dry)[5]		0.1	0.01	
	SO 0691	Cotton-seed[5]		0.05	0.01	
	OR 0691	Cotton-seed oil, edible [5]			0.002	
	PE 0112	Egg[3]		0.05(*)	0	0
	FB 0269	Grape[2]		0.1	0.04	0.06
	VL 0480	Kale[3]		0.01(*)	0.01	0.01
	VB 0405	Kohlrabi[3]		0.05	0.02	0.05
	FC 0204	Lemon[3]		0.2	0.01	0.04
	MM 0097	Meat of cattle, pigs and sheep[3]		0.05(*)	0	0

Pesticide (Codex reference number)	CCN	Commodity	Recommended MRL mg/kg New	Recommended MRL mg/kg Previous	STMR or STMR-P, mg/kg	HR or HR-P mg/kg
	ML 0106	Milks[3]		0.01(*)	0	
	FC 0004	Orange, sweet, sour [2,3]		0.2	0.01	0.04
	FP 0230	Pear[3]		0.05	0.01	0.04
	MF 0818	Pig fat[3]		0.05(*)	0	0
	VR 0589	Potato[4]	0.01(*)	0.05(*)	0.01	0.01
	PF 0111	Poultry fat[3]		0.05(*)	0	0
	PM 0110	Poultry meat[3]		0.05(*)	0	0
	GC 0650	Rye[4]	0.02(*)	0.05(*)	0.01	
	AS 0650	Rye straw and fodder, dry[4]	0.1	2		
	MF 0822	Sheep fat[3]		0.05(*)	0	0
	VR 0596	Sugar beet[4]	0.01(*)	0.05(*)	0.01	
	AV 0596	Sugar-beet leaf or top[4]	0.05	0.05(*)		
	GC 0654	Wheat[4]	0.02(*)	0.05(*)	0.01	
	AS 0654	Wheat straw and fodder, dry[4]	0.1	2		

Residue for compliance with MRLs and estimation of dietary intake in plant and animal commodities: sum of oxydemeton methyl, demeton-S-methyl and demeton-S-methylsulphon, expressed as oxydemeton methyl

Notes The definition of the residue and recommendations for MRLs are based only on use of oxydemeton methyl.

[1] Group ADI for demeton-S-methyl and related compounds

[2] The information provided precluded an estimate that the dietary would be below the ARfD for children aged ≤ 6 years

[3] STMRs/STMRPs or HRs/HRPs were estimated on the basis of the 1998 evaluation.

[4] STMRs/STMRPs or HRs/HRPs were estimated on the basis of new data submitted to the 2004 JMPR.

[5] Case 3 commodities do not require estimation of the HR for short-term intake assessment.

Pesticide (Codex reference number)	CCN	Commodity	New	Previous	STMR	HR
Paraquat** (057)	AM 0660	Almond hulls	0.01 (*)			
ADI: 0–0.005 mg/kg bw	FI 0030	Assorted tropical and subtropical fruits minus inedible peel (except passion fruit)	0.01 (*)		0.01	0.01
ARfD: 0.006 mg/kg bw	FB 0018	Berries and other small fruit	0.01 (*)		0	0
	MO 1280	Cattle kidney	W	0.5		
	FC 0001	Citrus fruit	0.02	–	0.01	0.02
	JF 0004	Orange juice			0	
	SO 0691	Cotton-seed	2	0.2	0.21	
	OC 0691	Cotton-seed oil, crude			0.01	
	OR 0691	Cotton-seed oil, edible	W	0.05 (*)		
	MO 0105	Edible offal (mammalian)	0.05		0.0007	0.033
	MO 0097	Edible offal of cattle, pigs and sheep	W	0.05 (*)		
	PE 0112	Egg	0.005 (*)	0.01 (*)	0	0
	AV 1051	Fodder beet leaves or tops	0.2 (dry wt)			
	VC 0045	Fruiting vegetables, cucurbits	0.02		0	0
	VO 0050	Fruiting vegetables, other than cucurbits	0.05		0.01	0.04
	JF 0448	Tomato juice			0	
		Tomato ketchup			0	
	DH 1100	Hops, dry	0.1	0.2	0.05	0.05
		Beer			0.0001	

Annex 1

Pesticide (Codex reference number)	CCN	Commodity	Recommended MRL mg/kg New	Recommended MRL mg/kg Previous	STMR or STMR-P, mg/kg	HR or HR-P mg/kg
	VL 0053	Leafy vegetables	0.07		0.025	0.05
	GC 0645	Maize	0.03	0.1	0.025	
	CF 1255	Maize flour	0.05		0.038	
		Maize germ			0.0075	
		Maize grits and meal			0.013	
	OC 0645	Maize oil, crude			0.006	
		Corn starch			0.006	
	AS 0645	Maize fodder	10 (dry wt.)			
	AF 0645	Maize forage	5 (dry wt.)			
	MM 0095	Meat (from mammals other than marine mammals)	0.005		0.0001	0.005
	MM 0097	Meat of cattle, pigs and sheep	W	0.05 (*)		
	ML 0106	Milks	0.005(*)	0.01 (*)	0.00008	
	FT 0305	Olive	0.1	1	0.05	0.1
	OC 0305	Olive oil, virgin			0.018	
	FI 0351	Passion fruit	W	0.2		
	MO 1284	Pig kidney	W	0.5		
	FP 0009	Pome fruits	0.01 (*)	–	0	0
	VR 0589	Potato	W	0.2		
		Potato crisps			0.02	
		Potato granules			0.05	
	PO 0111	Poultry, edible offal of	0.005 (*)		0	0
	PM 0110	Poultry meat	0.005 (*)		0	0
	VD 0070	Pulses	0.5		0.1	
	GC 0649	Rice	W	10		
	CM 1205	Rice, polished	W	0.5		
	VR 0075	Root and tuber vegetables	0.05		0.02	0.05
	MO 1288	Sheep kidney	W	0.5		
	GC 0651	Sorghum	0.03	0.5	0.025	
		Sorghum flour			0.004	
		Sorghum germ			0.013	
	AF 0651	Sorghum forage (green)	0.3 (dry wt)			
	AS 0651	Sorghum straw and fodder, dry	0.3 (dry wt)			
	VD 0541	Soya bean (dry)	W	0.1		
	AL 0541	Soya bean fodder	0.5 (dry wt)			
	AL 1265	Soya bean forage (green)	2 (dry wt)			
	OC 0541	Soya bean oil, crude			0.01	
	FS 0012	Stone fruit	0.01 (*)		0	0
	DF 0014	Prune			0	
	SO 0702	Sunflower seed	2	2	0.22	
	OC 0702	Sunflower seed oil, crude	W	0.05 (*)	0	
	OR 0702	Sunflower seed oil, edible	W	0.05 (*)		
	DT 1114	Tea, green, black (black, fermented and dried)	0.2		0.06	
	TN 0085	Tree nuts	0.05		0.01	0.05
	AO1 0002	Vegetables (except those listed)	W	0.05 (*)		

284   Annex 1

Pesticide (Codex reference number)	CCN	Commodity	Recommended MRL mg/kg		STMR or STMR-P, mg/kg	HR or HR-P mg/kg
			New	Previous		

Residue for compliance with MRLs and estimation of dietary intake in plant and animal commodities: paraquat cation

Phorate** (112)
ADI: 0–0.0007 mg/kg bw
ARfD: 0.003 mg/kg bw

Pirimicarb** (101)
ADI: 0–0.02 mg/kg bw
ARfD: 0.1 mg/kg bw

Pesticide	CCN	Commodity	New	Previous	STMR	HR
Pirimiphos-methyl (086)	MO 0105	Edible offal (mammalian)	0.01(*) [1]		0	0
ADI: 0–0.03 mg/kg bw	PE 0112	Egg	0.01	0.05	0	0.01
	MM 0095	Meat (from mammals other than marine mammals)	0.01(*) (fat) [1]	0.05(*)	Muscle: 0 Fat: 0	Muscle: 0 Fat: 0
	PM 0110	Poultry meat	0.01(*) [1]		Muscle: 0 Fat: 0	Muscle: 0 Fat: 0
	PM 0111	Poultry, edible offal of	0.01(*) [1]		0	0

[1] No residues expected from consumption of feed commodities with pirimiphos-methyl residues, as evaluated by JMPR.

Residue for compliance with MRLs and estimation of dietary intake in plant and animal commodities: pirimiphos-methyl. The residue is fat-soluble.

Pesticide	CCN	Commodity	New	Previous	STMR	HR
Prochloraz** (142)	FI 0326	Avocado	W [1]	5 (Po)		
ADI: 0–0.01 mg/kg bw	FI 0030	Assorted tropical and subtropical fruits minus inedible peel	7 (Po)		0.1	0.7
ARfD: 0.1 mg/kg bw	FI 0327	Banana	W [1]	5 (Po)		
	GC 0640	Barley	W [1]	0.5		
	AS 0640	Barley straw and fodder, dry	W [1]	15		
		Beer			0.01	
	MF 0812	Cattle fat	W [1]	0.5		
	MM 0812	Cattle meat	W [1]	0.1 (*)		
	MO 0812	Cattle, edible offal of	W [1]	5		
	GC 0080	Cereal grains	2		0.11	1.2
	FC 0001	Citrus fruits	10 (Po)		0.1	0.92
	SB 0716	Coffee beans	W	0.2		
	MO 0105	Edible offal (mammalian)	10		1.2	6.2
	PE 0112	Eggs	0.1		0.012	0.07
	SO 0693	Linseed	0.05 (*)		0.05	0.05
	FI 0345	Mango	W [1]	2 (Po)		
	MM 0095	Meat (from mammals other than marine mammals)	0.5 (fat)		0.02 (muscle) 0.06 (fat)	0.1 (muscle) 0.38 (fat)
	ML 0106	Milks	0.05 *	0.1*	0	0
	VO 0450	Mushroom	40	2	4.9	37
	GC 0647	Oat	W [1]	0.5		
	AS 0647	Oat straw and fodder, dry	W [1]	15		
	FC 0004	Orange, sweet, sour	W [1]	5 Po		

Annex 1

Pesticide (Codex reference number)	CCN	Commodity	Recommended MRL mg/kg New	Recommended MRL mg/kg Previous	STMR or STMR-P, mg/kg	HR or HR-P mg/kg
	FI 0350	Papaya	W [1]	1 Po		
	HS 0790	Pepper, black, white	10		5.1	5.1
	PM 0110	Poultry meat	0.05*		0.001 (muscle) 0.001 (fat)	0.005 (muscle) 0.007 (fat)
	PO 0111	Poultry, edible offal of	0.2		0.015	0.1
	SO 0495	Rape seed	0.7	0.5	0.1	0.48
		Rape-seed meal			0.06	
	OR 0495	Rape-seed oil, edible			0.06	
	GC 0650	Rye	W [1]	0.5		
	AS 0650	Rye straw and fodder, dry	W [1]	15		
	FS 0012	Stone fruit	W	0.05		
	AS 0081	Straw and fodder (dry) of cereal grain	40			
	SO 0702	Sunflower seed	0.5		0.1	0.32
		Sunflower seed meal			0.05	
	OR 0702	Sunflower seed oil, edible	1		0.06	
	GC 0654	Wheat	W [1]	0.5		
	CM 0654	Wheat bran, unprocessed	7		0.54	
	AS 0654	Wheat straw and fodder, dry	W [1]	15		
	CF 1211	Wheat flour			0.025	
	CP 1212	Wholemeal bread			0.14	

[1] Replaced by recommendation for wider group of commodities

Residue for compliance with MRLs and estimation of dietary intake in plant and animal commodities: sum of prochloraz and its metabolites containing the 2,4,6-trichlorophenol moiety, expressed as prochloraz.

The residue is fat-soluble

Propiconazole (160)

ADI: 0–0.07 mg/kg bw

ARfD: 0.3 mg/kg bw

Pesticide (Codex reference number)	CCN	Commodity	New	Previous	STMR	HR
Propineb	FP 0226	Apple	W	2		
ADI: 0–0.007mg/kg bw	FS 0013	Cherry	0.2	–	0.13	0.35
Interim[1] ARfD: 0.1 mg/kg bw	VC 0424	Cucumber	1	–	0.49	1.1
	MO 0105	Edible offal (mammalian)	0.05 (*)	–	0	0
	PE 0112	Egg	0.01 (*)	–	0	0
	FB 0269	Grape	W	2		
	MM 0095	Meat (from mammals other than marine mammals)	0.05 (*)	–	0	0
	VC 0046	Melon (except watermelon)	W	0.1 (*)		
	VA 0385	Onion, bulb	W	0.2 (*)		
	FP 0230	Pear	W	2		
	VO 0445	Peppers, sweet	7	–	1.6	13
	VR 0587	Potato	0.1	0.1 (*)	0.12	0.16
	PM 0110	Poultry meat	0.05 (*)	–	0	0
	PO 0111	Poultry, edible offal of	0.05 (*)	–	0	0

Pesticide (Codex reference number)	CCN	Commodity	Recommended MRL mg/kg New	Recommended MRL mg/kg Previous	STMR or STMR-P, mg/kg	HR or HR-P mg/kg
	VO 0448	Tomato	2	1	1	2.9

Residue for compliance with MRLs in plant and animal commodities: total dithiocarbamates, determined as CS_2, evolved during acid digestion and expressed as mg CS_2/kg

Residue for estimation of dietary intake in plant and animal commodities: sum of propineb and propylenethiourea

Notes All recommended MRLs are covered by the existing or recommended MRLs for dithiocarbamates.

[1] The WHO Core Assessment Group is occasionally asked by the FAO Panel of Experts to establish an ARfD for compounds that were not scheduled for toxicological evaluation, for use in acute dietary risk assessment. This value is derived ad hoc from data from previous meetings and is therefore called an interim ARfD. It can be used in dietary risk assessments until they are replaced by full evaluations, if this is considered necessary.

Pesticide (Codex reference number)	CCN	Commodity	Recommended MRL mg/kg New	Recommended MRL mg/kg Previous	STMR or STMR-P, mg/kg	HR or HR-P mg/kg
Pyraclostrobin* (210)	AM 0660	Almond hull	2		0.20	
ADI: 0–0.03 mg/kg bw	TN 0660	Almond	0.02(*)		0.02	0.02
ARfD: 0.05 mg/kg bw	FI 0327	Banana	0.02(*)		0.02	0.02
	GC 0640	Barley	0.5		0.03	
		Malt			0.03	
	VD 0071	Bean (dry)	0.2		0.02	0.10
	FB 0020	Blueberry	1		0.34	0.57
	VR 0577	Carrot	0.5		0.12	0.24
	FS 0013	Cherry	1		0.43	0.63
	FC 0001	Citrus fruit	1		0.19	0.51
	DF 0269	Dried grape (currant, raisin, sultana)	5		1.36	4.27
	MO 0095	Edible offal, mammalian	0.05(*)		0.008	0.037
	PE 0112	Egg	0.05(*)		0	0
	AV 1051	Fodder beet leaf or top	50			
	VA 0381	Garlic	0.05(*)		0.05	0.05
	FB 0269	Grape	2		0.44	1.38
	JF 0269	Grape juice			0.005	
		Wine			0.04	
	VD 0533	Lentil (dry)	0.5		0.13	
	GC 0645	Maize	0.02(*)		0.02	
	FI 0345	Mango	0.05(*)		0.05	0.05
	MM 0095	Meat (from mammals other than marine mammals)	0.5 (fat)		Muscle: 0.009 Fat: 0.063	Muscle: 0.044 Fat: 0.41
	ML 0106	Milks	0.03		0.01	
	GC 0647	Oat	0.5		0.17	
	VA 0385	Onion, bulb	0.2		0.02	0.09
	FI 0350	Papaya	0.05(*)		0.05	0.05
	FS 0247	Peach	0.5		0.15	0.31
	SO 0703	Peanut	0.02(*)		0.02	0.02
	AL 0697	Peanut fodder	50			
	VD 0072	Pea (dry)	0.3		0.07	
		Pea vines	40			
		Pea hay	30			
	TN 0672	Pecan	0.02(*)		0.02	0.02
	TN 0678	Pistachio nut	1		0.22	0.45
	FS 0014	Plum (including prune)	0.3		0.06	0.19

Pesticide (Codex reference number)	CCN	Commodity	Recommended MRL mg/kg New	Previous	STMR or STMR-P, mg/kg	HR or HR-P mg/kg
	VR 0589	Potato	0.02(*)		0.02	0.02
	PM 0110	Poultry meat	0.05(*) (fat)		Muscle 0 fat 0	Muscle 0 fat 0
	PO 0111	Poultry, edible offal of	0.05(*)		0	0
	VR 0494	Radish	0.5		0.08	0.3
	VL 0494	Radish leaf (including radish top)	20		9.9	15
	VC 0431	Squash, summer	0.3		0.15	0.18
	AS 0081	Straw and fodder (dry) of cereal grain	30		1.69	
	FB 0275	Strawberry	0.5		0.16	0.26
	VR 0596	Sugar beet	0.2		0.04	
	VO 0448	Tomato	0.3		0.12	0.21
	GC 0654	Wheat	0.2		0.02	
	CF 1211	Wheat flour			0.012	
	CF 1210	Wheat germ			0.016	

Residue for compliance with MRLs and estimation of dietary intake in plant and animal commodities: pyraclostrobin.

The residue is fat-soluble.

Spinosad (203)	FM 0812	Cattle milk fat	5			
ADI: 0–0.02 mg/kg bw	GC 0080	Cereal grain	1 (Po)		0.70	
ARfD: unnecessary	DF 0269	Dried grape (currant, raisin and sultana)	1		0.13	
	MO 0105	Edible offal (mammalian) [except cattle][2]	0.5		Liver: 0.064 Kidney: 0.032	
	FB 0269	Grape	0.5		0.084	
	GC 0645	Maize	W [1]	0.01(*)		
	MM 0095	Meat (from mammals other than marine mammals) [except cattle][2]	2 (fat)		Meat 0.01 Fat: 0.32	
	MM 0822	Sheep meat	W [1]	0.01(*) (fat)		
	MO 0822	Sheep, edible offal of	W [1]	0.01(*)		
	GC 0651	Sorghum	W [1]	1		
	CM 0654	Wheat bran, unprocessed	2		1.4	
		Grits			0.057	
	CF 1255	Maize flour			0.13	
	OC 0645	Maize oil, crude			0.77	
	CM 1206	Rice bran, unprocessed			0.55	
		Rice hull			2.0	
	CM 0649	Rice, husked (brown rice)			0.077	
	CM 1205	Rice, polished (white rice)			0.015	
	CF 1211	Wheat flour			0.18	
	CP 1211	White bread			0.098	
		Wine			0.027	

Residue for compliance with MRLs and estimation of dietary intake in plant and animal commodities: sum of spinosyn A and spinosyn D

The residue is fat-soluble, but residues in milk should be determined in whole milk.

[1] Replaced by recommendation for wider group of commodities.

Pesticide (Codex reference number)	CCN	Commodity	Recommended MRL mg/kg New	Recommended MRL mg/kg Previous	STMR or STMR-P, mg/kg	HR or HR-P mg/kg

[2] The recommendations are derived from a dairy cow feeding study and the corresponding animal dietary burden. They are extended to 'Edible offal (mammalian) [except cattle]' and 'Meat (from mammals other than marine mammals) [except cattle]' following the policy of the 2002 JMPR. The proposed MRLs for cattle meat (3 mg/kg, fat), cattle kidney (1 mg/kg) and cattle liver (2mg/kg), arising from the direct use of spinosad on cattle, should remain. They exceed the recommendations for mammalian meat and offal, and so require the restriction 'except cattle' to the general commodity descriptions.

Triadimefon ** (133)
ADI: 0–0.03 mg/kg bw
ARfD: 0.08 mg/kg bw

Triadimenol ** (168)
ADI: 0–0.03 mg/kg bw
ARfD: 0.08 mg/kg bw

Pesticide	CCN	Commodity	New	Previous	STMR	HR
Trifloxystrobin* (213)	AM 0660	Almond hulls[1]	3			
ADI: 0–0.04 mg/kg bw	DF 0226	Apple, dried			0.043	
ARfD: unnecessary	JF 0226	Apple juice			0.018	
		Apple sauce			0.053	
		Apple preserve			0.053	
		Apricot, canned			0.019	
	FI 0327	Banana	0.05		0.02	
	GC 0640	Barley	0.5		0.04	
	AS 0640	Barley straw and fodder (dry)[1]	7			
		Beer (residue arising from hops)			0.01	
	VB 0402	Brussels sprout	0.5		0.17	
	VR 0577	Carrot	0.1		0.035	
	VB 0041	Cabbage, head	0.5		0.17	
	VS 0624	Celery	1		0.18	
	FC 0001	Citrus fruit[2]	0.5		0.095	
	JF 0001	Citrus juice[2]			0.018	
	AB 0001	Citrus pulp, dry[1]	1			
		Citrus oil (orange)			12	
	DF 0269	Dried grape (raisin)	5		0.345	
	PE 0112	Egg	0.04(*)		0	
	VB 0042	Flowerhead brassica	0.5		0.17	
	VC 0045	Fruiting vegetables, cucurbits	0.3		0.095	
	FB 0269	Grape	3		0.15	
	JF 0269	Grape juice			0.036	
		Must			0.07	
		Wine			0.023	
	DH 1100	Hops, dry	40		9.95	
	MO 0098	Kidney of cattle, goats, pigs and sheep	0.04(*)		0.004	
	VA 0384	Leek	0.7		0.31	
	MO 0099	Liver of cattle, goats, pigs and sheep	0.05		0.008	
	GC 0645	Maize	0.02		0.02	

Pesticide (Codex reference number)	CCN	Commodity	Recommended MRL mg/kg New	Recommended MRL mg/kg Previous	STMR or STMR-P, mg/kg	HR or HR-P mg/kg
	AS 0645	Maize fodder[1]	10			
	MM 0095	Meat (from mammals other than marine mammals)	0.05 (fat)		Fat: 0.006 Muscle: 0	
	ML0106	Milks	0.02(*)		0	
		Nectarine, canned			0.019	
		Peach, canned			0.019	
	SO 0697	Peanut	0.02(*)		0	
	AL 0697	Peanut fodder[1]	5			
		Pear, dried			0.043	
		Pear juice			0.018	
		Pear preserve			0.053	
	VO 0445	Peppers, sweet	0.3		0.1	
	FP 0009	Pome fruit	0.7		0.11	
	VR 0589	Potato	0.02(*)		0.02	
		Potato chips			0.008	
		Potato flakes			0.008	
	PO 0111	Poultry, edible offal of	0.04(*)		0	
	PM 0110	Poultry meat	0.04(*) (fat)		Fat: 0 Muscle: 0	
		Prune, dried			0.57	
	GC 0649	Rice	5		0.16	
	CM 1205	Rice, polished			0.029	
	CM 1206	Rice bran, unprocessed[1]	7		0.25	
	AS 0649	Rice straw and fodder, dry[1]	10			
	FS 0012	Stone fruit	3		0.38	
	FB 0275	Strawberry	0.2		0.1	
		Strawberry, canned			0.029	
		Strawberry jam			0.062	
	VR 0596	Sugar beet	0.05		0.02	
		White sugar			0.0036	
	AB 0596	Sugar beet pulp, dry[1]	0.2			
	DM 0596	Sugar beet molasses[1]	0.1			
	AM 0596	Sugar beet leaves or tops[1]	15			
	VO 0448	Tomato	0.7		0.08	
		Tomato paste			0.13	
		Tomato puree			0.045	
	TN 0085	Tree nuts	0.02(*)		0	
	GC 0654	Wheat	0.2		0.02	
	CM 0654	Wheat bran, unprocessed[1]	0.5		0.062	
	CF 1211	Wheat flour			0.008	
	CF 1210	Wheat germ			0.013	
	CF 1212	Wheat wholemeal			0.01	
	CP 1212	Wheat wholemeal bread			0.005	
	AS 0654	Wheat straw and fodder, dry[1]	5			

Pesticide (Codex reference number)	CCN	Commodity	Recommended MRL mg/kg New	Recommended MRL mg/kg Previous	STMR or STMR-P, mg/kg	HR or HR-P mg/kg

Residue in plant commodities
    for compliance with MRLs: trifloxystrobin
    for estimation of dietary intake: sum of trifloxystrobin and [(*E,E*)-methoxyimino-{2-[1-(3-trifluoromethylphenyl)ethylideneaminooxymethyl]phenyl}acetic acid] (CGA 321113), expressed as trifloxystrobin.

Residue in animal commodities
    For compliance with MRLs and estimation of dietary intake:
    sum of trifloxystrobin and [(*E,E*)-methoxyimino-{2-[1-(3-trifluoromethylphenyl)ethylideneamino-oxymethyl]phenyl}acetic acid] (CGA 321113), expressed as trifloxystrobin.

The residue is fat-soluble.
[1] Expressed on dry weight
[2] STMR, STMR-P based on whole fruit residue data

Table 2. Recommended MRLs, STMR and HR values for spices

Pesticide	Group or subgroup of spices	Proposed MRL (mg/kg)	Median[2] (mg/kg)	High residue[2] (mg/kg)
Acephate	Entire group 028[3]	0.2(*)	0.2	0.2
Azinphos-methyl	Entire group 028[3]	0.5(*)	0.1	< 0.5
	Seeds	5		
Chlorpyrifos	Fruits or berries	1	0.09	3.6
	Roots or rhizomes	1		
Chlorpyrifos-methyl	Seeds	1		
	Fruits	0.3	0.77	2.9
	Roots or rhizomes	5		
Cypermethrin	Fruits or berries	0.1	0.11	0.12
	Roots or rhizomes	0.2		
	Seeds	5		
Diazinon	Fruits	0.1 (*)	0.19	3.6
	Roots or rhizomes	0.5		
Dichlorvos	Entire group 028[3]	0.1(*)	0.1	0.1
	Seeds	0.05(*)		
Dicofol	Fruits or berries	0.1	0.05	0.05
	Roots or rhizomes	0.1		
	Seeds	5		
Dimethoate	Fruits or berries	0.5	0.17	3
	Roots or rhizomes	0.1(*)		
Disulfoton	Entire group 028[3]	0.05(*)	0.05	0.05
	Seeds	1		
Endosulfan (total)	Fruits or berries	5	0.12	3.2
	Roots or rhizomes	0.5		
	Seeds	3		
Ethion	Fruits or berries	5	1.7	3.1
	Roots or rhizomes	0.3		
	Seeds	7	0.4	5.4
Fenitrothion	Fruits or berries	1		
	Roots or rhizomes	0.1(*)		

Annex 1

Pesticide	Group or subgroup of spices	Proposed MRL (mg/kg)	Median[2] (mg/kg)	High residue[2] (mg/kg)
Iprodion	Seeds	0.05(*)		
	Roots or rhizomes	0.1	0.05	0.05
Malathion	Seeds	2	0.48	0.86
	Fruits or berries	1		
	Roots or rhizomes	0.5		
Metalaxyl	Seeds	5	0.43	3.2
Methamidophos	Entire group 028[3]	0.1(*)	0.01	0.1
Mevinphos	Seeds	5		
	Fruits or berries	0.2(*)	0.05	2.9
	Roots or rhizomes	1		
Parathion	Seeds	0.1 (*)		
	Fruits or berries	0.2	0.1	0.1
	Roots or rhizomes	0.2		
Parathion-methyl	Seeds	5		
	Fruits or berries	5	0.43	3.4
	Roots or rhizomes	3		
Permethrin	Entire group 028[3]	0.05 (*)	0.05	0.05
Phenthoate	Seeds	7[c]	1.2	5.2
	Seeds sub-group	0.5	0.21	0.3
Phorate	Fruits or berries	0.1(*)		
	Roots or rhizomes	0.1(*)		
Phosalone	Seeds	2		
	Fruits	2	0.85	
	Roots or rhizomes	3		1.5
Pirimicarb	Seeds	5	0.14	3
Pirimiphos-methyl	Seeds subgroup	3	0.23	1.8
	Fruits subgroup	0.5		
	Seeds subgroup	0.1		
Quintozene	Fruits or berries	0.02	0.05	1.2
	Roots or rhizomes	2		
Vinclozoline	Entire spice group[3]	0.05 (*)	0.05	0.05

[1] The residue definitions remain the same as those recommended for the given pesticide in other plant commodities.
[2] Values recommended for calculation of short term and chronic intakes are indicated.
[3] The Group of A28 as modified by the 36th Session of CCPR

Table 3. Recommended maximum residue levels of residues in/on dried chili peppers[1]

Pesticide		Recommended MRL for dried chili pepper (mg/kg)	Notes
177	Abamectin	0.2	
95	Acephate	50	
2	Azinphos-methyl	10	
155	Benalaxyl	0.5	
47	Bromide ion	200	
8	Carbaryl	50	
72	Carbendazim (based on chilli peper)	20	

Pesticide		Recommended MRL for dried chili pepper (mg/kg)	Notes
81	Chlorothalonil	70	
17	Chlorpyrifos	20	
90	Chlorpyrifos-methyl	5	
157	Cyfluthrin	2	
67	Cyhexatin	5	
118	Cypermethrin	5	
169	Cyromazine	10	
22	Diazinon	0.5	
82	Dichlofluanid	20	
26	Dicofol	10	
27	Dimethoate	50	
87	Dinocap	2	
105	Dithiocarbamates	10	
106	Ethephon	50	
149	Ethoprophos	0.2	(a)
192	Fenarimol	5	
185	Fenpropathrin	10	
119	Fenvalerate	5	
206	Imidacloprid	1	
49	Malathion	1	
138	Metalaxyl	10	
100	Methamidophos	20	
94	Methomyl	10	(b)
209	Methoxyfenozide	20	
54	Monocrotophos	2	
126	Oxamyl	50	
120	Permethrin	10	
61	Phosphamidon	2	
62	Piperonyl butoxide	20	
101	Pirimicarb	20	
86	Pirimiphos-methyl	10	(c)
136	Procymidone	50	
171	Profenofos	50	
148	Propamocarb	10	
63	Pyrethrins	0.5	
64	Quintozene	0.1	
203	Spinosad	3	
189	Tebuconazole	5	
196	Tebufenozide	10	
162	Tolylfluanid	20	
133	Triadimefon	1	
168	Triadimenol	1	
159	Vinclozolin	30	

The residue definitions remain the same as those recommended for the given pesticide in other plant commodities.
(a) The 2004 JMPR recommended a new maximum reside level of 0.05 mg/kg for sweet pepper
(b) Withdrawn by the 2001 JMPR. The 2004 JMPR recommended a new maximum reside level of 0.7 mg/kg.
(c) Withdrawn by the 2003 JMPR

ANNEX 2

INDEX OF REPORTS AND EVALUATIONS OF PESTICIDES BY THE JMPR

Numbers in parentheses after the names of pesticides are Codex classification numbers. The abbreviations used are:
  T, evaluation of toxicology
  R, evaluation of residue and analytical aspects
  E, evaluation of effects on the environment

Abamectin (177)	1992 (T,R), 1994 (T,R), 1995 (T), 1997 (T,R), 2000 (R)
Acephate (095)	1976 (T,R), 1979 (R), 1981 (R), 1982 (T), 1984 (T,R), 1987 (T), 1988 (T), 1990 (T,R), 1991 (corr. to 1990 R evaluation), 1994 (R), 1996 (R), 2002 (T), 2003 (R), 2004 (corr. to 2003 report)
Acrylonitrile	1965 (T,R)
Aldicarb (117)	1979 (T,R), 1982 (T,R), 1985 (R), 1988 (R), 1990 (R), 1991 (corr. to 1990 evaluation), 1992 (T), 1993 (R), 1994 (R), 1996 (R), 2001 (R), 2002 (R)
Aldrin (001)	1965 (T), 1966 (T,R), 1967 (R), 1974 (R), 1975 (R), 1977 (T), 1990 (R), 1992 (R)
Allethrin	1965 (T,R)
Aminocarb (134)	1978 (T,R), 1979 (T,R)
Aminomethylphosphonic acid (AMPA, 198)	1997 (T,R)
Amitraz (122)	1980 (T,R), 1983 (R), 1984 (T,R), 1985 (R), 1986 (R), 1989 (R), 1990 (T,R), 1991 (R & corr. to 1990 R evaluation), 1998 (T)
Amitrole (079)	1974 (T,R), 1977 (T), 1993 (T,R), 1997 (T), 1998 (R)
Anilazine (163)	1989 (T,R), 1992 (R)
Azinphos-ethyl (068)	1973 (T,R), 1983 (R)
Azinphos-methyl (002)	1965 (T), 1968 (T,R), 1972 (R), 1973 (T), 1974 (R), 1991 (T,R), 1992 (corr. to 1991 report), 1993 (R), 1995 (R)
Azocyclotin (129)	1979 (R), 1981 (T), 1982 (R),1983 (R), 1985 (R), 1989 (T,R), 1991 (R), 1994 (T)
Benalaxyl (155)	1986 (R), 1987 (T), 1988 (R), 1992 (R), 1993 (R)
Bendiocarb (137)	1982 (T,R), 1984 (T,R), 1989 (R), 1990 (R)
Benomyl (069)	1973 (T,R), 1975 (T,R), 1978 (T,R), 1983 (T,R), 1988 (R), 1990 (R), 1994 (R), 1995 (T,E), 1998 (R)
Bentazone (172)	1991 (T,R), 1992 (corr. to 1991 report, Annex I), 1994 (R), 1995 (R), 1998 (T,R), 1999 (corr. to 1998 report), 2004(T)
BHC (technical-grade)	1965 (T), 1968 (T,R), 1973 (T,R) (see also Lindane)
Bifenthrin (178)	1992 (T,R), 1995 (R), 1996 (R), 1997 (R)
Binapacryl (003)	1969 (T,R), 1974 (R), 1982 (T), 1984 (R), 1985 (T,R)
Bioresmethrin (093)	1975 (R), 1976 (T,R), 1991 (T,R)
Biphenyl	See Diphenyl
Bitertanol (144)	1983 (T), 1984 (R), 1986 (R), 1987 (T), 1988 (R), 1989 (R), 1991 (R), 1998 (T), 1999 (R), 2002 (R)
Bromide ion (047)	1968 (R), 1969 (T,R), 1971 (R), 1979 (R), 1981 (R), 1983 (R), 1988 (T,R), 1989 (R), 1992 (R)

Bromomethane (052)	1965 (T,R), 1966 (T,R), 1967 (R), 1968 (T,R), 1971 (R), 1979 (R), 1985 (R), 1992 (R)
Bromophos (004)	1972 (T,R), 1975 (R), 1977 (T,R), 1982 (R), 1984 (R), 1985 (R)
Bromophos-ethyl (005)	1972 (T,R), 1975 (T,R), 1977 (R)
Bromopropylate (070)	1973 (T,R), 1993 (T,R)
Butocarboxim (139)	1983 (R), 1984 (T), 1985 (T), 1986 (R)
Buprofezin (173)	1991 (T,R), 1995 (R), 1996 (corr. to 1995 report.), 1999 (R)
sec-Butylamine (089)	1975 (T,R), 1977 (R), 1978 (T,R), 1979 (R), 1980 (R), 1981 (T), 1984 (T,R: withdrawal of temporary ADI, but no evaluation)
Cadusafos (174)	1991 (T,R), 1992 (R), 1992 (R)
Campheclor (071)	1968 (T,R), 1973 (T,R)
Captafol (006)	1969 (T,R), 1973 (T,R), 1974 (R), 1976 (R), 1977 (T,R), 1982 (T), 1985 (T,R), 1986 (corr. to 1985 report), 1990 (R), 1999 (acute Rf D)
Captan (007)	1965 (T), 1969 (T,R), 1973 (T), 1974 (R), 1977 (T,R), 1978 (T,R), 1980 (R), 1982 (T), 1984 (T,R), 1986 (R), 1987 (R and corr. to 1986 R evaluation), 1990 (T,R), 1991 (corr. to 1990 R evaluation), 1994 (R), 1995 (T), 1997 (R), 2000 (R), 2004 (T)
Carbaryl (008)	1965 (T), 1966 (T,R), 1967 (T,R), 1968 (R), 1969 (T,R), 1970 (R), 1973 (T,R), 1975 (R), 1976 (R), 1977 (R), 1979 (R), 1984 (R), 1996 (T), 2001 (T), 2002 (R)
Carbendazim (072)	1973 (T,R), 1976 (R), 1977 (T), 1978 (R), 1983 (T,R), 1985 (T,R), 1987 (R), 1988 (R), 1990 (R), 1994 (R), 1995 (T,E), 1998 (T,R), 2003 (R)
Carbofuran (096)	1976 (T,R), 1979 (T,R), 1980 (T), 1982 (T), 1991 (R), 1993 (R), 1996 (T), 1997 (R), 1999 (corr. to 1997 report), 2002 (T, R), 2003 (R) (See also carbosulfan), 2004 (R)
Carbon disulfide (009)	1965 (T,R), 1967 (R), 1968 (R), 1971 (R), 1985 (R)
Carbon tetrachloride (010)	1965 (T,R), 1967 (R), 1968 (T,R), 1971 (R), 1979 (R), 1985 (R)
Carbophenothion (011)	1972 (T,R), 1976 (T,R), 1977 (T,R), 1979 (T,R), 1980 (T,R), 1983 (R)
Carbosulfan (145)	1984 (T,R), 1986 (T), 1991 (R), 1992 (corr. to 1991 report), 1993 (R), 1997 (R), 1999 (R), 2002 (R), 2003 (T, R), 2004 (R, corr. to 2003 report)
Cartap (097)	1976 (T,R), 1978 (T,R), 1995 (T,R)
Chinomethionat (080)	1968 (T,R) (as oxythioquinox), 1974 (T,R), 1977 (T,R), 1981 (T,R), 1983 (R), 1984 (T,R), 1987 (T)
Chlorbenside	1965 (T)
Chlordane (012)	1965 (T), 1967 (T,R), 1969 (R), 1970 (T,R), 1972 (R), 1974 (R), 1977 (T,R), 1982 (T), 1984 (T,R), 1986 (T)
Chlordimeform (013)	1971 (T,R), 1975 (T,R), 1977 (T), 1978 (T,R), 1979(T), 1980(T), 1985(T), 1986 (R), 1987 (T)
Chlorfenson	1965 (T)
Chlorfenvinphos (014)	1971 (T,R), 1984 (R), 1994 (T), 1996 (R)
Chlormequat (015)	1970 (T,R), 1972 (T,R), 1976 (R), 1985 (R), 1994 (T,R), 1997 (T), 1999 (acute Rf D), 2000 (R)
Chlorobenzilate (016)	1965 (T), 1968 (T,R), 1972 (R), 1975 (R), 1977 (R), 1980 (T)
Chloropicrin	1965 (T,R)

Chloropropylate	1968 (T,R), 1972 (R)
Chlorothalonil (081)	1974 (T,R), 1977 (T,R), 1978 (R), 1979 (T,R), 1981 (T,R), 1983 (T,R), 1984 (corr. to 1983 report and T evaluation), 1985 (T,R), 1987 (T), 1988 (R), 1990 (T,R), 1991 (corr. to 1990 evaluation), 1992 (T), 1993 (R), 1997 (R)
Chlorpropham	1965 (T), 2000 (T), 2001 (R)
Chlorpyrifos (017)	1972 (T,R), 1974 (R), 1975 (R), 1977 (T,R), 1981 (R), 1982 (T,R), 1983 (R), 1989 (R), 1995 (R), 1999 (T), 2000 (R), 2004 (R)
Chlorpyrifos-methyl (090)	1975 (T,R), 1976 (R, Annex I only), 1979 (R), 1990 (R), 1991 (T,R), 1992 (T and corr. to 1991 report), 1993 (R), 1994 (R), 2001 (T)
Chlorthion	1965 (T)
Clethodim (187)	1994 (T,R), 1997 (R), 1999 (R), 2002 (R)
Clofentezine (156)	1986 (T,R), 1987 (R), 1989 (R), 1990 (R), 1992 (R)
Coumaphos (018)	1968 (T,R), 1972 (R), 1975 (R), 1978 (R), 1980 (T,R), 1983 (R), 1987 (T), 1990 (T,R)
Crufomate (019)	1968 (T,R), 1972 (R)
Cyanophenfos (091)	1975 (T,R), 1978 (T: ADI extended, but no evaluation), 1980, (T), 1982 (R), 1983 (T)
Cycloxydim (179)	1992 (T,R), 1993 (R)
Cyfluthrin (157)	1986 (R), 1987 (T and corr. to 1986 report), 1989 (R), 1990 (R), 1992 (R)
Cyhalothrin (146)	1984 (T,R), 1986 (R), 1988 (R)
Cyhexatin (tricyclohexyltin hydroxide) (067)	1970 (T,R), 1973 (T,R), 1974 (R), 1975 (R), 1977 (T), 1978 (T,R), 1980 (T), 1981 (T), 1982 (R), 1983 (R), 1985 (R), 1988 (T), 1989 (T), 1991 (T,R), 1992 (R), 1994 (T)
Cypermethrin (118)	1979 (T,R), 1981 (T,R), 1982 (R), 1983 (R), 1984 (R), 1985 (R), 1986 (R), 1987 (corr. to 1986 evaluation), 1988 (R), 1990 (R)
Cyprodinil (207)	2003 (T,R), 2004 (corr. to 2003 report)
Cyromazine (169)	1990 (T,R), 1991 (corr. to 1990 R evaluation), 1992 (R)
2,4-D (020)	1970 (T,R), 1971 (T,R), 1974 (T,R), 1975 (T,R), 1980 (R), 1985, (R), 1986 (R), 1987 (corr. to 1986 report, Annex I), 1996 (T), 1997 (E), 1998 (R), 2001 (R)
Daminozide (104)	1977 (T,R), 1983 (T), 1989 (T,R), 1991 (T)
DDT (021)	1965 (T), 1966 (T,R), 1967 (T,R),1968 (T,R), 1969 (T,R), 1978 (R), 1979 (T), 1980 (T), 1983 (T), 1984 (T), 1993 (R), 1994 (R), 1996 (R)
Deltamethrin (135)	1980 (T,R), 1981 (T,R), 1982 (T,R), 1984 (R), 1985 (R), 1986 (R), 1987 (R), 1988 (R), 1990 (R), 1992 (R), 2000 (T), 2002 (R)
Demeton (092)	1965 (T), 1967 (R), 1975 (R), 1982 (T)
Demeton-*S*-methyl (073)	1973 (T,R), 1979 (R), 1982 (T), 1984 (T,R), 1989 (T,R), 1992 (R), 1998 (R)
Demeton-*S*-methylsulphon (164)	1973 (T,R), 1982 (T), 1984 (T,R), 1989 (T,R), 1992 (R)
Dialifos (098)	1976 (T,R), 1982 (T), 1985 (R)

Diazinon (022)	1965 (T), 1966 (T), 1967 (R), 1968 (T,R), 1970 (T,R), 1975 (R), 1979 (R), 1993 (T,R), 1994 (R), 1996 (R), 1999 (R), 2001 (T)
1,2-Dibromoethane (023)	1965 (T,R), 1966 (T,R), 1967 (R), 1968 (R), 1971 (R), 1979 (R), 1985 (R)
Dicloran (083)	2003 (R)
Dichlorfluanid (082)	1969 (T,R), 1974 (T,R), 1977 (T,R), 1979 (T,R), 1981 (R),1982 (R), 1983 (T,R), 1985 (R)
1,2-Dichloroethane (024)	1965 (T,R), 1967 (R), 1971 (R), 1979 (R), 1985 (R)
Dichlorvos (025)	1965 (T,R), 1966 (T,R), 1967 (T,R), 1969 (R), 1970 (T,R), 1974 (R), 1977 (T), 1993 (T,R)
Dicloran (083)	1974 (T,R), 1977 (T,R), 1998 (T,R)
Dicofol (026)	1968 (T,R), 1970 (R), 1974 (R), 1992 (T,R), 1994 (R)
Dieldrin (001)	1965 (T), 1966 (T,R), 1967 (T,R), 1968 (R), 1969 (R), 1970, (T,R), 1974 (R), 1975 (R), 1977 (T), 1990 (R), 1992 (R)
Diflubenzuron (130)	1981 (T,R), 1983 (R), 1984 (T,R), 1985 (T,R), 1988 (R), 2001 (T), 2002 (R)
Dimethipin (151)	1985 (T,R), 1987 (T,R), 1988 (T,R), 1999 (T), 2001 (R), 2004 (T)
Dimethoate (027)	1965 (T), 1966 (T), 1967 (T,R), 1970 (R), 1973 (R in evaluation of formothion), 1977 (R), 1978 (R), 1983 (R) 1984 (T,R) 1986 (R), 1987 (T,R), 1988 (R), 1990 (R), 1991 (corr. to 1990 evaluation), 1994 (R), 1996 (T), 1998 (R), 2003 (T,R), 2004 (corr. to 2003 report)
Dimethrin	1965 (T)
Dinocap (087)	1969 (T,R), 1974 (T,R), 1989 (T,R), 1992 (R), 1998 (R), 1999 (R), 2000 (T), 2001 (R)
Dioxathion (028)	1968 (T,R), 1972 (R)
Diphenyl (029)	1966 (T,R), 1967 (T)
Diphenylamine (030)	1969 (T,R), 1976 (T,R), 1979 (R), 1982 (T), 1984 (T,R), 1998 (T), 2001 (R), 2003 (R)
Diquat (031)	1970 (T,R), 1972 (T,R), 1976 (R), 1977 (T,R), 1978 (R), 1994 (R)
Disulfoton (074)	1973 (T,R), 1975 (T,R), 1979 (R), 1981 (R), 1984 (R), 1991 (T,R), 1992 (corr. to 1991 report, Annex I), 1994 (R), 1996 (T), 1998 (R)
Dithianon (180)	1992 (T,R), 1995 (R), 1996 (corr. to 1995 report)
Dithiocarbamates (105)	1965 (T), 1967 (T,R), 1970 (T,R), 1983 (R propineb, thiram), 1984 (R propineb), 1985 (R), 1987 (T thiram), 1988 (R thiram), 1990 (R), 1991 (corr. to 1990 evaluation), 1992 (T thiram), 1993 (T,R), 1995 (R), 1996 (T,R ferbam, ziram;, R thiram), 2004 (R)
4,6-Dinitro-*ortho*-cresol (DNOC)	1965 (T)
Dodine (084)	1974 (T,R), 1976 (T,R), 1977 (R), 2000 (T), 2003 (R) 2004 (corr. to 2003 report)
Edifenphos (099)	1976 (T,R), 1979 (T,R), 1981 (T,R)
Endosulfan (032)	1965 (T), 1967 (T,R), 1968 (T,R), 1971 (R), 1974 (R), 1975 (R), 1982 (T), 1985 (T,R), 1989 (T,R), 1993 (R), 1998 (T)
Endrin (033)	1965 (T), 1970 (T,R), 1974 (R), 1975 (R), 1990 (R), 1992 (R)
Esfenvalerate (204)	2002 (T, R)
Ethephon (106)	1977 (T,R), 1978 (T,R), 1983 (R), 1985 (R), 1993 (T), 1994 (R), 1995 (T), 1997 (T), 2002 (T)

Ethiofencarb (107)	1977 (T,R), 1978 (R), 1981 (R), 1982 (T,R), 1983 (R)
Ethion (034)	1968 (T,R), 1969 (R), 1970 (R), 1972 (T,R), 1975 (R), 1982 (T), 1983 (R), 1985 (T), 1986 (T), 1989 (T), 1990 (T), 1994 (R)
Ethoprophos (149)	1983 (T), 1984 (R), 1987 (T), 1999 (T), 2004 (R)
Ethoxyquin (035)	1969 (T,R), 1998 (T), 1999 (R)
Ethylene dibromide	See 1,2-Dibromoethane
Ethylene dichloride	See 1,2-Dichloroethane
Ethylene oxide	1965 (T,R), 1968 (T,R), 1971 (R)
Ethylenethiourea (ETU) (108)	1974 (R), 1977 (T,R), 1986 (T,R), 1987 (R), 1988 (T,R), 1990 (R), 1993 (T,R)
Etofenprox (184)	1993 (T,R)
Etrimfos (123)	1980 (T,R), 1982 (T,R^1), 1986 (T,R), 1987 (R), 1988 (R), 1989 (R), 1990 (R)
Famoxadone (208)	2003 (T,R)
Fenamiphos (085)	1974 (T,R), 1977 (R), 1978 (R), 1980 (R), 1985 (T), 1987 (T), 1997 (T), 1999 (R), 2002 (T)
Fenarimol (192)	1995 (T,R,E), 1996 (R and corr. to 1995 report)
Fenbuconazole (197)	1997 (T,R)
Fenbutatin oxide (109)	1977 (T,R), 1979 (R), 1992 (T), 1993 (R)
Fenchlorfos (036)	1968 (T,R), 1972 (R), 1983 (R)
Fenitrothion (037)	1969 (T,R), 1974 (T,R), 1976 (R), 1977 (T,R), 1979 (R), 1982, (T) 1983 (R), 1984 (T,R), 1986 (T,R), 1987 (R and corr. to 1986 R evaluation), 1988 (T), 1989 (R), 2000 (T), 2003 (R), 2004 (R, corr. to 2003 report)
Fenpropathrin (185)	1993 (T,R)
Fenpropimorph (188)	1994 (T), 1995 (R), 1999 (R), 2001 (T), 2004 (T)
Fenpyroximate (193)	1995 (T,R), 1996 (corr. to 1995 report.), 1999 (R), 2004 (T)
Fensulfothion (038)	1972 (T,R), 1982 (T), 1983 (R)
Fenthion (039)	1971 (T,R), 1975 (T,R), 1977 (R), 1978 (T,R), 1979 (T), 1980 (T), 1983 (R), 1989 (R), 1995 (T,R,E), 1996 (corr. to 1995 report), 1997 (T), 2000 (R)
Fentin compounds (040)	1965 (T), 1970 (T,R), 1972 (R), 1986 (R), 1991 (T,R), 1993 (R), 1994 (R)
Fenvalerate (119)	1979 (T,R), 1981 (T,R), 1982 (T), 1984 (T,R), 1985 (R), 1986 (T,R), 1987 (R and corr. to 1986 report), 1988 (R), 1990 (R), 1991 (corr. to 1990 R evaluation)
Ferbam	See Dithiocarbamates, 1965 (T), 1967 (T,R), 1996 (T,R)
Fipronil	1997 (T), 2000 (T), 2001 (R)
Fipronil-desulfinyl	1997 (T)
Flucythrinate (152)	1985 (T,R), 1987 (R), 1988 (R), 1989 (R), 1990 (R), 1993 (R)
Fludioxinil ( )	2004 (T,R)
Flumethrin (195)	1996 (T,R)
Flusilazole (165)	1989 (T,R), 1990 (R), 1991 (R), 1993 (R), 1995 (T)
Flutolanil (205)	2002 (T, R)
Folpet (041)	1969 (T,R), 1973 (T), 1974 (R), 1982 (T), 1984 (T,R), 1986 (T), 1987 (R), 1990 (T,R), 1991 (corr. to 1990 R evaluation), 1993 (T,R), 1994 (R), 1995 (T), 1997 (R), 1998 (R), 1999(R) , 2002 (T), 2004 (T)
Formothion (042)	1969 (T,R), 1972 (R), 1973 (T,R), 1978 (R), 1998 (R)

Glufosinate-ammonium (175)	1991 (T,R), 1992 (corr. to 1991 report, Annex I), 1994 (R), 1998 (R), 1999 (T,R)
Glyphosate (158)	1986 (T,R), 1987 (R and corr. to 1986 report), 1988 (R), 1994 (R), 1997 (T,R), 2004 (T)
Guazatine (114)	1978 (T.R), 1980 (R), 1997 (T,R)
Haloxyfop (194)	1995 (T,R), 1996 (R and corr. to 1995 report), 2001 (R)
Heptachlor (043)	1965 (T), 1966 (T,R), 1967 (R), 1968 (R), 1969 (R), 1970 (T,R), 1974 (R), 1975 (R), 1977 (R), 1987 (R), 1991 (T,R), 1992 (corr. to 1991 report, Annex I), 1993 (R), 1994 (R)
Hexachlorobenzene (044)	1969 (T,R), 1973 (T,R), 1974 (T,R), 1978(T), 1985 (R)
Hexaconazole (170)	1990 (T,R), 1991 (R and corr. to 1990 R evaluation), 1993 (R)
Hexythiazox (176)	1991 (T,R), 1994 (R), 1998 (R)
Hydrogen cyanide (045)	1965 (T,R)
Hydrogen phosphide (046)	1965 (T,R), 1966 (T,R), 1967 (R), 1969 (R), 1971 (R)
Imazalil (110)	1977 (T,R), 1980 (T,R), 1984 (T,R), 1985 (T,R), 1986 (T), 1988 (R), 1989 (R), 1991 (T), 1994 (R), 2000 (T), 2001 (T)
Imidacloprid	2001 (T), 2002 (R)
Iprodione (111)	1977 (T,R), 1980 (R), 1992 (T), 1994 (R), 1995 (T), 2001 (R)
Isofenphos (131)	1981 (T,R), 1982 (T,R), 1984 (R), 1985 (R), 1986 (T,R), 1988 (R), 1992 (R)
Kresoxim-methyl (199)	1998 (T,R), 2001 (R)
Lead arsenate	1965 (T), 1968 (T,R)
Leptophos (088)	1974 (T,R), 1975 (T,R), 1978 (T,R)
Lindane (048)	1965 (T), 1966 (T,R), 1967 (R), 1968 (R), 1969 (R), 1970 (T,R, published as Annex VI to 1971 evaluations), 1973 (T,R), 1974 (R), 1975 (R), 1977 (T,R), 1978 (R), 1979 (R), 1989 (T,R), 1997 (T), 2002 (T), 2003 (R), 2004 (corr. to 2003 report)
Malathion (049)	1965 (T), 1966 (T,R), 1967 (corr. to 1966 R evaluation), 1968 (R), 1969 (R), 1970 (R), 1973 (R), 1975 (R), 1977 (R), 1984 (R), 1997 (T), 1999 (R), 2000 (R), 2003 (T), 2004 (R)
Maleic hydrazide (102)	1976 (T,R), 1977 (T,R), 1980 (T), 1984 (T,R), 1996 (T), 1998 (R)
Mancozeb (050)	1967 (T,R), 1970 (T,R), 1974 (R), 1977 (R), 1980 (T,R), 1993 (T,R)
Maneb	See Dithiocarbamates, 1965 (T), 1967 (T,R), 1987 (T), 1993 (T,R)
Mecarbam (124)	1980 (T,R), 1983 (T,R), 1985 (T,R), 1986 (T,R), 1987 (R)
Metalaxyl (138)	1982 (T,R), 1984 (R), 1985 (R), 1986 (R), 1987 (R), 1989 (R), 1990 (R), 1992 (R), 1995 (R)
Metalaxyl –M (212)	2002 (T), 2004 (R)

Methacrifos (125)	1980 (T,R), 1982 (T), 1986 (T), 1988 (T), 1990 (T,R), 1992 (R)
Methamidophos (100)	1976 (T,R), 1979 (R), 1981 (R), 1982 (T,R), 1984 (R), 1985 (T), 1989 (R), 1990 (T,R), 1994 (R), 1996 (R), 1997 (R), 2002 (T), 2003 (R), 2004 (R, corr. to 2003 report)
Methidathion (051)	1972 (T,R), 1975 (T,R), 1979 (R), 1992 (T,R), 1994 (R), 1997 (T)
Methiocarb (132)	1981 (T,R), 1983 (T,R), 1984 (T), 1985 (T), 1986 (R), 1987 (T,R), 1988 (R), 1998 (T), 1999 (R)
Methomyl (094)	1975 (R), 1976 (R), 1977 (R), 1978 (R), 1986 (T,R), 1987 (R), 1988 (R), 1989 (T,R), 1990 (R), 1991 (R), 2001 (T,R), 2004 (R)
Methoprene (147)	1984 (T,R), 1986 (R), 1987 (T and corr. to 1986 report), 1988 (R), 1989 (R), 2001 (T)
Methoxychlor	1965 (T), 1977 (T)
Methoxyfenozide (209)	2003 (T,R), 2004 (corr. to 2003 report)
Methyl bromide (052)	See Bromomethane
Metiram (186)	1993 (T), 1995 (R)
Mevinphos (053)	1965 (T), 1972 (T,R), 1996 (T), 1997 (E,R), 2000 (R)
MGK 264	1967 (T,R)
Monocrotophos (054)	1972 (T,R), 1975 (T,R), 1991 (T,R), 1993 (T), 1994 (R)
Myclobutanil (181)	1992 (T,R), 1997 (R), 1998 (R)
Nabam	See Dithiocarbamates, 1965 (T), 1976 (T,R)
Nitrofen (140)	1983 (T,R)
Omethoate (055)	1971 (T,R), 1975 (T,R), 1978 (T,R), 1979 (T), 1981 (T,R), 1984 (R), 1985 (T), 1986 (R), 1987 (R), 1988 (R), 1990 (R), 1998 (R)
Organomercury compounds	1965 (T), 1966 (T,R), 1967 (T,R)
Oxamyl (126)	1980 (T,R), 1983 (R), 1984 (T), 1985 (T,R), 1986 (R), 2002 (T,R)
Oxydemeton-methyl (166)	1965 (T, as demeton-*S*-methyl sulfoxide), 1967 (T), 1968 (R), 1973 (T,R), 1982 (T), 1984 (T,R), 1989 (T,R), 1992 (R), 1998 (R), 1999 (corr. to 1992 report), 2002 (T), 2004 (R)
Oxythioquinox	See Chinomethionat
Paclobutrazol (161)	1988 (T,R), 1989 (R)
Paraquat (057)	1970 (T,R), 1972 (T,R), 1976 (T,R), 1978 (R), 1981 (R), 1982 (T), 1985 (T), 1986 (T), 2003 (T), 2004 (R)
Parathion (058)	1965 (T), 1967 (T,R), 1969 (R), 1970 (R), 1984 (R), 1991 (R), 1995 (T,R), 1997 (R), 2000 (R)
Parathion-methyl (059)	1965 (T), 1968 (T,R), 1972 (R), 1975 (T,R), 1978 (T,R), 1979 (T), 1980 (T), 1982 (T), 1984 (T,R), 1991 (R), 1992 (R), 1994 (R), 1995 (T), 2000 (R), 2003 (R)
Penconazole (182)	1992 (T,R), 1995 (R)
Permethrin (120)	1979 (T,R), 1980 (R), 1981 (T,R), 1982 (R), 1983 (R), 1984 (R), 1985 (R), 1986 (T,R), 1987 (T), 1988 (R), 1989 (R), 1991 (R), 1992 (corr. to 1991 report), 1999 (T)
2-Phenylphenol (056)	1969 (T,R), 1975 (R), 1983 (T), 1985 (T,R), 1989 (T), 1990 (T,R), 1999 (T,R), 2002 (R)
Phenothrin (127)	1979 (R), 1980 (T,R), 1982 (T), 1984 (T), 1987 (R), 1988 (T,R)
Phenthoate (128)	1980 (T,R), 1981 (R), 1984 (T)

Phorate (112)	1977 (T,R), 1982 (T), 1983 (T), 1984 (R), 1985 (T), 1990 (R), 1991 (R), 1992 (R), 1993 (T), 1994 (T), 1996 (T), 2004 (T)
Phosalone (060)	1972 (T,R), 1975 (R), 1976 (R), 1993 (T), 1994 (R), 1997 (T), 1999 (R), 2001 (T)
Phosmet (103)	1976 (R), 1977 (corr. to 1976 R evaluation), 1978 (T,R), 1979 (T,R), 1981 (R), 1984 (R), 1985 (R), 1986 (R), 1987 (R and corr. to 1986 R evaluation), 1988 (R), 1994 (T), 1997 (R), 1998 (T), 2002 (R), 2003(R)
Phosphine	See Hydrogen phosphide
Phosphamidon (061)	1965 (T), 1966 (T), 1968 (T,R), 1969 (R), 1972 (R), 1974 (R), 1982 (T), 1985 (T), 1986 (T)
Phoxim (141)	1982 (T), 1983 (R), 1984 (T,R), 1986 (R), 1987 (R), 1988 (R)
Piperonyl butoxide (062)	1965 (T,R), 1966 (T,R), 1967 (R), 1969 (R), 1972 (T,R), 1992 (T,R), 1995 (T), 2001 (R), 2002 (R)
Pirimicarb (101)	1976 (T,R), 1978 (T,R), 1979 (R), 1981 (T,R), 1982 (T), 1985 (R), 2004 (T)
Pirimiphos-methyl (086)	1974 (T,R), 1976 (T,R), 1977 (R), 1979 (R), 1983 (R), 1985 (R), 1992 (T), 1994 (R), 2003 (R), 2004 (R, corr. to 2003 report)
Prochloraz (142)	1983 (T,R), 1985 (R), 1987 (R), 1988 (R), 1989 (R), 1990 (R), 1991 (corr. to 1990 report, Annex I, and R evaluation), 1992 (R), 2001 (T), 2004 (R)
Procymidone(136)	1981 (R), 1982 (T), 1989 (T,R), 1990 (R), 1991 (corr. to 1990 Annex I), 1993 (R), 1998 (R)
Profenofos (171)	1990 (T,R), 1992 (R), 1994 (R), 1995 (R)
Propamocarb (148)	1984 (T,R), 1986 (T,R), 1987 (R)
Propargite (113)	1977 (T,R), 1978 (R), 1979 (R), 1980 (T,R), 1982 (T,R), 1999 (T), 2002 (R)
Propham (183)	1965 (T), 1992 (T,R)
Propiconazole (160)	1987 (T,R), 1991 (R), 1994 (R), 2004 (T)
Propineb	1977 (T,R), 1980 (T), 1983 (T), 1984 (R), 1985 (T,R), 1993 (T,R), 2004 2004 (R)
Propoxur (075)	1973 (T,R), 1977 (R), 1981 (R), 1983 (R), 1989 (T), 1991 (R), 1996 (R)
Propylenethiourea (PTU, 150)	1993 (T,R), 1994 (R), 1999 (T)
Pyraclostrobin (210)	2003 (T), 2004 (R)
Pyrazophos (153)	1985 (T,R), 1987 (R), 1992 (T,R), 1993 (R)
Pyrethrins (063)	1965 (T), 1966 (T,R), 1967 (R), 1968 (R), 1969 (R), 1970 (T), 1972 (T,R), 1974 (R), 1999 (T), 2000 (R), 2003 (T,R)
Pyriproxyfen (200)	1999 (R,T), 2000 (R), 2001 (T)
Quintozene (064)	1969 (T,R) 1973 (T,R), 1974 (R), 1975 (T,R), 1976 (Annex I, corr. to 1975 R evaluation), 1977 (T,R), 1995 (T,R), 1998 (R)
Spinosad (203)	2001 (T,R, 2004 (R)
2,4,5-T (121)	1970 (T,R), 1979 (T,R), 1981 (T)
Tebuconazole (189)	1994 (T,R), 1996 (corr. to Annex II of 1995 report), 1997 (R)
Tebufenozide (196)	1996 (T,R), 1997 (R), 1999 (R), 2001 (T,R), 2003(T)
Tecnazine (115)	1974 (T,R), 1978 (T,R), 1981 (R), 1983 (T), 1987 (R), 1989 (R), 1994 (T,R)
Teflubenzuron (190)	1994 (T), 1996 (R)

Terbufos (167)	1989 (T,R), 1990 (T,R), 2003 (T)
Thiabendazole (065)	1970 (T,R), 1971 (R), 1972 (R), 1975 (R), 1977 (T,R), 1979 (R), 1981 (R), 1997 (R), 2000 (R)
Thiodicarb (154)	1985 (T,R), 1986 (T), 1987 (R), 1988 (R), 2000 (T), 2001 (R)
Thiometon (076)	1969 (T,R), 1973 (T,R), 1976 (R), 1979 (T,R), 1988 (R)
Thiophanate-methyl (077)	1973 (T,R), 1975 (T,R), 1977 (T), 1978 (R), 1988 (R), 2002 (R)
	1990 (R), 1994 (R), 1995 (T,E), 1998 (T,R)
Thiram (105)	See Dithiocarbamates, 1965 (T), 1967 (T,R), 1970 (T,R), 1974 (T), 1977 (T), 1983 (R), 1984 (R), 1985 (T,R), 1987 (T), 1988 (R), 1989 (R), 1992 (T), 1996 (R)
Tolclofos-methyl (191)	1994 (T,R) 1996 (corr. to Annex II of 1995 report)
Tolylfluanid (162)	1988 (T,R), 1990 (R), 1991 (corr. to 1990 report), 2002 (T,R), 2003 (R)
Toxaphene	See Camphechlor
Triadimefon (133)	1979 (R), 1981 (T,R), 1983 (T,R), 1984 (R), 1985 (T,R), 1986 (R), 1987 (R and corr. to 1986 R evaluation), 1988 (R), 1989 (R), 1992 (R), 1995 (R), 2004 (T)
Triadimenol (168)	1989 (T,R), 1992 (R), 1995 (R), 2004 (T)
Triazolylalanine	1989 (T,R)
Triazophos (143)	1982 (T), 1983 (R), 1984 (corr. to 1983 report, Annex I), 1986 (T,R), 1990 (R), 1991 (T and corr. to 1990 R evaluation), 1992 (R), 1993 (T,R), 2002 (T)
Trichlorfon (066)	1971 (T,R), 1975 (T,R), 1978 (T,R), 1987 (R)
Trichloronat	1971 (T,R)
Trichloroethylene	1968 (R)
Tricyclohexyltin hydroxide	See Cyhexatin
Trifloxystrobin (213)	2004 (T, R)
Triforine (116)	1977 (T), 1978 (T, R), 1997 (T)
Triphenyltin compounds	See Fentin compounds
Vamidothion (078)	1973 (T,R), 1982 (T), 1985 (T,R), 1987 (R), 1988 (T), 1990 (R), 1992 (R)
Vinclozolin (159)	1986 (T,R), 1987 (R and corr. to 1986 report and R evaluation), 1988 (T,R), 1989 (R), 1990 (R), 1992 (R), 1995 (T)
Zineb (105)	See Dithiocarbamates, 1965 (T), 1967 (T,R), 1993 (T)
Ziram (105)	See Dithiocarbamates, 1965 (T), 1967 (T,R), 1996 (T,R)

ANNEX 3

INTERNATIONAL ESTIMATED DAILY INTAKES OF PESTICIDE RESIDUES

The following tables give details of the international estimated daily intakes of the pesticides evaluated by the Meeting for the five GEMS/ Food diets and show the ratios of the estimated intakes to the corresponding ADIs.

(*) at or about the LOQ

The ranges of the ratios of intake:ADI for all the compounds evaluated are tabulated in Section 3.

Diet: g/person per day; intake: µg/person

CHLORPYRIFOS (017)

International estimated daily intake   ADI = 0–0.01 mg/kg bw

Codex code	Commodity	STMR or STMR-P (mg/kg)	Mid-East Diet	Mid-East Intake	Far-East Diet	Far-East Intake	African Diet	African Intake	Latin American Diet	Latin American Intake	European Diet	European Intake
TN 0660	Almond	0.05	0.5	0.0	0.0	0.0	0.0	0.0	0.1	0.0	1.8	0.1
JF 0226	Apple juice	0.027	4.5	0.1	0	0.0	0	0.0	0.3	0.0	3.8	0.1
FI 0327	Banana	0.01	8.3	0.1	26.2	0.3	21.0	0.2	102.3	1.0	22.8	0.2
VB 0400	Broccoli	0.02	0.5	0.0	1.0	0.0	0.0	0.0	1.1	0.0	2.7	0.1
VB 0403	Cabbage, Savoy	0.15	0.1	0.0	0.1	0.0	0.1	0.0	0.1	0.0	0.1	0.0
VR 0577	Carrot	0.025	2.8	0.1	2.5	0.1	0.0	0.0	6.3	0.2	22.0	0.6
MM 0812	Cattle meat	0.02	14.6	0.3	2.7	0.1	10.4	0.2	30.0	0.6	63.3	1.3
MO 1280	Cattle, kidney	0.01	0.1	0.0	0.0	0.0	0.1	0.0	0.2	0.0	0.2	0.0
MO 1281	Cattle, liver	0.01	0.2	0.0	0	0.0	0.1	0.0	0.3	0.0	0.4	0.0
VB 0404	Cauliflower	0.01	1.3	0.0	1.5	0.0	0	0.0	0.3	0.0	13	0.1
VL 0467	Chinese cabbage, type pe-tsai	0.18	ND	–	ND	–	ND	–	ND	–	ND	–
FC 0001	Citrus fruit	0.08	47.1	3.8	6.3	0.5	5.1	0.4	54.6	4.4	44.6	3.6
SB 0716	Coffee beans	0.01	5.3	0.1	0.4	0.0	0.0	0.0	3.6	0.0	7.9	0.1
VP 0526	Common bean (green pods or immature seeds)	0.01	3.5	0.0	0.8	0.0	0.0	0.0	4.0	0.0	12.0	0.1
OR 0691	Cotton-seed oil, edible	0.01	3.8	0.0	0.5	0.0	0.5	0.0	0.5	0.0	0.0	0.0
PE 0112	Eggs	0.001	14.6	0.0	13.1	0.0	3.7	0.0	11.9	0.0	37.6	0.0
FB 0269	Grapes (fresh, wine, dried)	0.085	16.1	1.4	1.0	0.1	0.0	0.0	1.6	0.1	16.1	1.4
GC 0645	Maize (fresh, flour)	0.015	48.3	0.7	31.2	0.5	106.2	1.6	41.8	0.6	8.8	0.1
OR 0645	Maize oil, edible	0.03	1.8	0.1	0.0	0.0	0.3	0.0	0.5	0.0	1.3	0.0
ML 0107	Milk of cattle, goats and sheep	0.005	110.3	0.6	23.9	0.1	41.3	0.2	160.1	0.8	289.3	1.4
VA 0385	Onion, bulb	0.04	23.0	0.9	11.5	0.5	7.3	0.3	13.8	0.6	27.8	1.1
JF 0004	Orange juice	0.007	7.3	0.1	0.0	0.0	0.0	0.0	0.3	0.0	4.5	0.0
FS 0247	Peach	0.042	1.3	0.1	0.3	0.0	0.0	0.0	0.4	0.0	6.3	0.3

Annex 3

Codex code	Commodity	STMR or STMR-P (mg/kg)	Mid-East	Far-East	African	Latin American	European
VP 0063	Peas (green pods and immature seeds)	0.01	5.5	2.0	0.0	0.8	14.0	0.1
TN 0672	Pecan	0.05	0.0	0.0	0.0	0.0	0.3	0.0
VO 0445	Peppers, sweet (including pim(i)ento)	0.38	3.3	2.0	5.3	2.3	10.3	3.9
MO 0818	Pig, edible offal of	0	0.0	1.0	0.0	1.0	5.0	0.0
MM 0818	Pig meat	0.001	0.0	27.2	2.6	10.5	75.8	0.1
FS 0014	Plum (fresh, prunes)	0.04	1.8	0.5	0.0	0.0	4.3	0.2
FP 0009	Pome fruits	0.17	10.8	7.5	0.3	6.5	51.3	8.7
VR 0589	Potato	0.51	59.0	19.2	20.6	40.8	240.8	122.8
PM 0110	Poultry meat	0.001	31.0	13.2	5.5	25.3	53.0	0.1
PO 0111	Poultry, edible offal of	0	0.1	0.1	0.1	0.4	0.4	0.0
CM 1206	Rice bran, unprocessed	0.22	ND	ND	ND	ND	ND	–
CM 0649	Rice, husked	0.016	0.0	1.8	34.7	21.0	2.5	0.0
CM 1205	Rice, polished	0.008	48.8	277.5	68.8	65.5	9.3	0.1
MM 0822	Sheep meat	0.02	13.5	0.7	2	3	10.3	0.2
MO 0822	Sheep, edible offal of	0.01	1.3	0.0	0.5	0.0	1.3	0.0
GC 0651	Sorghum	0.04	2.0	9.7	26.6	0.0	0.0	0.0
OR 0541	Soya bean oil, refined	0.004	1.3	1.7	3.0	14.5	4.3	0.0
FB 0275	Strawberry	0.09	0.0	0.0	0.0	0.0	5.3	0.5
VR 0596	Sugar-beet	0.015	0.5	0.0	0.0	0.3	2.0	0.0
VO 0447	Sweet corn (corn-on-the-cob)	0.01	0.0	0.0	4.4	0.0	8.3	0.1
DT 1114	Tea, green, black (black, fermented and dried)	0.34	2.3	1.2	0.5	0.5	2.3	0.8
VO 0448	Tomato (fresh)	0.13	44.1	5.7	14.6	25.5	34.9	4.5
JF 0448	Tomato juice	0.026	0.3	0.0	0.0	0.0	2.0	0.1
	Tomato paste	0.026	5.8	0.2	0.3	0.0	4.0	0.1
TN 0678	Walnuts	0.05	0.0	0.0	0.0	0.0	0.5	0.0
CM 0654	Wheat bran, unprocessed	0.03	ND	ND	ND	ND	ND	–
CF 1211	Wheat flour	0.002	323.0	114.0	28.3	112.0	175.8	0.4
	Total intake (µg/person)		49.7	18.1	19.9	36.0	153.3	
	Body weight per region (kg bw)		60	55	60	60	60	
	ADI (µg/person)		600	550	600	600	600	
	% ADI		8.3%	3.3%	3.3%	6.0%	25.6%	
	Rounded % ADI		8%	3%	3%	6%	30%	

Annex 3

ETHOPROPHOS (149) International estimated daily intake ADI = 0–0.0004 mg/kg bw

Codex Code	Commodity	STMR or STMR-P (mg/kg)	Mid-East Diet	Mid-East Intake	Far-East Diet	Far-East Intake	African Diet	African Intake	Latin American Diet	Latin American Intake	European Diet	European Intake
FI 0327	Banana	0.02	8.3	0.2	26.2	0.5	21.0	0.4	102.3	2.0	22.8	0.5
VC 0424	Cucumber	0.01	2.4	0.0	2.3	0.0	0.0	0.0	4.2	0.0	4.5	0.0
MO 0105	Edible offal (mammalian)	0	4.2	0.0	1.4	0.0	2.8	0.0	6.1	0.0	12.4	0.0
MM 0095	Meat from mammals other than marine mammals: 20% as fat	0	7.4	0.0	6.6	0.0	4.8	0.0	9.4	0.0	31.1	0.0
MM 0095	Meat from mammals other than marine mammals: 80% as muscle	0	29.6	0.0	26.2	0.0	19.0	0.0	37.6	0.0	124.4	0.0
VC 0046	Melon, except watermelon	0.005	16.0	0.1	2.0	0.0	0.0	0.0	2.8	0.0	18.3	0.1
ML 0106	Milks	0	116.9	0.0	32.1	0.0	41.8	0.0	160.1	0.0	289.3	0.0
VO 0445	Peppers, sweet (including pim(i)ento)	0.005	3.3	0.0	2.0	0.0	5.3	0.0	2.3	0.0	10.3	0.1
VR 0589	Potato	0.01	59.0	0.6	19.2	0.2	20.6	0.2	40.8	0.4	240.8	2.4
GS 0659	Sugar cane	0.02	18.5	0.4	7.3	0.1	15.9	0.3	3.5	0.1	0.0	0.0
VR 0508	Sweet potato	0.01	1.5	0.0	81.3	0.8	14.3	0.1	13.8	0.1	1.3	0.0
VO 0448	Tomato (fresh, juice, paste, peeled)	0.005	81.5	0.4	7.0	0.0	16.5	0.1	25.5	0.1	66.6	0.3
	Total intake (µg/person)			1.7		1.8		1.2		2.9		3.4
	Body weight per region (kg bw)			60		55		60		60		60
	ADI (µg/person)			24		22		24		24		24
	% ADI			7.0%		8.0%		5.0%		11.9%		14.2%
	Rounded % ADI			7%		8%		5%		10%		10%

Annex 3

FENITROTHION (37)   International estimated daily intake   ADI = 0–0.005 mg/kg bw

Codex code	Commodity	STMR or STMR-P (mg/kg)	Mid-East Diet	Mid-East Intake	Far-East Diet	Far-East Intake	African Diet	African Intake	Latin American Diet	Latin American Intake	European Diet	European Intake
FP 0226	Apple	0.04	7.5	0.3	4.7	0.2	0.3	0.0	5.5	0.2	40.0	1.6
GC 0640	Barley (beer only)	1	ND	–	ND	–	ND	–	ND	–	ND	–
GC 0640	Barley (fresh)	5	1.0	5.0	3.5	17.5	1.8	9.0	6.5	32.5	19.8	99.0
GC 0641	Buckwheat	5	0.0	0.0	1.0	5.0	0.0	0.0	0.0	0.0	0.0	0.0
MO 0105	Edible offal (mammalian)	0	4.2	0.0	1.4	0.0	2.8	0.0	6.1	0.0	12.4	0.0
PE 0112	Eggs	0	14.6	0.0	13.1	0.0	3.7	0.0	11.9	0.0	37.6	0.0
GC 0645	Maize (fresh, flour)	5	48.3	241.5	31.2	156.0	106.2	531.0	41.8	209.0	8.8	44.0
MM 0095	Meat from mammals other than marine mammals: 20% as fat	0	7.4	0.0	6.6	0.0	4.8	0.0	9.4	0.0	31.1	0.0
MM 0095	Meat from mammals other than marine mammals: 80% as muscle	0	29.6	0.0	26.2	0.0	19.0	0.0	37.6	0.0	124.4	0.0
ML 0106	Milks	0	116.9	0.0	32.1	0.0	41.8	0.0	160.1	0.0	289.3	0.0
GC 0646	Millet	5	2.5	12.5	9.3	46.5	51.8	259.0	0.0	0.0	0.0	0.0
GC 0647	Oats	5	0.0	0.0	0.0	0.0	0.2	1.0	0.8	4.0	2.0	10.0
PM 0110	Poultry meat: 10% as fat	0	3.1	0.0	1.3	0.0	0.6	0.0	2.5	0.0	5.3	0.0
PM 0110	Poultry meat: 90% as muscle	0	27.9	0.0	11.9	0.0	5.0	0.0	22.8	0.0	47.7	0.0
CM 1206	Rice bran, unprocessed	36	ND	–	ND	–	ND	–	ND	–	ND	–
CM 0649	Rice, husked and cooked	0.55	0.0	0.0	1.8	1.0	34.7	19.1	21.0	11.6	2.5	1.4
CM 1205	Rice, polished and cooked	0.2	48.8	9.8	277.5	55.5	68.8	13.8	65.5	13.1	9.3	1.9
CF 1250	Rye flour	1.175	0.0	0.0	1.0	1.2	0.0	0.0	0.0	0.0	1.5	1.8
GC 0651	Sorghum	5	2.0	10.0	9.7	48.5	26.6	133.0	0.0	0.0	0.0	0.0
GC 0653	Triticale	5	0.0	0.0	1.0	5.0	0.0	0.0	0.0	0.0	0.0	0.0
CM 0654	Wheat bran, unprocessed	19.75	ND	–	ND	–	ND	–	ND	–	ND	–
	Wheat bulgur wholemeal	1.175	0.3	0.4	0.0	0.0	0.0	0.0	0.0	0.0	0.0	0.0
	Wheat macaroni	1.175	1.0	1.2	0.3	0.4	0.0	0.0	2.8	3.3	1.3	1.5
	Wheat pastry	1.175	3.0	3.5	0.5	0.6	0.0	0.0	2.0	2.4	1.0	1.2
CP 1211	White bread	0.5	215.3	107.7	76.0	38.0	18.9	9.5	37.3	18.7	117.2	58.6
CP 1212	Wholemeal bread	1.9	107.7	204.6	38.0	72.2	9.4	17.9	74.7	141.9	58.6	111.3
	Total intake (µg/person)			596.4		447.5		993.2		436.6		332.2
	Body weight per region (kg bw)			60		55		60		60		60
	ADI (µg/person)			300		275		300		300		300
	% ADI			198.8%		162.7%		331.1%		145.5%		110.7%
	Rounded % ADI			200%		160%		330%		150%		110%

Annex 3

FLUDIOXONIL (211) International estimated daily intake ADI = 0–0.4 mg/kg bw

Codex code	Commodity	STMR or STMR-P (mg/kg)	Mid-East Diet	Mid-East Intake	Far-East Diet	Far-East Intake	African Diet	African Intake	Latin American Diet	Latin American Intake	European Diet	European Intake
FC 0001	Citrus fruit	1.1	47.1	51.8	6.3	6.9	5.1	5.6	54.6	60.1	44.6	49.1
VD 0071	Beans (dry)	0.02	2.3	0.0	4.8	0.1	0.0	0.0	13.0	0.3	3.5	0.1
VP 0061	Beans except broad and soya beans (green pods and immature seeds)	0.04	3.9	0.2	0.9	0.0	0.0	0.0	4.4	0.2	13.2	0.5
VP 0062	Beans, shelled (immature seeds)	0.02	0.1	0.0	0.1	0.0	0.1	0.0	0.1	0.0	0.1	0.0
FB 0264	Blackberry	1	0.0	0.0	0.0	0.0	0.0	0.0	0.0	0.0	0.0	0.0
FB 0020	Blueberry	0.6	0.0	0.0	0.0	0.0	0.0	0.0	0.0	0.0	0.5	0.3
VB 0400	Broccoli	0.23	0.5	0.1	1.0	0.2	0.0	0.0	1.1	0.3	2.7	0.6
	Cabbages (head and leafy Brassicas, kohlrabi)	0.24	5.0	1.2	9.7	2.3	0.0	0.0	10.5	2.5	26.8	6.4
VR 0577	Carrot	0.2	2.8	0.6	2.5	0.5	0.0	0.0	6.3	1.3	22.0	4.4
GC 0080	Cereal grains	0.02	429.9	8.6	450.8	9.0	318.3	6.4	252.4	5.0	221.9	4.4
OR 0691	Cotton-seed oil, edible	0.05	3.8	0.2	0.5	0.0	0.5	0.0	0.5	0.0	0.0	0.0
VC 0424	Cucumber	0.06	2.4	0.1	2.3	0.1	0.0	0.0	4.2	0.2	4.5	0.3
FB 0266	Dewberry, including boysenberry and loganberry	1	0.0	0.0	0.0	0.0	0.0	0.0	0.0	0.0	0.0	0.0
DH 0170	Dried herbs	22	ND	–	ND	–	ND	–	ND	–	ND	–
MO 0105	Edible offal (mammalian)	0	4.2	0.0	1.4	0.0	2.8	0.0	6.1	0.0	12.4	0.0
VO 0440	Egg plant	0.06	6.3	0.4	3.0	0.2	0.7	0.0	6.0	0.4	2.3	0.1
PE 0112	Eggs	0	14.6	0.0	13.1	0.0	3.7	0.0	11.9	0.0	37.6	0.0
FB 0269	Grapes (fresh, wine, excluding dried grapes)	0.28	15.8	4.4	1.0	0.3	0.0	0.0	1.3	0.4	13.8	3.9
DF 0269	Grapes, dried (currants, raisins and sultanas)	0.31	0.3	0.1	0.0	0.0	0.0	0.0	0.3	0.1	2.3	0.7
HH 0720	Herbs		ND	–	ND	–	ND	–	ND	–	ND	–
FI 0341	Kiwi fruit	7.2	0.0	0.0	0.0	0.0	1.9	13.7	0.1	0.7	1.5	10.8
VL 0482	Lettuce, head	2.7	2.3	6.2	0.0	0.0	0.0	0.0	5.8	15.7	22.5	60.8
MM 0095	Meat from mammals other than marine mammals	0	37.0	0.0	32.8	0.0	23.8	0.0	47.0	0.0	155.5	0.0
VC 0046	Melon, except watermelon	0.02	16.0	0.3	2.0	0.0	0.0	0.0	2.8	0.1	18.3	0.4
ML 0106	Milks	0	116.9	0.0	32.1	0.0	41.8	0.0	160.1	0.0	289.3	0.0
VL 0485	Mustard greens	1.2	0.1	0.1	0.1	0.1	0.1	0.1	0.1	0.1	0.1	0.1
VA 0385	Onion, bulb	0.04	23.0	0.9	11.5	0.5	7.3	0.3	13.8	0.6	27.8	1.1
FP 0230	Pear	0.21	3.3	0.7	2.8	0.6	0.0	0.0	1.0	0.2	11.3	2.4
VD 0072	Peas (dry)	0.02	0.5	0.0	1.7	0.0	5.1	0.1	1.3	0.0	1.8	0.0
VP 0063	Peas (green pods and immature seeds)	0.04	5.5	0.2	2.0	0.1	0.0	0.0	0.8	0.0	14.0	0.6
VP 0064	Peas, shelled (immature seeds)	0.02	4.0	0.1	0.5	0.0	0.0	0.0	0.2	0.0	10.1	0.2

Annex 3

Codex code	Commodity	STMR or STMR-P (mg/kg)	Mid-East Diet	Mid-East Intake	Far-East Diet	Far-East Intake	African Diet	African Intake	Latin American Diet	Latin American Intake	European Diet	European Intake
VO 0445	Peppers, sweet (including pim(i)ento)	0.18	3.3	0.6	2.0	0.4	5.3	1.0	2.3	0.4	10.3	1.9
TN 0675	Pistachio nut	0.05	0.3	0.0	0.0	0.0	0.0	0.0	0.0	0.0	0.0	0.0
VR 0589	Potato	0.01	59.0	0.6	19.2	0.2	20.6	0.2	40.8	0.4	240.8	2.4
PM 0110	Poultry meat	0	31.0	0.0	13.2	0.0	5.5	0.0	25.3	0.0	53.0	0.0
PO 0111	Poultry, edible offal of	0	0.1	0.0	0.1	0.0	0.1	0.0	0.4	0.0	0.4	0.0
DF 0014	Prunes	0.96	0.0	0.0	0.0	0.0	0.0	0.0	0.0	0.0	0.5	0.5
OR 0495	Rape-seed oil, edible	0.02	4.5	0.1	2.7	0.1	0.0	0.0	0.3	0.0	7.3	0.1
FB 0272	Raspberry, red, black	1	0.0	0.0	0.0	0.0	0.0	0.0	0.0	0.0	0.5	0.5
VA 0389	Spring onion	0.59	0.0	0.0	2.0	1.2	1.5	0.9	4.0	2.4	1.0	0.6
VC 0431	Squash, summer	0.06	10.5	0.6	2.2	0.1	0.0	0.0	14.0	0.8	3.5	0.2
FS 0012	Stone fruit	0.8	7.3	5.8	1.0	0.8	0.0	0.0	0.8	0.6	23.3	18.6
FB 0275	Strawberry	0.27	0.0	0.0	0.0	0.0	0.0	0.0	0.0	0.0	5.3	1.4
VO 0447	Sweet corn (corn-on-the-cob)	0.01	0.0	0.0	0.0	0.0	4.4	0.0	0.0	0.0	8.3	0.1
VO 0448	Tomato (fresh)	0.12	44.1	5.3	5.7	0.7	14.6	1.8	25.5	3.1	34.9	4.2
JF 0448	Tomato juice	0.026	0.3	0.0	0.0	0.0	0.0	0.0	0.0	0.0	2.0	0.1
	Tomato paste	0.17	5.8	1.0	0.2	0.0	0.3	0.1	0.1	0.1	4.0	0.7
VL 0473	Watercress	1.2	0.1	0.1	0.1	0.1	0.1	0.1	0.1	0.1	0.1	0.1
	Wine only	0.01	0.5	0.0	0.0	0.0	0.8	0.0	19.8	0.2	97.8	1.0
	Total intake (µg/person)			90.5		24.6		30.3		96.1		179.5
	Body weight per region (kg bw)			60		55		60		60		60
	ADI (µg/person)			24 000		22 000		24 000		24 000		24 000
	% AD			0.4%		0.1%		0.1%		0.4%		0.7%
	Rounded % ADI			0%		0%		0%		0%		1%

Annex 3

GLYPHOSATE (158) ADI = 0–1 mg/kg bw

Codex code	Commodity	MRL (mg/kg)	Mid-East Diet	Mid-East Intake	Far-East Diet	Far-East Intake	African Diet	African Intake	Latin American Diet	Latin American Intake	European Diet	European Intake
GC 0640	Barley (fresh)	20	1.0	20.0	3.5	70.0	1.8	36.0	6.5	130.0	19.8	396.0
VD 0071	Beans (dry)	2	2.3	4.6	4.8	9.6	0.0	0.0	13.0	26.0	3.5	7.0
MM 0812	Cattle meat	0.1	14.6	1.5	2.7	0.3	10.4	1.0	30.0	3.0	63.3	6.3
ML 0812	Cattle milk	0.1	79.5	8.0	23.2	2.3	35.8	3.6	159.3	15.9	287.0	28.7
MO 0812	Cattle, edible offal of	2	2.5	5.0	0.3	0.6	1.8	3.6	5.0	10.0	6.0	12.0
SO 0691	Cotton-seed	10	0.0	0.0	0.0	0.0	0.0	0.0	0.0	0.0	0.0	0.0
OC 0691	Cotton-seed oil, crude	0.05	3.8	0.2	0.5	0.0	0.5	0.0	0.5	0.0	0.0	0.0
OR 0691	Cotton-seed oil, edible	0.05	3.8	0.2	0.5	0.0	0.5	0.0	0.5	0.0	0.0	0.0
PE 0112	Eggs	0.1	14.6	1.5	13.1	1.3	3.7	0.4	11.9	1.2	37.6	3.8
FI 0341	Kiwi fruit	0.1	0.0	0.0	0.0	0.0	1.9	0.2	0.1	0.0	1.5	0.2
GC 0645	Maize (fresh, flour)	1	48.3	48.3	31.2	31.2	106.2	106.2	41.8	41.8	8.8	8.8
GC 0647	Oats	20	0.0	0.0	0.0	0.0	0.2	4.0	0.8	16.0	2.0	40.0
VD 0072	Peas (dry)	5	0.5	2.5	1.7	8.5	5.1	25.5	1.3	6.5	1.8	9.0
MO 0818	Pig, edible offal of	1	0.0	0.0	1.0	1.0	0.0	0.0	1.0	1.0	5.0	5.0
MM 0818	Pigmeat	0.1	0.0	0.0	27.2	2.7	2.6	0.3	10.5	1.1	75.8	7.6
PM 0110	Poultry meat	0.1	31.0	3.1	13.2	1.3	5.5	0.6	25.3	2.5	53.0	5.3
SO 0495	Rape-seed	10	0.0	0.0	0.0	0.0	0.0	0.0	0.0	0.0	0.0	0.0
GC 0649	Rice	0.1	48.8	4.9	279.3	27.9	103.4	10.3	86.5	8.7	11.8	1.2
GC 0651	Sorghum	20	2.0	40.0	9.7	194.0	26.6	532.0	0.0	0.0	0.0	0.0
VD 0541	Soya bean (dry)	20	4.5	90.0	2.0	40.0	0.5	10.0	0.0	0.0	0.0	0.0
VP 0541	Soya bean (immature seeds)	0.2	0.1	0.0	0.1	0.0	0.1	0.0	0.0	0.0	0.0	0.0
VO 0447	Sweet corn (corn-on-the-cob)	0.1	0.0	0.0	0.0	0.0	4.4	0.4	0.0	0.0	8.3	0.8
GC 0654	Wheat	5	4.3	21.5	0.8	4.0	0.0	0.0	4.8	24.0	2.2	11.0
CM 0654	Wheat bran, unprocessed	20	ND	–	ND	–	ND	–	ND	–	ND	–
CF 1211	Wheat flour	0.5	323.0	161.5	114.0	57.0	28.3	14.2	112.0	56.0	175.8	87.9
CF 1212	Wheat wholemeal	5	ND	–	ND	–	ND	–	ND	–	ND	–

	Mid-East	Far-East	African	Latin American	European
Total intake (μg/person)	412.7	451.8	748.3	343.7	630.5
Body weight per region (kg bw)	60	55	60	60	60
ADI (μg/person)	60 000	55 000	60 000	60 000	60 000
% ADI	0.7%	0.8%	1.2%	0.6%	1.1%
Rounded % ADI	1%	1%	1%	1%	1%

Annex 3

MALATHION (49)

International estimated daily intake

ADI = 0–0.3 mg/kg bw

Codex code	Commodity	STMR or STMR-P (mg/kg)	Mid-East Diet	Mid-East Intake	Far-East Diet	Far-East Intake	African Diet	African Intake	Latin American Diet	Latin American Intake	European Diet	European Intake
FP 0226	Apple	0.11	7.5	0.8	4.7	0.5	0.3	0.0	5.5	0.6	40.0	4.4
VS 0621	Asparagus	0.305	0.0	0.0	0.0	0.0	0.0	0.0	0.0	0.0	1.5	0.5
VD 0071	Beans (dry)	0.36	2.3	0.8	4.8	1.7	0.0	0.0	13.0	4.7	3.5	1.3
VP 0061	Beans except broad and soya bean (green pods and immature seeds)	0.31	3.9	1.2	0.9	0.3	0.0	0.0	4.4	1.4	13.2	4.1
FB 0020	Blueberry	2.27	0.0	0.0	0.0	0.0	0.0	0.0	0.0	0.0	0.5	1.1
FC 0001	Citrus fruit	0.02	47.1	0.9	6.3	0.1	5.1	0.1	54.6	1.1	44.6	0.9
OR 0691	Cotton-seed oil, edible	3.06	3.8	11.6	0.5	1.5	0.5	1.5	0.5	1.5	0.0	0.0
VC 0424	Cucumber	0.02	2.4	0.0	2.3	0.0	0.0	0.0	4.2	0.1	4.5	0.1
FB 0269	Grapes (fresh, wine, dried)	0.16	16.1	2.6	1.0	0.2	0.0	0.0	1.6	0.3	16.1	2.6
GC 0645	Maize (fresh, flour)	0.01	48.3	0.5	31.2	0.3	106.2	1.1	41.8	0.4	8.8	0.1
VL 0485	Mustard greens	0.07	0.1	0.0	0.1	0.0	0.1	0.0	0.1	0.0	0.1	0.0
VA 0385	Onion, bulb	0.23	23.0	5.3	11.5	2.6	7.3	1.7	13.8	3.2	27.8	6.4
VO 0051	Peppers	0.01	3.4	0.0	2.1	0.0	5.4	0.1	2.4	0.0	10.4	0.1
VL 0502	Spinach	0.35	0.5	0.2	0.0	0.0	0.0	0.0	0.3	0.1	2.0	0.7
VA 0389	Spring onion	0.52	0.0	0.0	2.0	1.0	1.5	0.8	4.0	2.1	1.0	0.5
FB 0275	Strawberry	0.25	0.0	0.0	0.0	0.0	0.0	0.0	0.0	0.0	5.3	1.3
VO 0447	Sweet corn (corn-on-the-cob)	0.01	0.0	0.0	0.0	0.0	4.4	0.0	0.0	0.0	8.3	0.1
VO 0448	Tomato (fresh)	0.25	44.1	11.0	5.7	1.4	14.6	3.7	25.5	6.4	34.9	8.7
JF 0448	Tomato juice	0	0.3	0.0	0.0	0.0	0.0	0.0	0.0	0.0	2.0	0.0
VR 0506	Turnip, garden	0.05	0.5	0.0	0.0	0.0	0.0	0.0	0.3	0.0	2.0	0.1
GC 0654	Wheat	0.04	327.3	13.1	114.8	4.6	28.3	1.1	116.8	4.7	178.0	7.1
	Total intake (μg/person)			48.2		14.4		10.1		26.5		40.1
	Body weight per region (kg bw)			60		55		60		60		60
	ADI (μg/person)			18 000		16 500		18 000		18 000		18 000
	% ADI			0.3%		0.1%		0.1%		0.1%		0.2%
	Rounded % ADI			0%		0%		0%		0%		0%

Annex 3

METALAXYL (138) and METALAXYL-M (212) Theoretical maximum daily intake ADI = 0–0.08 mg/kg bw

Codex code	Commodity	MRL (mg/kg)	Mid-East Diet	Mid-East Intake	Far-East Diet	Far-East Intake	African Diet	African Intake	Latin American Diet	Latin American Intake	European Diet	European Intake
VS 0621	Asparagus	0.05	0.0	0.0	0.0	0.0	0.0	0.0	0.0	0.0	1.5	0.1
FI 0326	Avocado	0.2	0.0	0.0	0.0	0.0	0.2	0.0	3.3	0.7	1.0	0.2
VB 0400	Broccoli	0.5	0.5	0.3	1.0	0.5	0.0	0.0	1.1	0.6	2.7	1.4
VB 0402	Brussels sprouts	0.2	0.5	0.1	1.0	0.2	0.0	0.0	1.1	0.2	2.7	0.5
VB 0403	Cabbage, Savoy	0.5	0.1	0.1	0.1	0.1	0.1	0.1	0.1	0.1	0.1	0.1
VB 0041	Cabbages, head (see Savoy cabbage)		ND	–	ND	–	ND	–	ND	–	ND	–
SB 0715	Cacao bean	0.2	0.5	0.1	0.0	0.0	0.0	0.0	1.3	0.3	3.1	0.6
VR 0577	Carrot	0.05	2.8	0.1	2.5	0.1	0.0	0.0	6.3	0.3	22.0	1.1
VB 0404	Cauliflower	0.5	1.3	0.7	1.5	0.8	0	0.0	0.3	0.2	13	6.5
GC 0080	Cereal grain	0.05	429.9	21.5	450.8	22.5	318.3	15.9	252.4	12.6	221.9	11.1
FC 0001	Citrus fruit	5	47.1	235.5	6.3	31.5	5.1	25.5	54.6	273.0	44.6	223.0
SO 0691	Cotton-seed	0.05	0.0	0.0	0.0	0.0	0.0	0.0	0.0	0.0	0.0	0.0
VC 0424	Cucumber	0.5	2.4	1.2	2.3	1.1	0.0	0.0	4.2	2.1	4.5	2.3
VC 0425	Gherkin	0.5	2.4	1.2	2.3	1.1	0.0	0.0	4.2	2.1	4.5	2.3
FB 0269	Grapes (fresh, wine, dried)	1	16.1	16.1	1.0	1.0	0.0	0.0	1.6	1.6	16.1	16.1
DH 1100	Hops, dry	10	0.1	1.0	0.1	1.0	0.1	1.0	0.1	1.0	0.1	1.0
VL 0482	Lettuce, head	2	2.3	4.6	0.0	0.0	0.0	0.0	5.8	11.6	22.5	45.0
VC 0046	Melon, except watermelon	0.2	16.0	3.2	2.0	0.4	0.0	0.0	2.8	0.6	18.3	3.7
VA 0385	Onion, bulb	2	23.0	46.0	11.5	23.0	7.3	14.6	13.8	27.6	27.8	55.6
SO 0697	Peanut	0.1	0.3	0.0	0.2	0.0	2.3	0.2	0.3	0.0	3.0	0.3
VP 0064	Peas, shelled (immature seeds)	0.05	4.0	0.2	0.5	0.0	0.0	0.0	0.2	0.0	10.1	0.5
VO 0051	Peppers	1	3.4	3.4	2.1	2.1	5.4	5.4	2.4	2.4	10.4	10.4
FP 0009	Pome fruits	1	10.8	10.8	7.5	7.5	0.3	0.3	6.5	6.5	51.3	51.3
VR 0589	Potato	0.05	59.0	3.0	19.2	1.0	20.6	1.0	40.8	2.0	240.8	12.0
FB 0272	Raspberry, red, black	0.2	0.0	0.0	0.0	0.0	0.0	0.0	0.0	0.0	0.5	0.1
VD 0541	Soya bean (dry)	0.05	4.5	0.2	2.0	0.1	0.5	0.0	0.0	0.0	0.0	0.0
VL 0502	Spinach	2	0.5	1.0	0.0	0.0	0.0	0.0	0.3	0.6	2.0	4.0
VC 0431	Squash, summer	0.2	10.5	2.1	2.2	0.4	0.0	0.0	14.0	2.8	3.5	0.7
VR 0596	Sugar-beet	0.05	0.5	0.0	0.0	0.0	0.0	0.0	0.3	0.0	2.0	0.1
OC 0702	Sunflower seed oil, crude	0.05	9.3	0.5	0.5	0.0	0.3	0.0	0.8	0.0	8.5	0.4
SO 0702	Sunflower seed, consumed fresh	0.05	1.0	0.1	0.0	0.0	0.6	0.0	0.0	0.0	0.0	0.0
VO 0448	Tomato (fresh, juice, paste, peeled)	0.5	81.5	40.8	7.0	3.5	16.5	8.3	25.5	12.8	66.6	33.3

Annex 3

Codex code	Commodity	MRL (mg/kg)	Mid-East Diet	Mid-East Intake	Far-East Diet	Far-East Intake	African Diet	African Intake	Latin American Diet	Latin American Intake	European Diet	European Intake
VC 0432	Watermelon	0.2	49.3	9.9	9.5	1.9	0.0	0.0	5.5	1.1	7.8	1.6
VC 0433	Winter squash	0.2	1.5	0.3	0.3	0.1	0.0	0.0	2.0	0.4	0.5	0.1
	Total intake (µg/person)			403.8		99.9		72.4		363.0		485.2
	Body weight per region (kg bw)			60		55		60		60		60
	ADI (µg/person)			4800		4400		4800		4800		4800
	% ADI			8.4%		2.3%		1.5%		7.6%		10.1%
	Rounded % ADI			8%		2%		2%		8%		10%

METHOMYL (94)

International estimated daily intake ADI = 0–0.02 mg/kg bw

Codex code	Commodity	STMR or STMR-P (mg/kg)	Mid-East Diet	Mid-East Intake	Far-East Diet	Far-East Intake	African Diet	African Intake	Latin American Diet	Latin American Intake	European Diet	European Intake
FP 0226	Apple	0.41	7.5	3.1	4.7	1.9	0.3	0.1	5.5	2.3	40.0	16.4
VS 0621	Asparagus	0.33	0.0	0.0	0.0	0.0	0.0	0.0	0.0	0.0	1.5	0.5
GC 0640	Barley (fresh)	2[a]	1.0	2.0	3.5	7.0	1.8	3.6	6.5	13.0	19.8	39.6
–	Beans (dry), including broad beans	0.02	6.8	0.1	6.8	0.1	0.0	0.0	13.5	0.3	4.3	0.1
VP 0061	Beans except broad bean and soya bean (green pods and immature seeds)	0.055	3.9	0.2	0.9	0.0	0.0	0.0	4.4	0.2	13.2	0.7
VB 0040	Brassica vegetables (flowerhead, head and leafy Brassicas, kohlrabi)	1.3	6.3	8.2	11.2	14.6	0.0	0.0	10.8	14.0	39.8	51.7
VS 0624	Celery	0.66	0.5	0.3	0.0	0.0	0.0	0.0	0.3	0.2	2.0	1.3
FC 0001	Citrus fruit	0.034	47.1	1.6	6.3	0.2	5.1	0.2	54.6	1.9	44.6	1.5
OR 0691	Cotton-seed oil, edible	0.006	3.8	0.0	0.5	0.0	0.5	0.0	0.5	0.0	0.0	0.0
MO 0105	Edible offal (mammalian)	0	4.2	0.0	1.4	0.0	2.8	0.0	6.1	0.0	12.4	0.0
PE 0840	Chicken eggs	0	14.5	0.0	13.0	0.0	3.6	0.0	11.8	0.0	37.5	0.0
VC 0045	Fruiting vegetables, cucurbits	0.02	80.5	1.6	18.2	0.4	0.0	0.0	30.5	0.6	38.5	0.8
FB 0269	Grapes (fresh, wine, excluding dried grapes)	0.86	15.8	13.6	1.0	0.9	0.0	0.0	1.3	1.1	13.8	11.9
VL 0053	Leafy vegetables	1.4	7.8	10.9	9.7	13.6	0.7	1.0	16.5	23.1	51.7	72.4
GC 0645	Maize (fresh, flour)	0.02	48.3	1.0	31.2	0.6	106.2	2.1	41.8	0.8	8.8	0.2
OR 0645	Maize oil, edible	0.004	1.8	0.0	0.0	0.0	0.3	0.0	0.5	0.0	1.3	0.0
MM 0095	Meat from mammals other than marine mammals	0	37.0	0.0	32.8	0.0	23.8	0.0	47.0	0.0	155.5	0.0
ML 0106	Milks	0	116.9	0.0	32.1	0.0	41.8	0.0	160.1	0.0	289.3	0.0
FS 0245	Nectarine	0.05	1.3	0.1	0.3	0.0	0.0	0.0	0.4	0.0	6.3	0.3
GC 0647	Oats	0.02[a]	0.0	0.0	0.0	0.0	0.2	0.0	0.8	0.0	2.0	0.0
VA 0385	Onion, bulb	0.068	23.0	1.6	11.5	0.8	7.3	0.5	13.8	0.9	27.8	1.9

Annex 3

Codex code	Commodity	STMR or STMR-P (mg/kg)	Mid-East Diet	Mid-East Intake	Far-East Diet	Far-East Intake	African Diet	African Intake	Latin American Diet	Latin American Intake	European Diet	European Intake
JF 0004	Orange juice	0.004	7.3	0.0	0.0	0.0	0.0	0.0	0.3	0.0	4.5	0.0
FS 0247	Peach	0.05	1.3	0.1	0.3	0.0	0.0	0.0	0.4	0.0	6.3	0.3
FP 0230	Pear	0.09	3.3	0.3	2.8	0.3	0.0	0.0	1.0	0.1	11.3	1.0
VP 0063	Peas (green pods and immature seeds)	0.46	5.5	2.5	2.0	0.9	0.0	0.0	0.8	0.4	14.0	6.4
VO 0051	Peppers	0.105	3.4	0.4	2.1	0.2	5.4	0.6	2.4	0.3	10.4	1.1
FS 0014	Plum (fresh, prune)	0.08	1.8	0.1	0.5	0.0	0.0	0.0	0.0	0.0	4.3	0.3
VR 0589	Potato	0	59.0	0.0	19.2	0.0	20.6	0.0	40.8	0.0	240.8	0.0
PO 0111	Poultry, edible offal of	0	0.1	0.0	0.1	0.0	0.1	0.0	0.4	0.0	0.4	0.0
PM 0110	Poultry meat	0	31.0	0.0	13.2	0.0	5.5	0.0	25.3	0.0	53.0	0.0
VD 0541	Soya bean (dry)	0.04	4.5	0.2	2.0	0.1	0.5	0.0	0.0	0.0	0.0	0.0
OR 0541	Soya bean oil, refined	0.04	1.3	0.1	1.7	0.1	3.0	0.1	14.5	0.6	4.3	0.2
VO 0447	Sweet corn (corn-on-the-cob)	0.065	0.0	0.0	0.0	0.0	4.4	0.3	0.0	0.0	8.3	0.5
VO 0448	Tomato (fresh)	0.16	44.1	7.1	5.7	0.9	14.6	2.3	25.5	4.1	34.9	5.6
	Tomato paste	0.007	5.8	0.0	0.2	0.0	0.3	0.0	0.0	0.0	4.0	0.0
CM 0654	Wheat bran, unprocessed	0.27	ND	-	ND	-	ND	-	ND	-	ND	-
CF 1211	Wheat flour	0.003	323.0	1.0	114.0	0.3	28.3	0.1	112.0	0.3	175.8	0.5
CF 1210	Wheat germ	0.13	0.1	0.0	0.1	0.0	0.0	0.0	0.1	0.0	0.1	0.0
	Total intake (µg/person)			56.0		43.0		10.9		64.2		215.4
	Body weight per region (kg bw)			60		55		60		60		60
	ADI (µg/person)			1200		1100		1200		1200		1200
	% ADI			4.7%		3.9%		0.9%		5.4%		18.0%
	Rounded % ADI			5%		4%		1%		5%		20%

Annex 3

OXYDEMETON METHYL International estimated daily intake ADI = 0–0.02 mg/kg bw

Codex code	Commodity	STMR or STMR-P (mg/kg)	Mid-East Diet	Mid-East Intake	Far East Diet	Far East Intake	African Diet	African Intake	Latin American Diet	Latin American Intake	European Diet	European Intake
FC 0204	Lemon	0.01	1.9	0.0	0.2	0.0	0.0	0.0	5.4	0.1	2.4	0.0
FC 0004	Orange, sweet, sour (including orange-like hybrids)	0.01	31.5	0.3	4.0	0.0	4.8	0.0	31.0	0.3	29.8	0.3
FP 0226	Apple	0.01	7.5	0.1	4.7	0.0	0.3	0.0	5.5	0.1	40.0	0.4
JF 0226	Apple juice	0.01	4.5	0.0	0	0.0	0	0.0	0.3	0.0	3.8	0.0
FP 0230	Pear	0.01	3.3	0.0	2.8	0.0	0.0	0.0	1.0	0.0	11.3	0.1
FB 0269	Grapes (fresh, wine, dried)	0.04	16.1	0.6	1.0	0.0	0.0	0.0	1.6	0.1	16.1	0.6
VR 0589	Potato	0.01	59.0	0.6	19.2	0.2	20.6	0.2	40.8	0.4	240.8	2.4
VR 0596	Sugar-beet	0.01	0.5	0.0	0.0	0.0	0.0	0.0	0.3	0.0	2.0	0.0
VB 0403	Cabbage, Savoy	0.03	0.1	0.0	0.1	0.0	0.1	0.0	0.1	0.0	0.1	0.0
VB 0404	Cauliflower	0.01	1.3	0.0	1.5	0.0	0	0.0	0.3	0.0	13	0.1
VB 0405	Kohlrabi	0.02	0.1	0.0	0.1	0.0	0.1	0.0	0.1	0.0	0.1	0.0
VL 0480	Kale	0.01	0.5	0.0	0.0	0.0	0.0	0.0	0.3	0.0	2.0	0.0
VD 0526	Common bean (dry)	0.01	0.1	0.0	0.1	0.0	0.1	0.0	0.1	0.0	0.1	0.0
SO 0691	Cotton-seed	0.01	0.0	0.0	0.0	0.0	0.0	0.0	0.0	0.0	0.0	0.0
OR 0691	Cotton-seed oil, edible	0.002	3.8	0.0	0.5	0.0	0.5	0.0	0.5	0.0	0.0	0.0
GC 0640	Barley (fresh)	0.01	1.0	0.0	3.5	0.0	1.8	0.0	6.5	0.1	19.8	0.2
GC 0650	Rye	0.01	0.0	0.0	1.0	0.0	0.0	0.0	0.0	0.0	1.5	0.0
GC 0654	Wheat	0.01	327.3	3.3	114.8	1.1	28.3	0.3	116.8	1.2	178.0	1.8
MM 0097	Meat of cattle, pigs and sheep	0	28.1	0.0	30.6	0.0	15.0	0.0	43.5	0.0	149.4	0.0
MF 0812	Cattle fat	0	0.3	0.0	0.3	0.0	0.3	0.0	1.5	0.0	0.0	0.0
MF 0818	Pig fat	0	ND	–	ND	–	ND	–	ND	–	ND	–
MF 0822	Sheep fat	0	0.1	0.0	0.1	0.0	0.1	0.0	0.1	0.0	0.1	0.0
PM 0110	Poultry meat	0	31.0	0.0	13.2	0.0	5.5	0.0	25.3	0.0	53.0	0.0
PF 0111	Poultry, fat	0	3.1	0.0	1.3	0.0	0.6	0.0	2.5	0.0	5.3	0.0
ML 0106	Milks	0	116.9	0.0	32.1	0.0	41.8	0.0	160.1	0.0	289.3	0.0
PE 0112	Eggs	0	14.6	0.0	13.1	0.0	3.7	0.0	11.9	0.0	37.6	0.0
	Total intake (µg/person)			5.0		1.6		0.6		2.2		6.1
	Body weight per region (kg bw)			60		55		60		60		60
	ADI (µg/person)			18		16.5		18		18		18
	% ADI			28.0%		9.5%		3.1%		11.9%		33.8%
	Rounded % ADI			30%		9%		3%		10%		30%

Annex 3

PARAQUAT (057) International estimated daily intake ADI = 0–0.005 mg/kg bw

Codex code	Commodity	STMR or STMR-P (mg/kg)	Mid-East Diet	Mid-East Intake	Far-East Diet	Far-East Intake	African Diet	African Intake	Latin American Diet	Latin American Intake	European Diet	European Intake
FI 0326	Avocado	0.01	0.0	0.0	0.0	0.0	0.2	0.0	3.3	0.0	1.0	0.0
FI 0327	Banana	0.01	8.3	0.1	26.2	0.3	21.0	0.2	102.3	1.0	22.8	0.2
FB 0018	Berry and other small fruits	0	16.0	0.0	1.0	0.0	0.0	0.0	1.6	0.0	23.5	0.0
FC 0001	Citrus fruit	0.01	47.1	0.5	6.3	0.1	5.1	0.1	54.6	0.5	44.6	0.4
OC 0691	Cotton-seed oil, crude	0.01	3.8	0.0	0.5	0.0	0.5	0.0	0.5	0.0	0.0	0.0
MO 0105	Edible offal (mammalian)	0.0007	4.2	0.0	1.4	0.0	2.8	0.0	6.1	0.0	12.4	0.0
PE 0112	Eggs	0	14.6	0.0	13.1	0.0	3.7	0.0	11.9	0.0	37.6	0.0
VO 0050	Fruiting vegetables other than cucurbits	0.01	92.3	0.9	12.6	0.1	27.0	0.3	33.9	0.3	91.6	0.9
VC 0045	Fruiting vegetables, cucurbits	0	80.5	0.0	18.2	0.0	0.0	0.0	30.5	0.0	38.5	0.0
DH 1100	Hops, dry	0.05	0.1	0.0	0.1	0.0	0.1	0.0	0.1	0.0	0.1	0.0
FI 0341	Kiwi fruit	0.01	0.0	0.0	0.0	0.0	1.9	0.0	0.1	0.0	1.5	0.0
VL 0053	Leafy vegetables	0.025	7.8	0.2	9.7	0.2	0.7	0.0	16.5	0.4	51.7	1.3
CF 1255	Maize flour	0.038	31.8	1.2	31.2	1.2	106.2	4.0	40.3	1.5	8.8	0.3
GC 0645	Maize (fresh)	0.025	16.5	0.4	0.0	0.0	0.0	0.0	1.5	0.0	0.0	0.0
OC 0645	Maize oil, crude	0.006	1.8	0.0	0.0	0.0	0.3	0.0	0.5	0.0	1.3	0.0
FI 0345	Mango	0.01	2.3	0.0	5.3	0.1	3.4	0.0	6.3	0.1	0.0	0.0
MM 0095	Meat from mammals other than marine mammals	0.0001	37.0	0.0	32.8	0.0	23.8	0.0	47.0	0.0	155.5	0.0
ML 0106	Milks	0.00008	116.9	0.0	32.1	0.0	41.8	0.0	160.1	0.0	289.3	0.0
OC 0305	Olive oil, crude	0.018	1.5	0.0	0.0	0.0	0.0	0.0	0.0	0.0	7.8	0.1
FT 0305	Olives	0.05	1.3	0.1	0.0	0.0	0.0	0.0	0.3	0.0	2.8	0.1
FI 0350	Papaya	0.01	0.0	0.0	0.2	0.0	0.0	0.0	5.3	0.1	0.0	0.0
FI 0351	Passion fruit	0.01	0.0	0.0	0.0	0.0	1.9	0.0	0.1	0.0	1.5	0.0
FI 0353	Pineapple (fresh, canned)	0.01	0.8	0.0	10.2	0.1	3.1	0.0	15.8	0.2	3.3	0.0
FI 0354	Plantain	0.01	0.0	0.0	0.0	0.0	41.3	0.4	56.5	0.6	0.0	0.0
FP 0009	Pome fruits	0	10.8	0.0	7.5	0.0	0.3	0.0	6.5	0.0	51.3	0.0
PM 0110	Poultry meat	0	31.0	0.0	13.2	0.0	5.5	0.0	25.3	0.0	53.0	0.0
PO 0111	Poultry, edible offal of	0	0.1	0.0	0.1	0.0	0.1	0.0	0.4	0.0	0.4	0.0
VD 0070	Pulses	0.1	18.9	1.9	14.5	1.5	17.5	1.8	20.3	2.0	9.4	0.9
VR 0075	Roots and tubers	0.02	61.8	1.2	108.5	2.2	321.3	6.4	159.3	3.2	242.0	4.8
GC 0651	Sorghum	0.025	2.0	0.1	9.7	0.2	26.6	0.7	0.0	0.0	0.0	0.0
OC 0541	Soya bean oil, crude	0.01	1.3	0.0	1.7	0.0	3.0	0.0	14.5	0.1	4.3	0.0
FS 0012	Stone fruits	0	7.3	0.0	1.0	0.0	0.0	0.0	0.8	0.0	23.3	0.0
OC 0702	Sunflower seed oil, crude	0	9.3	0.0	0.5	0.0	0.3	0.0	0.8	0.0	8.5	0.0
SO 0702	Sunflower seed, consumed fresh	0.22	1.0	0.2	0.0	0.0	0.6	0.1	0.0	0.0	0.0	0.0

Annex 3

Codex code	Commodity	STMR or STMR-P (mg/kg)	Mid-East Diet	Mid-East Intake	Far-East Diet	Far-East Intake	African Diet	African Intake	Latin American Diet	Latin American Intake	European Diet	European Intake
DT 1114	Tea, green, black (black, fermented and dried)	0.06	2.3	0.1	1.2	0.1	0.5	0.0	0.5	0.0	2.3	0.1
TN 0085	Tree nuts	0.01	1.1	0.0	13.5	0.1	4.5	0.0	17.8	0.2	4.6	0.0
	Total intake (µg/person)			7.0		6.1		14.2		10.4		9.6
	Body weight per region (kg bw)			60		55		60		60		60
	ADI (µg/person)			300		275		300		300		300
	% ADI			2.3%		2.2%		4.7%		3.5%		3.2%
	Rounded % ADI			2%		2%		5%		3%		3%

PHORATE (112)

International estimated daily intake   ADI = 0–0.0007 mg/kg bw

Codex code	Commodity	MRL (mg/kg)	Mid-East Diet	Mid-East Intake	Far-East Diet	Far-East Intake	African Diet	African Intake	Latin American Diet	Latin American Intake	European Diet	European Intake
VP 0526	Common bean (green pods or immature seeds)	0.1	3.5	0.4	0.8	0.1	0.0	0.0	4.0	0.4	12.0	1.2
SO 0691	Cotton-seed	0.05	0.0	0.0	0.0	0.0	0.0	0.0	0.0	0.0	0.0	0.0
PE 0112	Eggs	0.05	14.6	0.7	13.1	0.7	3.7	0.2	11.9	0.6	37.6	1.9
GC 0645	Maize (fresh, flour)	0.05	48.3	2.4	31.2	1.6	106.2	5.3	41.8	2.1	8.8	0.4
MM 0095	Meat from mammals other than marine mammals	0.05	37.0	1.9	32.8	1.6	23.8	1.2	47.0	2.4	155.5	7.8
ML 0106	Milks	0.05	116.9	5.8	32.1	1.6	41.8	2.1	160.1	8.0	289.3	14.5
SO 0697	Peanut	0.1	0.3	0.0	0.2	0.0	2.3	0.2	0.3	0.0	3.0	0.3
OC 0697	Peanut oil, crude	0.05	0.0	0.0	1.8	0.1	3.5	0.2	0.5	0.0	1.8	0.1
OR 0697	Peanut oil, edible	0.05	0.0	0.0	1.8	0.1	3.5	0.2	0.5	0.0	1.8	0.1
VR 0589	Potato	0.2	59.0	11.8	19.2	3.8	20.6	4.1	40.8	8.2	240.8	48.2
GC 0651	Sorghum	0.05	2.0	0.1	9.7	0.5	26.6	1.3	0.0	0.0	0.0	0.0
VP 0541	Soya bean (immature seeds)	0.05	0.1	0.0	0.1	0.0	0.1	0.0	0.0	0.0	0.0	0.0
VR 0596	Sugar-beet	0.05	0.5	0.0	0.0	0.0	0.0	0.0	0.3	0.0	2.0	0.1
VO 0447	Sweet corn (corn-on-the-cob)	0.05	0.0	0.0	0.0	0.0	4.4	0.2	0.0	0.0	8.3	0.4
GC 0654	Wheat	0.05	327.3	16.4	114.8	5.7	28.3	1.4	116.8	5.8	178.0	8.9
	Total intake (µg/person)			39.5		15.8		16.4		27.5		83.8
	Body weight per region (kg bw)			60		55		60		60		60
	ADI (µg/person)			42		38.5		42		42		42
	% ADI			94.1%		41.1%		39.2%		65.6%		199.6%
	Rounded % ADI			90%		40%		40%		70%		200%

Annex 3

PIRIMICARB (101)

Theoretical maximum daily intake

ADI = 0–0.02 mg/kg bw

Codex code	Commodity	MRL (mg/kg)	Mid-East Diet	Mid-East Intake	Far-East Diet	Far-East Intake	African Diet	African Intake	Latin American Diet	Latin American Intake	European Diet	European Intake
FC 0001	Citrus fruit	0.05	47.1	2.4	6.3	0.3	5.1	0.3	54.6	2.7	44.6	2.2
GC 0640	Barley (fresh)	0.05	1.0	0.1	3.5	0.2	1.8	0.1	6.5	0.3	19.8	1.0
VP 0062	Beans, shelled (immature seeds)	0.1	0.1	0.0	0.1	0.0	0.1	0.0	0.1	0.0	0.1	0.0
VR 0574	Beetroot	0.05	0.5	0.0	0.0	0.0	0.0	0.0	0.3	0.0	2.0	0.1
VB 0400	Broccoli	1	0.5	0.5	1.0	1.0	0.0	0.0	1.1	1.1	2.7	2.7
VB 0402	Brussels sprouts	1	0.5	0.5	1.0	1.0	0.0	0.0	1.1	1.1	2.7	2.7
VB 0403	Cabbage, Savoy	1	0.1	0.1	0.1	0.1	0.1	0.1	0.1	0.1	0.1	0.1
VB 0041	Cabbage, head (see Savoy cabbage)		ND	–	ND	–	ND	–	ND	–	ND	–
VB 0404	Cauliflower	1	1.3	1.3	1.5	1.5	0	0.0	0.3	0.3	13	13.0
VS 0624	Celery	1	0.5	0.5	0.0	0.0	0.0	0.0	0.3	0.3	2.0	2.0
VP 0526	Common bean (green pods or immature seeds)	1	3.5	3.5	0.8	0.8	0.0	0.0	4.0	4.0	12.0	12.0
SO 0691	Cotton-seed	0.05	0.0	0.0	0.0	0.0	0.0	0.0	0.0	0.0	0.0	0.0
VC 0424	Cucumber	1	2.4	2.4	2.3	2.3	0.0	0.0	4.2	4.2	4.5	4.5
FB 0278	Currant, black	0.5	0.0	0.0	0.0	0.0	0.0	0.0	0.0	0.0	0.0	0.0
VO 0440	Eggplant	1	6.3	6.3	3.0	3.0	0.7	0.7	6.0	6.0	2.3	2.3
PE 0112	Eggs	0.05	14.6	0.7	13.1	0.7	3.7	0.2	11.9	0.6	37.6	1.9
VL 0476	Endive	1	0.5	0.5	0.0	0.0	0.0	0.0	0.3	0.3	2.0	2.0
VC 0425	Gherkin	1	2.4	2.4	2.3	2.3	0.0	0.0	4.2	4.2	4.5	4.5
VB 0405	Kohlrabi	0.5	0.1	0.1	0.1	0.1	0.1	0.1	0.1	0.1	0.1	0.1
VA 0384	Leek	0.5	0.5	0.3	0.0	0.0	0.0	0.0	0.3	0.2	2.0	1.0
VL 0482	Lettuce, head	1	2.3	2.3	0.0	0.0	0.0	0.0	5.8	5.8	22.5	22.5
MM 0095	Meat from mammals other than marine mammals	0.05	37.0	1.9	32.8	1.6	23.8	1.2	47.0	2.4	155.5	7.8
ML 0106	Milks	0.05	116.9	5.8	32.1	1.6	41.8	2.1	160.1	8.0	289.3	14.5
GC 0647	Oats	0.05	0.0	0.0	0.0	0.0	0.2	0.0	0.8	0.0	2.0	0.1
VA 0385	Onion, bulb	0.5	23.0	11.5	11.5	5.8	7.3	3.7	13.8	6.9	27.8	13.9
FC 0004	Orange, sweet, sour (including orange-like hybrids)	0.5	31.5	15.8	4.0	2.0	4.8	2.4	31.0	15.5	29.8	14.9
HH 0740	Parsley	1	0.1	0.1	0.1	0.1	0.1	0.1	0.1	0.1	0.1	0.1
VR 0588	Parsnip	0.05	0.5	0.0	0.0	0.0	0.0	0.0	0.3	0.0	2.0	0.1
FS 0247	Peach	0.5	1.3	0.6	0.3	0.1	0.0	0.0	0.4	0.2	6.3	3.1
VP 0063	Peas (green pods and immature seeds)	0.2	5.5	1.1	2.0	0.4	0.0	0.0	0.8	0.2	14.0	2.8
TN 0672	Pecan	0.05	0.0	0.0	0.0	0.0	0.1	0.0	0.1	0.0	0.3	0.0
VO 0444	Peppers, chili	2	0.1	0.2	0.1	0.2	0.1	0.2	0.1	0.2	0.1	0.2
VO 0445	Peppers, sweet (including pim(i)ento)	1	3.3	3.3	2.0	2.0	5.3	5.3	2.3	2.3	10.3	10.3
FS 0014	Plum (fresh, prunes)	0.5	1.8	0.9	0.5	0.3	0.0	0.0	0.0	0.0	4.3	2.2
FP 0009	Pome fruit	1	10.8	10.8	7.5	7.5	0.3	0.3	6.5	6.5	51.3	51.3
VR 0589	Potato	0.05	59.0	3.0	19.2	1.0	20.6	1.0	40.8	2.0	240.8	12.0

317

Annex 3

VR 0494	Radish	0.05	0.5	0.0	0.0	0.0	0.3	0.0	2.0	0.1
SO 0495	Rape-seed	0.2	0.0	0.0	0.0	0.0	0.0	0.0	0.0	0.0
FB 0272	Raspberry, red, black	0.5	0.0	0.0	0.0	0.0	0.0	0.0	0.5	0.3
VL 0502	Spinach	1	0.5	0.5	0.0	0.0	0.3	0.3	2.0	2.0
FB 0275	Strawberry	0.5	0.0	0.0	0.0	0.0	0.0	0.0	5.3	2.7
VR 0596	Sugar-beet	0.05	0.0	0.0	0.0	0.0	0.3	0.0	2.0	0.1
VO 0447	Sweet corn (corn-on-the-cob)	0.05	0.0	0.0	0.0	0.2	0.0	0.0	8.3	0.4
VO 0448	Tomato (fresh, juice, paste, peeled)	1	81.5	81.5	7.0	16.5	25.5	25.5	66.6	66.6
VR 0506	Turnip, garden	0.05	0.5	0.0	0.0	0.0	0.3	0.0	2.0	0.1
VL 0473	Watercress	1	0.1	0.1	0.1	0.1	0.1	0.1	0.1	0.1
GC 0654	Wheat	0.05	327.3	16.4	114.8	5.7	28.3	1.4	116.8	5.8	178.0	8.9

Total intake (μg/person)	177.3	48.5	35.9		107.3	289.1
Body weight per region (kg bw)	60	55	60		60	60
ADI (μg/person)	1200	1100	1200		1200	1200
% ADI	14.8%	4.4%	3.0%		8.9%	24.1%
Rounded % ADI	10%	4%	3%		9%	20%

Annex 3

PROCHLORAZ (142) International estimated daily intake ADI = 0–0.01 mg/kg bw

Codex code	Commodity	STMR or STMR-P (mg/kg)	Mid-East Diet	Mid-East Intake	Far-East Diet	Far-East Intake	African Diet	African Intake	Latin American Diet	Latin American Intake	European Diet	European Intake
FC 0001	Citrus fruit	0.1	47.1	4.7	6.3	0.6	5.1	0.5	54.6	5.5	44.6	4.5
FI 0326	Avocado	0.1	0.0	0.0	0.0	0.0	0.2	0.0	3.3	0.3	1.0	0.1
FI 0327	Banana	0.1	8.3	0.8	26.2	2.6	21.0	2.1	102.3	10.2	22.8	2.3
GC 0640	Barley (beer only)	0.01	ND	–	ND	–	ND	–	ND	–	ND	–
GC 0080	Cereal grain	0.11	0.0	0.0	298.8	32.9	280.6	30.9	65.7	7.2	0.0	0.0
FI 0332	Custard apple	0.1	ND	–	ND	–	ND	–	ND	–	ND	–
MO 0105	Edible offal (mammalian)	1.2	4.2	5.0	1.4	1.7	2.8	3.4	6.1	7.3	12.4	14.9
PE 0112	Eggs	0.012	14.6	0.2	13.1	0.2	3.7	0.0	11.9	0.1	37.6	0.5
FI 0335	Feijoa	0.1	ND	–	ND	–	ND	–	ND	–	ND	–
FI 0336	Guava	0.1	ND	–	ND	–	ND	–	ND	–	ND	–
FI 0338	Jackfruit	0.1	ND	–	ND	–	ND	–	ND	–	ND	–
FI 0341	Kiwi fruit	0.1	0.0	0.0	0.0	0.0	1.9	0.2	0.1	0.0	1.5	0.2
SO 0693	Linseed	0.05	0.0	0.0	0.0	0.0	0.0	0.0	0.0	0.0	0.0	0.0
FI 0345	Mango	0.1	2.3	0.2	5.3	0.5	3.4	0.3	6.3	0.6	0.0	0.0
MM 0095	Meat from mammals other than marine mammals: 20% as fat	0.06	7.4	0.4	6.6	0.4	4.8	0.3	9.4	0.6	31.1	1.9
		0.02	29.6	0.6	26.2	0.5	19.0	0.4	37.6	0.8	124.4	2.5
ML 0106	Milks	0	116.9	0.0	32.1	0.0	41.8	0.0	160.1	0.0	289.3	0.0
VO 0450	Mushroom	4.9	0.3	1.5	0.5	2.5	0.0	0.0	0.0	0.0	4.0	19.6
FI 0350	Papaya	0.1	0.0	0.0	0.2	0.0	0.0	0.0	5.3	0.5	0.0	0.0
FI 0351	Passion fruit	0.1	0.0	0.0	0.0	0.0	1.9	0.2	0.1	0.0	1.5	0.2
HS 0790	Pepper (black, white)	5.1	0.3	1.5	0.0	0.0	0.0	0.0	0.0	0.0	0.0	0.0
FI 0352	Persimmon, American	0.1	ND	–	ND	–	ND	–	ND	–	ND	–
FI 0353	Pineapple (fresh, canned)	0.1	0.8	0.1	10.2	1.0	3.1	0.3	15.8	1.6	3.3	0.3
FI 0354	Plantain	0.1	0.0	0.0	0.0	0.0	41.3	4.1	56.5	5.7	0.0	0.0
PM 0110	Poultry meat: 10% as fat	0.001	3.1	0.0	1.3	0.0	0.6	0.0	2.5	0.0	5.3	0.0
PM 0110	Poultry meat: 90% as muscle	0.001	27.9	0.0	11.9	0.0	5.0	0.0	22.8	0.0	47.7	0.0
PO 0111	Poultry, edible offal of	0.015	0.1	0.0	0.1	0.0	0.1	0.0	0.4	0.0	0.4	0.0
FI 0358	Rambutan	0.1	ND	–	ND	–	ND	–	ND	–	ND	–
SO 0495	Rape-seed	0.1	0.0	0.0	0.0	0.0	0.0	0.0	0.0	0.0	0.0	0.0
OR 0495	Rape-seed oil, edible	0.06	4.5	0.3	2.7	0.2	0.3	0.0	0.3	0.0	7.3	0.4
OR 0702	Sunflower seed oil, edible	0.06	9.3	0.6	0.5	0.0	0.3	0.0	0.8	0.0	8.5	0.5
SO 0702	Sunflower seed, consumed fresh	0.1	1.0	0.1	0.0	0.0	0.6	0.1	0.0	0.0	0.0	0.0
CM 0654	Wheat bran, unprocessed	0.54	ND	–	ND	–	ND	–	ND	–	ND	–

Annex 3

Codex code	Commodity	STMR or STMR-P (mg/kg)	Mid-East Diet	Mid-East Intake	Far-East Diet	Far-East Intake	African Diet	African Intake	Latin American Diet	Latin American Intake	European Diet	European Intake
CF 1211	Wheat flour	0.025	323.0	8.1	114.0	2.9	28.3	0.7	112.0	2.8	175.8	4.4
CP 1212	Wholemeal bread	0.14	107.7	15.1	38.0	5.3	9.4	1.3	74.7	10.5	58.6	8.2
	Total intake (µg/person)			39.2		51.3		44.8		53.8		60.4
	Body weight per region (kg bw)			60		55		60		60		60
	ADI (µg/person)			600		550		600		600		600
	% ADI			6.5%		9.3%		7.5%		9.0%		10.1%
	Rounded % ADI			7%		9%		7%		9%		10%

Annex 3

PROPICONAZOLE (160) International estimated daily intake ADI = 0–0.0700 mg/kg bw

Codex code	Commodity	STMR or STMR-P (mg/kg)	Mid-East Diet	Mid-East Intake	Far-East Diet	Far-East Intake	African Diet	African Intake	Latin American Diet	Latin American Intake	European Diet	European Intake
TN 0660	Almond	0.05	0.5	0.0	0.0	0.0	0.0	0.0	0.1	0.0	1.8	0.1
FI 0327	Banana	0.1	8.3	0.8	26.2	2.6	21.0	2.1	102.3	10.2	22.8	2.3
GC 0640	Barley (fresh)	0.05	1.0	0.1	3.5	0.2	1.8	0.1	6.5	0.3	19.8	1.0
SB 0716	Coffee bean	0.1	5.3	0.5	0.4	0.0	0.0	0.0	3.6	0.4	7.9	0.8
MO 0105	Edible offal (mammalian)	0.05	4.2	0.2	1.4	0.1	2.8	0.1	6.1	0.3	12.4	0.6
PE 0112	Eggs	0.05	14.6	0.7	13.1	0.7	3.7	0.2	11.9	0.6	37.6	1.9
FB 0269	Grapes (fresh, wine, dried)	0.5	16.1	8.1	1.0	0.5	0.0	0.0	1.6	0.8	16.1	8.1
FI 0345	Mango	0.05	2.3	0.1	5.3	0.3	3.4	0.2	6.3	0.3	0.0	0.0
MM 0095	Meat from mammals other than marine mammals	0.05	37.0	1.9	32.8	1.6	23.8	1.2	47.0	2.4	155.5	7.8
ML 0106	Milks	0.01	116.9	1.2	32.1	0.3	41.8	0.4	160.1	1.6	289.3	2.9
GC 0647	Oats	0.05	0.0	0.0	0.0	0.0	0.2	0.0	0.8	0.0	2.0	0.1
SO 0703	Peanut, whole	0.1	0.0	0.0	4.0	0.4	5.5	0.6	1.3	0.1	0.3	0.0
TN 0672	Pecan	0.05	0.0	0.0	0.0	0.0	0.0	0.0	0.0	0.0	0.3	0.0
PM 0110	Poultry meat	0.05	31.0	1.6	13.2	0.7	5.5	0.3	25.3	1.3	53.0	2.7
SO 0495	Rape-seed	0.05	0.0	0.0	0.0	0.0	0.0	0.0	0.0	0.0	0.0	0.0
SO 0650	Rye	0.05	0.0	0.0	1.0	0.1	0.0	0.0	0.0	0.0	1.5	0.1
FS 0012	Stone fruits	1	7.3	7.3	1.0	1.0	0.0	0.0	0.8	0.8	23.3	23.3
VR 0596	Sugar-beet	0.05	0.5	0.0	0.0	0.0	0.0	0.0	0.3	0.0	2.0	0.1
GS 0659	Sugar-cane	0.05	18.5	0.9	7.3	0.4	15.9	0.8	3.5	0.2	0.0	0.0
GC 0654	Wheat	0.05	327.3	16.4	114.8	5.7	28.3	1.4	116.8	5.8	178.0	8.9

	Mid-East	Far-East	African	Latin American	European
Total intake (µg/person)	39.7	14.5	7.3	25.2	60.5
Body weight per region (kg bw)	60	55	60	60	60
ADI (µg/person)	4200	3850	4200	4200	4200
% ADI	0.9%	0.4%	0.2%	0.6%	1.4%
Rounded % ADI	1%	0%	0%	1%	1%

Annex 3

PROPINEB International estimated daily intake ADI = 0–0.007 mg/kg bw

Codex code	Commodity	STMR or STMR-P (mg/kg)	Mid-East Diet	Mid-East Intake	Far-East Diet	Far-East Intake	African Diet	African Intake	Latin American Diet	Latin American Intake	European Diet	European Intake
FS 0013	Cherry	0.13	0.0	0.0	0.0	0.0	0.0	0.0	0.0	0.0	3.0	0.4
VC 0424	Cucumber	0.49	2.4	1.2	2.3	1.1	0.0	0.0	4.2	2.0	4.5	2.2
MO 0105	Edible offal (mammalian)	0	4.2	0.0	1.4	0.0	2.8	0.0	6.1	0.0	12.4	0.0
PE 0112	Eggs	0	14.6	0.0	13.1	0.0	3.7	0.0	11.9	0.0	37.6	0.0
MM 0095	Meat from mammals other than marine mammals: 20% as fat	0	7.4	0.0	6.6	0.0	4.8	0.0	9.4	0.0	31.1	0.0
MM 0095	Meat from mammals other than marine mammals: 80% as muscle	0	29.6	0.0	26.2	0.0	19.0	0.0	37.6	0.0	124.4	0.0
ML 0106	Milks	0	116.9	0.0	32.1	0.0	41.8	0.0	160.1	0.0	289.3	0.0
VO 0445	Peppers, sweet (including pim(i)ento)	1.6	3.3	5.3	2.0	3.2	5.3	8.5	2.3	3.7	10.3	16.5
VR 0589	Potato	0.12	59.0	7.1	19.2	2.3	20.6	2.5	40.8	4.9	240.8	28.9
PM 0110	Poultry meat: 10% as fat	0	3.1	0.0	1.3	0.0	0.6	0.0	2.5	0.0	5.3	0.0
PM 0110	Poultry meat: 90% as muscle	0	27.9	0.0	11.9	0.0	5.0	0.0	22.8	0.0	47.7	0.0
PO 0111	Poultry, edible offal of	0	0.1	0.0	0.1	0.0	0.1	0.0	0.4	0.0	0.4	0.0
VO 0448	Tomato (fresh, juice, paste, peeled)	1	81.5	81.5	7.0	7.0	16.5	16.5	25.5	25.5	66.6	66.6
JF 0448	Tomato juice	0.182	0.3	0.1	0.0	0.0	0.0	0.0	0.0	0.0	2.0	0.4
	Tomato paste	1.86	5.8	10.8	0.2	0.4	0.3	0.6	0.0	0.0	4.0	7.4
	Total intake (μg/person)			105.9		14.0		28.0		36.1		122.4
	Body weight per region (kg bw)			60		55		60		60		60
	ADI (μg/person)			420		385		420		420		420
	% ADI			25.2%		3.6%		6.7%		8.6%		29.1%
	Rounded % ADI			30%		4%		7%		9%		30%

Annex 3

PYRACLOSTROBIN (210) International estimated daily intake ADI = 0–0.03 mg/kg bw

Codex code	Commodity	STMR or STMR-P (mg/kg)	Mid-East Diet	Mid-East Intake	Far-East Diet	Far-East Intake	African Diet	African Intake	Latin American Diet	Latin American Intake	European Diet	European Intake
FC 0001	Citrus fruit	0.19	47.1	8.9	6.3	1.2	5.1	1.0	54.6	10.4	44.6	8.5
TN 0660	Almond	0.02	0.5	0.0	0.0	0.0	0.0	0.0	0.1	0.0	1.8	0.0
FI 0327	Banana	0.02	8.3	0.2	26.2	0.5	21.0	0.4	102.3	2.0	22.8	0.5
GC 0640	Barley (fresh)	0.03	1.0	0.0	3.5	0.1	1.8	0.1	6.5	0.2	19.8	0.6
VD 0071	Beans (dry)	0.02	2.3	0.0	4.8	0.1	0.0	0.0	13.0	0.3	3.5	0.1
FB 0020	Blueberry	0.34	0.0	0.0	0.0	0.0	0.0	0.0	0.0	0.0	0.5	0.2
VR 0577	Carrot	0.12	2.8	0.3	2.5	0.3	0.0	0.0	6.3	0.8	22.0	2.6
FS 0013	Cherry	0.43	0.0	0.0	0.0	0.0	0.0	0.0	0.0	0.0	3.0	1.3
MO 0105	Edible offal (mammalian)	0.008	4.2	0.0	1.4	0.0	2.8	0.0	6.1	0.0	12.4	0.1
PE 0112	Eggs	0	14.6	0.0	13.1	0.0	3.7	0.0	11.9	0.0	37.6	0.0
VA 0381	Garlic	0.05	2.0	0.1	2.2	0.1	0.0	0.0	0.5	0.0	3.0	0.2
FB 0269	Grapes (fresh, wine, excluding dried grapes)	0.44	15.8	7.0	1.0	0.4	0.0	0.0	1.3	0.6	13.8	6.1
DF 0269	Grapes, dried (currants, raisins and sultanas)	1.36	0.3	0.4	0.0	0.0	0.0	0.0	0.3	0.4	2.3	3.1
VD 0533	Lentil (dry)	0.13	2.8	0.4	0.7	0.1	0.0	0.0	0.0	0.0	2.3	0.3
GC 0645	Maize (fresh)	0.02	16.5	0.3	0.0	0.0	0.0	0.0	1.5	0.0	0.0	0.0
FI 0345	Mango	0.05	2.3	0.1	5.3	0.3	3.4	0.2	6.3	0.3	0.0	0.0
MM 0095	Meat from mammals other than marine mammals: 20% as fat	0.063	7.4	0.5	6.6	0.4	4.8	0.3	9.4	0.6	31.1	2.0
MM 0095	Meat from mammals other than marine mammals: 80% as muscle	0.009	29.6	0.3	26.2	0.2	19.0	0.2	37.6	0.3	124.4	1.1
ML 0106	Milks	0.01	116.9	1.2	32.1	0.3	41.8	0.4	160.1	1.6	289.3	2.9
GC 0647	Oats	0.17	0.0	0.0	0.0	0.0	0.2	0.0	0.8	0.1	2.0	0.3
FI 0350	Onion, dry	0.02	23.0	0.5	9.5	0.2	5.8	0.1	9.8	0.2	26.8	0.5
	Papaya	0.05	0.0	0.0	0.2	0.0	0.0	0.0	5.3	0.3	0.0	0.0
	Peach and nectarine	0.15	2.5	0.4	0.5	0.1	0.0	0.0	0.8	0.1	12.5	1.9
SO 0697	Peanut	0.02	0.3	0.0	0.2	0.0	2.3	0.0	0.3	0.0	3.0	0.1
VD 0072	Peas (dry)	0.07	0.5	0.0	1.7	0.1	5.1	0.4	1.3	0.1	1.8	0.1
TN 0672	Pecan	0.02	0.0	0.0	0.0	0.0	0.0	0.0	0.0	0.0	0.3	0.0
TN 0675	Pistachio nut	0.22	0.3	0.1	0.0	0.0	0.0	0.0	0.0	0.0	0.0	0.0
FS 0014	Plum (fresh)	0.06	1.8	0.1	0.5	0.0	0.0	0.0	0.0	0.0	3.8	0.2
VR 0589	Potato	0.02	59.0	1.2	19.2	0.4	20.6	0.4	40.8	0.8	240.8	4.8
PM 0110	Poultry meat	0	31.0	0.0	13.2	0.0	5.5	0.0	25.3	0.0	53.0	0.0
PO 0111	Poultry, edible offal of	0	0.1	0.0	0.1	0.0	0.1	0.0	0.4	0.0	0.4	0.0
PF 0111	Poultry, fats	0	3.1	0.0	1.3	0.0	0.6	0.0	2.5	0.0	5.3	0.0
VR 0494	Radish	0.08	0.5	0.0	0.0	0.0	0.0	0.0	0.3	0.0	2.0	0.2
VC 0431	Squash, summer	0.15	10.5	1.6	2.2	0.3	0.0	0.0	14.0	2.1	3.5	0.5
FB 0275	Strawberry	0.16	0.0	0.0	0.0	0.0	0.0	0.0	0.0	0.0	5.3	0.8

323

Annex 3

Codex code	Commodity	STMR or STMR-P (mg/kg)	Mid-East Diet	Mid-East Intake	Far-East Diet	Far-East Intake	African Diet	African Intake	Latin American Diet	Latin American Intake	European Diet	European Intake
VO 0448	Tomato (fresh)	0.12	44.1	5.3	5.7	0.7	14.6	1.8	25.5	3.1	34.9	4.2
	Total intake (µg/person)			33.0		7.3		5.6		26.7		49.5
GC 0654	Wheat	0.02	4.3	0.1	0.8	0.0	0.0	0.0	4.8	0.1	2.2	0.0
CM 0654	Wheat bran. unprocessed	0.012	ND	–	ND	–	ND	–	ND	–	ND	–
CF 1211	Wheat flour	0.012	323.0	3.9	114.0	1.4	28.3	0.3	112.0	1.3	175.8	2.1
CF 1210	Wheat germ	0.016	0.1	0.0	0.1	0.0	0.0	0.0	0.1	0.0	0.1	0.0
	Wine only	0.04	0.5	0.0	0.0	0.0	0.8	0.0	19.8	0.8	97.8	3.9
	Body weight per region (kg bw)			60		55		60		60		60
	ADI (µg/person)			1800		1650		1800		1800		1800
	% ADI			1.8%		0.4%		0.3%		1.5%		2.7%
	Rounded % ADI			2%		0%		0%		1%		3%

Annex 3

SPINOSAD (203) International estimated daily intake ADI = 0–0.02 mg/kg bw

Codex code	Commodity	STMR or STMR-P (mg/kg)	Mid-East Diet	Mid-East Intake	Far-East Diet	Far-East Intake	African Diet	African Intake	Latin American Diet	Latin American Intake	European Diet	European Intake
TN 0660	Almond	0.01	0.5	0.0	0.0	0.0	0.0	0.0	0.1	0.0	1.8	0.0
FP 0226	Apple	0.0165	7.5	0.1	4.7	0.1	0.3	0.0	5.5	0.1	40.0	0.7
JF 0226	Apple juice	0.0013	4.5	0.0	0	0.0	0	0.0	0.3	0.0	3.8	0.0
	Brassica vegetables (flowerhead, head and leafy Brassicas, kohlrabi)	0.27	6.3	1.7	11.2	3.0	0.0	0.0	10.8	2.9	39.8	10.7
MM 0812	Cattle meat	0.078	14.6	1.1	2.7	0.2	10.4	0.8	30.0	2.3	63.3	4.9
ML 0812	Cattle milk	0.65	79.5	51.7	23.2	15.1	35.8	23.3	159.3	103.5	287.0	186.6
MO 1280	Cattle kidney	0.31	0.1	0.0	0.0	0.0	0.1	0.0	0.2	0.1	0.2	0.1
MO 1281	Cattle liver	0.66	0.2	0.1	0	0.0	0.1	0.1	0.3	0.2	0.4	0.3
VS 0624	Celery	0.97	0.5	0.5	0.0	0.0	0.0	0.0	0.3	0.3	2.0	1.9
GC 0080	Cereal grains, excluding maize flour, polished rice, wheat flour	0.7	26.3	18.4	28.1	19.7	115.0	80.5	34.6	24.2	28.0	19.6
FC 0001	Citrus fruit	0.01	47.1	0.5	6.3	0.1	5.1	0.1	54.6	0.5	44.6	0.4
OC 0691	Cotton-seed oil, crude	0.0018	3.8	0.0	0.5	0.0	0.5	0.0	0.5	0.0	0.0	0.0
OR 0691	Cotton-seed oil, edible	0.002	3.8	0.0	0.5	0.0	0.5	0.0	0.5	0.0	0.0	0.0
MO 0105	Edible offal (mammalian), excluding cattle	0.064	1.7	0.1	1.1	0.1	1.0	0.1	1.1	0.1	6.4	0.4
PE 0112	Eggs	0.01	14.6	0.1	13.1	0.1	3.7	0.0	11.9	0.1	37.6	0.4
VC 0045	Fruiting vegetables. cucurbits	0.046	80.5	3.7	18.2	0.8	0.0	0.0	30.5	1.4	38.5	1.8
FB 0269	Grapes (fresh, wine, excluding dried grapes)	0.084	15.8	1.3	1.0	0.1	0.0	0.0	1.3	0.1	13.8	1.1
DF 0269	Grapes, dried (currants, raisins and sultanas)	0.13	0.3	0.0	0.0	0.0	0.0	0.0	0.3	0.0	2.3	0.3
FI 0341	Kiwi fruit	0.02	0.0	0.0	0.0	0.0	1.9	0.0	0.1	0.0	1.5	0.0
VL 0053	Leafy vegetables	1.9	7.8	14.8	9.7	18.4	0.7	1.3	16.5	31.4	51.7	98.2
VP 0060	Legume vegetables	0.041	9.9	0.4	3.1	0.1	0.1	0.0	5.6	0.2	28.4	1.2
CF 1255	Maize flour	0.13	31.8	4.1	31.2	4.1	106.2	13.8	40.3	5.2	8.8	1.1
OC 0645	Maize oil, crude	0.77	1.8	1.4	0.0	0.0	0.3	0.2	0.5	0.4	1.3	1.0
MM 0095	Meat from mammals other than marine mammals: 20% as fat, excluding cattle	0.32	4.5	1.4	6.0	1.9	2.7	0.9	3.4	1.1	18.4	5.9
MM 0095	Meat from mammals other than marine mammals: 80% as muscle, excluding cattle	0.01	17.9	0.2	24.1	0.2	10.7	0.1	13.6	0.1	73.8	0.7
JF 0004	Orange juice	0.0072	7.3	0.1	0.0	0.0	0.0	0.0	0.3	0.0	4.5	0.0
VO 0051	Peppers	0.056	3.4	0.2	2.1	0.1	5.4	0.3	2.4	0.1	10.4	0.6
VR 0589	Potato	0	59.0	0.0	19.2	0.0	20.6	0.0	40.8	0.0	240.8	0.0
PM 0110	Poultry meat	0.01	31.0	0.3	13.2	0.1	5.5	0.1	25.3	0.3	53.0	0.5
CM 1206	Rice bran. unprocessed	0.55	ND	–	ND	–	ND	–	ND	–	ND	–

325

Annex 3

Codex code	Commodity	STMR or STMR-P (mg/kg)	Mid-East Diet	Mid-East Intake	Far-East Diet	Far-East Intake	African Diet	African Intake	Latin American Diet	Latin American Intake	European Diet	European Intake
CM 0649	Rice, husked	0.077	0.0	0.0	1.8	0.1	34.7	2.7	21.0	1.6	2.5	0.2
CM 1205	Rice, polished	0.015	48.8	0.7	277.5	4.2	68.8	1.0	65.5	1.0	9.3	0.1
VD 0541	Soya bean (dry)	0	4.5	0.0	2.0	0.0	0.5	0.0	0.0	0.0	0.0	0.0
FS 0012	Stone fruits	0.0265	7.3	0.2	1.0	0.0	0.0	0.0	0.8	0.0	23.3	0.6
VO 0447	Sweet corn (corn-on-the-cob)	0.01	0.0	0.0	0.0	0.0	0.0	0.0	0.0	0.0	8.3	0.1
VO 0448	Tomato (fresh)	0.03	44.1	1.3	5.7	0.2	14.6	0.4	25.5	0.8	34.9	1.0
JF 0448	Tomato juice	0.0075	0.3	0.0	0.0	0.0	0.0	0.0	0.0	0.0	2.0	0.0
	Tomato paste	0.059	5.8	0.3	0.2	0.0	0.3	0.0	0.0	0.0	4.0	0.2
CM 0654	Wheat bran. unprocessed	1.4	ND	-	ND	-	ND	-	ND	-	ND	-
CF 1211	Wheat flour	0.18	323.0	58.1	114.0	20.5	28.3	5.1	112.0	20.2	175.8	31.6
CP 1211	White bread	0.098	215.3	21.1	76.0	7.4	18.9	1.9	37.3	3.7	117.2	11.5
	Wine only	0.027	0.5	0.0	0.0	0.0	0.8	0.0	19.8	0.5	97.8	2.6
	Total intake (µg/person)			184.2		96.8		132.7		202.5		386.7
	Body weight per region (kg bw)			60		55		60		60		60
	ADI (µg/person)			1200		1100		1200		1200		1200
	% ADI			15.4%		8.8%		11.1%		16.9%		32.2%
	Rounded % ADI			20%		9%		10%		20%		30%

Annex 3

TRIADIMEFON (133)

Estimated theoretical maximum daily intake

ADI = 0–0.03 mg/kg bw

Codex code	Commodity	MRL (mg/kg)	Mid-East Diet	Mid-East Intake	Far-East Diet	Far-East Intake	African Diet	African Intake	Latin American Diet	Latin American Intake	European Diet	European Intake
FP 0009	Pome fruit	0.5	10.8	5.4	7.5	3.8	0.3	0.2	6.5	3.3	51.3	25.7
GC 0640	Barley (fresh)	0.5	1.0	0.5	3.5	1.8	1.8	0.9	6.5	3.3	19.8	9.9
VD 0524	Chick-pea (dry)	0.05	3.3	0.2	2.5	0.1	0.0	0.0	0.0	0.0	1.0	0.1
SB 0716	Coffee bean	0.05	5.3	0.3	0.4	0.0	0.0	0.0	3.6	0.2	7.9	0.4
FB 0021	Currant, red, black, white	0.2	0.0	0.0	0.0	0.0	0.0	0.0	0.0	0.0	0.3	0.1
PE 0112	Eggs	0.05	14.6	0.7	13.1	0.7	3.7	0.2	11.9	0.6	37.6	1.9
VC 0045	Fruiting vegetables, cucurbits	0.1	80.5	8.1	18.2	1.8	0.0	0.0	30.5	3.1	38.5	3.9
FB 0269	Grapes (fresh. Wine, dried)	0.5	16.1	8.1	1.0	0.5	0.0	0.0	1.6	0.8	16.1	8.1
DH 1100	Hops, dry	10	0.1	1.0	0.1	1.0	0.1	1.0	0.1	1.0	0.1	1.0
FI 0345	Mango	0.05	2.3	0.1	5.3	0.3	3.4	0.2	6.3	0.3	0.0	0.0
MM 0095	Meat from mammals other than marine mammals	0.05	37.0	1.9	32.8	1.6	23.8	1.2	47.0	2.4	155.5	7.8
ML 0106	Milks	0.05	116.9	5.8	32.1	1.6	41.8	2.1	160.1	8.0	289.3	14.5
GC 0647	Oats	0.1	0.0	0.0	0.0	0.0	0.2	0.0	0.8	0.1	2.0	0.2
VA 0387	Onion, Welsh	0.05	ND	–	ND	–	ND	–	ND	–	ND	–
VP 0063	Peas (green pods and immature seeds)	0.1	5.5	0.6	2.0	0.2	0.0	0.0	0.8	0.1	14.0	1.4
VO 0445	Peppers. sweet (including pim(i)ento)	0.1	3.3	0.3	2.0	0.2	5.3	0.5	2.3	0.2	10.3	1.0
FI 0353	Pineapple (fresh)	2	0.0	0.0	9.3	18.6	2.6	5.2	15.5	31.0	1.3	2.6
PM 0110	Poultry meat	0.05	31.0	1.6	13.2	0.7	5.5	0.3	25.3	1.3	53.0	2.7
FB 0272	Raspberry, red, black	1	0.0	0.0	0.0	0.0	0.0	0.0	0.0	0.0	0.5	0.5
GC 0650	Rye	0.1	0.0	0.0	1.0	0.1	0.0	0.0	0.0	0.0	1.5	0.2
VA 0389	Spring onion	0.05	0.0	0.0	2.0	0.1	1.5	0.1	4.0	0.2	1.0	0.1
FB 0275	Strawberry	0.1	0.0	0.0	0.0	0.0	0.0	0.0	0.0	0.0	5.3	0.5
VR 0596	Sugar-beet	0.1	0.5	0.1	0.0	0.0	0.0	0.0	0.3	0.0	2.0	0.2
VO 0448	Tomato (fresh. juice. paste. peeled)	0.2	81.5	16.3	7.0	1.4	16.5	3.3	25.5	5.1	66.6	13.3
GC 0654	Wheat	0.1	327.3	32.7	114.8	11.5	28.3	2.8	116.8	11.7	178.0	17.8

Total intake (µg/person)	83.5	45.9	17.9	72.5	113.5
Body weight per region (kg bw)	60	55	60	60	60
ADI (µg/person)	1800	1650	1800	1800	1800
% ADI	4.6%	2.8%	1.0%	4.0%	6.3%
Rounded % ADI	5%	3%	1%	4%	6%

Annex 3

TRIADIMENOL (168) Theoretical maximum daily intake ADI = 0–0.03 mg/kg bw

Codex code	Commodity	MRL (mg/kg)	Mid-East Diet	Mid-East Intake	Far-East Diet	Far-East Intake	African Diet	African Intake	Latin American Diet	Latin American Intake	European Diet	European Intake
FP 0009	Pome fruit	0.5	10.8	5.4	7.5	3.8	0.3	0.2	6.5	3.3	51.3	25.7
VS 0620	Artichoke, globe	1	2.3	2.3	0.0	0.0	0.0	0.0	0.0	0.0	5.5	5.5
FI 0327	Banana	0.2	8.3	1.7	26.2	5.2	21.0	4.2	102.3	20.5	22.8	4.6
GC 0640	Barley (fresh)	0.5	1.0	0.5	3.5	1.8	1.8	0.9	6.5	3.3	19.8	9.9
VD 0524	Chick-pea (dry)	0.05	3.3	0.2	2.5	0.1	0.0	0.0	0.0	0.0	1.0	0.1
SB 0716	Coffee bean	0.1	5.3	0.5	0.4	0.0	0.0	0.0	3.6	0.4	7.9	0.8
FB 0021	Currants, red, black, white	0.5	0.0	0.0	0.0	0.0	0.0	0.0	0.0	0.0	0.3	0.2
PE 0112	Eggs	0.05	14.6	0.7	13.1	0.7	3.7	0.2	11.9	0.6	37.6	1.9
VC 0045	Fruiting vegetables. cucurbits	2	80.5	161.0	18.2	36.4	0.0	0.0	30.5	61.0	38.5	77.0
FB 0269	Grapes (fresh, wine, dried)	2	16.1	32.2	1.0	2.0	0.0	0.0	1.6	3.2	16.1	32.2
FI 0345	Mango	0.05	2.3	0.1	5.3	0.3	3.4	0.2	6.3	0.3	0.0	0.0
MM 0095	Meat from mammals other than marine mammals	0.05	37.0	1.9	32.8	1.6	23.8	1.2	47.0	2.4	155.5	7.8
ML 0106	Milks	0.01	116.9	1.2	32.1	0.3	41.8	0.4	160.1	1.6	289.3	2.9
GC 0647	Oats	0.2	0.0	0.0	0.0	0.0	0.2	0.0	0.8	0.2	2.0	0.4
VA 0387	Onion, Welsh	0.05	ND	-	ND	-	ND	-	ND	-	ND	-
VP 0063	Peas (green pods and immature seeds)	0.1	5.5	0.6	2.0	0.2	0.0	0.0	0.8	0.1	14.0	1.4
VO 0445	Peppers, sweet (including pim(i)ento)	0.1	3.3	0.3	2.0	0.2	5.3	0.5	2.3	0.2	10.3	1.0
FI 0353	Pineapple (fresh, canned, juice, dried)	1	ND	-	ND	-	ND	-	ND	-	ND	-
PM 0110	Poultry meat	0.05	31.0	1.6	13.2	0.7	5.5	0.3	25.3	1.3	53.0	2.7
FB 0272	Raspberry, red, black	0.5	0.0	0.0	0.0	0.0	0.0	0.0	0.0	0.0	0.5	0.3
GC 0650	Rye	0.2	0.0	0.0	1.0	0.2	0.0	0.0	0.0	0.0	1.5	0.3
VA 0389	Spring onion	0.05	0.0	0.0	2.0	0.1	1.5	0.1	4.0	0.2	1.0	0.1
FB 0275	Strawberry	0.1	0.0	0.0	0.0	0.0	0.0	0.0	0.0	0.0	5.3	0.5
VR 0596	Sugar-beet	0.1	0.5	0.1	0.0	0.0	0.0	0.0	0.3	0.0	2.0	0.2
VO 0448	Tomato (fresh, juice, paste, peeled)	0.5	81.5	40.8	7.0	3.5	16.5	8.3	25.5	12.8	66.6	33.3
GC 0654	Wheat	0.2	327.3	65.5	114.8	23.0	28.3	5.7	116.8	23.4	178.0	35.6
	Total intake (µg/person)			316.3		80.0		22.1		134.5		244.1
	Body weight per region (kg bw)			60		55		60		60		60
	ADI (µg/person)			1800		1650		1800		1800		1800
	% ADI			17.6%		4.8%		1.2%		7.5%		13.6%
	Rounded % ADI			20%		5%		1%		7%		10%

Annex 3

TRIFLOXYSTROBIN (213) International estimated daily intake ADI = 0–0.04 mg/kg bw

Codex code	Commodity	STMR or STMR-P (mg/kg)	Mid-East Diet	Mid-East Intake	Far-East Diet	Far-East Intake	African Diet	African Intake	Latin American Diet	Latin American Intake	European Diet	European Intake
FI 0327	Banana	0.02	8.3	0.2	26.2	0.5	21.0	0.4	102.3	2.0	22.8	0.5
GC 0640	Barley (fresh)	0.04	1.0	0.0	3.5	0.1	1.8	0.1	6.5	0.3	19.8	0.8
VB 0400	Broccoli	0.17	0.5	0.1	1.0	0.2	0.0	0.0	1.1	0.2	2.7	0.5
VB 0402	Brussels sprouts	0.17	0.5	0.1	1.0	0.2	0.0	0.0	1.1	0.2	2.7	0.5
VB 0403	Cabbage, Savoy	0.17	0.1	0.0	0.1	0.0	0.1	0.0	0.1	0.0	0.1	0.0
VR 0577	Carrot	0.035	2.8	0.1	2.5	0.1	0.0	0.0	6.3	0.2	22.0	0.8
VB 0404	Cauliflower	0.17	1.3	0.2	1.5	0.3	0	0.0	0.3	0.1	13	2.2
VS 0624	Celery	0.18	0.5	0.1	0.0	0.0	0.0	0.0	0.3	0.1	2.0	0.4
FC 0001	Citrus fruit	0.095	47.1	4.5	6.3	0.6	5.1	0.5	54.6	5.2	44.6	4.2
MO 0097	Edible offal of cattle, pigs and sheep	0.008	3.8	0.0	1.3	0.0	2.3	0.0	6.0	0.0	12.3	0.1
PE 0112	Eggs	0	14.6	0.0	13.1	0.0	3.7	0.0	11.9	0.0	37.6	0.0
VC 0045	Fruiting vegetables. cucurbits	0.095	80.5	7.6	18.2	1.7	0.0	0.0	30.5	2.9	38.5	3.7
FB 0269	Grapes (fresh, wine, excluding dried grapes)	0.15	15.8	2.4	1.0	0.2	0.0	0.0	1.3	0.2	13.8	2.1
DF 0269	Grapes, dried (currants, raisins and sultanas)	0.345	0.3	0.1	0.0	0.0	0.0	0.0	0.3	0.1	2.3	0.8
DH 1100	Hops, dry	9.95	0.1	1.0	0.1	1.0	0.1	1.0	0.1	1.0	0.1	1.0
VA 0384	Leek	0.31	0.5	0.2	0.0	0.0	0.0	0.0	0.3	0.1	2.0	0.6
GC 0645	Maize (fresh. flour)	0.02	48.3	1.0	31.2	0.6	106.2	2.1	41.8	0.8	8.8	0.2
MM 0095	Meat from mammals other than marine mammals: 20% as fat	0.006	7.4	0.0	6.6	0.0	4.8	0.0	9.4	0.0	31.1	0.0
MM 0095	Meat from mammals other than marine mammals: 80% as muscle	0	29.6	0.0	26.2	0.0	19.0	0.0	37.6	0.0	124.4	0.0
ML 0106	Milks	0	116.9	0.0	32.1	0.0	41.8	0.0	160.1	0.0	289.3	0.0
JF 0004	Orange juice	0.018	7.3	0.1	0.0	0.0	0.0	0.0	0.3	0.0	4.5	0.1
SO 0697	Peanut	0	0.3	0.0	0.2	0.0	2.3	0.0	0.3	0.0	3.0	0.0
VO 0445	Peppers, sweet (including pim(i)ento)	0.1	3.3	0.3	2.0	0.2	5.3	0.5	2.3	0.2	10.3	1.0
FP 0009	Pome fruit	0.11	10.8	1.2	7.5	0.8	0.3	0.0	6.5	0.7	51.3	5.6
VR 0589	Potato	0.02	59.0	1.2	19.2	0.4	20.6	0.4	40.8	0.8	240.8	4.8
PM 0110	Poultry meat: 10% as fat	0	3.1	0.0	1.3	0.0	0.6	0.0	2.5	0.0	5.3	0.0
PM 0110	Poultry meat: 90% as muscle	0	27.9	0.0	11.9	0.0	5.0	0.0	22.8	0.0	47.7	0.0
PO 0111	Poultry, edible offal of	0	0.1	0.0	0.1	0.0	0.1	0.0	0.4	0.0	0.4	0.0
GC 0649	Rice	0.16	48.8	7.8	279.3	44.7	103.4	16.5	86.5	13.8	11.8	1.9
FS 0012	Stone fruit	0.38	7.3	2.8	1.0	0.4	0.0	0.0	0.8	0.3	23.3	8.9
FB 0275	Strawberry	0.1	0.0	0.0	0.0	0.0	0.0	0.0	0.0	0.0	5.3	0.5
VR 0596	Sugar-beet	0.02	0.5	0.0	0.0	0.0	0.0	0.0	0.3	0.0	2.0	0.0
VO 0448	Tomato (fresh)	0.08	44.1	3.5	5.7	0.5	14.6	1.2	25.5	2.0	34.9	2.8
	Tomato paste	0.13	5.8	0.8	0.2	0.0	0.3	0.0	0.0	0.0	4.0	0.5

Annex 3

Codex code	Commodity	STMR or STMR-P (mg/kg)	Mid-East Diet	Mid-East Intake	Far-East Diet	Far-East Intake	African Diet	African Intake	Latin American Diet	Latin American Intake	European Diet	European Intake
TN 0085	Tree nuts	0	1.1	0.0	13.5	0.0	4.5	0.0	17.8	0.0	4.6	0.0
GC 0654	Wheat	0.02	327.3	6.5	114.8	2.3	28.3	0.6	116.8	2.3	178.0	3.6
CF 1210	Wheat germ	0.013	0.1	0.0	0.1	0.0	0.0	0.0	0.1	0.0	0.1	0.0
	Total intake (µg/person)			41.8		54.7		23.4		33.7		48.0
	Body weight per region (kg bw)			60		55		60		60		60
	ADI (µg/person)			2400		2200		2400		2400		2400
	% ADI			1.7%		2.5%		1.0%		1.4%		2.0%
	Rounded % ADI			2%		2%		1%		1%		2%

ANNEX 4
INTERNATIONAL ESTIMATES OF SHORT-TERM DIETARY INTAKES OF PESTICIDE RESIDUES

CARBOFURAN (96)

International estimate of short-term intake for ARfD = 0.009 mg/kg bw (9 µg/kg bw)
Maximum % ARfD: 50%

General population

Codex code	Commodity	STMR or STMR-P (mg/kg)	HR or HR-P (mg/kg)	Country	Body weight (kg)	Large portion (g/person)	Unit weight (g)	Unit weight Country	Unit weight, edible portion (g)	Varia-bility factor	Case	IESTI (µg/kg bw/day)	% ARfD rounded
FC 0001	Citrus fruits	–	0.05										–
FI 0327	Banana	–	0.1	SAF	55.7	613	708	USA	481	3	2a	2.83	30%
VC 4199	Cantaloupe	–	0.13	USA	65.0	606	552	USA	276	3	2a	2.32	30%
MF 0812	Cattle fat	–	0.05	USA	65.0	60	–	–	ND		1	0.05	1%
SB 0716	Coffee beans	0.1	–	NLD	63.0	66	–	–	ND		3	0.10	1%
VC 0424	Cucumber	–	0.29	NLD	63.0	313	301	USA	286	3	2a	4.07	50%
MO 0096	Edible offal of cattle goats, horses, pigs and sheep	–	0.05	FRA	62.3	277	–	–	ND		1	0.22	2%
MF 0814	Goat fat	–	0.05	USA	65.0	18	–	–	ND		1	0.01	0%
FC 0203	Grapefruit	–	0.05	JPN	52.6	947	340	SWE	167	3	2a	1.22	10%
FC 0204	Lemon	–	0.05	FRA	62.3	115	173	SWE	92	3	2a	0.24	3%
FC 0206	Mandarin	–	0.05	JPN	52.6	409	168	USA	124	3	2a	0.62	7%
MM 0096	Meat of cattle goats, horses, pigs and sheep	–	0.05	AUS	67.0	520	–	–	ND	3	1	0.39	4%
ML 0106	Milks	0.05	–	USA	65.0	2466	–	–	ND		3	1.90	20%
FC 0004	Orange, sweet, sour (including orange-like hybrids)	–	0.05	USA	65.0	564	251	SWE	178	3	2a	0.71	8%
MF 0818	Pig fat	–	0.05	AUS	67.0	144	–	–	ND		1	0.11	1%
VR 0589	Potato	–	0.05	NLD	63.0	687	216	UNK	216	3	2a	0.89	10%
SO 0699	Safflower seed	0.1	–			ND	–	–	ND		3	ND	–
FC 0005	Shaddocks or pomelos (including Shaddock-like hybrids)	–	0.05	USA	65.0	448	210	FRA	126	3	2a	0.54	6%
VC 0431	Squash, summer	–	0.26	FRA	62.3	343	300	FRA	270	3	2a	3.68	40%
GS 0659	Sugar-cane	0.1	–	SAF	55.7	89	–	–	ND				–
VO 0447	Sweet corn (corn-on-the-cob)	–	0.08	USA	65.0	367	215	UNK	125	3	2a	0.76	8%

331

Annex 4

CARBOFURAN (96)

International estimate of short-term intake for Children up to 6 years

Acute RfD= 0.009 mg/kg bw; 9 µg/kg bw
Maximum % ARfD: 100%

Codex code	Commodity	STMR or STMR-P (mg/kg)	HR or HR-P (mg/kg)	Large portion diet Country	Body weight (kg)	Large portion (g/person)	Unit weight (g)	Unit weight Country	Unit weight, edible portion (g)	Variability factor	Case	IESTI (µg/kg bw/day)	% ARfD rounded
FC 0001	Citrus fruit	–	0.05										
FI 0327	Banana	–	0.1	JPN	15.9	312	708	USA	481	3	2b	5.88	70%
VC 4199	Cantaloupe	–	0.13	USA	15.0	270	552	USA	276	3	2b	7.01	80%
MF 0812	Cattle fat	–	0.05	USA	15.0	27	–	–	ND		1	0.09	1%
SB 0716	Coffee bean	0.1	–	NLD	17.0	19	–	–	ND		3	0.11	1%
VC 0424	Cucumber	–	0.29	NLD	17.0	162	301	USA	286	3	2b	8.29	90%
MO 0096	Edible offal of cattle goats, horses, pigs and sheep	–	0.05	FRA	17.8	203	–	–	ND		1	0.57	6%
MF 0814	Goat fat	–	0.05	USA	15.0	3	–	–	ND		1	0.01	0%
FC 0203	Grapefruit	–	0.05	FRA	17.8	381	340	SWE	167	3	2a	2.01	20%
FC 0204	Lemon	–	0.05	JPN	15.9	88	173	SWE	92	3	2b	0.83	9%
FC 0206	Mandarin	–	0.05	JPN	15.9	353	168	USA	124	3	2a	1.89	20%
MM 0096	Meat of cattle goats, horses, pigs and sheep	–	0.05	AUS	19.0	261	–	–	ND		1	0.69	8%
ML 0106	Milks	0.05	–	USA	15.0	1286	–	–	ND		3	4.29	50%
FC 0004	Orange, sweet, sour (including orange-like hybrids)	–	0.05	UNK	14.5	495	251	SWE	178	3	2a	2.94	30%
MF 0818	Pig fat	–	0.05	FRA	17.8	85	–	–	ND		1	0.24	3%
VR 0589	Potato	–	0.05	SAF	14.2	300	216	UNK	216	3	2a	2.58	30%
SO 0699	Safflower seed	0.1	–	–	–	ND	–	–	ND		3	ND	–
FC 0005	Shaddocks or pomelos (including Shaddock-like hybrids)	–	0.05	FRA	17.8	381	210	FRA	126	3	2a	1.78	20%
VC 0431	Squash, summer	–	0.26	AUS	19.0	219	300	FRA	270	3	2b	8.99	100%
GS 0659	Sugar-cane	0.1	–	SAF	14.2	60	–	–	ND	ND	ND	ND	–
VO 0447	Sweet corn (corn-on-the-cob)	–	0.08	UNK	14.5	161	215	UNK	125	3	2a	2.26	30%

Annex 4

CHLORPYRIFOS (17)

International estimate of short-term intake for **General population**

Acute RfD= 0.1 mg/kg bw; 100 µg/kg bw
Maximum % ARfD: 10%

Codex code	Commodity	STMR or STMR-P (mg/kg)	HR or HR-P (mg/kg)	Country	Body weight (kg)	Large portion (g/person)	Unit weight (g)	Country	Unit weight, edible portion (g)	Variability factor	Case	IESTI (µg/kg bw/day)	% ARfD rounded
VR 0589	Potato	–	0.86	NLD	63.0	687	200	FRA	160	3	2a	13.74	10%
OR 0691	Cotton-seed oil, edible	0.01	–	USA	65.0	9	–	–	ND	ND	3	0.00	0%
CM 1206	Rice bran, unprocessed	0.22	–	AUS	67.0	50	–	–	ND	ND	3	0.16	0%
CM 0649	Rice, husked	0.016	–	JPN	52.6	319	–	–	ND	ND	3	0.10	0%
CM 1205	Rice, polished	0.008	–	JPN	52.6	402	–	–	ND	ND	3	0.06	0%
OR 0541	Soya bean oil, refined	0.004	–	USA	65.0	98	–	–	ND	ND	3	0.01	0%
DT 0171	Teas (tea and herb teas)	0.34	–	–	–	ND	–	–	ND	ND	3	ND	–

CHLORPYRIFOS (17)

International estimate of short-term intake for **Children up to 6 years**

Acute RfD= 0.1 mg/kg bw; 100 µg/kg bw
Maximum % ARfD: 40%

Codex code	Commodity	STMR or STMR-P (mg/kg)	HR or HR-P (mg/kg)	Country	Body weight (kg)	Large portion (g/person)	Unit weight (g)	Country	Unit weight, edible portion (g)	Variability factor	Case	IESTI (µg/kg bw/day)	% ARfD rounded
VR 0589	Potato	–	0.86	SAF	14.2	300	200	FRA	160	3	2a	37.53	40%
OR 0691	Cotton-seed oil, edible	0.01	–	USA	15.0	6	–	–	ND	ND	3	0.00	0%
CM 1206	Rice bran, unprocessed	0.22	–	USA	15.0	3	–	–	ND	ND	3	0.05	0%
CM 0649	Rice, husked	0.016	–	FRA	17.8	223	–	–	ND	ND	3	0.20	0%
CM 1205	Rice, polished	0.008	–	JPN	15.9	199	–	–	ND	ND	3	0.10	0%
OR 0541	Soya bean oil, refined	0.004	–	USA	15.0	35	–	–	ND	ND	3	0.01	0%
DT 0171	Teas (tea and herb teas)	0.34	–	–	–	ND	–	–	ND	ND	3	ND	–

Annex 4

DIMETHIPIN (151) International estimate of short-term intake for **General population** Acute RfD: 0.2 mg/kg bw; 200 µg/kg bw Maximum ARfD: % : 0%

Codex Code	Commodity	STMR or STMR-P (mg/kg)	HR or HR-P (mg/kg)	Country	Body weight (kg)	Large portion (g/person)	Unit weight (g)	Country	Unit weight, edible portion (g)	Varia-bility factor	Case	IESTI (µg/kg bw/day)	% ARfD rounded
VR 0589	Potato	–	0.02	NLD	63.0	687	216	UNK	216	3	2a	0.36	0%
PE 0840	Chicken egg	–	0	FRA	62.3	219	–	–	ND	ND	1	0.00	0%
OR 0691	Cotton-seed oil, edible	0.02	–	USA	65.0	9	–	–	ND	ND	3	0.00	0%
MO 0105	Edible offal (mammalian)	–	0	FRA	62.3	277	–	–	ND	ND	1	0.00	0%
MM 0095	Meat from mammals other than marine mammals: 20% as fat	–	0	AUS	67.0	104	–	–	ND	ND	1	0.00	0%
MM 0095	Meat from mammals other than marine mammals: 80% as muscle	–	0	AUS	67.0	417	–	–	ND	ND	1	0.00	0%
ML 0106	Milks	0	–	USA	65.0	2466	–	–	ND	ND	3	0.00	0%
PM 0110	Poultry meat: 10% as fat	–	0	AUS	67.0	43	–	–	ND	ND	1	0.00	0%
PM 0110	Poultry meat: 90% as muscle	–	0	AUS	67.0	388	–	–	ND	ND	1	0.00	0%
PO 0111	Poultry, edible offal of	–	0	USA	65.0	248	–	–	ND	ND	1	0.00	0%
OR 0495	Rape-seed oil, edible	0.1	–	AUS	67.0	65	–	–	ND	ND	3	0.10	0%
SO 0702	Sunflower seed	0.01	–	USA	65.0	193	–	–	ND	ND	3	0.03	0%

DIMETHIPIN (151) International estimate of short-term intake for **Children up to 6 years** Acute RfD: 0.2 mg/kg bw; 200 µg/kg bw Maximum % ARfD: 1%

Codex Code	Commodity	STMR or STMR-P (mg/kg)	HR or HR-P (mg/kg)	Country	Body weight (kg)	Large portion (g/person)	Unit weight (g)	Country	Unit weight, edible portion (g)	Varia-bility factor	Case	IESTI (µg/kg bw/day)	% ARfD rounded
VR 0589	Potato	–	0.02	SAF	14.2	300	216	UNK	216	3	2a	1.03	1%
PE 0840	Chicken eggs	–	0	FRA	17.8	134	–	–	ND	ND	1	0.00	0%
OR 0691	Cotton-seed oil, edible	0.02	–	USA	15.0	6	–	–	ND	ND	3	0.01	0%
MO 0105	Edible offal (mammalian)	–	0	FRA	17.8	203	–	–	ND	ND	1	0.00	0%
MM 0095	Meat from mammals other than marine mammals: 20% as fat	–	0	AUS	19.0	52	–	–	ND	ND	1	0.00	0%

Annex 4

Codex code	Commodity		Country							
MM 0095	Meat from mammals other than marine mammals: 80% as muscle	–	AUS	19.0	208	–	–	ND	ND	1	0.00	0%
ML 0106	Milks	0	USA	15.0	1286	–	–	ND	ND	3	0.00	0%
PM 0110	Poultry meat: 10% as fat	–	AUS	19.0	22	–	–	ND	ND	1	0.00	0%
PM 0110	Poultry meat: 90% as muscle	–	AUS	19.0	201	–	–	ND	ND	1	0.00	0%
PO 0111	Poultry, edible offal of	–	USA	15.0	37	–	–	ND	ND	1	0.00	0%
OR 0495	Rape-seed oil, edible	0.1	AUS	19.0	18	–	–	ND	ND	3	0.10	0%
SO 0702	Sunflower seed	0.01	USA	15.0	24	–	–	ND	ND	3	0.02	0%

ETHOPROPHOS (149)

International estimate of short-term intake for **General population**

Acute RfD= 0.05 mg/kg bw; 50 µg/kg bw
Maximum % ARfD: 1%

Codex code	Commodity	STMR or STMR-P (mg/kg)	HR or HR-P (mg/kg)	Large portion diet Country	Body weight (kg)	Large portion (g/person)	Unit weight (g)	Unit weight Country	Unit weight, edible portion (g)	Varia-bility factor	Case	IESTI (µg/kg bw/day)	% ARfD rounded
FI 0327	Banana	–	0.02	SAF	55.7	613	900	FRA	612	3	2a	0.66	1%
VC 0424	Cucumber	–	0.01	NLD	63.0	313	400	FRA	360	3	2b	0.15	0%
MO 0105	Edible offal (mammalian)	–	0	FRA	62.3	277	–	–	ND	ND	1	0.00	0%
MM 0095	Meat from mammals other than marine mammals: 20% as fat	–	0	AUS	67.0	104	–	–	ND	ND	1	0.00	0%
MM 0095	Meat from mammals other than marine mammals: 80% as muscle	–	0	AUS	67.0	417	–	–	ND	ND	1	0.00	0%
VC 0046	Melon, except watermelon	–	0.012	USA	65.0	655	700	JPN	700	3	2b	0.36	1%
ML 0106	Milks	0	–	USA	65.0	2466	–	–	ND	ND	3	0.00	0%
VO 0445	Peppers, sweet (including pim(i)ento)	–	0.044	FRA	62.3	207	172	UNK	160	3	2a	0.37	1%
VR 0589	Potato	–	0.03	NLD	63.0	687	216	UNK	216	3	2a	0.53	1%
GS 0659	Sugar-cane	–	0.02	SAF	55.7	89	–	–	ND	ND	ND	ND	–
VR 0508	Sweet potato	–	0.03	USA	65.0	536	130	USA	105	3	2a	0.34	1%
VO 0448	Tomato (fresh, juice, paste, peeled)	–	0.01	USA	65.0	391	150	JPN	150	3	2a	0.11	0%

335

Annex 4

ETHOPROPHOS (149)

International estimate of short-term intake for Children up to 6 years

Acute RfD = 0.050 mg/kg bw; 50 µg/kg bw
Maximum % ARfD: 3%

Codex code	Commodity	STMR or STMR-P (mg/kg)	HR or HR-P (mg/kg)	Large portion diet Country	Body weight (kg)	Large portion (g/person)	Unit weight (g)	Unit weight Country	Unit weight, edible portion (g)	Variability factor	Case	IESTI (µg/kg bw/day)	% ARfD rounded
FI 0327	Banana	–	0.02	JPN	15.9	312	900	FRA	612	3	2b	1.18	2%
VC 0424	Cucumber	–	0.01	NLD	17.0	162	400	FRA	360	3	2b	0.29	1%
MO 0105	Edible offal (mammalian)	–	0	FRA	17.8	203	–	–	ND	ND	1	0.00	0%
MM 0095	Meat from mammals other than marine mammals: 20% as fat	–	0	AUS	19.0	52	–	–	ND	ND	1	0.00	0%
MM 0095	Meat from mammals other than marine mammals: 80% as muscle	–	0	AUS	19.0	208	–	–	ND	ND	1	0.00	0%
VC 0046	Melon, except watermelon	–	0.012	AUS	19.0	413	700	FRA	420	3	2b	0.78	2%
ML 0106	Milks	0	–	USA	15.0	1286	–	–	ND	ND	3	0.00	0%
VO 0445	Peppers, sweet (including pim(i)ento)	–	0.044	AUS	19.0	60	172	UNK	160	3	2b	0.42	1%
VR 0589	Potato	–	0.03	SAF	14.2	300	216	UNK	216	3	2a	1.55	3%
GS 0659	Sugar-cane	–	0.02	SAF	14.2	60	–	–	ND	ND	ND	ND	–
VR 0508	Sweet potato	–	0.03	USA	15.0	166	130	USA	105	3	2a	0.75	2%
VO 0448	Tomato (fresh, juice, paste, peeled)	–	0.01	USA	15.0	159	150	JPN	150	3	2a	0.31	1%

Annex 4

FENITROTHION (37)

International estimate of short-term intake for **General population**

Acute RfD = 0.04 mg/kg bw (40 µg/kg bw)
Maximum % ARfD: 100%

Codex code	Commodity	STMR or STMR-P (mg/kg)	HR or HR-P (mg/kg)	Country	Body weight (kg)	Large portion (g/person)	Unit weight (g)	Country	Unit weight, edible portion (g)	Variability factor	Case	IESTI (µg/kg bw/day)	% ARfD rounded
FP 0226	Apple	–	0.41	USA	65.0	1348	200	JPN	200	3	2a	11.03	30%
GC 0640	Barley (beer only)	–	1.52	AUS	67.0	528	–	–	ND	ND	1	11.98	30%
GC 0640	Barley (fresh)	–	7.6	–	–	ND	–	–	ND	ND	1	ND	–
GC 0641	Buckwheat	–	7.6	NLD	63.0	117	–	–	ND	ND	1	14.14	40%
PE 0840	Chicken eggs	–	0	FRA	62.3	219	–	–	ND	ND	1	0.00	0%
MO 0105	Edible offal (mammalian)	–	0	FRA	62.3	277	–	–	ND	ND	1	0.00	0%
GC 0645	Maize (fresh, flour, oil)	–	7.6	FRA	62.3	260	–	–	ND	ND	1	31.69	80%
MM 0095	Meat from mammals other than marine mammals: 20% as fat	–	0	AUS	67.0	104	–	–	ND	ND	1	0.00	0%
MM 0095	Meat from mammals other than marine mammals: 80% as muscle	–	0	AUS	67.0	417	–	–	ND	ND	1	0.00	0%
ML 0106	Milks	0	–	USA	65.0	2466	–	–	ND	ND	3	0.00	0%
GC 0646	Millet	–	7.6	AUS	67.0	101	–	–	ND	ND	1	11.40	30%
GC 0647	Oats	–	7.6	FRA	62.3	305	–	–	ND	ND	1	37.24	90%
PM 0110	Poultry meat: 10% as fat	–	0	AUS	67.0	43	–	–	ND	ND	1	0.00	0%
PM 0110	Poultry meat: 90% as muscle	–	0	AUS	67.0	388	–	–	ND	ND	1	0.00	0%
CM 1206	Rice bran, unprocessed	–	54.72	AUS	67.0	50	–	–	ND	ND	1	40.82	100%
CM 0649	Rice, husked and cooked	–	0.836	JPN	52.6	319	–	–	ND	ND	1	5.07	10%
CM 1205	Rice, polished and cooked	–	0.304	JPN	52.6	402	–	–	ND	ND	1	2.32	6%
CF 1250	Rye flour	–	1.786	FRA	62.3	115	–	–	ND	ND	1	3.29	8%
GC 0651	Sorghum	–	7.6	USA	65.0	18	–	–	ND	ND	1	2.05	5%
GC 0653	Triticale	–	7.6	–	–	ND	–	–	ND	ND	1	ND	–
CM 0654	Wheat bran, unprocessed	–	30.02	USA	65.0	80	–	–	ND	ND	1	36.92	90%
	Wheat bulgur wholemeal	–	1.786	–	–	ND	–	–	ND	ND	1	ND	–
CF 1211	Wheat flour	–	1.786	USA	65.0	365	–	–	ND	ND	1	10.04	30%
	Wheat macaroni	–	1.786	–	–	ND	–	–	ND	ND	1	ND	–
	Wheat pastry	–	1.786	–	–	ND	–	–	ND	ND	1	ND	–
CP 1211	White bread	–	0.76	SAF	55.7	479	–	–	ND	ND	1	6.54	20%
CP 1212	Wholemeal bread	–	2.888	SAF	55.7	395	–	–	ND	ND	1	20.50	50%

Annex 4

FENITROTHION (37)

International estimate of short-term intake for

Children up to 6 years

Acute RfD = 0.04 mg/kg bw; 40 µg/kg bw

Maximum % ARfD: 160%

Codex code	Commodity	STMR or STMR-P (mg/kg)	HR or HR-P (mg/kg)	Country	Body weight (kg)	Large portion (g/person)	Unit weight (g)	Country	Unit weight, edible portion (g)	Variability factor	Case	IESTI (µg/kg bw/day)	% ARfD rounded
FP 0226	Apple	–	0.41	USA	15.0	679	200	JPN	200	3	2a	29.49	70%
GC 0640	Barley (beer only)	–	1.52	AUS	19.0	12	–	–	ND	ND	1	0.94	2%
GC 0640	Barley (fresh)	–	7.6	–	–	ND	–	–	ND	ND	1	ND	–
GC 0641	Buckwheat	–	7.6	NLD	17.0	59	–	–	ND	ND	1	26.30	70%
PE 0840	Chicken eggs	–	0	FRA	17.8	134	–	–	ND	ND	1	0.00	0%
MO 0105	Edible offal (mammalian)	–	0	FRA	17.8	203	–	–	ND	ND	1	0.00	0%
GC 0645	Maize (fresh, flour, oil)	–	7.6	FRA	17.8	148	–	–	ND	ND	1	63.31	160%
MM 0095	Meat from mammals other than marine mammals: 20% as fat	–	0	AUS	19.0	52	–	–	ND	ND	1	0.00	0%
MM 0095	Meat from mammals other than marine mammals: 80% as muscle	–	0	AUS	19.0	208	–	–	ND	ND	1	0.00	0%
ML 0106	Milks	0	–	USA	15.0	1286	–	–	ND	ND	3	0.00	0%
GC 0646	Millet	–	7.6	–	–	ND	–	–	ND	ND	1	ND	–
GC 0647	Oats	–	7.6	USA	15.0	62	–	–	ND	ND	1	31.54	80%
PM 0110	Poultry meat: 10% as fat	–	0	AUS	19.0	22	–	–	ND	ND	1	0.00	0%
PM 0110	Poultry meat: 90% as muscle	–	0	AUS	19.0	201	–	–	ND	ND	1	0.00	0%
CM 1206	Rice bran, unprocessed	–	54.72	USA	15.0	3	–	–	ND	ND	1	11.49	30%
CM 1649	Rice, husked	–	0.836	FRA	17.8	223	–	–	ND	ND	1	10.45	30%
CM 1205	Rice, polished	–	0.304	JPN	15.9	199	–	–	ND	ND	1	3.80	9%
CF 1250	Rye flour	–	1.786	USA	15.0	18	–	–	ND	ND	1	2.11	5%
GC 0651	Sorghum	–	7.6	–	–	ND	–	–	ND	ND	1	ND	–
GC 0653	Triticale	–	7.6	–	–	ND	–	–	ND	ND	1	ND	–
CM 0654	Wheat bran, unprocessed	–	30.02	USA	15.0	30	–	–	ND	ND	1	59.44	150%
	Wheat bulgur wholemeal	–	1.786	–	–	ND	–	–	ND	ND	1	ND	–
CF 1211	Wheat flour	–	1.786	AUS	19.0	194	–	–	ND	ND	1	18.27	50%
	Wheat macaroni	–	1.786	–	–	ND	–	–	ND	ND	1	ND	–
	Wheat pastry	–	1.786	–	–	ND	–	–	ND	ND	1	ND	–
CP 1211	White bread	–	0.76	SAF	14.2	270	–	–	ND	ND	1	14.44	40%
CP 1212	Wholemeal bread	–	2.888	SAF	14.2	240	–	–	ND	ND	1	48.81	120%

Annex 4

FENPROPIMORPH (188)

International estimate of short-term intake for **General population**

Acute RfD = 0.2 mg/kg bw; 200 µg/kg bw
Maximum % ARfD: 7%

Codex code	Commodity	STMR or STMR-P (mg/kg)	HR or HR-P (mg/kg)	Country	Body weight (kg)	Large portion (g/person)	Unit weight (g)	Country	Unit weight, edible portion (g)	Variability factor	Case	IESTI (µg/kg bw/day)	% ARfD rounded
FI 0327	Banana	–	0.43	SAF	55.7	613	900	FRA	612	3	2a	14.18	7%
GC 0640	Barley (beer only)	0.05	–	AUS	67.0	528	–	–	ND	ND	3	0.39	0%
GC 0640	Barley (fresh)	0.05	–	–	–	ND	–	–	ND	ND	3	ND	–
GC 0640	Barley (fresh, flour, beer)	0.05	–	NLD	63.0	378	–	–	ND	ND	3	0.30	0%
PE 0840	Chicken eggs	–	0	FRA	62.3	219	–	–	ND	ND	1	0.00	0%
MO 0098	Kidney of cattle (g)oats, pigs and sheep	–	0.026	USA	65.0	788	–	–	ND	ND	1	0.32	0%
MO 0099	Liver of cattle (g)oats, pigs and sheep	–	0.22	USA	65.0	380	–	–	ND	ND	1	1.28	1%
MM 0095	Meat from mammals other than marine mammals: 20% as fat	–	0.006	AUS	67.0	104	–	–	ND	ND	1	0.01	0%
MM 0095	Meat from mammals other than marine mammals: 80% as muscle	–	0.009	AUS	67.0	417	–	–	ND	ND	1	0.06	0%
ML 0106	Milks	0.004	–	USA	65.0	2466	–	–	ND	ND	3	0.15	0%
GC 0647	Oats	0.05	–	FRA	62.3	305	–	–	ND	ND	3	0.25	0%
PM 0110	Poultry meat: 10% as fat	–	0	AUS	67.0	43	–	–	ND	ND	1	0.00	0%
PM 0110	Poultry meat: 90% as muscle	–	0	AUS	67.0	388	–	–	ND	ND	1	0.00	0%
PO 0111	Poultry, edible offal of	–	0	USA	65.0	248	–	–	ND	ND	1	0.00	0%
GC 0650	Rye	0.05	–	NLD	63.0	77	–	–	ND	ND	3	0.06	0%
VR 0596	Sugar beet	0.05	–	–	–	ND	–	–	ND	ND	ND	ND	–
GC 0654	Wheat	0.05	–	USA	65.0	383	–	–	ND	ND	3	0.29	0%

Annex 4

FENPROPIMORPH (188)

International estimate of short-term intake for Children up to 6 years

Acute RfD= 0.2 mg/kg bw; 200 μg/kg bw
Maximum % ARfD: 10%

Codex code	Commodity	STMR or STMR-P (mg/kg)	HR or HR-P (mg/kg)	Country	Body weight (kg)	Large portion (g/person)	Unit weight (g)	Country	Unit weight, edible portion (g)	Variability factor	Case	IESTI (μg/kg bw/day)	% ARfD rounded
FI 0327	Banana	–	0.43	JPN	15.9	312	900	FRA	612	3	2b	25.30	10%
GC 0640	Barley (beer only)	0.05	–	AUS	19.0	12	–	–	ND	ND	3	0.03	0%
GC 0640	Barley (fresh)	0.05	–	–	–	ND	–	–	ND	ND	3	ND	–
GC 0640	Barley (fresh, flour, beer)	0.05	–	AUS	19.0	14	–	–	ND	ND	3	0.04	0%
PE 0840	Chicken eggs	–	0	FRA	17.8	134	–	–	ND	ND	1	0.00	0%
MO 0098	Kidney of cattle (g)oats, pigs and sheep	–	0.026	USA	15.0	187	–	–	ND	ND	1	0.32	0%
MO 0099	Liver of cattle (g)oats, pigs and sheep	–	0.22	FRA	17.8	203	–	–	ND	ND	1	2.51	1%
MM 0095	Meat from mammals other than marine mammals: 20% as fat	–	0.006	AUS	19.0	52	–	–	ND	ND	1	0.02	0%
MM 0095	Meat from mammals other than marine mammals: 80% as muscle	–	0.009	AUS	19.0	208	–	–	ND	ND	1	0.10	0%
ML 0106	Milks	0.004	–	USA	15.0	1286	–	–	ND	ND	3	0.34	0%
GC 0647	Oats	0.05	–	USA	15.0	62	–	–	ND	ND	3	0.21	0%
PM 0110	Poultry meat: 10% as fat	–	0	AUS	19.0	22	–	–	ND	ND	1	0.00	0%
PM 0110	Poultry meat: 90% as muscle	–	0	AUS	19.0	201	–	–	ND	ND	1	0.00	0%
PO 0111	Poultry, edible offal of	–	0	USA	15.0	37	–	–	ND	ND	1	0.00	0%
GC 0650	Rye	0.05	–	NLD	17.0	37	–	–	ND	ND	3	0.11	0%
VR 0596	Sugar beet	0.05	–	–	–	ND	–	–	ND	ND	ND	ND	–
GC 0654	Wheat	0.05	–	USA	15.0	151	–	–	ND	ND	3	0.50	0%

Annex 4

FENPYROXIMATE (193)

International estimate of short-term intake for **General population**

Acute RfD = 0.01 mg/kg bw; 10 µg/kg bw
Maximum % ARfD 120 %

Codex code	Commodity	STMR or STMR-P (mg/kg)	HR or HR-P (mg/kg)	Country	Body weight (kg)	Large portion (g/person)	Unit weight (g)	Country	Unit weight, edible portion (g)	Variability factor	Case	IESTI (µg/kg bw/day)	% ARfD rounded
FP 0226	Apple	–	0.18	USA	65.0	1348	200	JPN	200	3	2a	4.84	50%
JF 0226	Apple juice	0.04	–	–	–	ND	–	–	ND	ND	ND	ND	–
MM 0812	Cattle meat: 20% as fat	–	0	AUS	67.0	93	–	–	ND	ND	1	0.00	0%
MM 0812	Cattle meat: 80% as muscle	–	0	AUS	67.0	374	–	–	ND	ND	1	0.00	0%
MO 0812	Cattle, edible offal of	–	0	SAF	55.7	524	–	–	ND	ND	1	0.00	0%
FB 0269	Grapes (fresh, dried, excluding wine)	–	0.57	AUS	67.0	513	456	SWE	438	3	2a	11.81	120%
DH 1100	Hops, dry	4.4	–	USA	65.0	6	–	–	ND	ND	3	0.40	4%
ML 0106	Milks	0.002	–	USA	65.0	2466	–	–	ND	ND	3	0.08	1%
FC 0004	Orange, sweet, sour (including orange-like hybrids)	–	0.09	USA	65.0	564	200	JPN	200	3	2a	1.34	10%
	Wine only	0.005	–	AUS	67.0	1131	–	–	ND	ND	3	0.08	1%

FENPYROXIMATE (193)

International estimate of short-term intake for **Children up to 6 years**

Acute RfD = 0.01 mg/kg bw; 10 µg/kg bw
Maximum %ArfD: 310%

Codex code	Commodity	STMR or STMR-P (mg/kg)	HR or HR-P (mg/kg)	Country	Body weight (kg)	Large portion (g/person)	Unit weight (g)	Country	Unit weight, edible portion (g)	Variability factor	Case	IESTI (µg/kg bw/day)	% ARfD rounded
FP 0226	Apple	–	0.18	USA	15.0	679	200	JPN	200	3	2a	12.95	130%
JF 0226	Apple juice	0.04	–	–	–	ND	–	–	ND	ND	ND	ND	–
MM 0812	Cattle meat: 20% as fat	–	0	AUS	19.0	48	–	–	ND	ND	1	0.00	0%
MM 0812	Cattle meat: 80% as muscle	–	0	AUS	19.0	190	–	–	ND	ND	1	0.00	0%
MO 0812	Cattle, edible offal of	–	0	FRA	17.8	203	–	–	ND	ND	1	0.00	0%
FB 0269	Grapes (fresh, dried, excluding wine)	–	0.57	AUS	19.0	342	456	SWE	438	3	2b	30.78	310%
DH 1100	Hops, dry	4.4	–	JPN	15.9	0	–	–	ND	ND	3	0.13	1%
ML 0106	Milks	0.002	–	USA	15.0	1286	–	–	ND	ND	3	0.17	2%
FC 0004	Orange, sweet, sour (including orange-like hybrids)	–	0.09	UNK	14.5	495	200	JPN	200	3	2a	5.56	60%
	Wine only	0.005	–	AUS	19.0	4	–	–	ND	ND	3	0.00	0%

341

Annex 4

MALATHION (49)

International estimate of short-term intake for **General population**

Acute RfD= 2. mg/kg bw; 2000 µg/kg bw
Maximum % ARfD: 4%

Codex code	Commodity	STMR or STMR-P (mg/kg)	HR or HR-P (mg/kg)	Country	Body weight (kg)	Large portion (g/person)	Unit weight (g)	Country	Unit weight, edible portion (g)	Variability factor	Case	IESTI (µg/kg bw/day)	% ARfD rounded
FP 0226	Apple	–	0.37	USA	65.0	1348	162	SWE	149	3	2a	9.37	0%
FC 0203	Grapefruit	–	0.22	JPN	52.6	947	340	SWE	167	3	2a	5.35	0%
FB 0269	Grapes (fresh, wine, dried)	–	2.6	AUS	67.0	1004	456	SWE	438	3	2a	72.95	4%
FC 0204	Lemon	–	0.22	FRA	62.3	115	173	SWE	92	3	2a	1.05	0%
FC 0206	Mandarin	–	0.22	JPN	52.6	409	133	UNK	100	3	2a	2.54	0%
FC 0004	Orange, sweet, sour (including orange-like hybrids)	–	0.22	USA	65.0	564	251	SWE	178	3	2a	3.12	0%

MALATHION (49)

International estimate of short-term intake for **Children up to 6 years**

ARfD = 2.0 mg/kg bw (2000 µg/kg bw)
Maximum % ARfD: 10%

Codex code	Commodity	STMR or STMR-P (mg/kg)	HR or HR-P (mg/kg)	Country	Body weight (kg)	Large portion (g/person)	Unit weight (g)	Country	Unit weight, edible portion (g)	Variability factor	Case	IESTI (µg/kg bw/day)	% ARfD rounded
FP 0226	Apple	–	0.37	USA	15.0	679	162	SWE	149	3	2a	24.10	1%
FC 0203	Grapefruit	–	0.22	FRA	17.8	381	340	SWE	**167**	3	2a	8.83	0%
FB 0269	Grapes (fresh, wine, dried)	–	2.6	JPN	15.9	388	456	SWE	438	3	2b	190.24	10%
FC 0204	Lemon	–	0.22	JPN	15.9	88	173	SWE	92	3	2b	3.67	0%
FC 0206	Mandarin	–	0.22	JPN	15.9	353	133	UNK	100	3	2a	7.65	0%
FC 0004	Oranges, sweet, sour (including orange-like hybrids)	–	0.22	UNK	14.5	495	251	SWE	178	3	2a	12.92	1%

Annex 4

METHOMYL (94)

International estimate of short-term intake for **General population**

ARfD = 0.02 mg/kg bw; 20 µg/kg bw

Maximum % ARfD: 20%

Codex code	Commodity	STMR or STMR-P (mg/kg)	HR or HR-P (mg/kg)	Large portion diet Country	Body weight (kg)	Large portion (g/person)	Unit weight (g)	Unit weight Country	Unit weight, edible portion (g)	Variability factor	Case	IESTI (µg/kg bw/day)	% ARfD rounded
VO 0051	Peppers	0.10	0.44	FRA	62.3	207	172	UNK	160	3	2a	3.72	20%

METHOMYL (94)

International estimate of short-term intake for **Children up to 6 years**

ARfD = 0.02 mg/kg bw; 20 µg/kg bw

Maximum % ARfD: 20%

Codex code	Commodity	STMR or STMR-P (mg/kg)	HR or HR-P (mg/kg)	Large portion diet Country	Body weight (kg)	Large portion (g/person)	Unit weight (g)	Unit weight Country	Unit weight, edible portion (g)	Variability factor	Case	IESTI (µg/kg bw/day)	% ARfD rounded
VO 0051	Peppers	0.10	0.44	AUS	19.0	60	172	UNK	160	3	2b	4.17	20%

OXYDEMETON METHYL (166)

International estimate of short-term intake for **General population**

ARfD = 0.002 mg/kg bw (2 µg/kg bw)

Maximum % ARfD: 80%

Codex code	Commodity	STMR or STMR-P (mg/kg)	HR or HR-P (mg/kg)	Large portion diet Country	Body weight (kg)	Large portion (g/person)	Unit weight (g)	Unit weight Country	Unit weight, edible portion (g)	Variability factor	Case	IESTI (µg/kg bw/day)	% ARfD rounded
FP 0226	Apple	–	0.04	USA	65.0	1348	162	SWE	149	3	2a	1.01	50%
JF 0226	Apple juice	0.01	–	–	–	ND	–	–	ND	ND	ND	ND	–
GC 0640	Barley (fresh, flour, beer)	0.01	–	NLD	63.0	378	–	–	ND	ND	3	0.06	3%
VB 0041	Cabbages, head	–	0.05	SAF	55.7	362	771	UNK	540	3	2b	0.98	50%

343

Annex 4

Codex code	Commodity	STMR or STMR-P (mg/kg)	HR or HR-P (mg/kg)	Large portion diet Country	Body weight (kg)	Large portion (g/person)	Unit weight (g)	Unit weight Country	Unit weight, edible portion (g)	Variability factor	Case	IESTI (µg/kg bw/day)	% ARfD rounded
MF 0812	Cattle fat	–	0	USA	65.0	60	–	–	ND	ND	1	0.00	0%
VB 0404	Cauliflower (head)	–	0.01	UNK	70.1	579	1733	UNK	780	3	2b	0.25	10%
VD 0526	Common bean (dry)	0.01	–	FRA	62.3	283	–	–	ND	ND	3	0.05	2%
PE 0112	Eggs	–	0	–	–	ND	–	–	ND	ND	1	ND	–
FB 0269	Grapes (fresh, wine, dried)	–	0.06	AUS	67.0	1004	456	SWE	438	3	2a	1.68	80%
VL 0480	Kale	–	0.01	NLD	63.0	337	–	–	ND	ND	ND	ND	–
VB 0405	Kohlrabi	–	0.05	NLD	63.0	283	135	USA	99	3	2a	0.38	20%
FC 0204	Lemon	–	0.04	FRA	62.3	115	173	SWE	92	3	2a	0.19	10%
MM 0097	Meat of cattle, pigs and sheep	–	0	AUS	67.0	520	–	–	ND	ND	1	0.00	0%
ML 0106	Milks	0	–	USA	65.0	2466	–	–	ND	ND	3	0.00	0%
FC 0004	Orange, sweet, sour (including orange-like hybrids)	–	0.04	USA	65.0	564	251	SWE	178	3	2a	0.57	30%
FP 0230	Pear	–	0.04	USA	65.0	693	187	UNK	170	3	2a	0.64	30%
MF 0818	Pig fat	–	0	AUS	67.0	144	–	–	ND	ND	1	0.00	0%
VR 0589	Potato	–	0.01	NLD	63.0	687	216	UNK	216	3	2a	0.18	9%
PM 0110	Poultry meat	–	0	AUS	67.0	431	–	–	ND	ND	1	0.00	0%
PF 0111	Poultry, fat	–	0	FRA	62.3	46	–	–	ND	ND	1	0.00	0%
GC 0650	Rye	0.01	–	NLD	63.0	77	–	–	ND	ND	3	0.01	1%
MF 0822	Sheep fat	–	0	USA	65.0	54	–	–	ND	ND	1	0.00	0%
VR 0596	Sugar beet	0.01	–	–	–	ND	–	–	ND	ND	ND	ND	–
GC 0654	Wheat	0.01	–	USA	65.0	383	–	–	ND	ND	3	0.06	3%

Annex 4

OXYDEMETON-METHYL (166)

International estimate of short-term intake for Children up to 6 years

ARfD = 0.002 mg/kg bw; 2 µg/kg bw
Maximum % ARfD: 220%

Codex code	Commodity	STMR or STMR-P (mg/kg)	HR or HR-P (mg/kg)	Large portion diet Country	Body weight (kg)	Large portion (g/person)	Unit weight (g)	Unit weight Country	Unit weight, edible portion (g)	Variability factor	Case	IESTI (µg/kg bw/day)	% ARfD rounded
FP 0226	Apple	–	0.04	USA	15.0	679	162	SWE	149	3	2a	2.60	130%
JF 0226	Apple juice	0.01	–	–	–	ND	–	–	ND	ND	ND	ND	–
GC 0640	Barley (fresh, flour, beer)	0.01	–	AUS	19.0	14	–	–	ND	ND	3	0.01	0%
VB 0041	Cabbage, head	–	0.05	SAF	14.2	220	771	UNK	ND	3	2b	2.33	120%
MF 0812	Cattle fat	–	0	USA	15.0	27	–	–	ND	ND	1	0.00	0%
VB 0404	Cauliflower (head)	–	0.01	NLD	17.0	209	1733	UNK	780	3	2b	0.37	20%
VD 0526	Common bean (dry)	0.01	–	FRA	17.8	209	–	–	ND	ND	3	0.12	6%
PE 0112	Eggs	–	0	–	–	ND	–	–	ND	ND	1	ND	–
FB 0269	Grapes (fresh, wine, dried)	–	0.06	JPN	15.9	388	456	SWE	438	3	2b	4.39	220%
VL 0480	Kale	–	0.01	NLD	17.0	149	–	–	ND	ND	ND	ND	–
VB 0405	Kohlrabi	–	0.05	–	–	ND	135	USA	99	3	ND	ND	–
FC 0204	Lemon	–	0.04	JPN	15.9	88	173	SWE	92	3	2b	0.67	30%
MM 0097	Meat of cattle, pigs and sheep	–	0	AUS	19.0	261	–	–	ND	ND	1	0.00	0%
ML 0106	Milks	0	–	USA	15.0	1286	–	–	ND	ND	3	0.00	0%
FC 0004	Orange, sweet, sour (including orange-like hybrids)	–	0.04	UNK	14.5	495	251	SWE	178	3	2a	2.35	120%
FP 0230	Pear	–	0.04	UNK	14.5	279	187	UNK	170	3	2a	1.71	90%
MF 0818	Pig fat	–	0	FRA	17.8	85	–	–	ND	ND	1	0.00	0%
VR 0589	Potato	–	0.01	SAF	14.2	300	216	UNK	216	3	2a	0.52	30%
PM 0110	Poultry meat	–	0	AUS	19.0	224	–	–	ND	ND	1	0.00	0%
PF 0111	Poultry, fat	–	0	FRA	17.8	20	–	–	ND	ND	1	0.00	0%
GC 0650	Rye	0.01	–	NLD	17.0	37	–	–	ND	ND	3	0.02	1%
MF 0822	Sheep fat	–	0	USA	15.0	28	–	–	ND	ND	1	0.00	0%
VR 0596	Sugar-beet	0.01	–	–	–	ND	–	–	ND	ND	ND	ND	–

345

Annex 4

PARAQUAT (057): International estimate of short-term intake for **General population**

ARfD = 0.006 mg/kg bw/day (6 μg/kg bw/day)
Maximum % ARfD: 20%

Codex code	Commodity	STMR or STMR-P (mg/kg)	HR or HR-P (mg/kg)	Large portion diet Country	Body weight (kg)	Large portion (g/person)	Unit weight (g)	Unit weight Country	Unit weight, edible portion (g)	Variability factor	Case	IESTI (μg/kg bw/day)	% ARfD rounded
TN 0660	Almond	–	0.05	JPN	52.6	74	–	–	ND	ND	1	0.07	1%
FP 0226	Apple	–	0	USA	65.0	1348	138	USA	127	3	2a	0.00	0%
FI 0327	Banana	–	0.01	SAF	55.7	613	708	USA	481	3	2a	0.28	5%
VR 0577	Carrot	–	0.05	NLD	63.0	335	250	JPN	250	3	2a	0.66	10%
ML 0812	Cattle milk	0.00008	–	NLD	63.0	2515	–	–	ND	ND	3	0.00	0%
PE 0840	Chicken eggs	–	0	FRA	62.3	219	–	–	ND	ND	1	0.00	0%
VD 0524	Chick-pea (dry)	0.1	–	FRA	62.3	203	–	–	ND	ND	3	0.33	5%
VD 0526	Common bean (dry)	0.1	–	FRA	62.3	283	–	–	ND	ND	3	0.45	8%
SO 0691	Cotton-seed	0.21	–	USA	65.0	3	–	–	ND	ND	3	0.01	0%
OC 0691	Cotton-seed oil, crude†	0.01	–	USA	65.0	9	–	–	ND	ND	3	0.00	0%
MO 0105	Edible offal (mammalian)	–	0.033	FRA	62.3	277	–	–	ND	ND	1	0.15	2%
FB 0269	Grapes (fresh, wine, dried)	–	0	AUS	67.0	1004	125	FRA	118	3	2a	0.00	0%
FC 0203	Grapefruit	–	0.02	JPN	52.6	947	256	USA	125	3	2a	0.46	8%
FI 0336	Guava	–	0.01	AUS	67.0	450	90	USA	87	3	2a	0.09	2%
TN 0666	Hazelnut	–	0.05	AUS	67.0	70	–	–	ND	ND	1	0.05	1%
DH 1100	Hops, dry	0.05	–	USA	65.0	6	–	–	ND	ND	3	0.00	0%
FI 0341	Kiwi fruit	–	0.01	NLD	63.0	355	76	USA	74	3	2a	0.08	1%
FC 0204	Lemon	–	0.02	FRA	62.3	115	108	USA	72	3	2a	0.08	1%
VL 0482	Lettuce, head	–	0.05	USA	65.0	213	539	USA	512	3	2b	0.49	8%
VL 0483	Lettuce, leaf	–	0.05	NLD	63.0	152	10	USA	10	1	1	0.12	2%
TN 0669	Macadamia nuts	–	0.05	USA	65.0	107	–	–	ND	ND	1	0.08	1%
CF 1255	Maize flour	0.038	–	AUS	67.0	90	–	–	ND	ND	3	0.05	1%
GC 0645	Maize (fresh, flour, oil)	0.025	–	FRA	62.3	260	–	–	ND	ND	3	0.10	2%
OC 0645	Maize oil, crude†	0.006	–	NLD	63.0	43	–	–	ND	ND	3	0.00	0%
MM 0095	Meat from mammals other than marine mammals	–	0.005	AUS	67.0	521	–	–	ND	ND	1	0.04	1%
FS 0245	Nectarine	–	0	USA	65.0	590	136	USA	125	3	2a	0.00	0%
FT 0305	Olive	–	0.1	NLD	63.0	63	–	–	ND	ND	ND	ND	–
OC 0305	Olive oil, virgin	0.018	–	–	–	ND	–	–	ND	ND	3	ND	–

346

Annex 4

Codex code	Commodity	STMR or STMR-P (mg/kg)	HR or HR-P (mg/kg)	Large portion diet Country	Body weight (kg)	Large portion (g/person)	Unit weight (g)	Unit weight Country	Unit weight, edible portion (g)	Varia-bility factor	Case	IESTI (µg/kg bw/day)	% ARfD rounded
FC 0004	Orange, sweet, sour (including orange-like hybrids)	–	0.02	USA	65.0	564	131	USA	96	3	2a	0.23	4%
FS 0247	Peach	–	0	SAF	55.7	685	98	USA	85	3	2a	0.00	0%
FP 0230	Pear	–	0	USA	65.0	693	166	USA	151	3	2a	0.00	0%
VD 0072	Peas (dry)	0.1	–	FRA	62.3	445	–	–	ND	ND	3	0.71	10%
VO 0445	Peppers, sweet (including pim(i)ento)	–	0.04	FRA	62.3	207	119	USA	98	3	2a	0.26	4%
VR 0589	Potato	–	0.05	NLD	63.0	687	122	USA	99	3	2a	0.70	10%
PM 0110	Poultry meat	–	0	AUS	67.0	431	–	–	ND	ND	1	0.00	0%
FB 0275	Strawberry	–	0	FRA	62.3	346	14	FRA	13	1	1	0.00	0%
GC 0651	Sorghum	0.025	–	USA	65.0	18	–	–	ND	ND	3	0.01	0%
VD 0541	Soya bean (dry)	0.1	–	JPN	52.6	159	–	–	ND	ND	3	0.30	5%
OC 0541	Soya bean oil, crude†	0.01	–	USA	65.0	98	–	–	ND	ND	3	0.02	0%
VL 0502	Spinach (bunch)	–	0.05	NLD	63.0	820	340	USA	245	3	2a	1.04	20%
SO 0702	Sunflower seed	0.22	–	USA	65.0	193	–	–	ND	ND	3	0.65	10%
OC 0702	Sunflower seed oil, crude†	0	–	FRA	62.3	61	–	–	ND	ND	3	0.00	0%
DT 1114	Tea, green, black (black, fermented and dried)	0.01	–	JPN	52.6	16	–	–	ND	ND	3	0.00	0%
VO 0448	Tomato (fresh, juice, paste, peeled)	–	0.04	USA	65.0	391	123	USA	123	3	2a	0.39	7%
TN 0085	Tree nuts	–	0.05	JPN	52.6	107	–	–	ND	ND	1	0.10	2%
VL 0506	Turnip greens	–	0.05	USA	65.0	353	800	JPN	800	3	2b	0.81	10%

†, calculated from information on consumption of edible oils

Annex 4

PARAQUAT (057) : International estimate of short-term intake for Children up to 6 years

ARfD = 0.006 mg/kg bw; 6 µg/kg bw
Maximum % ARfD: 50%

Codex code	Commodity	STMR or STMR-P (mg/kg)	HR or HR-P (mg/kg)	Country	Body weight (kg)	Large portion (g/person)	Unit weight (g)	Unit weight Country	Unit weight, edible portion (g)	Variability factor	Case	IESTI (µg/kg bw/day)	% ARfD rounded
TN 0660	Almond	–	0.05	FRA	17.8	31	–	–	ND	ND	1	0.09	1%
FP 0226	Apple	–	0	USA	15.0	679	138	USA	127	3	2a	0.00	0%
FI 0327	Banana	–	0.01	JPN	15.9	312	708	USA	481	3	2b	0.59	10%
VR 0577	Carrot	–	0.05	FRA	17.8	205	250	JPN	250	3	2b	1.73	30%
ML 0812	Cattle milk	0.00008	–	AUS	19.0	1450	–	–	ND	ND	3	0.01	0%
PE 0840	Chicken eggs	–	0	FRA	17.8	134	–	–	ND	ND	1	0.00	0%
VD 0524	Chick-pea (dry)	0.1	–	USA	15.0	34	–	–	ND	ND	3	0.23	4%
VD 0526	Common bean (dry)	0.1	–	FRA	17.8	209	–	–	ND	ND	3	1.18	20%
SO 0691	Cotton-seed	0.21	–	USA	15.0	1	–	–	ND	ND	3	0.01	0%
OC 0691	Cotton-seed oil, crude†	0.01	–	USA	15.0	6	–	–	ND	ND	3	0.00	0%
MO 0105	Edible offal (mammalian)	–	0.033	FRA	17.8	203	–	–	ND	ND	1	0.38	6%
FB 0269	Grapes (fresh, wine, dried)	–	0	JPN	15.9	388	125	FRA	118	3	2a	0.00	0%
FC 0203	Grapefruit	–	0.02	FRA	17.8	381	256	USA	125	3	2a	0.71	10%
FI 0336	Guava	–	0.01	AUS	19.0	34	90	USA	87	3	2b	0.05	1%
TN 0666	Hazelnuts	–	0.05	NLD	17.0	11	–	–	ND	ND	1	0.03	1%
DH 1100	Hops, dry	0.05	–	JPN	15.9	0	–	–	ND	ND	3	0.00	0%
FI 0341	Kiwi fruit	–	0.01	JPN	15.9	162	76	USA	74	3	2a	0.19	3%
FC 0204	Lemon	–	0.02	JPN	15.9	88	108	USA	72	3	2a	0.29	5%
VL 0482	Lettuce, head	–	0.05	NLD	17.0	84	539	USA	512	3	2b	0.74	10%
VL 0483	Lettuce, leaf	–	0.05	NLD	17.0	102	10	USA	10	1	1	0.30	5%
TN 0669	Macadamia nuts	–	0.05	–	–	ND	–	–	ND	ND	1	ND	–
CF 1255	Maize flour	0.038	–	AUS	19.0	60	–	–	ND	ND	3	0.12	2%
GC 0645	Maize (fresh, flour, oil)	0.025	–	FRA	17.8	148	–	–	ND	ND	3	0.21	3%
OC 0645	Maize oil, crude†	0.006	–	FRA	17.8	21	–	–	ND	ND	3	0.01	0%
MM 0095	Meat from mammals other than marine mammals	–	0.005	AUS	19.0	261	–	–	ND	ND	1	0.07	1%
FS 0245	Nectarine	–	0	AUS	19.0	302	136	USA	125	3	2a	0.00	0%
FT 0305	Olive	–	0.1	FRA	17.8	49	–	–	ND	ND	ND	ND	–
OC 0305	Olive oil, virgin	0.018	–	–	–	ND	–	–	ND	ND	3	ND	–

Annex 4

FC 0004	Orange, sweet, sour (including orange-like hybrids)	–	0.02	UNK	14.5	495	131	USA	96	3	2a	0.95	20%
FS 0247	Peach	–	0	AUS	19.0	315	98	USA	85	3	2a	0.00	0%
FP 0230	Pear	–	0	UNK	14.5	279	166	USA	151	3	2a	0.00	0%
VD 0072	Peas (dry)	0.1	–	FRA	17.8	107	–	–	ND	ND	3	0.60	10%
VO 0445	Peppers, sweet (including pim(i)ento)	–	0.04	AUS	19.0	60	119	USA	98	3	2b	0.38	6%
VR 0589	Potato	–	0.05	SAF	14.2	300	122	USA	99	3	2a	1.75	30%
PM 0110	Poultry meat	–	0	AUS	19.0	224	–	–	ND	ND	1	0.00	0%
GC 0651	Sorghum	0.025	–	–	–	ND	–	–	ND	ND	3	ND	–
VD 0541	Soya bean (dry)	0.1	–	JPN	15.9	88	–	–	ND	ND	3	0.56	9%
OC 0541	Soya bean oil, crude†	0.01	–	USA	15.0	35	–	–	ND	ND	3	0.02	0%
VL 0502	Spinach (bunch)	–	0.05	SAF	14.2	420	340	USA	245	3	2a	3.20	50%
FB 0275	Strawberry	–	0	AUS	19.0	176	14	FRA	13	1	1	0.00	0%
SO 0702	Sunflower seed	0.22	–	USA	15.0	24	–	–	ND	ND	3	0.35	6%
OC 0702	Sunflower seed oil, crude†	0	–	FRA	17.8	37	–	–	ND	ND	3	0.00	0%
DT 1114	Tea, green, black (black, fermented and dried)	0.01	–	JPN	15.9	10	–	–	ND	ND	3	0.01	0%
VO 0448	Tomato (fresh, juice, paste, peeled)	–	0.04	USA	15.0	159	123	USA	123	3	2a	1.08	20%
TN 0085	Tree nuts	–	0.05	AUS	19.0	28	–	–	ND	ND	1	0.07	1%
VL 0506	Turnip greens	–	0.05	USA	15.0	90	800	JPN	800	3	2b	0.90	10%

†, calculated from information on consumption of edible oils

Annex 4

International estimate of short-term intake for **General population**

ARfD = 0.1 mg/kg bw; 100 μg/kg bw
Maximum % ARfD: 130%

Codex code	Commodity	STMR or STMR-P (mg/kg)	HR or HR-P (mg/kg)	Large portion diet Country	Body weight (kg)	Large portion (g/person)	Unit weight (g)	Unit weight Country	Unit weight, edible portion (g)	Variability factor	Case	IESTI (μg/kg bw/day)	% ARfD rounded
FI 0326	Avocado	–	0.7	FRA	62.3	260	300	FRA	180	3	2a	6.96	7%
FI 0327	Banana	–	0.7	SAF	55.7	613	720	JPN	720	3	2b	23.10	20%
GC 0640	Barley (fresh, flour, beer)	0.11	–	NLD	63.0	378	–	–	ND	ND	3	0.66	1%
CM 0081	Bran, unprocessed of cereal grain (except buckwheat, canihua, quinoa)	0.54	–	AUS	67.0	37	–	–	ND	ND	3	0.30	0%
PE 0840	Chicken eggs	–	0.07	FRA	62.3	219	–	–	ND	ND	1	0.25	0%
FI 0332	Custard apple	–	0.7	AUS	67.0	654	–	–	ND	ND	1	6.83	7%
MO 0105	Edible offal (mammalian)	–	6.2	FRA	62.3	277	–	–	ND	ND	1	27.53	30%
FI 0335	Feijoa	–	0.7	AUS	67.0	120	–	–	ND	ND	1	1.25	1%
FC 0203	Grapefruit	–	0.92	JPN	52.6	947	400	JPN	400	3	2a	30.55	30%
FI 0336	Guava	–	0.7	AUS	67.0	450	90	USA	87	3	2a	6.53	7%
FI 0338	Jackfruit	–	0.7	AUS	67.0	348	–	–	ND	ND	1	3.63	4%
FI 0341	Kiwi fruit	–	0.7	NLD	63.0	355	120	JPN	120	3	2a	6.61	7%
FC 0204	Lemon	–	0.92	FRA	62.3	115	173	SWE	92	3	2a	4.41	4%
FC 0205	Lime	–	0.92	AUS	67.0	590	67	USA	56	3	2a	9.64	10%
SO 0693	Linseed	0.05	–	NLD	63.0	21	–	–	ND	ND	3	0.02	0%
GC 0645	Maize (fresh, flour, oil)	0.11	–	FRA	62.3	260	–	–	ND	ND	3	0.46	0%
FC 0206	Mandarin	–	0.92	JPN	52.6	409	168	USA	124	3	2a	11.50	10%
FC 0003	Mandarins (including Mandarin-like hybrids)	–	0.92	USA	65.0	394	–	–	ND	ND	1	5.58	6%
MM 0095	Meat from mammals other than marine mammals: 20% as fat	–	0.38	AUS	67.0	104	–	–	ND	ND	1	0.59	1%
ML 0106	Milks	0	–	USA	65.0	2466	–	–	ND	ND	3	0.00	0%
GC 0646	Millet	0.11	–	AUS	67.0	101	–	–	ND	ND	3	0.17	0%
VO 0450	Mushroom	–	37	FRA	62.3	219	21	UNK	20	1	1	129.87	130%
GC 0647	Oats	0.11	–	FRA	62.3	305	–	–	ND	ND	3	0.54	1%
FC 0004	Orange, sweet, sour (including orange-like hybrids)	–	0.92	USA	65.0	564	200	JPN	200	3	2a	13.65	10%

Annex 4

Codex code	Commodity	STMR or STMR-P (mg/kg)	HR or HR-P (mg/kg)	Large portion diet Country	Body weight (kg)	Large portion (g/person)	Unit weight (g)	Unit weight Country	Unit weight, edible portion (g)	Variability factor	Case	IESTI (µg/kg bw/day)	% ARfD rounded
FI 0350	Papaya	–	0.7	USA	65.0	567	250	JPN	250	3	2a	11.49	10%
FI 0351	Passion fruit	–	0.7	JPN	52.6	554	–	–	ND	ND	1	7.37	7%
HS 0790	Pepper (black, white)	5.1	–	–	–	ND	–	–	ND	ND	3	ND	–
FI 0352	Persimmon, American	–	0.7	AUS	67.0	672	122	SWE	102	3	2a	9.16	9%
FI 0353	Pineapple (fresh, canned, juice, dried)	–	0.7	JPN	52.6	371	700	FRA	420	3	2b	14.83	10%
FI 0354	Plantain	–	0.7	AUS	67.0	160	–	–	ND	ND	1	1.67	2%
PM 0110	Poultry meat: 10% as fat	–	0.007	AUS	67.0	43	–	–	ND	ND	1	0.00	0%
PM 0110	Poultry meat: 90% as muscle	–	0.005	AUS	67.0	388	–	–	ND	ND	1	0.03	0%
PO 0111	Poultry, edible offal of	–	0.1	USA	65.0	248	–	–	ND	ND	1	0.38	0%
FI 0358	Rambutan	–	0.7	AUS	67.0	562	–	–	ND	ND	1	5.87	6%
OR 0495	Rape-seed oil, edible	0.06	–	AUS	67.0	65	–	–	ND	ND	3	0.06	0%
GC 0649	Rice	0.11	–	FRA	62.3	312	–	–	ND	ND	3	0.55	1%
GC 0650	Rye	0.11	–	NLD	63.0	77	–	–	ND	ND	3	0.13	0%
FC 0005	Shaddocks or pomelos (including Shaddock-like hybrids)	–	0.92	USA	65.0	448	210	FRA	126	3	2a	9.91	10%
GC 0651	Sorghum	0.11	–	USA	65.0	18	–	–	ND	ND	3	0.03	0%
SO 0702	Sunflower seed	0.1	–	USA	65.0	193	–	–	ND	ND	3	0.30	0%
OR 0702	Sunflower seed oil, edible	0.06	–	FRA	62.3	61	–	–	ND	ND	3	0.06	0%
FC 4031	Tangelo	–	0.92	AUS	67.0	114	–	–	ND	ND	3	ND	–
GC 0653	Triticale	0.11	–	–	–	ND	–	–	ND	ND	3	ND	–
GC 0654	Wheat	0.11	–	USA	65.0	383	–	–	ND	ND	3	0.65	1%
CM 0654	Wheat bran, unprocessed	0.54	–	USA	65.0	80	–	–	ND	ND	3	0.66	1%
CF 1211	Wheat flour	0.025	–	USA	65.0	365	–	–	ND	ND	3	0.14	0%
CP 1212	Wholemeal bread	0.14	–	SAF	55.7	395	–	–	ND	ND	3	0.99	1%

Annex 4

PROCHLORAZ (142)

International estimate of short-term intake for Children up to 6 years

ARfD = 0.1 mg/kg bw ; 100 µg/kg bw
Maximum % ARfD: 150%

Codex code	Commodity	STMR or STMR-P (mg/kg)	HR or HR-P (mg/kg)	Large portion diet Country	Body weight (kg)	Large portion (g/person)	Unit weight (g)	Unit weight Country	Unit weight, edible portion (g)	Varia-bility factor	Case	IESTI (µg/kg bw/day)	% ARfD rounded
FI 0326	Avocado	–	0.7	USA	15.0	131	300	FRA	180	3	2b	18.27	20%
FI 0327	Banana	–	0.7	JPN	15.9	312	900	FRA	612	3	2b	41.18	40%
GC 0640	Barley (fresh, flour, beer)	0.11	–	AUS	19.0	14	–	–	ND	ND	3	0.08	0%
CM 0081	Bran, unprocessed of cereal grain (except buckwheat, canihua, quinoa)	0.54	–	AUS	19.0	13	–	–	ND	ND	3	0.36	0%
PE 0840	Chicken eggs	–	0.07	FRA	17.8	134	–	–	ND	ND	1	0.53	1%
FI 0332	Custard apple	–	0.7	–	–	ND	–	–	ND	ND	1	ND	–
MO 0105	Edible offal (mammalian)	–	6.2	FRA	17.8	203	–	–	ND	ND	1	70.62	70%
FI 0335	Feijoa	–	0.7	–	–	ND	–	–	ND	ND	1	ND	–
FC 0203	Grapefruit	–	0.92	FRA	17.8	381	400	JPN	400	3	2b	59.15	60%
FI 0336	Guava	–	0.7	AUS	19.0	34	90	USA	87	3	2b	3.80	4%
FI 0338	Jackfruit	–	0.7	–	–	ND	–	–	ND	ND	1	ND	–
FI 0341	Kiwi fruit	–	0.7	JPN	15.9	162	76	USA	74	3	2a	13.63	10%
FC 0204	Lemon	–	0.92	JPN	15.9	88	173	SWE	92	3	2b	15.35	20%
FC 0205	Lime	–	0.92	AUS	19.0	26	67	USA	56	3	2b	3.75	4%
SO 0693	Linseed	0.05	–	–	–	ND	–	–	ND	ND	3	ND	–
GC 0645	Maize (fresh, flour, oil)	0.11	–	FRA	17.8	148	–	–	ND	ND	3	0.92	1%
FC 0206	Mandarin	–	0.92	JPN	15.9	353	168	USA	124	3	2a	34.83	30%
FC 0003	Mandarin (including mandarin-like hybrids)	–	0.92	USA	15.0	205	–	–	ND	ND	1	12.56	10%
FI 0345	Mango	–	0.7	AUS	19.0	207	339	SWE	234	3	2b	22.89	20%
MM 0095	Meat from mammals other than marine mammals: 20% as fat	–	0.38	AUS	19.0	52	–	–	ND	ND	1	1.04	1%
MM 0095	Meat from mammals other than marine mammals: 80% as muscle	–	0.1	AUS	19.0	208	–	–	ND	ND	1	1.10	1%
ML 0106	Milks	0	–	USA	15.0	1286	–	–	ND	ND	3	0.00	0%
GC 0646	Millet	0.11	–	–	–	ND	–	–	ND	ND	3	ND	–
VO 0450	Mushroom	–	37	FRA	17.8	71	21	UNK	20	1	1	148.00	150%
GC 0647	Oats	0.11	–	USA	15.0	62	–	–	ND	ND	3	0.46	0%
FC 0004	Orange, sweet, sour (including orange-like hybrids)	–	0.92	UNK	14.5	495	200	JPN	200	3	2a	56.79	60%

Annex 4

Codex code	Commodity	STMR or STMR-P (mg/kg)	HR or HR-P (mg/kg)	Large portion diet Country	Body weight (kg)	Large portion (g/person)	Unit weight (g)	Unit weight Country	Unit weight, edible portion (g)	Variability factor	Case	IESTI (μg/kg bw/day)	% ARfD rounded
FI 0350	Papaya	–	0.7	USA	15.0	240	346	SWE	232	3	2a	32.84	30%
FI 0351	Passion fruit	–	0.7	JPN	15.9	167	–	–	ND	ND	1	7.37	7%
HS 0790	Pepper (black, white)	5.1	–	–	–	ND	–	–	ND	ND	3	ND	–
FI 0352	Persimmon, American	–	0.7	–	–	ND	122	SWE	ND	3	ND	ND	–
FI 0353	Pineapple (fresh, canned, juice, dried)	–	0.7	JPN	15.9	216	700	FRA	420	3	2b	28.58	30%
FI 0354	Plantain	–	0.7	–	–	ND	–	–	ND	ND	1	ND	–
PM 0110	Poultry meat: 10% as fat	–	0.007	AUS	19.0	22	–	–	ND	ND	1	0.01	0%
PM 0110	Poultry meat: 90% as muscle	–	0.005	AUS	19.0	201	–	–	ND	ND	1	0.05	0%
PO 0111	Poultry, edible offal of	–	0.1	USA	15.0	37	–	–	ND	ND	1	0.25	0%
FI 0358	Rambutan	–	0.7	AUS	19.0	20	–	–	ND	ND	1	0.74	1%
OR 0495	Rape-seed oil, edible	0.06	–	AUS	19.0	18	–	–	ND	ND	3	0.06	0%
GC 0649	Rice	0.11	–	FRA	17.8	223	–	–	ND	ND	3	1.38	1%
GC 0650	Rye	0.11	–	NLD	17.0	37	–	–	ND	ND	3	0.24	0%
FC 0005	Shaddocks or pomelos (including Shaddock-like hybrids)	–	0.92	FRA	17.8	381	210	FRA	126	3	2a	32.74	30%
GC 0651	Sorghum	0.11	–	–	–	ND	–	–	ND	ND	3	ND	–
SO 0702	Sunflower seed	0.1	–	USA	15.0	24	–	–	ND	ND	3	0.16	0%
OR 0702	Sunflower seed oil, edible	0.06	–	FRA	17.8	37	–	–	ND	ND	3	0.12	0%
FC 4031	Tangelo	–	0.92	–	–	ND	–	–	ND	ND	ND	ND	–
GC 0653	Triticale	0.11	–	–	–	ND	–	–	ND	ND	3	ND	–
GC 0654	Wheat	0.11	–	USA	15.0	151	–	–	ND	ND	3	1.11	1%
CM 0654	Wheat bran, unprocessed	0.54	–	USA	15.0	30	–	–	ND	ND	3	1.07	1%
CF 1211	Wheat flour	0.025	–	AUS	19.0	194	–	–	ND	ND	3	0.26	0%
CP 1212	Wholemeal bread	0.14	–	SAF	14.2	240	–	–	ND	ND	3	2.37	2%

Annex 4

PROPINEB International estimate of short-term intake for **General population**

ARfD = 0.1 mg/kg bw ; 100 µg/kg bw
Maximum % ARfD: 110%

Codex code	Commodity	STMR or STMR-P (mg/kg)	HR or HR-P (mg/kg)	Large portion diet Country	Body weight (kg)	Large portion (g/person)	Unit weight (g)	Unit weight Country	Unit weight, edible portion (g)	Variability factor	Case	IESTI (µg/kg bw/day)	% ARfD rounded
FS 0013	Cherry	–	0.35	FRA	62.3	375	12	UNK	10	1	1	2.11	2%
PE 0840	Chicken eggs	–	0	FRA	62.3	219	–	–	ND	ND	1	0.00	0%
VC 0424	Cucumber	–	1.1	NLD	63.0	313	400	FRA	360	3	2b	16.40	20%
MO 0105	Edible offal (mammalian)	–	0	FRA	62.3	277	–	–	ND	ND	1	0.00	0%
MM 0095	Meat from mammals other than marine mammals: 20% as fat	–	0	AUS	67.0	104	–	–	ND	ND	1	0.00	0%
MM 0095	Meat from mammals other than marine mammals: 80% as muscle	–	0	AUS	67.0	417	–	–	ND	ND	1	0.00	0%
ML 0106	Milks	0	–	USA	65.0	2466	–	–	ND	ND	3	0.00	0%
VO 0445	Peppers, sweet (including pim(i)ento)	–	13	FRA	62.3	207	172	UNK	160	3	2a	110.05	110%
VR 0589	Potato	–	0.16	NLD	63.0	687	216	UNK	216	3	2a	2.84	3%
PM 0110	Poultry meat: 10% as fat	–	0	AUS	67.0	43	–	–	ND	ND	1	0.00	0%
PM 0110	Poultry meat: 90% as muscle	–	0	AUS	67.0	388	–	–	ND	ND	1	0.00	0%
PO 0111	Poultry, edible offal of	–	0	USA	65.0	248	–	–	ND	ND	1	0.00	0%
VO 0448	Tomato (fresh, juice, paste, peeled)	–	2.9	USA	65.0	391	150	JPN	150	3	2a	30.81	30%
JF 0448	Tomato juice	0.18	–	–	–	ND	–	–	ND	ND	3	ND	–
	Tomato paste	1.86	–	–	–	ND	–	–	ND	ND	ND	ND	–

Annex 4

PROPINEB

International estimate of short-term intake for **Children up to 6 years**

ARfD = 0.1 mg/kg bw ; 100 µg/kg bw
Maximum % ARfD: 120%

Codex code	Commodity	STMR or STMR-P (mg/kg)	HR or HR-P (mg/kg)	Country	Body weight (kg)	Large portion (g/person)	Unit weight (g)	Country	Unit weight, edible portion (g)	Variability factor	Case	IESTI (µg/kg bw/day)	% ARfD rounded
FS 0013	Cherry	–	0.35	FRA	17.8	297	12	UNK	10	1	1	5.83	6%
PE 0840	Chicken eggs	–	0	FRA	17.8	134	–	–	ND	ND	1	0.00	0%
VC 0424	Cucumber	–	1.1	NLD	17.0	162	400	FRA	360	3	2b	31.45	30%
MO 0105	Edible offal (mammalian)	–	0	FRA	17.8	203	–	–	ND	ND	1	0.00	0%
MM 0095	Meat from mammals other than marine mammals: 20% as fat	–	0	AUS	19.0	52	–	–	ND	ND	1	0.00	0%
MM 0095	Meat from mammals other than marine mammals: 80% as muscle	–	0	AUS	19.0	208	–	–	ND	ND	1	0.00	0%
ML 0106	Milks	0	–	USA	15.0	1286	–	–	ND	ND	3	0.00	0%
VO 0445	Peppers, sweet (including pim(i)ento)	–	13	AUS	19.0	60	172	UNK	160	3	2b	123.24	120%
VR 0589	Potato	–	0.16	SAF	14.2	300	216	UNK	216	3	2a	8.24	8%
PM 0110	Poultry meat: 10% as fat	–	0	AUS	19.0	22	–	–	ND	ND	1	0.00	0%
PM 0110	Poultry meat: 90% as muscle	–	0	AUS	19.0	201	–	–	ND	ND	1	0.00	0%
PO 0111	Poultry, edible offal of	–	0	USA	15.0	37	–	–	ND	ND	1	0.00	0%
VO 0448	Tomato (fresh, juice, paste, peeled)	–	2.9	USA	15.0	159	150	JPN	150	3	2a	88.74	90%
FJ 0448	Tomato juice	0.18	–	–	–	ND	–	–	ND	ND	3	ND	–
	Tomato paste	1.86	–	–	–	ND	–	–	ND	ND	ND	ND	–

355

Annex 4

PYRACLOSTROBIN (210)

International estimate of short-term intake for **General population**

Acute RfD = 0.05 mg/kg bw; 50µg/kg bw
Maximum % of ARfD 30%

Codex code	Commodity	STMR or STMR-P (mg/kg)	HR or HR-P (mg/kg)	Large portion diet Country	Body weight (kg)	Large portion (g/person)	Unit weight (g)	Unit weight Country	Unit weight, edible portion (g)	Varia-bility factor	Case	IESTI (µg/kg bw/day)	% ArfD rounded
TN 0660	Almond	–	0.02	JPN	52.6	74	–	–	ND	ND	1	0.03	0%
FI 0327	Banana	–	0.02	SAF	55.7	613	900	FRA	612	3	2a	0.66	1%
GC 0640	Barley (fresh, flour, beer)	0.04	–	NLD	63.0	378	–	–	ND	ND	3	0.24	0%
VD 0071	Beans (dry)	0.02	–	FRA	62.3	255	–	–	ND	ND	3	0.08	0%
FB 0020	Blueberry	–	0.57	AUS	67.0	158	–	–	ND	ND	ND	ND	–
VR 0577	Carrot	–	0.24	NLD	63.0	335	100	FRA	89	3	2a	1.95	4%
FS 0013	Cherry	–	0.63	FRA	62.3	375	5	FRA	4	1	1	3.79	8%
SB 0716	Coffee bean	0.03	–	NLD	63.0	66	–	–	ND	ND	3	0.03	0%
MO 0105	Edible offal (mammalian)	–	0.037	FRA	62.3	277	–	–	ND	ND	1	0.16	0%
PE 0112	Eggs	–	0	–	–	ND	–	–	ND	ND	1	ND	–
VA 0381	Garlic	–	0.05	FRA	62.3	22	–	–	ND	ND	ND	ND	–
FB 0269	Grapes (fresh, dried, excluding wine)	–	1.38	AUS	67.0	513	150	JPN	150	3	2a	16.75	30%
DF 0269	Grapes, dried (currants, raisins and sultanas)	–	4.27	FRA	62.3	135	–	–	ND	ND	1	9.27	20%
VR 0583	Horseradish	–	0.3	FRA	62.3	493	–	–	ND	ND	ND	ND	–
VD 0533	Lentil (dry)	0.13	–	FRA	62.3	435	–	–	ND	ND	3	0.91	2%
GC 0645	Maize (fresh)	0.02	–	–	–	ND	–	–	ND	ND	3	ND	–
GC 0645	Maize (fresh, flour, oil)	0.02	–	FRA	62.3	260	–	–	ND	ND	3	0.08	0%
FI 0345	Mango	–	0.05	FRA	62.3	567	207	USA	139	3	2a	0.68	1%
MM 0095	Meat from mammals other than marine mammals 20% as fat	–	0.41	AUS	67.0	104	–	–	ND	ND	1	0.64	1%
MM 0095	Meat from mammals other than marine mammals 80% as fat	–	0.044	AUS	67.0	417	–	–	ND	ND	1	0.27	1%
ML0106	Milks	0.01	–	USA	65.0	2466	–	–	ND	ND	3	0.38	1%
GC 0647	Oats	0.17	–	FRA	62.3	305	–	–	ND	ND	3	0.83	2%
VA 0385	Onion, bulb	–	0.09	FRA	62.3	306	165	UNK	150	3	2a	0.88	2%
FC 0004	Orange, sweet, sour (including orange-like hybrids)	–	0.51	USA	65.0	564	190	FRA	137	3	2a	6.57	10%
FI 0350	Papaya	–	0.05	USA	65.0	567	250	JPN	250	3	2a	0.82	2%
VD 4511	Pea (dry) = field pea (dry)	0.17	–	NLD	63.0	252	–	–	ND	ND	3	0.68	1%

356

Annex 4

Codex code	Commodity	STMR or STMR-P (mg/kg)	HR or HR-P (mg/kg)	Large portion diet Country	Body weight (kg)	Large portion (g/person)	Unit weight (g)	Unit weight Country	Unit weight, edible portion (g)	Variability factor	Case	IESTI (µg/kg bw/day)	% ArfD rounded
FS 0247	Peach	–	0.31	SAF	55.7	685	150	JPN	150	3	2a	5.48	10%
SO 0697	Peanut	0.02	–	FRA	62.3	161	–	–	ND	ND	3	0.05	0%
TN 0672	Pecan	–	0.02	AUS	67.0	23	–	–	ND	ND	1	0.01	0%
TN 0675	Pistachio nut	–	0.45	AUS	67.0	300	–	–	ND	ND	1	2.02	4%
FS 0014	Plums (fresh, prunes)	–	0.19	USA	65.0	413	59	UNK	55	3	2a	1.53	3%
VR 0589	Potato	0.02	0.02	NLD	63.0	687	200	FRA	160	3	2a	0.32	1%
PM 0110	Poultry meat	–	0	AUS	67.0	431	–	–	ND	ND	1	0.00	0%
PO 0111	Poultry, edible offal of	–	0	USA	65.0	248	–	–	ND	ND	1	0.00	0%
PF 0111	Poultry, fats	–	0	FRA	62.3	46	–	–	ND	ND	1	0.00	0%
VC 0431	Squash, summer	–	0.18	FRA	62.3	343	196	USA	186	3	2a	2.07	4%
FB 0275	Strawberry	–	0.26	FRA	62.3	346	14	FRA	13	1	1	1.44	3%
VO 0448	Tomato (fresh, juice, paste, peeled)	–	0.21	USA	65.0	391	85	UNK	85	3	2a	1.81	4%
GC 0654	Wheat	0.02	–	USA	65.0	383	–	–	ND	ND	3	0.12	0%
CM 0654	Wheat bran, unprocessed	0.012	–	USA	65.0	80	–	–	ND	ND	3	0.01	0%
CF 1211	Wheat flour	0.012	–	USA	65.0	365	–	–	ND	ND	3	0.07	0%
CF 1210	Wheat germ	0.016	–	FRA	62.3	207	–	–	ND	ND	3	0.05	0%
	Wine only	0.04	–	AUS	67.0	1131	–	–	ND	ND	3	0.68	1%

For citrus, it was noted that no residues occurred in pulp, so the acute exposure was 0.
For grapes, it was noted that the consumption took in account dried grapes.

Annex 4

PYRACLOSTROBIN (210)

International estimate of short-term intake for r
Acute RfD=0.05 mg/kg bw; 50µg/kg bw

Children up to 6 years

Codex code	Commodity	STMR or STMR-P (mg/kg)	HR or HR-P (mg/kg)	Country	Body weight (kg)	Large portion (g/person)	Unit weight (g)	Unit weight Country	Unit weight, edible portion (g)	Variability factor	Case	IESTI (µg/kg bw/day)	% ARfD rounded
TN 0660	Almond	–	0.02	FRA	17.8	31	–	–	ND	ND	1	0.04	0%
FI 0327	Banana	–	0.02	JPN	15.9	312	900	FRA	612	3	2b	1.18	2%
GC 0640	Barley (fresh, flour, beer)	0.04	–	AUS	19.0	14	–	–	ND	ND	3	0.03	0%
VD 0071	Beans (dry)	0.02	–	FRA	17.8	209	–	–	ND	ND	3	0.24	0%
FB 0020	Blueberry	–	0.57	FRA	17.8	138	–	–	ND	ND	ND	ND	–
VR 0577	Carrot	–	0.24	FRA	17.8	205	100	FRA	89	3	2a	5.16	10%
FS 0013	Cherry	–	0.63	FRA	17.8	297	5	FRA	4	1	1	10.50	20%
SB 0716	Coffee bean	0.03	–	NLD	17.0	19	–	–	ND	ND	3	0.03	0%
MO 0105	Edible offal (mammalian)	–	0.037	FRA	17.8	203	–	–	ND	ND	1	0.57	1%
PE 0112	Eggs	–	0	–	–	ND	–	–	ND	ND	1	ND	–
VA 0381	Garlic	–	0.05	FRA	17.8	30	–	–	ND	ND	ND	ND	–
FB 0269	Grapes (fresh, dried, excluding wine)	–	1.38	AUS	19.0	342	150	JPN	150	3	2a	46.63	90%
DF 0269	Grapes. dried (currants, raisins and sultanas)	–	4.27	USA	15.0	59	–	–	ND	ND	1	16.87	30%
VR 0583	Horseradish	–	0.3	USA	15.0	127	–	–	ND	ND	ND	ND	–
VD 0533	Lentil (dry)	0.13	–	FRA	17.8	127	–	–	ND	ND	3	0.93	2%
GC 0645	Maize (fresh)	0.02	–	–	–	ND	–	–	ND	ND	3	ND	–
GC 0645	Maize (fresh, flour, oil)	0.02	–	FRA	17.8	148	–	–	ND	ND	3	0.17	0%
FI 0345	Mango	–	0.05	AUS	19.0	207	207	USA	139	3	2a	1.27	3%
MM 0095	Meat from mammals other than marine mammals 20% as fat	–	0.41	AUS	19.0	52	–	–	ND	ND	1	1.12	2%
MM 0095	Meat from mammals other than marine mammals 80% as fat	–	0.044	AUS	19.0	208	–	–	ND	ND	1	0.48	1%
ML 0106	Milks	0.01	–	USA	15.0	1286	–	–	ND	ND	3	0.86	2%
GC 0647	Oats	0.17	–	USA	15.0	62	–	–	ND	ND	3	0.71	1%
VA 0385	Onion, bulb	–	0.09	FRA	17.8	127	165	UNK	150	3	2b	1.93	4%
FC 0004	Orange, sweet, sour (including orange-like hybrids)	–	0.51	UNK	14.5	495	190	FRA	137	3	2a	27.03	50%
FI 0350	Papaya	–	0.05	USA	15.0	240	250	JPN	250	3	2b	2.40	5%
VD 4511	Pea (dry) = field pea (dry)	0.17	–	–	–	ND	–	–	ND	ND	3	ND	–
FS 0247	Peach	–	0.31	AUS	19.0	315	150	JPN	150	3	2a	10.04	20%

Annex 4

Codex code	Commodity	STMR or STMR-P (mg/kg)	HR or HR-P (mg/kg)	Large portion diet Country	Body weight (kg)	Large portion (g/person)	Unit weight (g)	Unit weight Country	Unit weight, edible portion (g)	Variability factor	Case	IESTI (μg/kg bw/day)	% ARfD rounded
SO 0697	Peanut	0.02	–	USA	15.0	78	–	–	ND	ND	3	0.10	0%
TN 0672	Pecan	–	0.02	AUS	19.0	22	–	–	ND	ND	1	0.02	0%
TN 0675	Pistachio nut	–	0.45	AUS	19.0	63	–	–	ND	ND	1	1.48	3%
FS 0014	Plums (fresh, prunes)	–	0.19	FRA	17.8	254	59	UNK	55	3	2a	3.90	8%
VR 0589	Potato	0.02	0.02	SAF	14.2	300	200	FRA	160	3	2a	0.87	2%
PM 0110	Poultry meat	–	0	AUS	19.0	224	–	–	ND	ND	1	0.00	0%
PO 0111	Poultry, edible offal of	–	0	USA	15.0	37	–	–	ND	ND	1	0.00	0%
PF 0111	Poultry, fat	–	0	FRA	17.8	20	–	–	ND	ND	1	0.00	0%
VC 0431	Squash, summer	–	0.18	AUS	19.0	219	196	USA	186	3	2a	5.60	10%
FB 0275	Strawberry	–	0.26	AUS	19.0	176	14	FRA	13	1	1	2.41	5%
VO 0448	Tomato (fresh, juice, paste, peeled)	–	0.21	USA	15.0	159	85	UNK	85	3	2a	4.61	9%
GC 0654	Wheat	0.02	–	USA	15.0	151	–	–	ND	ND	3	0.20	0%
CM 0654	Wheat bran, unprocessed	0.012	–	USA	15.0	30	–	–	ND	ND	3	0.02	0%
CF 1211	Wheat flour	0.012	–	AUS	19.0	194	–	–	ND	ND	3	0.12	0%
CF 1210	Wheat germ	0.016	–	USA	15.0	8	–	–	ND	ND	3	0.01	0%
	Wine only	0.04	–	AUS	19.0	4	–	–	ND	ND	3	0.01	0%

For citrus, it was noted that no residues were found in pulp.
Raisin included dried grapes.

ANNEX 5

ESTIMATION OF DAILY INTAKE OF PESTICIDES IN AND ON SPICES

In view of the nature of data derived from monitoring and the lack of detailed information on spice consumption, several assumptions and approximations had to be made for calculating the daily intake of residues of pesticides in or on spices:

- As the subgroups of spices used in the GEM/Food diets did not coincide with those agreed upon by the CCPR at its Thirty-sixth session, consumption of the entire spices group was used in calculating long- and short-term intake.

- The ratio of the number of samples containing detectable residues and all samples taken from a given commodity–pesticide combination was used to reflect the proportion of commodities that were treated with or exposed to the pesticide. The factor derived was used in calculating the IEDI.

- The IEDI was calculated only from the residue levels detected in a particular spice commodity–pesticide combination that made the greatest contribution to intake of any of the subgroups.

- The IESTI was calculated when an ARfD value was available.

- The consumption of 'dried chili peppers' was estimated to be about 10% of the combined consumption of fresh sweet and chili peppers (VO 0051, VO 0444, VO 0445).

- For dried chili peppers, only TMDIs could be calculated, as the only residue value available was one derived by extrapolation from the MRLs of fresh peppers. The IESTI was also calculated from estimated maximum residue levels.

- The residue level in a composite sample taken from a spice shipment containing several lots represents the average level in the mixed commodity. This value provides information similar to that of the median residue level from supervised trials. Therefore, the highest residue levels observed in composite samples were used to calculate short-term intake, instead of the median value which would have been used for results derived from supervised trials.

- Estimates were made for spices and chili pepper independently.

The Meeting evaluated residues of 28 pesticides from monitoring data and estimated maximum reside levels for 47 pesticides in dried chili peppers on the basis of MRLs established for fresh sweet and chili peppers. The intakes from spices and chili were compared with the existing ADI and ARfD values only; intakes arising from other uses of the compounds were not considered.

Long-term intake

Spices

The results of intake calculations showed that the IEDI for spices was less than 1% in all diets. The Meeting concluded that spice consumption would not change the risk assessment based on all other uses of the compounds.

Chilli peppers

The calculated TMDIs for dried chilli peppers were less than 5% of the ADI in any GEMS/Food diet for abamectin, benalaxyl, cyluthrin, cypermethrin, diazinon, dichlofulanid, dinocap, dithiocarbamates, fenpropathrin, fenvalerate, imidacoprid, metalaxyl, methoxyfenozide, permethrin, piperonyl butoxide, propamocarb, pyrethrins, quintozene, spinosad, tebuconazole, tolylfluanid, triadimefon and triadimenol. The Meeting concluded that consumption of dried chilli pepper would not change the risk assessment based on all other uses of the compounds.

The residues of acephate, azinphos-methyl, carbaryl, carbendazim, chlorothalonil, chlorpyrifos, chlorpyrifos-methyl, cyhexatin, cyromazine, dicofol, dimethoate, dithiocarbamates, ethephon, ethoprophos, fenarimol, fenpropathrin, methamidophos, methomyl, monocrotophos, oxamyl, phosphamidon, procymidone, profenofos, tebufenozide and vinclozolin in dried chili pepper contribute more than 5% of the ADI. The Meeting concluded that a complete assessment of the long-term intake of these compounds should be carried out, taking into account the residues derived from all other uses.

Use of dimethoate on peppers exceeded the ADI (430%) on chili pepper alone. The Meeting could not conclude that its use would not present a health public concern.

Short-term intake

The IESTI values for spices ranged from < 1% to 170% (mevinphos).

For chili peppers, intake values at or above the ARfD were calculated for dimethoate (120%), methamidophos (100%) and oxamyl (270%).

The intakes of both spices and chili pepper (expressed in mg/kg bw per day) were similar for children and for the general population.

The Meeting concluded that short-term intake of pesticide residues, other than those listed above, deriving from consumption of spices and dried chili pepper is unlikely to present a public health concern.

Summary of results of IEDI and IESTI calculations for spices

Compound	Long-term intake (% ADI)[a]					Short-term intake (% ArfD)[a]	
	Mid-East (0.043 g/kg bw/day)	Far-East (0.055 g/kg bw/day)	African (0.030 g/kg bw/day)	Latin American (0.008 g/kg bw/day)	European (0.008 g/kg bw/day)	Adult (1.75 g/kg bw/day)	Children (1.67 g/kg bw/day)
Acephate	0	0	0	0	0	0	0
Azinphos-methyl	0	0	0	0	0		
Chlorpyrifos	0	0	0	0	0	5	5
Chlorpyrifos-methyl	0	0	0	0	0		
Cypermethrin	0	0	0	0	0		
Diazinon	0	0	0	0	0	20	20
Dichlorvos	0	0	0	0	0		
Dicofol	0	0	0	0	0		
Dimethoate	0	0	0	0	0	25	25
Disulfoton	0	0	0	0	0	5	5
Endosulfan	0	0	0	0	0	30	25
Ethion	0	0	0	0	0		
Fenitrothion	0	0	0	0	0	25	25
Iprodion	0	0	0	0	0		
Malathion	0	0	0	0	0	0	0
Metalaxyl	0	0	0	0	0		
Methamidophos	0	0	0	0	0	0	0
Mevinphos	0	0	0	0	0	170	160
Parathion	0	0	0	0	0		
Parathion-methyl	0	0	0	0	0	20	20
Permethrin	0	0	0	0	0		
Phenthoate	0	0	0	0	0		
Phorate	0	0	0	0	0		

Annex 5

Compound	Long-term intake (% ADI)[a]					Short-term intake (% ArfD)[a]	
	Mid-East (0.043 g/kg bw/day)	Far-East (0.055 g/kg bw/day)	African (0.030 g/kg bw/day)	Latin American (0.008 g/kg bw/day)	European (0.008 g/kg bw/day)	Adult (1.75 g/kg bw/day)	Children (1.67 g/kg bw/day)
Phosalone	0	0	0	0	0	0	0
Pirimicarb	0	0	0	0	0		
Pirimiphos-methyl	0	0	0	0	0		
Quintozene	0	0	0	0	0		
Vinclozolin	0	0	0	0	0		

[a] Rounded figures

Summary of results of calculations of TMDI (% ADI) and IESTI (% ARfD) from MRLs proposed for dried chili peppers

Compound	Long-term intake (% ADI)[a]					Short-term intake (% ArfD)[a]	
	Mid-East (0.057 g/kg bw/day)	Far-East (0.038 g/kg bw/day)	African (0.090 g/kg bw/day)	Latin American (0.040 g/kg bw/day)	European (0.173 g/kg bw/day)	Adults (0.472 g/kg bw/day)	Children (0.477 g/kg bw/day)
Abamectin	1	0	1	0	2		
Acephate	30	20	45	20	90	50	50
Azinphos-methyl	10	10	20	10	35		
Benalaxyl	0	0	0	0	0		
Carbaryl	35	20	60	25	110	10	10
Carbendazim	5	5	5	5	10		
Chlorothalonil	10	10	20	10	40		
Chlorpyrifos	10	10	20	10	35	10	10
Chlorpyrifos-methyl	5	0	5	0	10		
Cyfluthrin	0	0	0	0	0		
Cyhexatin	0	5	5	5	10		
Cypermethrin	0	0	0	0	0		
Cyromazine	5	0	5	0	10		
Diazinon	0	0	0	0	0	0	0
Dichlofluanid	0	0	0	0	1		
Dicofol	30	20	45	20	90		
Dimethoate	140	90	230	100	430	120	120
Dinocap	0	0	0	0	0		
Dithiocarbamates	0	0	5	0	5		
Ethephon	5	5	10	5	20	50	50
Ethoprophos	5	0	5	0	10	0	0
Fenarimol	5	0	5	0	10		
Fenpropathrin	5	0	5	0	5		
Fenvalerate	0	0	0	0	5		
Imidacloprid	0	0	0	0	0	0	0
Metalaxyl	0	0	0	0	0		
Methamidophos	30	20	50	20	90	90	100

Compound	Long-term intake (% ADI)[a]					Short-term intake (% ArfD)[a]	
	Mid-East (0.057 g/kg bw/day)	Far-East (0.038 g/kg bw/day)	African (0.090 g/kg bw/day)	Latin American (0.040 g/kg bw/day)	European (0.173 g/kg bw/day)	Adults (0.472 g/kg bw/day)	Children (0.477 g/kg bw/day)
Methomyl	5	0	5	0	10	20	20
Methoxyfenozide	0	0	0	5	0	0	0
Monocrotophos	20	10	30	10	60		
Oxamyl	30	20	50	20	100	260	270
Permethrin	0	0	0	0	5		
Phosphamidon	20	20	40	20	70		
Piperonyl butoxide	0	0	0	0	0		
Procymidone	0	0	5	0	10		
Profenofos	30	20	50	20	90		
Propamocarb	0	0	0	0	2		
Pyrethrins	0	0	0	0	0	0	0
Quintozene	0	0	0	0	0		
Spinosad	0	0	0	0	5		
Tebuconazole	0	0	0	0	5		
Tebufenozide	5	0	5	0	10		
Tolylfluanid	0	0	0	0	5	0	0
Triadimefon	0	0	0	0	0		
Triadimenol	0	0	0	0	0		
Vinclozolin	20	10	30	10	50		

[a] Rounded figures

ANNEX 6

REPORTS AND OTHER DOCUMENTS RESULTING FROM PREVIOUS JOINT MEETINGS OF THE FAO PANEL OF EXPERTS ON PESTICIDE RESIDUES IN FOOD AND THE ENVIRONMENT AND WHO EXPERT GROUPS ON PESTICIDE RESIDUES

1. Principles governing consumer safety in relation to pesticide residues. Report of a meeting of a WHO Expert Committee on Pesticide Residues held jointly with the FAO Panel of Experts on the Use of Pesticides in Agriculture. FAO Plant Production and Protection Division Report, No. PL/1961/11; WHO Technical Report Series, No. 240, 1962.

2. Evaluation of the toxicity of pesticide residues in food. Report of a Joint Meeting of the FAO Committee on Pesticides in Agriculture and the WHO Expert Committee on Pesticide Residues. FAO Meeting Report, No. PL/1963/13; WHO/Food Add./23, 1964.

3. Evaluation of the toxicity of pesticide residues in food. Report of the Second Joint Meeting of the FAO Committee on Pesticides in Agriculture and the WHO Expert Committee on Pesticide Residues. FAO Meeting Report, No. PL/1965/10; WHO/Food Add./26.65, 1965.

4. Evaluation of the toxicity of pesticide residues in food. FAO Meeting Report, No. PL/1965/10/1; WHO/Food Add./27.65, 1965.

5. Evaluation of the hazards to consumers resulting from the use of fumigants in the protection of food. FAO Meeting Report, No. PL/1965/10/2; WHO/Food Add./28.65, 1965.

6. Pesticide residues in food. Joint report of the FAO Working Party on Pesticide Residues and the WHO Expert Committee on Pesticide Residues. FAO Agricultural Studies, No. 73; WHO Technical Report Series, No. 370, 1967.

7. Evaluation of some pesticide residues in food. FAO/PL:CP/15; WHO/Food Add./67.32, 1967.

8. Pesticide residues. Report of the 1967 Joint Meeting of the FAO Working Party and the WHO Expert Committee. FAO Meeting Report, No. PL:1967/M/11; WHO Technical Report Series, No. 391, 1968.

9. 1967 Evaluations of some pesticide residues in food. FAO/PL:1967/M/11/1; WHO/Food Add./68.30, 1968.

10. Pesticide residues in food. Report of the 1968 Joint Meeting of the FAO Working Party of Experts on Pesticide Residues and the WHO Expert Committee on Pesticide Residues. FAO Agricultural Studies, No. 78; WHO Technical Report Series, No. 417, 1968.

11. 1968 Evaluations of some pesticide residues in food. FAO/PL:1968/M/9/1; WHO/Food Add./69.35, 1969.

12. Pesticide residues in food. Report of the 1969 Joint Meeting of the FAO Working Party of Experts on Pesticide Residues and the WHO Expert Group on Pesticide Residues. FAO Agricultural Studies, No. 84; WHO Technical Report Series, No. 458, 1970.

13. 1969 Evaluations of some pesticide residues in food. FAO/PL:1969/M/17/1; WHO/Food Add./70.38, 1970.

14. Pesticide residues in food. Report of the 1970 Joint Meeting of the FAO Working Party of Experts on Pesticide Residues and the WHO Expert Committee on Pesticide Residues. FAO Agricultural Studies, No. 87; WHO Technical Report Series, No. 4574, 1971.

15. 1970 Evaluations of some pesticide residues in food. AGP:1970/M/12/1; WHO/Food Add./71.42, 1971.

16. Pesticide residues in food. Report of the 1971 Joint Meeting of the FAO Working Party of Experts on Pesticide Residues and the WHO Expert Committee on Pesticide Residues. FAO Agricultural Studies, No. 88; WHO Technical Report Series, No. 502, 1972.

17. 1971 Evaluations of some pesticide residues in food. AGP:1971/M/9/1; WHO Pesticide Residue Series, No. 1, 1972.

18. Pesticide residues in food. Report of the 1972 Joint Meeting of the FAO Working Party of Experts on Pesticide Residues and the WHO Expert Committee on Pesticide Residues. FAO Agricultural Studies, No. 90; WHO Technical Report Series, No. 525, 1973.

19. 1972 Evaluations of some pesticide residues in food. AGP:1972/M/9/1; WHO Pesticide Residue Series, No. 2, 1973.

20. Pesticide residues in food. Report of the 1973 Joint Meeting of the FAO Working Party of Experts on Pesticide Residues and the WHO Expert Committee on Pesticide Residues. FAO Agricultural Studies, No. 92; WHO Technical Report Series, No. 545, 1974.

21. 1973 Evaluations of some pesticide residues in food. FAO/AGP/1973/M/9/1; WHO Pesticide Residue Series, No. 3, 1974.

22. Pesticide residues in food. Report of the 1974 Joint Meeting of the FAO Working Party of Experts on Pesticide Residues and the WHO Expert Committee on Pesticide Residues. FAO Agricultural Studies, No. 97; WHO Technical Report Series, No. 574, 1975.

23. 1974 Evaluations of some pesticide residues in food. FAO/AGP/1974/M/11; WHO Pesticide Residue Series, No. 4, 1975.

24. Pesticide residues in food. Report of the 1975 Joint Meeting of the FAO Working Party of Experts on Pesticide Residues and the WHO Expert Committee on Pesticide Residues. FAO Plant Production and Protection Series, No. 1; WHO Technical Report Series, No. 592, 1976.

25. 1975 Evaluations of some pesticide residues in food. AGP:1975/M/13; WHO Pesticide Residue Series, No. 5, 1976.

26. Pesticide residues in food. Report of the 1976 Joint Meeting of the FAO Panel of Experts on Pesticide Residues and the Environment and the WHO Expert Group on Pesticide Residues. FAO Food and Nutrition Series, No. 9; FAO Plant Production and Protection Series, No. 8; WHO Technical Report Series, No. 612, 1977.

27. 1976 Evaluations of some pesticide residues in food. AGP:1976/M/14, 1977.

28. Pesticide residues in food—1977. Report of the Joint Meeting of the FAO Panel of Experts on Pesticide Residues and Environment and the WHO Expert Group on Pesticide Residues. FAO Plant Production and Protection Paper 10 Rev, 1978.

29. Pesticide residues in food: 1977 evaluations. FAO Plant Production and Protection Paper 10 Suppl., 1978.

30. Pesticide residues in food—1978. Report of the Joint Meeting of the FAO Panel of Experts on Pesticide Residues and Environment and the WHO Expert Group on Pesticide Residues. FAO Plant Production and Protection Paper 15, 1979.

31. Pesticide residues in food: 1978 evaluations. FAO Plant Production and Protection Paper 15 Suppl., 1979.

32. Pesticide residues in food—1979. Report of the Joint Meeting of the FAO Panel of Experts on Pesticide Residues in Food and the Environment and the WHO Expert Group on Pesticide Residues. FAO Plant Production and Protection Paper 20, 1980.

33. Pesticide residues in food: 1979 evaluations. FAO Plant Production and Protection Paper 20 Suppl., 1980

34. Pesticide residues in food—1980. Report of the Joint Meeting of the FAO Panel of Experts on Pesticide Residues in Food and the Environment and the WHO Expert Group on Pesticide Residues. FAO Plant Production and Protection Paper 26, 1981.

35. Pesticide residues in food: 1980 evaluations. FAO Plant Production and Protection Paper 26 Suppl., 1981.

36. Pesticide residues in food—1981. Report of the Joint Meeting of the FAO Panel of Experts on Pesticide Residues in Food and the Environment and the WHO Expert Group on Pesticide Residues. FAO Plant Production and Protection Paper 37, 1982.

37. Pesticide residues in food: 1981 evaluations. FAO Plant Production and Protection Paper 42, 1982.

38. Pesticide residues in food—1982. Report of the Joint Meeting of the FAO Panel of Experts on Pesticide Residues in Food and the Environment and the WHO Expert Group on Pesticide Residues. FAO Plant Production and Protection Paper 46, 1982.

39. Pesticide residues in food: 1982 evaluations. FAO Plant Production and Protection Paper 49, 1983.

40. Pesticide residues in food—1983. Report of the Joint Meeting of the FAO Panel of Experts on Pesticide Residues in Food and the Environment and the WHO Expert Group on Pesticide Residues. FAO Plant Production and Protection Paper 56, 1985.

41. Pesticide residues in food: 1983 evaluations. FAO Plant Production and Protection Paper 61, 1985.

42. Pesticide residues in food—1984. Report of the Joint Meeting on Pesticide Residues. FAO Plant Production and Protection Paper 62, 1985.

43. Pesticide residues in food—1984 evaluations. FAO Plant Production and Protection Paper 67, 1985.

44. Pesticide residues in food—1985. Report of the Joint Meeting of the FAO Panel of Experts on Pesticide Residues in Food and the Environment and a WHO Expert Group on Pesticide Residues. FAO Plant Production and Protection Paper 68, 1986.

45. Pesticide residues in food—1985 evaluations. Part I. Residues. FAO Plant Production and Protection Paper 72/1, 1986.

46. Pesticide residues in food—1985 evaluations. Part II. Toxicology. FAO Plant Production and Protection Paper 72/2, 1986.

47. Pesticide residues in food—1986. Report of the Joint Meeting of the FAO Panel of Experts on Pesticide Residues in Food and the Environment and a WHO Expert Group on Pesticide Residues. FAO Plant Production and Protection Paper 77, 1986.

48. Pesticide residues in food—1986 evaluations. Part I. Residues. FAO Plant Production and Protection Paper 78, 1986.

49. Pesticide residues in food—1986 evaluations. Part II. Toxicology. FAO Plant Production and Protection Paper 78/2, 1987.

50. Pesticide residues in food—1987. Report of the Joint Meeting of the FAO Panel of Experts on Pesticide Residues in Food and the Environment and a WHO Expert Group on Pesticide Residues. FAO Plant Production and Protection Paper 84, 1987.

51. Pesticide residues in food—1987 evaluations. Part I. Residues. FAO Plant Production and Protection Paper 86/1, 1988.

52. Pesticide residues in food—1987 evaluations. Part II. Toxicology. FAO Plant Production and Protection Paper 86/2, 1988.

53. Pesticide residues in food—1988. Report of the Joint Meeting of the FAO Panel of Experts on Pesticide Residues in Food and the Environment and a WHO Expert Group on Pesticide Residues. FAO Plant Production and Protection Paper 92, 1988.

54. Pesticide residues in food—1988 evaluations. Part I. Residues. FAO Plant Production and Protection Paper 93/1, 1988.

55. Pesticide residues in food—1988 evaluations. Part II. Toxicology. FAO Plant Production and Protection Paper 93/2, 1989.

56. Pesticide residues in food—1989. Report of the Joint Meeting of the FAO Panel of Experts on Pesticide Residues in Food and the Environment and a WHO Expert Group on Pesticide Residues. FAO Plant Production and Protection Paper 99, 1989.

57. Pesticide residues in food—1989 evaluations. Part I. Residues. FAO Plant Production and Protection Paper 100, 1990.

58. Pesticide residues in food—1989 evaluations. Part II. Toxicology. FAO Plant Production and Protection Paper 100/2, 1990.

59. Pesticide residues in food—1990. Report of the Joint Meeting of the FAO Panel of Experts on Pesticide Residues in Food and the Environment and a WHO Expert Group on Pesticide Residues. FAO Plant Production and Protection Paper 102, Rome, 1990.

60. Pesticide residues in food—1990 evaluations. Part I. Residues. FAO Plant Production and Protection Paper 103/1, Rome, 1990.

61. Pesticide residues in food—1990 evaluations. Part II. Toxicology. World Health Organization, WHO/PCS/91.47, Geneva, 1991.

62. Pesticide residues in food—1991. Report of the Joint Meeting of the FAO Panel of Experts on Pesticide Residues in Food and the Environment and a WHO Expert Group on Pesticide Residues. FAO Plant Production and Protection Paper 111, Rome, 1991.

63. Pesticide residues in food—1991 evaluations. Part I. Residues. FAO Plant Production and Protection Paper 113/1, Rome, 1991.

64. Pesticide residues in food—1991 evaluations. Part II. Toxicology. World Health Organization, WHO/PCS/92.52, Geneva, 1992.

65. Pesticide residues in food—1992. Report of the Joint Meeting of the FAO Panel of Experts on Pesticide Residues in Food and the Environment and a WHO Expert Group on Pesticide Residues. FAO Plant Production and Protection Paper 116, Rome, 1993.

66. Pesticide residues in food—1992 evaluations. Part I. Residues. FAO Plant Production and Protection Paper 118, Rome, 1993.

67. Pesticide residues in food—1992 evaluations. Part II. Toxicology. World Health Organization, WHO/PCS/93.34, Geneva, 1993.

68. Pesticide residues in food—1993. Report of the Joint Meeting of the FAO Panel of Experts on Pesticide Residues in Food and the Environment and a WHO Expert Group on Pesticide Residues. FAO Plant Production and Protection Paper 122, Rome, 1994.

69. Pesticide residues in food—1993 evaluations. Part I. Residues. FAO Plant Production and Protection Paper 124, Rome, 1994.

70. Pesticide residues in food—1993 evaluations. Part II. Toxicology. World Health Organization, WHO/PCS/94.4, Geneva, 1994.

71. Pesticide residues in food—1994. Report of the Joint Meeting of the FAO Panel of Experts on Pesticide Residues in Food and the Environment and a WHO Expert Group on Pesticide Residues. FAO Plant Production and Protection Paper 127, Rome, 1995.

72. Pesticide residues in food—1994 evaluations. Part I. Residues. FAO Plant Production and Protection Paper 131/1 and 131/2 (2 volumes), Rome, 1995.

73. Pesticide residues in food—1994 evaluations. Part II. Toxicology. World Health Organization, WHO/PCS/95.2, Geneva, 1995.

74. Pesticide residues in food—1995. Report of the Joint Meeting of the FAO Panel of Experts on Pesticide Residues in Food and the Environment and the Core Assessment Group. FAO Plant Production and Protection Paper 133, Rome, 1996.

75. Pesticide residues in food—1995 evaluations. Part I. Residues. FAO Plant Production and Protection Paper 137, 1996.

76. Pesticide residues in food—1995 evaluations. Part II. Toxicological and Environmental. World Health Organization, WHO/PCS/96.48, Geneva, 1996.

77. Pesticide residues in food—1996. Report of the Joint Meeting of the FAO Panel of Experts on Pesticide Residues in Food and the Environment and the WHO Core Assessment Group. FAO Plant Production and Protection Paper, 140, 1997.

78. Pesticide residues in food—1996 evaluations. Part I. Residues. FAO Plant Production and Protection Paper, 142, 1997.

79. Pesticide residues in food—1996 evaluations. Part II. Toxicological. World Health Organization, WHO/PCS/97.1, Geneva, 1997.

80. Pesticide residues in food—1997. Report of the Joint Meeting of the FAO Panel of Experts on Pesticide Residues in Food and the Environment and the WHO Core Assessment Group. FAO Plant Production and Protection Paper, 145, 1998.

81. Pesticide residues in food—1997 evaluations. Part I. Residues. FAO Plant Production and Protection Paper, 146, 1998.

82. Pesticide residues in food—1997 evaluations. Part II. Toxicological and Environmental. World Health Organization, WHO/PCS/98.6, Geneva, 1998.

83. Pesticide residues in food—1998. Report of the Joint Meeting of the FAO Panel of Experts on Pesticide Residues in Food and the Environment and the WHO Core Assessment Group. FAO Plant Production and Protection Paper, 148, 1999.

84. Pesticide residues in food—1998 evaluations. Part I. Residues. FAO Plant Production and Protection Paper, 152/1 and 152/2 (two volumes).

85. Pesticide residues in food—1998 evaluations. Part II. Toxicological and Environmental. World Health Organization, WHO/PCS/99.18, Geneva, 1999.

86. Pesticide residues in food—1999. Report of the Joint Meeting of the FAO Panel of Experts on Pesticide Residues in Food and the Environment and the WHO Core Assessment Group. FAO Plant Production and Protection Paper, 153, 1999.

87. Pesticide residues in food—1999 evaluations. Part I. Residues. FAO Plant Production and Protection Paper, 157, 2000.

88. Pesticide residues in food—1999 evaluations. Part II. Toxicological. World Health Organization, WHO/PCS/00.4, Geneva, 2000.

89. Pesticide residues in food—2000. Report of the Joint Meeting of the FAO Panel of Experts on Pesticide Residues in Food and the Environment and the WHO Core Assessment Group. FAO Plant Production and Protection Paper, 163, 2001.

90. Pesticide residues in food—2000 evaluations. Part I. Residues. FAO Plant Production and Protection Paper, 165, 2001.

91. Pesticide residues in food—2000 evaluations. Part II. Toxicological. World Health Organization, WHO/PCS/01.3, 2001.

92. Pesticide residues in food—2001. Report of the Joint Meeting of the FAO Panel of Experts on Pesticide Residues in Food and the Environment and the WHO Core Assessment Group. FAO Plant Production and Protection Paper, 167, 2001.

93. Pesticide residues in food—2001 evaluations. Part I. Residues. FAO Plant Production and Protection Paper, 171, 2002.

94. Pesticide residues in food—2001 evaluations. Part II. Toxicological. World Health Organization, WHO/PCS/02.1, 2002.

95. Pesticide residues in food—2002. Report of the Joint Meeting of the FAO Panel of Experts on Pesticide Residues in Food and the Environment and the WHO Core Assessment Group. FAO Plant Production and Protection Paper, 172, 2002.

96. Pesticide residues in food—2002 evaluations. Part I. Residues. FAO Plant Production and Protection Paper, 175/1 and 175/2 (two volumes).

97. Pesticide residues in food—2002 evaluations. Part II. Toxicological. World Health Organization, WHO/PCS, 2003.

98. Pesticide residues in food—2003. Report of the Joint Meeting of the FAO Panel of Experts on Pesticide Residues in Food and the Environment and the WHO Core Assessment Group. FAO Plant Production and Protection Paper, 176, 2004.

99. Pesticide residues in food—2003 evaluations. Part I. Residues. FAO Plant Production and Protection Paper, 177, 2004.

100. Pesticide residues in food—2003 evaluations. Part II. Toxicological. World Health Organization, WHO/PCS, 2004.

ANNEX 7

CORRECTIONS TO THE REPORT OF THE 2003 MEETING

Changes are shown in bold. Only significant factual errors and omissions are listed.

Table of contents
p. iii   **4.3 Dimethoate (027) (T,R)**

Section 3
p. 17   Table 1, Dodine, **change** ADI 0–0.2 to **ADI 0–0.1** mg/kg bw
p. 20   Table 2 Acephate, **change** ArfD 0.003 to ARfD **0.05** mg/kg bw

Section 4
p. 36   para 3, line 5, **change** 190–630% to **110–630%**
p. 40   Maximum burden table, footnote 4, **change** 2 mg/kg to **13 mg/kg** as benomyl

4.3 Carbosulfan/carbofuran
p. 47   para 2, last sentence, **change** Carbosulfan was less than 0.02 mg/kg to **0.002** mg/kg
p. 48   Potatoes, para 6, **insert**: The only measurable value of 0.02 mg/kg was taken as the HR.
p. 51   lines 1–2. **change** '… an STMR of 0.217 mg/kg for carbofuran in sugar beet leaves or tops.' to '… an STMR of 0.217 mg/kg **(dry wt)** for carbofuran in sugar beet leaves or tops.'
pp. 51–52, Maximum burden table, column , row 3: **change** Sugar beet to Sugar beet **leaves or tops**
                            row 4: **change** Dry citrus to Dry citrus **pulp**
p.52   STMR burden table, column 1, row 3: **change** Sugar beet to Sugar beet **leaves or tops**
         Penultimate para, **change to** The **ruminant dietary burden of carbofuran** established by the **1997 JMPR (Report, p. 50)** was based on a diet containing 80% of alfalfa fodder; there **were** few animal feed items.

4.4 Cyprodinil
p. 61   lines 2–3, **change** 2-anilino-4-(2-hydroxypropyl)-6-methylpyrimidinamin-5-ol to **2-anilino-4-(2-hydroxypropyl)-6-methylpyrimidin-5-ol**

4.6 Dimethoate
p. 79   Plant metabolism: last para. **change** The plant metabolites of omethoate to **The plant metabolites of III, XI, XII, and XX**
p. 81   para 3, lines 1–2, **change** Residues in whole lemon fruit at a 21-day PHI were 0.76 and 1.10 mg/kg dimethoate and 0.07 and 0.11 mg/kg omethoate. to **Residues in whole lemon fruit at a 21- or 28-day PHI were 0.76 and 1.10 mg/kg dimethoate and 0.10 and 0.11 mg/kg omethoate.**
         para 6, lines 3–4, **change** The dimethoate residues … are in rank order 0.15, 0.2, 0.29, 0.35, 0.37 (2), 0.41, 0.48, 0.52, 0.65, 0.76, 0.77, 0.83, 0.85, 1.04, 1.1, 1.17, 1.34, 1.48, 1.50 and 3.1 mg/kg. to **The dimethoate residues … are in rank order 0.02, 0.06, 0.11, 0.15, 0.2, 0.21, 0.29, 0.35, 0.37 (2), 0.41, 0.48, 0.52, 0.65, 0.76, 0.77 (2), 0.83, 0.85, 0.97, 1.04, 1.1, 1.17, 1.33, 1.34, 1.48, 1.50 and 3.1 mg/kg.**
         para 8, lines 3–4, **chamge** For the purpose of estimating STMR-Ps for processed commodities, the STMRs are 0.76 mg/kg for dimethoate and 0.035 mg/kg for omethoate. to **For the purpose of estimating STMR-Ps for processed commodities, the STMRs are 0.71 mg/kg for dimethoate and 0.08 mg/kg for omethoate.**
p. 82   Olives, line 6., **change** In Italy the rate is 0.028–0.56 kg ai/hl …. to **In Italy the rate is 0.028–0.056 kg ai/hl ….**
p. 86   last para, **change** The residues in wheat straw from northern Europe at PHI 28 days (or at earliest commercial harvest) in ranked order were <0.002 (3), <0.01 (2), 0.02, 0.05 and 0.07 mg/kg of dimethoate, and <0.002 (5), <0.01 (3) mg/kg of omethoate. to **The residues in**

374 Annex 7

wheat straw from northern Europe at PHI 28 days (or at earliest commercial harvest) in ranked order were <0.01 (3), 0.01, 0.02, 0.05 and 0.07 mg/kg of dimethoate, and <0.002, <0.01 (6) mg/kg of omethoate.

p. 87 para 6, line 3 and following table, **change** ... STMRs for citrus fruits (dimethoate 0.76, Omethoate 0.08 mg/kg) ... to ... **STMRs for citrus fruits (dimethoate 0.71, Omethoate 0.08 mg/kg)** ...

The estimated processing factors and STMR-Ps for orange juice, dry orange pulp, processed olive products and wheat products are summarized below. **Replace the following entries.**

Processed commodity	Processing factor		STMR of RAC, mg/kg		STMR-P[1], mg/kg
	Dimethoate	Omethoate	Dimethoate	Omethoate	
Orange juice	0.14[2]	0.21[2]	**0.71**	**0.08**	**0.27**
Orange pulp, dry	2.1[2]	1.7[2]	**0.71**	**0.08**	**1.6**[3]
Olive, processed	0.21	0.12	0.04	0.22	**0.272**

4.8 Dodine
p. 94 Table, Farm animal dietary burden. **Replace the following entries.**

Commodity	STMR-P (mg/kg)	Group	Dry matter %	Residue on dry basis mg/kg	Percent of diet		Residue contribution, mg/kg	
					Beef cattle	Dairy cattle	Beef cattle	Dairy cattle
Apple **pomace, wet**	8.69	AB	40	21.7	40	20	8.7	4.4

4.10 Fenitrothion
p. 117 para 4, lines 5–6. **Change** ...0.089 and 0.11 for white bread (mean 0.010) ...to...**0.089 and 0.11 for white bread (mean 0.10)** ...

para 5, lines 3–4. **Change**... and STMR-Ps of ... 0.05 mg/kg in white bread ... to ... **and STMR-Ps of ... 0.50 mg/kg in white bread** ...

p. 118 last 2 lines.
... (Transfer factor = residue level in egg or tissue ÷ feeding level in ~~diet~~ **metabolism study**) at ~~this feeding level~~ **the level of the dietary burden**.

4.12 Lindane
p. 128 Mean farm animal dietary burden. **Replace the following entries.**

Commodity Group	Residue mg/kg	Basis	% dry matter	Residue on dry wt mg/kg	Chosen diets, %			Residue contribution, mg/kg		
					Beef cattle	Dairy cattle	Poultry	Beef cattle	Dairy cattle	Poultry
Wheat hay	AS	**0.01**	STMR	88	**0.011**	25			**0.003**		
						Mean dietary burden			0.011	0.014	0.005

4.13 Methamidophos
p. 141 para 3, last sentence. **Change** ... the existing CXLs of 0.01 (*) mg/kg for cattle fat, cattle meat, goat meat, goat fat, sheep meat, sheep fat and milks. to ... **the existing CXLs of 0.01 (*) mg/kg for cattle fat, cattle meat, goat meat, goat fat, pig meat, pig fat, sheep meat, sheep fat and milks.**

para 5, last line. **Change** ... with the exception of cabbage (head) and tomatoes. to ... **with the exception of broccoli, cauliflower, apples, sweet peppers, cabbage (head) and tomatoes.**

Annex 7

4.14 Methoxyfenozide

p. 165   second para. **Change** ... the acute RfD varied from 0% to 100% for the general population ... to **... the acute RfD varied from 0% to 10% for the general population ...**

4.18 Pirimiphos-methyl

p. 181   para 3, lines 1–2. **Change** Since no data were reported on the storage stability of pirimiphos-methyl or its metabolites in animal tissues, milk or eggs, the Meeting concluded...to **Since no data were reported on the storage stability of pirimiphos-methyl or its metabolites in animal tissues or eggs, Meeting concluded...**

Annex 1

Replace the following entries.

Pesticide (Codex reference no.)	ADI, mg/kg bw	CCN	Commodity	Recommended MRL, mg/kg New	Recommended MRL, mg/kg Previous	STMR or STMR-P, mg/kg	HR or HR-P mg/kg
Carbofuran[1] (096)	0–0.002	AF 0645	Maize forage	**0.5 (dry wt)**		**0.13 (dry wt)**	
		AV 0596	Sugar beet leaves or tops	0.7	W	0.217 **(dry wt)**	
Carbosulfan** (145)	0–0.01	AF 0645	Maize forage	0.05* **(dry wt)**	–	0	
		AV 0596	Sugar beet leaves or tops	0.05* **(dry wt)**			
Dimethoate[1] (027)		VO 0448	**Tomato**	**W**	**2**		
Fenitrothion**(037)[1]		CP 1211	White bread	W	0.2 PoP	**0.50**	
Methamidophos* (100)	0–0.004	AM1051	Fodder beet leaves or tops	30 **(dry wt)**		9.1 **(dry wt)**	26.5 **(dry wt)**
		AV 0596	Sugar beet leaves or tops	30 **(dry wt)**	1	9.1 **(dry wt)**	26.5 **(dry wt)**
Methoxyfenozide* (209)	0–0.1	AM 0660	Almond hulls	50 **(dry wt)**		13 **(fresh wt)**	
		SO 0691	Cotton seed	7		**0.39**	
		AS 0645	Maize fodder	60 (dry wt)		8.2 **(fresh wt)**	
		AF 0645	Maize forage	50 (dry wt)		4.5 **(fresh wt)**	
		TN 0085	Tree nuts	0.1		**0.021**	0.074
Pyraclostrobin * **(210)**							

FAO TECHNICAL PAPERS

FAO PLANT PRODUCTION AND PROTECTION PAPERS

1	Horticulture: a select bibliography, 1976 (E)	29	Sesame: status and improvement, 1981 (E)
2	Cotton specialists and research institutions in selected countries, 1976 (E)	30	Palm tissue culture, 1981 (C E)
3	Food legumes: distribution, adaptability and biology of yield, 1977 (E F S)	31	An eco-climatic classification of intertropical Africa, 1981 (E)
4	Soybean production in the tropics, 1977 (C E F S)	32	Weeds in tropical crops: selected abstracts, 1981 (E)
4 Rev.1	Soybean production in the tropics (first revision), 1982 (E)	32 Sup.1	Weeds in tropical crops: review of abstracts, 1982 (E)
5	Les systèmes pastoraux sahéliens, 1977 (F)	33	Plant collecting and herbarium development, 1981 (E)
6	Pest resistance to pesticides and crop loss assessment – Vol. 1, 1977 (E F S)	34	Improvement of nutritional quality of food crops, 1981 (C E)
6/2	Pest resistance to pesticides and crop loss assessment – Vol. 2, 1979 (E F S)	35	Date production and protection, 1982 (Ar E)
6/3	Pest resistance to pesticides and crop loss assessment – Vol. 3, 1981 (E F S)	36	El cultivo y la utilización del tarwi – *Lupinus mutabilis* Sweet, 1982 (S)
7	Rodent pest biology and control – Bibliography 1970-74, 1977 (E)	37	Pesticide residues in food 1981 – Report, 1982 (E F S)
8	Tropical pasture seed production, 1979 (E F** S**)	38	Winged bean production in the tropics, 1982 (E)
9	Food legume crops: improvement and production, 1977 (E)	39	Seeds, 1982 (E/F/S)
10	Pesticide residues in food, 1977 – Report, 1978 (E F S)	40	Rodent control in agriculture, 1982 (Ar C E F S)
10 Rev.	Pesticide residues in food 1977 – Report, 1978 (E)	41	Rice development and rainfed rice production, 1982 (E)
10 Sup.	Pesticide residues in food 1977 – Evaluations, 1978 (E)	42	Pesticide residues in food 1981 – Evaluations, 1982 (E)
11	Pesticide residues in food 1965-78 – Index and summary, 1978 (E F S)	43	Manual on mushroom cultivation, 1983 (E F)
12	Crop calendars, 1978 (E/F/S)	44	Improving weed management, 1984 (E F S)
13	The use of FAO specifications for plant protection products, 1979 (E F S)	45	Pocket computers in agrometeorology, 1983 (E)
14	Guidelines for integrated control of rice insect pests, 1979 (Ar C E F S)	46	Pesticide residues in food 1982 – Report, 1983 (E F S)
15	Pesticide residues in food 1978 – Report, 1979 (E F S)	47	The sago palm, 1983 (E F)
15 Sup.	Pesticide residues in food 1978 – Evaluations, 1979 (E)	48	Guidelines for integrated control of cotton pests, 1983 (Ar E F S)
16	Rodenticides: analyses, specifications, formulations, 1979 (E F S)	49	Pesticide residues in food 1982 – Evaluations, 1983 (E)
17	Agrometeorological crop monitoring and forecasting, 1979 (C E F S)	50	International plant quarantine treatment manual, 1983 (C E)
18	Guidelines for integrated control of maize pests, 1979 (C E)	51	Handbook on jute, 1983 (E)
19	Elements of integrated control of sorghum pests, 1979 (E F S)	52	The palmyrah palm: potential and perspectives, 1983 (E)
20	Pesticide residues in food 1979 – Report, 1980 (E F S)	53/1	Selected medicinal plants, 1983 (E)
20 Sup.	Pesticide residues in food 1979 – Evaluations, 1980 (E)	54	Manual of fumigation for insect control, 1984 (C E F S)
21	Recommended methods for measurement of pest resistance to pesticides, 1980 (E F)	55	Breeding for durable disease and pest resistance, 1984 (C E)
22	China: multiple cropping and related crop production technology, 1980 (E)	56	Pesticide residues in food 1983 – Report, 1984 (E F S)
23	China: development of olive production, 1980 (E)	57	Coconut, tree of life, 1984 (E S)
24/1	Improvement and production of maize, sorghum and millet – Vol. 1. General principles, 1980 (E F)	58	Economic guidelines for crop pest control, 1984 (E F S)
24/2	Improvement and production of maize, sorghum and millet – Vol. 2. Breeding, agronomy and seed production, 1980 (E F)	59	Micropropagation of selected rootcrops, palms, citrus and ornamental species, 1984 (E)
25	*Prosopis tamarugo*: fodder tree for arid zones, 1981 (E F S)	60	Minimum requirements for receiving and maintaining tissue culture propagating material, 1985 (E F S)
26	Pesticide residues in food 1980 – Report, 1981 (E F S)	61	Pesticide residues in food 1983 – Evaluations, 1985 (E)
26 Sup.	Pesticide residues in food 1980 – Evaluations, 1981 (E)	62	Pesticide residues in food 1984 – Report, 1985 (E F S)
27	Small-scale cash crop farming in South Asia, 1981 (E)	63	Manual of pest control for food security reserve grain stocks, 1985 (C E)
28	Second expert consultation on environmental criteria for registration of pesticides, 1981 (E F S)	64	Contribution à l'écologie des aphides africains, 1985 (F)
		65	Amélioration de la culture irriguée du riz des petits fermiers, 1985 (F)
		66	Sesame and safflower: status and potentials, 1985 (E)
		67	Pesticide residues in food 1984 – Evaluations, 1985 (E)
		68	Pesticide residus in food 1985 – Report, 1986 (E F S)
		69	Breeding for horizontal resistance to wheat diseases, 1986 (E)
		70	Breeding for durable resistance in perennial crops, 1986 (E)

71	Technical guideline on seed potato micropropagation and multiplication, 1986 (E)
72/1	Pesticide residues in food 1985 – Evaluations – Part I: Residues, 1986 (E)
72/2	Pesticide residues in food 1985 – Evaluations – Part II: Toxicology, 1986 (E)
73	Early agrometeorological crop yield assessment, 1986 (E F S)
74	Ecology and control of perennial weeds in Latin America, 1986 (E S)
75	Technical guidelines for field variety trials, 1993 (E F S)
76	Guidelines for seed exchange and plant introduction in tropical crops, 1986 (E)
77	Pesticide residues in food 1986 – Report, 1986 (E F S)
78	Pesticide residues in food 1986 – Evaluations – Part I: Residues, 1986 (E)
78/2	Pesticide residues in food 1986 – Evaluations – Part II: Toxicology, 1987 (E)
79	Tissue culture of selected tropical fruit plants, 1987 (E)
80	Improved weed management in the Near East, 1987 (E)
81	Weed science and weed control in Southeast Asia, 1987 (E)
82	Hybrid seed production of selected cereal, oil and vegetable crops, 1987 (E)
83	Litchi cultivation, 1989 (E S)
84	Pesticide residues in food 1987 – Report, 1987 (E F S)
85	Manual on the development and use of FAO specifications for plant protection products, 1987 (E** F S)
86/1	Pesticide residues in food 1987 – Evaluations – Part I: Residues, 1988 (E)
86/2	Pesticide residues in food 1987 – Evaluations – Part II: Toxicology, 1988 (E)
87	Root and tuber crops, plantains and bananas in developing countries – challenges and opportunities, 1988 (E)
88	*Jessenia* and *Oenocarpus*: neotropical oil palms worthy of domestication, 1988 (E S)
89	Vegetable production under arid and semi-arid conditions in tropical Africa, 1988 (E F)
90	Protected cultivation in the Mediterranean climate, 1990 (E F S)
91	Pastures and cattle under coconuts, 1988 (E S)
92	Pesticide residues in food 1988 – Report, 1988 (E F S)
93/1	Pesticide residues in food 1988 – Evaluations – Part I: Residues, 1988 (E)
93/2	Pesticide residues in food 1988 – Evaluations – Part II: Toxicology, 1989 (E)
94	Utilization of genetic resources: suitable approaches, agronomical evaluation and use, 1989 (E)
95	Rodent pests and their control in the Near East, 1989 (E)
96	*Striga* – Improved management in Africa, 1989 (E)
97/1	Fodders for the Near East: alfalfa, 1989 (Ar E)
97/2	Fodders for the Near East: annual medic pastures, 1989 (Ar E F)
98	An annotated bibliography on rodent research in Latin America 1960-1985, 1989 (E)
99	Pesticide residues in food 1989 – Report, 1989 (E F S)
100	Pesticide residues in food 1989 – Evaluations – Part I: Residues, 1990 (E)
100/2	Pesticide residues in food 1989 – Evaluations – Part II: Toxicology, 1990 (E)
101	Soilless culture for horticultural crop production, 1990 (E)
102	Pesticide residues in food 1990 – Report, 1990 (E F S)
103/1	Pesticide residues in food 1990 – Evaluations – Part I: Residues, 1990 (E)
104	Major weeds of the Near East, 1991 (E)
105	Fundamentos teórico-prácticos del cultivo de tejidos vegetales, 1990 (S)
106	Technical guidelines for mushroom growing in the tropics, 1990 (E)
107	*Gynandropsis gynandra* (L.) Briq. – a tropical leafy vegetable – its cultivation and utilization, 1991 (E)
108	Carambola cultivation, 1993 (E S)
109	Soil solarization, 1991 (E)
110	Potato production and consumption in developing countries, 1991 (E)
111	Pesticide residues in food 1991 – Report, 1991 (E)
112	Cocoa pest and disease management in Southeast Asia and Australasia, 1992 (E)
113/1	Pesticide residues in food 1991 – Evaluations – Part I: Residues, 1991 (E)
114	Integrated pest management for protected vegetable cultivation in the Near East, 1992 (E)
115	Olive pests and their control in the Near East, 1992 (E)
116	Pesticide residues in food 1992 – Report, 1993 (E F S)
117	Quality declared seed, 1993 (E F S)
118	Pesticide residues in food 1992 – Evaluations – Part I: Residues, 1993 (E)
119	Quarantine for seed, 1993 (E)
120	Weed management for developing countries, 1993 (E S)
120/1	Weed management for developing countries, Addendum 1, 2004 (E S)
121	Rambutan cultivation, 1993 (E)
122	Pesticide residues in food 1993 – Report, 1993 (E F S)
123	Rodent pest management in eastern Africa, 1994 (E)
124	Pesticide residues in food 1993 – Evaluations – Part I: Residues, 1994 (E)
125	Plant quarantine: theory and practice, 1994 (Ar)
126	Tropical root and tuber crops – Production, perspectives and future prospects, 1994 (E)
127	Pesticide residues in food 1994 – Report, 1994 (E)
128	Manual on the development and use of FAO specifications for plant protection products – Fourth edition, 1995 (E F S)
129	Mangosteen cultivation, 1995 (E)
130	Post-harvest deterioration of cassava – A biotechnology perspective, 1995 (E)
131/1	Pesticide residues in food 1994 – Evaluations – Part I: Residues, Volume 1, 1995 (E)
131/2	Pesticide residues in food 1994 – Evaluations – Part I: Residues, Volume 2, 1995 (E)
132	Agro-ecology, cultivation and uses of cactus pear, 1995 (E)
133	Pesticide residues in food 1995 – Report, 1996 (E)
134	(Number not assigned)
135	Citrus pest problems and their control in the Near East, 1996 (E)
136	El pepino dulce y su cultivo, 1996 (S)
137	Pesticide residues in food 1995 – Evaluations – Part I: Residues, 1996 (E)
138	Sunn pests and their control in the Near East, 1996 (E)
139	Weed management in rice, 1996 (E)
140	Pesticide residues in food 1996 – Report, 1997 (E)
141	Cotton pests and their control in the Near East, 1997 (E)
142	Pesticide residues in food 1996 – Evaluations – Part I: Residues, 1997 (E)

143	Management of the whitefly-virus complex, 1997 (E)	
144	Plant nematode problems and their control in the Near East region, 1997 (E)	
145	Pesticide residues in food 1997 – Report, 1998 (E)	
146	Pesticide residues in food 1997 – Evaluations – Part I: Residues, 1998 (E)	
147	Soil solarization and integrated management of soilborne pests, 1998 (E)	
148	Pesticide residues in food 1998 – Report, 1999 (E)	
149	Manual on the development and use of FAO specifications for plant protection products – Fifth edition, including the new procedure, 1999 (E)	
150	Restoring farmers' seed systems in disaster situations, 1999 (E)	
151	Seed policy and programmes for sub-Saharan Africa, 1999 (E F)	
152/1	Pesticide residues in food 1998 – Evaluations – Part I: Residues, Volume 1, 1999 (E)	
152/2	Pesticide residues in food 1998 – Evaluations – Part I: Residues, Volume 2, 1999 (E)	
153	Pesticide residues in food 1999 – Report, 1999 (E)	
154	Greenhouses and shelter structures for tropical regions, 1999 (E)	
155	Vegetable seedling production manual, 1999 (E)	
156	Date palm cultivation, 1999 (E)	
156 Rev.1	Date palm cultivation, 2002 (E)	
157	Pesticide residues in food 1999 – Evaluations – Part I: Residues, 2000 (E)	
158	Ornamental plant propagation in the tropics, 2000 (E)	
159	Seed policy and programmes in the Near East and North Africa, 2000	
160	Seed policy and programmes for Asia and the Pacific, 2000 (E)	
161	Silage making in the tropics with particular emphasis on smallholders, 2000 (E S)	
162	Grassland resource assessment for pastoral systems, 2001, (E)	
163	Pesticide residues in food 2000 – Report, 2001 (E)	
164	Seed policy and programmes in Latin America and the Caribbean, 2001 (E S)	
165	Pesticide residues in food 2000 – Evaluations – Part I, 2001 (E)	
166	Global report on validated alternatives to the use of methyl bromide for soil fumigation, 2001 (E)	
167	Pesticide residues in food 2001 – Report, 2001 (E)	
168	Seed policy and programmes for the Central and Eastern European countries, Commonwealth of Independent States and other countries in transition, 2001 (E)	
169	Cactus (*Opuntia* spp.) as forage, 2003 (E S)	
170	Submission and evaluation of pesticide residues data for the estimation of maximum residue levels in food and feed, 2002 (E)	
171	Pesticide residues in food 2001 – Evaluations – Part I, 2002 (E)	
172	Pesticide residues in food, 2002 – Report, 2002 (E)	
173	Manual on development and use of FAO and WHO specifications for pesticides, 2002 (E S)	
174	Genotype x environment interaction – Challenges and opportunities for plant breeding and cultivar recommendations, 2002 (E)	
175/1	Pesticide residues in food 2002 – Evaluations – Part 1: Residues – Volume 1 (E)	
175/2	Pesticide residues in food 2002 – Evaluations – Part 1: Residues – Volume 2 (E)	
176	Pesticide residues in food 2003 – Report, 2004 (E)	
177	Pesticide residues in food 2003 – Evaluations – Part 1: Residues, 2004 (E)	
178	Pesticide residues in food 2004 – Report, 2004 (E)	

Availability: November 2004

Ar	–	Arabic	Multil –	Multilingual
C	–	Chinese	*	Out of print
E	–	English	**	In preparation
F	–	French		
P	–	Portuguese		
S	–	Spanish		

The FAO Technical Papers are available through the authorized FAO Sales Agents or directly from Sales and Marketing Group, FAO, Viale delle Terme di Caracalla, 00100 Rome, Italy.